Crystal Chemistry of
Condensed Phosphates

Crystal Chemistry of Condensed Phosphates

A. Durif

Laboratory of Crystallography
CNRS
Grenoble, France

PLENUM PRESS • NEW YORK AND LONDON

Library of Congress Cataloging-in-Publication Data

Durif, A.
    Crystal chemistry of condensed phosphates / A. Durif.
      p.   cm.
    Includes bibliographical references (p.   -   ) and index.
    ISBN 0-306-44878-5
    1. Phosphates. 2. Crystallography.   I. Title.
  QD181.P1D88   1995
  546'.71224--dc20   95-2312
                                                                 CIP

ISBN 0-306-44878-5

© 1995 Plenum Press, New York
A Division of Plenum Publishing Corporation
233 Spring Street, New York, N. Y. 10013

10 9 8 7 6 5 4 3 2 1

Printed in the United States of America

To Professor André Guinier
who suggested I write this book

Preface

My first contact with condensed phosphates occurred during the early 1950s, when I was preparing some starting materials for the syntheses of silicate-like compounds. I was both surprised and intrigued by their very unusual behavior. I attributed these feelings to my total ignorance of this field, but during the following years the idea persisted in my mind that much was not clearly understood in this domain of chemistry. Unfortunately, at this time condensed phosphates had a bad reputation among chemists and crystallographers. They were said to be unstable, difficult to synthesize in a reproducible manner, impossible to crystallize, and so on. In addition, the literature in which all these "qualities" were documented was confusing, and various classifications mixing existing and hypothetical compounds were employed. For a crystallographer who admits the existence of a compound only when it is allocated a unit cell and a space group, this area of chemistry was a jungle.

In 1956, the classic book *Phosphorus and Its Compounds* by John R. van Wazer was published. Gathering and ordering clearly all the existing knowledge in this field, this book contributed fundamentally to my decision, a few years after its publication, to begin investigations in the field of condensed phosphates. Several research groups larger than mine were already exploring the same domain, in Berlin (E. Thilo), Moskow (I. V. Tananaev), St. Louis (E. G. Griffith), and elsewhere, but in fact the research activities in these various laboratories were complementary, and so we never had the feeling that we were in competition but, that on the contrary, we were contributing to a collective effort.

During this very productive period, two books dealing with phosphorus chemistry, *The Structural Chemistry of Phosphorus* (1974) and *Phosphorus* (1978), were published by D. E. C. Corbridge. They are today indispensable tools for a chemist or a crystallographer, but they came too early. Written before the huge proliferation of X-ray diffraction facilities, they do not include some fundamental results obtained by this technique.

The amount of results obtained by the early 1980s was sufficient to permit the formulation of a coherent classification of condensed phosphates. Some classes of compounds that had been purely speculative were proved to exist whereas hypothetical classes were discarded. At this stage, the need for a compilation of these results, dispersed in hundreds of articles, was evident. Thus, when Professor A. Guinier suggested that I undertake the task of collecting these results in a single volume, began work on this book. However, my ideas about what to include in this book were altered by the advent of large computerized data banks, as many of the numerical data that I had included in the first draft—atomic coordinates, for instance—can now be easily obtained from a data bank. I have been very slow in performing the necessary revisions, but this has turned out to be fortunate, because I have thus been able to include in the present version some very fundamental results obtained in the past three years, mainly in the domain of higher ring anions.

During this work, I received encouragement and valuable criticism from several German and Russian colleagues. Our discussions demonstrating the need for a book on this subject contributed greatly to my progress in writing this book. However, the major assistance came from Dr. M. T. Averbuch-Pouchot. This book could not have been written without her constant critical and valuable assistance.

A. Durif

Grenoble, France

Contents

Introduction

Condensed phosphates, the subject of this book, constitute an important part, probably the most intricate, of phosphate chemistry. This class of phosphates developed over a long period of time and, even today, is still relatively poorly understood as compared, for example, with the body of knowledge collected on condensed silicates.

But first let us try to explain what we commonly call a *phosphate*. A general and rather abstract definition can be given by saying that phosphates are salts of both the monophosphoric acid H_3PO_4 and its various condensed or polymeric forms. The corresponding anions have varied geometries but one common feature: they are all built up by pentavalent phosphorus atoms surrounded by more or less distorted tetrahedra made of four oxygen atoms. This definition, perhaps too restrictive suffers from many exceptions. For instance in current nomenclature used in the chemical literature, compounds that include in their anionic entities various substituted tetrahedra, such as PO_3H (phosphites), PO_3F (fluorophosphates) and PO_3S (thiophosphates) are considered to be phosphates or more precisely *substituted phosphates*.

A very broad definition was given by Corbridge[1] who wrote that the word *phosphates* must apply to any compound of pentavalent phosphorus whose anion includes a P–O bond. This broader definition includes substituted compounds (the various substituted tetrahedra mentioned above), and furthermore some compounds in which the phosphorus coordination is larger than four. Finally Corbridge concluded that the term "phosphates" must be used for compounds in which the phosphorus atoms are surrounded by a tetrahedron of four oxygen atoms (normal phosphates) or by tetrahedra in which one or several oxygen atoms are substituted by other atoms (substituted phosphates).

In this book we will discuss repeatedly the PO_4 tetrahedron. In order to give an idea of both interatomic distances and bond angles in this entity, we present in Table 1.1 the main geometrical features of a typical PO_4 tetrahedron as observed in a sodium monophosphate, $Na_3PO_4 \cdot 1/2H_2O$.

In a very elementary way, one can roughly classify phosphates into three main groups—monophosphates, condensed phosphates and oxyphosphates.

Monophosphates are salts derived from the normal H_3PO_4 acid. These salts are characterized by a very simple anionic entity, an isolated PO_4^{3-} group consisting of a central phosphorus atom surrounded by four oxygen atoms located at the corners of an almost regular tetrahedron. The geometry of this anion is described in Table 1.1.

As *condensed phosphates* are the subject of the present book they will be described in a very detailed way in the following pages. Let us simply say, at this stage that, they correspond to various anionic entities built from corner-sharing PO_4 tetrahedra. In this family of salts the O/P ratio in the anion is in the following range:

Table 1.1. Main Geometrical Features of a PO_4 Tetrahedron as Observed in $Na_3PO_4 \cdot 1/2H_2O$ by Averbuch-Pouchot and Durif.[a,b]

P	O(1)	O(2)	O(3)	O(4)
O(1)	<u>1.540(2)</u>	2.521(2)	2.498(2)	2.524(2)
O(2)	110.2(1)	<u>1.533(2)</u>	2.513(2)	2.515(2)
O(3)	108.2(1)	109.6(1)	<u>1.542(2)</u>	2.517(2)
O(4)	109.8(1)	109.6(1)	109.3(1)	<u>1.544(2)</u>

[a]Ref. 2.

[b]The four P–O values are given along the diagonal, part of this table, the six O–O distances above the diagonal and the six O–P–O angles below the diagonal. Estimated standard deviations are given in parentheses. Distances are in angstroms, and angles are in decimal degrees.

$$5/2 < O/P < 4$$

Oxyphosphates, for a time erroneously named "basic phosphates," include in their atomic arrangements some oxygen atoms not belonging to the anionic entity. In all the examples of such compounds reported to date, the phosphoric anion is an isolated PO_4 tetrahedron. Accordingly they can be defined, for the time, being phosphates having a global formula corresponding to an O/P ratio > 4.

1.1. Condensed Phosphates: Definition and Classification

First, let us explain what is commonly meant by a condensed phosphate. We can very simply say, before proceeding into the subject in greater detail, that any phosphoric anion in which there is a P–O–P bond is a condensed phosphoric anion. Another simple way to define this term is to say that any phosphoric anion corresponding to a formula characterized by a P/O ratio larger than 1/4 is a condensed one. It may be noted that this last definition is a direct consequence of the first one, since the basic unit of all phosphoric anions is the PO_4 tetrahedron.

These P–O–P bonds can be obtained by various processes; one of the simplest is the reorganization of two molecules of a monohydrogenmonophosphate after the elimination of a water molecule:

$$\begin{array}{ccc} & O & & O & & O & O \\ & \| & & \| & & \| & \| \\ O-P-O-H & + & H-O-P-O & \longrightarrow & O-P-O-P-O & +H_2O \\ & \| & & \| & & \| & \| \\ & O & & O & & O & O \end{array}$$

This scheme of condensation illustrates the most common way to prepare, for instance, tetrasodium diphosphate:

$$2Na_2HPO_4 \rightarrow Na_4P_2O_7 + H_2O$$

This condensation phenomenon can generate a great number of phosphoric anions with various geometries which we will examine now in greater detail.

In the present state of the crystal chemistry of condensed phosphates, one observes, for the anions, three very different types of condensation geometries. The first one

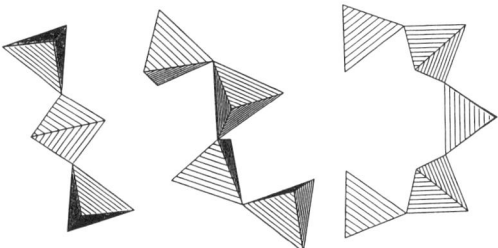

Figure 1.1. Some examples of polyphosphate anions.

corresponds to a progressive linear linkage of PO_4 tetrahedra sharing one or two of their oxygen atoms. Some examples of such condensed anions are presented in Fig. 1.1. The corresponding phosphates are usually called *polyphosphates*. As can easily be deduced from their geometries the general formula for this type of anions is given by

$$(P_nO_{3n+1})^{(n+2)-} \qquad (1.1)$$

n being the number of tetrahedra in the anionic entity. Phosphorus atoms belonging to PO_4 tetrahedra sharing two of their oxygen atoms with neighboring PO_4 tetrahedra are usually called "internal" phosphorus atoms, the others being known as "terminal" phosphorus atoms.

As it can be deduced from formula (1.1) when n becomes very large the ratio O/P $\rightarrow$ 3 and the geometry of the anion is that of an infinite chain. Generally, this type of anion is represented by the formula

$$(PO_3)_n{}^{n-} \qquad (1.2)$$

but in the chemical formulas of the corresponding salts, the degree of condensation (n) is omitted, and one simply writes, for instance, $NaPO_3$, for long-chain sodium polyphosphate. The geometry of such a chain is shown in Fig 1.2.

The second type of condensation is a cyclic one, leading to the formation of ring anions of the general formula

$$(P_nO_{3n})^{n-} \qquad (1.3)$$

At present phosphoric rings are known for $n = 3, 4, 5, 6, 8, 10,$ and 12. The corresponding phosphates are named *cyclophosphates*. Figure 1.3 illustrates some ring anions.

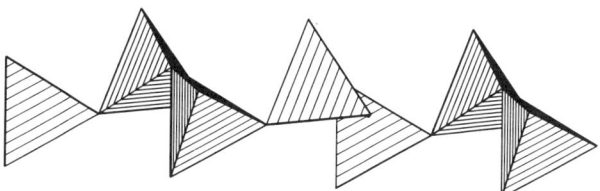

Figure 1.2. An infinite $(PO_3)_n$ chain.

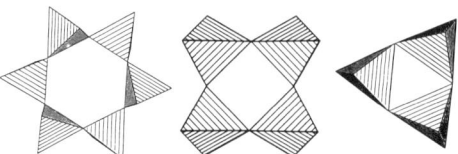

Figure 1.3. Some examples of ring-anion geometries.

In these first two types of condensation, one PO_4 tetrahedron shares one or two of its oxygen atoms with the neighboring PO_4 groups. The situation is different for the third type of condensation, observed in a class of P_2O_5–rich phosphates, called *ultraphosphates*. Here, in the anion, some PO_4 tetrahedra share three of their oxygen atoms with the neighboring PO_4 groups. This type of condensation leads to various anion geometries: finite groups, infinite ribbons, infinite layers, or three-dimensional networks. The phosphorus atom of a PO_4 tetrahedron sharing three of its oxygen atoms with adjacent tetrahedra is termed a "branching" phosphorus atom. The representation of an ultraphosphate anion is depicted in Fig. 1.4.

A chemical definition of these salts can be given by saying that their anions are richer in P_2O_5 than the end-term anions of the classical condensed phosphates, long-chain polyphosphates, or cyclophosphates, whose anionic entities are characterized by an O/P ratio of 3. On the basis of this definition the general anionic formula for this group of compounds can be written as follow, n and m being integers:

$$n[PO_3]^- + mP_2O_5 \quad \text{or} \quad [P_{(2m+n)}O_{(5m+3n)}]^{n-} \tag{1.4}$$

This very general formula corresponds to an infinite number of branches, depending on the values of m and n. At present the only unambiguously characterized ultraphosphate anions are those corresponding to $m = 1$, that is anions of general formula

$$[P_{(n+2)}O_{(3n+5)}]^{n-} \tag{1.5}$$

Such anions are now well known for $n = 2, 3, 4,$ and 6.

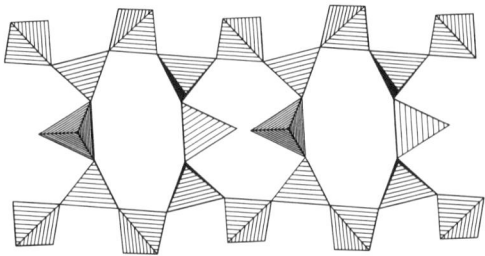

Figure 1.4. An ultraphosphate anion as observed in SmP_5O_{14}.

In this category of phosphates, no correlation exists between the chemical formula of the anion and its geometry. A given anionic formula can correspond to several types of condensation. For instance, an ultraphosphate anion of formula $P_5O_{14}^{3-}$ can build infinite ribbons in SmP_5O_{14} or a three-dimensional framework in TbP_5O_{14}.

It may appear surprising that the classification of phosphates presented above is based more on geometry than on chemistry, but it is now well recognized that any classification of condensed phosphates must be based on the geometry of the anions. We justify this statement with a very simple example. When NaH_2PO_4 is heated above 823 K, it undergoes the following reaction:

$$NaH_2PO_4 \rightarrow NaPO_3 + H_2O$$

and the phosphate obtained, $NaPO_3$, is a long-chain polyphosphate. However, the reaction is run for some hours at 823 K it must be written

$$3NaH_2PO_4 \rightarrow Na_3P_3O_9 + 3H_2O$$

and in this case the salt obtained, $Na_3P_3O_9$, is a cyclotriphosphate.

The global formula of these two phosphates is identical, but their chemical properties are fundamentally different. For instance, the second one is very water soluble while the first one is not.

1.2. Nomenclature of Condensed Phosphates

Through the years, the nomenclature used for condensed phosphates has often been very confusing. In this section, we simply report what is today the commonly accepted nomenclature of condensed phosphates.

Phosphates containing anions corresponding to the first type of condensation geometry are generally called *polyphosphates*, but with some variations depending on the degree of condensation of the anion. As mentioned above, in these salts the general formula of the anionic entity is $(P_nO_{3n+1})^{(n+2)-}$. For small degrees of condensation ($n < 20$), the usual name is *oligophosphates*, and the presently accepted nomenclature is as follows

For $n = 2$ (the P_2O_7 anion), the corresponding salts have traditionally been called "pyrophosphates," but the recommended name is *diphosphates*.
For $n = 3$ (the P_3O_{10} anion), the presently accepted name is *triphosphates*, but "tripolyphosphates" is still often erroneously used.
For $n = 4$ and 5 (the P_4O_{13} and P_5O_{16} anions), the designations *tetraphosphates* and *pentaphosphates* are now preferred to the previously used "tetrapolyphosphates" and "pentapolyphosphates."
When n is very large, the formula of the anion becomes, as we mentioned above, close to $(PO_3)_n$ and its geometry is that of an infinite chain. Phosphates containing this type of anion are commonly called *long-chain polyphosphates*.

Phosphates containing cyclic anions of general formula P_nO_{3n} have for a long time been called "metaphosphates" (trimetaphosphates, tetrametaphosphates, etc.). The present usage is to employ more descriptive names: *cyclotriphosphates* for phosphates containing P_3O_9 anions; *cyclotetraphosphates, cyclopentaphosphates*, etc. for phosphates containing larger ring anions (P_4O_{12}, P_5O_{15}, etc.).

For *ultraphosphates,* corresponding to the third type of condensation, no systematic or logical nomenclature exists. This lack is a source of frequent confusions, since some phosphates of this type have been erroneously designated by physicists or crystallographers according to the number of phosphorus atoms in the formula unit. Thus, some rare-earth ultraphosphates, with LnP_5O_{14} as general formula are commonly known as "pentaphosphates," a designation now used for the oligophosphates containing a P_5O_{16} anion. The restricted number of well-characterized ultraphosphates can probably explain this absence of any coherent nomenclature and classification for this kind of compounds, which still have not been extensively investigated.

At the beginning of this chapter, we described the present classification of phosphates. In light of the very rapid development of phosphate chemistry, this classification is now, in many respects, not satisfactory. In the present work, as we are dealing only with condensed phosphates, we will not discuss this more general aspect of phosphate chemistry. Insofar as condensed phosphates are concerned the present classification is, except in the case of the ultraphosphates, fairly satisfactory.

1.3. Brief Historical Survey of Condensed Phosphates

The present knowledge about the geometry of condensed phosphoric anions is the fruition of a long period of confusion lasting more than one and half century. Apparently the first observation of the condensation of phosphoric acid was reported by Berzelius[3] in 1816, when he described the changes in the properties of this acid after ignition. He was also the first to prepare a condensed phosphate by heating Na_2HPO_4 at "red heat." He had discovered the latter compound earlier and proved by his analysis that $2Na_2O \cdot P_2O_5$ was the final product of its calcination.

In 1827 Clark[4] extended these experiments and demonstrated that the compound prepared by Berzelius, $Na_4P_2O_7$, was a distinct chemical entity and not simply dehydrated Na_2HPO_4. Clark seems also to have been the first to distinguish between the easily removable water (crystallization water) and water that is more difficult to eliminate, resulting from the condensation of the anionic part of the salt. It was at this time a dilemma, because, according to the theory of Lavoisier, salts were considered to be combinations of basic and acid oxides and were classified on the basis of the ratio of basic to acid oxides, all the water content of a salt being considered to be crystallization water. Clark also demonstrated that hydrated sodium diphosphate and the corresponding arsenate are isomorphous.

Some years later, in 1833, Graham[5] repeated the Clark's experiments on the behavior of Na_2HPO_4, confirming Clark's results and extended the same kind of investigations to NaH_2PO_4. When studying the thermal behavior of this salt, he observed first the formation of an acidic pyrophosphate and at higher temperatures the formation of a new salt, which he named "metaphosphate."

$$2NaH_2PO_4 \xrightarrow{-H_2O} Na_2H_2P_2O_7 \xrightarrow{-H_2O} 2NaPO_3$$

After a long series of very careful experiments, he came to the conclusion that several types of water may exist in salts and suggested that a water molecule can be substituted for a base. Thus he explained the various sodium monophosphates by the following scheme:

$$P_2O_5 \cdot 3Na_2O = 2Na_3PO_4$$

$$P_2O_5 \cdot 2Na_2O \cdot H_2O = 2Na_2HPO_4$$

$$P_2O_5 \cdot Na_2O \cdot 2H_2O = 2NaH_2PO_4$$

$$P_2O_5 \cdot 3H_2O = 2H_3PO_4$$

Then he proposed a classification of phosphates into three groups, the orthophosphates, the pyrophosphates, and the metaphosphates, corresponding to the following three phosphoric acids:

$$P_2O_5 \cdot 3H_2O \text{ or } 2H_3PO_4 \text{ (orthophosphoric acid)}$$

$$P_2O_5 \cdot 2H_2O \text{ or } H_4P_2O_7 \text{ (pyrophosphoric acid)}$$

$$P_2O_5 \cdot H_2O \text{ or } 2HPO_3 \text{ (metaphosphoric acid)}$$

This outstanding proposal, quasi-prophetic at the time, caused nevertheless some confusion in the chemical literature for more than a century, because in fact the term "metaphosphates" covers a great variety of possible degrees of anionic condensation that were for a long time difficult to estimate. In spite of the deficiencies of this classification scheme, which became more and more evident over the years, it was so firmly established that, with a very small number of exceptions, it was the only one discussed in classical textbooks of chemistry up to 1950.

One of the most important results of the Graham's investigations was the discovery during his study of the thermal behavior of NaH_2PO_4 what he considered to be three forms of sodium "metaphosphate," $NaPO_3$. One of these forms was not water-soluble, a second one was very soluble and the third one appeared as a water-soluble glass. For a long time, the first form, which was in fact, long-chain crystalline sodium polyphosphate, has been known as the *Maddrell salt*. The readily water-soluble form was later identified as sodium cyclotriphosphate, whereas the water-soluble glass, obtained by rapid cooling of melted $NaPO_3$ and for a long time known as *Graham's salt*, is now recognized as being a mixture of polyphosphates of various chain lengths.

Graham's experiments stimulated a large number of investigations throughout past century. Most of these investigations involved attempts to determine the degree of condensation of the various forms of sodium metaphosphate and of its derivatives. For most of this period, during which chemical analysis was the only means by which to characterize a compound, the degree of condensation was guessed rather than being established through the determination of the formula of precipitated double salts.

In 1847 Maddrell[6] prepared several "metaphosphates" at relatively high temperature (589 K). From his detailed analyses, one can conclude that he was the first to characterize clearly, in a good crystalline state, a series of $M(PO_3)_2$ compounds (M = Mg, Ni, Co, Cu, Mn, Ca, Sr, and Ba) and also some derivatives with trivalent cations, $Cr(PO_3)_3$ and $Al(PO_3)_3$. All these salts were later recognized as cyclotetraphosphates or long-chain phosphates.

During approximately the same period (1847–1850), Fleitmann and Henneberg[7–9] made an important contribution to this field. They confirmed the existence of water-soluble forms of the sodium "metaphosphate," mainly the trimeric and tetrameric forms that are now called sodium cyclotri-, and cyclotetraphosphate, $Na_3P_3O_9$ and $Na_4P_4O_{12}$, and from these two materials, they prepared a great number of salts and double salts, such as $Na_2K_2P_4O_{12}\cdot2H_2O$, $Ag_3P_3O_9\cdot H_2O$, $Cu_2P_4O_{12}$, $Ba_3(P_3O_9)_2\cdot6H_2O$, $BaNaP_3O_9\cdot4H_2O$, $Cu_2P_4O_{12}\cdot8H_2O$ all of which have recently been characterized as cyclotri- or cyclotetraphosphates. Using the copper salt that had recently been discovered by Maddrell, they were the first to prepare sodium cyclotetraphosphate by an exchange reaction with sodium sulfide:

$$Cu_2P_4O_{12} + 2Na_2S \rightarrow Na_4P_4O_{12} + 2\,CuS$$

Up to about 1950, this reaction was the only way to prepare convenient amounts of sodium cyclotetraphosphate. In addition, it should be noticed that, with very few exceptions, all the chemical formulas reported by these two authors proved to be correct. In view of the confusion in the nomenclature at the time and the lack of methods for the determination of the condensation states of the anions, the amount of results that these authors produced seems almost incredible to a modern specialist in this field. In addition, during these investigations, they also characterized several phosphates, such as $Na_6P_4O_{13}$, $Mg_3P_4O_{13}$, and $Ag_{12}P_{10}O_{31}$, whose chemical formulas could not be explained by the Graham classification and thus suspected the existence of phosphoric anions, such as $P_4O_{13}^{6-}$ and $P_{10}O_{31}^{12-}$, that were not included in this classification. In many historical surveys of this field, Fleitmann and Henneberg are often credited with the condensation scheme leading to the general formula of polyphosphoric acids:

$$n\mathrm{H}_3\mathrm{PO}_4 - (\mathrm{n}-1)\mathrm{H}_2\mathrm{O} \rightarrow \mathrm{H}_{n+2}\mathrm{P}_n\mathrm{O}_{3n+1}$$

After a thorough investigation of the early chemical literature we think that it is difficult to support this assertion. It seems that it was not until the work of Schwarz, in 1895[10] that this formulation was clearly established, based upon, according to Schwarz, the "discovery" of $Na_6P_4O_{13}$ by Fleitmann and Henneberg. It was later on proved by Thilo and Ratz[11] that this material cannot be obtained in a crystalline state.

In 1875 Lindbom[12] repeated some of the work on the preparation of cyclotriphosphates that had been reported by Fleitmann and Henneberg and discussed their investigations, which he considered to be the main contribution to the knowledge of the various forms of metaphosphoric acid. Lindbom also described a chemical preparation for sodium cyclotriphosphate better than that reported by Fleitmann and Henneberg and leading to an almost pure salt (95%). In addition to preparing most of the cyclotriphosphates previously characterized by Fleitmann and Henneberg, he also prepared number of new ones. From the formulas of the double salts that he obtained, he came to the conclusion that the corresponding acid is a triacid and that the various forms of metaphosphoric acid must be explained in terms of different types of polymerization. He proposed for the first time in the history of condensed phosphates, a cyclic representation for cyclotriphosphoric acid.

In his famous "Inaugural Dissertation" (1880), Glatzel[13] described the investigation of more than 80 condensed phosphates, which he classified into "dimetaphosphates" and "tetrametaphosphates." Most of the compounds that he described were later proved to be cyclotetraphosphates or long-chain polyphosphates. Glatzel was also the first to postulate a cyclic conformation for the anion of cyclotetraphosphates.

Sabatier[14–15] investigated the transformation of aqueous solutions of metaphosphoric acid into monophosphoric acid as a function of concentration and temperature. He began from the assumption that this acid was a hexamer. As one of the processes he described for the preparation of this acid is the slow addition of P_4O_{10} to ice water, one can presume that he, in fact, investigated the hydrolysis rate of $H_4P_4O_{12}$.

Tammann in 1890 and 1892[16,17] examined carefully the various forms of sodium "metaphosphate" and several of their derivatives. He tried to determine their molecular weights from conductiometric or cryoscopic measurements in order to establish their degrees of condensation. During his investigations, he suspected the existence of a hexameric form.

In 1895, Schwarz,[10] in an article entitled "On a new polyphosphosphoric acid $H_5P_3O_{10}$ and some of its salts," made an important contribution to the chemistry of oligophosphates. He reported a scheme first suggested by Fleitmann and Henneberg to explain the formation of some phosphates with $P_4O_{13}^{6-}$ and $P_{10}O_{31}^{-12}$ anions that they had discovered during their investigations:

$$6H_2O + 2P_2O_5 \rightarrow 4H_3PO_4 \text{ (orthophosphoric acid)}$$

$$6H_2O + 3P_2O_5 \rightarrow 3H_4P_2O_7 \text{ (pyrophosphoric acid)}$$

$$6H_2O + 4P_2O_5 \rightarrow 2H_6P_4O_{13} \text{ (tetraphosphoric acid)}$$

$$6H_2O + 5P_2O_5 \rightarrow H_{12}P_{10}O_{31} \text{ (decaphosphoric acid)}$$

$$6H_2O + 6P_2O_5 \rightarrow 12HPO_3 \text{ (metaphosphoric acid)}$$

However, although this scheme, satisfactorily explained the results of Fleitmann and Henneberg, it could not account for the existence of the new polyphosphoric acid $H_5P_3O_{10}$ discovered by Schwarz. Schwarz described a number of crystalline derivatives of this acid, including $Na_5P_3O_{10}$, $CoNa_3P_3O_{10}\cdot12H_2O$, $Co_2NaP_3O_{10}$, $CoNa_3P_3O_{10}\cdot12H_2O$, Zn_2NaP_3-$O_{10}\cdot19/2H_2O$, $Ca_5(P_3O_{10})_2$, $Ba_5(P_3O_{10})_2$, and $Cu_5(P_3O_{10})_2\cdot13H_2O$. In addition, he confirmed the existence of the sodium tetraphosphate, $Na_6P_4O_{13}$, described by Fleitmann and Henneberg and characterized the corresponding lead salt, $Pb_3P_4O_{13}$. Schwarz proposed for the conformations of these various anions a representation corresponding to a linear linkage of PO_4 groups, in perfect accordance with the present results of structural analysis.

Three years later, Stange,[18] using Schwarz's results, produced some new triphosphates, $FeNa_3P_3O_{10}\cdot11/2H_2O$, $MgNa_3P_3O_{10}\cdot13H_2O$, $MnNa_3P_3O_{10}\cdot12H_2O$ and $CuNa_3P_3$-$O_{10}\cdot12H_2O$, all later identified as belonging to a series of isomorphous dodecahydrates.

To conclude our survey of last century let us examine the famous book *Chemische Krystallographie* by Groth.[19] Published in 1906, its aim was to report all previous optical morphological studies. Throughout its three volumes, several thousand such studies are reported and discussed, but among them are no more than five dealing with condensed phosphates, all of these being diphosphates (pyrophosphates). This lack of data can probably be mainly attributed to the difficulties of crystallizing these compounds, but the confusion in this field in the absence of any coherent nomenclature or classification was probably also a contributing factor.

At the very beginning of the century, two noteworthy investigations were reported. In 1900, Knorre[20] published "Beiträge zur Kenntnis der Metaphosphate." In this long article, he first presented a careful and critical survey of all the previous work in the field of condensed phosphates and then reported the results of his own investigations, mainly directed toward the optimization of the preparation of sodium cyclotriphosphate and the formation of double salts. He also carefully investigated the thermally induced transformation of various alkali monophosphates to condensed salts and the problem of the insoluble sodium metaphosphate. Although no new fundamental results were presented in this article, it is noteworthy in that it clarified some aspects of condended phosphate chemistry. Soon after, in 1903, Warschauer[21] published a detailed report of his investigations. He described improved processes for the preparation of $Cu_2P_4O_{12}$, $Ba(PO_3)_2$, and $Pb(PO_3)_2$ and tried to isolate the corresponding phosphoric acids from these salts by the action of H_2S in order to evaluate their degrees of condensation by various physical methods. He came to the conclusion that some of the "dimetaphosphates" previously described by Glatzel were, in fact, "tetrametaphosphates." During his investigations, he produced single crystals of $(NH_4)_4P_4O_{12}$ and reported a morphological study of this species. As in many articles of this period, a very good and critical survey of all previous work is included.

By and large the first 35 years of this century were a period of stagnation and lack of interest in this area of chemistry. This period seems to have been mainly characterized by a good number of fruitless attempts to produce hypothetical monometaphosphates and dimetaphosphates. Moreover, on the basis of the chemical literature of this period, one can suspect that the procedures used last century for the preparation of basic starting materials such as sodium "trimetaphosphate" or "tetrametaphosphates" had been forgotten. The following quotation from Pascal (1923)[22] summarizes well the common viewpoint during this period:

> There are few chapters in the history of inorganic chemistry in which the amount of confusion has exceeded that in the chemistry of metaphosphates. Each new study tends to invalidate the results of some previous study and to produce findings that are difficult to classify. These problems are compounded by the fact that the accepted nomenclature is not always employed by authors in a manner consistent with the degree of polymerization postulated for the various metaphosphoric acids.

Pascal denied most of the results obtained last century and claimed for the characterization of monometaphosphates.

At the end of this period, nevertheless, some significant contributions appeared. One of the major ones is the publication by Boullé[23] in 1938 of a procedure for the preparation of water-soluble cyclotriphosphates. This procedure, which is described in detail in Chapter 4, has been extensively used for the syntheses of cyclotriphosphates and has been extended to the preparation of cyclohexaphosphates and many other classes of condensed phosphates. It still remains a key process for the chemical preparation of these salts.

During the same period, Bonneman-Bémia[24] performed a very careful and systematic investigation of triphosphate chemistry. Using modern methods of synthesis and characterization, he confirmed the existence of all previously reported triphosphates and described a good number of new compounds in this field. He was the first to investigate $Na_5P_3O_{10}.6H_2O$ by X-ray diffraction.

A true resurgence of condensed phosphate chemistry occurred during the early fifties and can be attributed to several almost simultaneous advances, such as for instance the availability of various techniques that made it possible to measure more accurately the degree of polymerization, the discovery of paper chromatography, and the development of X-ray structural analysis. Since this period, the development of knowledge in this field of chemistry has been so fast that comprehensive survey cannot be given here. We simply describe in this section the main steps in this development during the past 40 years; the most recent developments in the various areas of condensed phosphate chemistry are discussed in the corresponding chapters of this book.

During this period, one of the most significant advances was the discovery of the possibilities offered by paper or thin-layer chromatography by Westman and Scott[25] and Ebel.[26] Through the applications of these techniques which made it possible to establish easily the degree of condensation of the polymeric anions, many previously reported species were shown to be mixtures of various condensed phosphates, and well-controlled reproducible preparations started to be elaborated.

Having recognized immediately the possibilities offered by these techniques for a better understanding of this field of chemistry, Thilo and his group in Berlin undertook a systematic investigation of condensed phosphates. A huge number of fundamental results were published by this group and remain today the basic chemical foundation for any kind of investigation in this field. Combining a thorough knowledge of chemistry, chromatography, and X-ray analysis, this group can be considered as major contributors to the modern chemistry of condensed phosphates.

During the same period Van Wazer and Griffith of the Monsanto Company in the United States conducted a large number investigations mainly concerning the chemical behavior of condensed phosphates, the elaboration of reproducible well-controlled procedures for their preparation, and the construction of some basic phase-equilibrium diagrams. Van Wazer also produced two volumes[27] reviewing the state of phosphorus chemistry at the time. This classical book is still now the best basic reference for any newcomer to this field of chemistry.

Some years later, several Soviet laboratories took a great interest to condensed phosphate chemistry. The leading laboratory was the Kurnakov Institute in Moscow, where the group led by Tananaev conducted a gigantic number of investigations of multicomponent systems based on classical methods of chemistry or by flux methods. Various other Soviet groups produced a large number of phase-equilibrium diagrams and investigated various multicomponent systems. In many cases, the Soviet studies were complemented by X-ray structural determinations, so one can say that a good part of our knowledge in the domain of the structural chemistry of condensed phosphates is the fruit of these investigations.

X-ray structural analysis has played a fundamental rôle in the development of condensed phosphate chemistry. As was discussed above, through the use of various techniques and of a great deal of imagination, the geometrical conformations of the condensed anions were guessed but they were not proved. However, despite the possibilities offered by the development of X-ray diffraction analysis, the application of this very powerful tool in this field was very long in coming. Although a satisfactory classification of silicates, based on numerous structural data, was made possible as early as 1940 through the investigations of the Bragg's school, it was not until 1935 that Levi and Peyronel in solving the crystal structure of $Zr_2P_2O_7$,[28] provided the first description of the geometrical features of a diphosphate anion. Two years later, Pauling and Sherman[29] determined the crystal structure of aluminum cyclotetraphosphate, $Al_4(P_4O_{12})_3$.

The first attempt to solve the crystal structure of a cyclotriphosphate ($Na_3P_3O_9 \cdot 6H_2O$) was made in 1942 by Caglioti *et al.*[30] but it was not completely successful, and it was not until 1956, when the crystal structure of $RbPO_3$ was determined by Corbridge,[31] that the geometrical configuration of a long-chain anion was obtained.

Since that date, corresponding approximately to the beginning of the huge development of facilities for X-ray diffraction analysis, a great number of structural characterizations have been performed, making it possible to confirm previous assumptions and to develop a coherent scheme for the classification of condensed phosphates. We simply enumerate in Table 1.2 what we consider to be the main steps in the advancement of knowledge about the structure of these salts. Each reported study corresponds to the first geometrical characterization of a given condensed anion.

Several well-documented surveys of the development of the condensed phosphate chemistry have been published during the past 40 years, including those by Topley,[42] Van Wazer,[27] Thilo,[43] and Kalliney.[44] Unfortunately, these reviews were all published before

Table 1.2. Chronology of the Main Steps in the Development of the Structural Chemistry of Condensed Phosphates[a]

Anion	Phosphate	Author(s)	Year	Ref.
P_2O_7	$Zr_2P_2O_7$	Levi and Peyronnel	1935	28
P_4O_{12}	$Al_4(P_4O_{12})_3$	Pauling and Sherman	1937	29
$(PO_3)_n$	$RbPO_3$	Corbridge	1956	31
P_3O_{10}	$Na_5P_3O_{10}$	Davies and Corbridge	1956	32
P_3O_9	$LiK_2P_3O_9 \cdot H_2O$	Eanes & Ondik	1962	33
P_6O_{18}	$Na_6P_6O_{18} \cdot 6H_2O$	Jost	1965	34
P_5O_{14}	LnP_5O_{14}	Bagieu-Beucher and Tranqui	1970	35
P_5O_{15}	$Na_4(NH_4)P_5O_{15} \cdot 4H_2O$	Jost	1972	36
P_8O_{24}	$Cu_3(NH_4)_2P_8O_{24}$	Laügt and Guitel	1975	37
P_4O_{13}	$Si(NH_4)_2P_4O_{13}$	Durif, Averbuch-Pouchot and Guitel	1976	38
P_5O_{16}	$Mg_2Na_3P_5O_{16}$	Smolin, Shepelev, Domanskii and Majling	1978	39
$P_{12}O_{36}$	$V_3Cs_3P_{12}O_{36}$	Lavrov, Voitenk and Tselebrovskaya	1981	41
$P_{10}O_{30}$	$Ba_2Zn_3P_{10}O_{30}$	Bagieu-Beucher, Durif and Guitel	1982	40

[a]The investigations cited in this table represent the first structural characterization of the anions listed in the first column; the second column contains the formula of the investigated phosphate.

the huge expansion of X-ray structural analysis, and so, cannot cover many fundamental results obtained through this technique.

The history of the development of ultraphosphate chemistry, which has been even still more erratic and confused than that of classical condensed phosphates, will be reported separately in Chapter 6, which is devoted to these salts.

1.4. Organization of This Book and General Comments

In the five chapters that follow, we review individually each of the major classes of condensed phosphates. Several of these chapters are subdivided into sections dealing with various families of compounds within a class. In reviewing each class or family, we begin with a short introduction dealing with the historical development of the subject, followed by a survey of the present state of chemistry in the field. This survey is not intended to be exhaustive, but we have attempted to report as completely as possible what is known today about *crystalline materials* characterized with a good degree of certainty. This section includes tables containing the main crystallographic data that are available for the materials described.

When general methods of preparation exist, they are described in a separate section entitled "Chemical Preparation of "..." However, unconventional methods of preparation are described in the section entitled "Present State of... Chemistry."

An important section in our review of each class of family, entitled "Atomic Arrangements of," is devoted to the description of a selection of atomic arrangements that we consider as representative. This is followed by detailed study of the various geometries and internal symmetries of the anions.

Some abbreviations will be commonly used throughout this book: S.G. for space group, Z for the number of formula units in the unit cell, and mp. for melting point. The units used are angströms for lengths, decimal degrees for angles, and Kelvin degrees for temperatures.

Some of the data that are indispensable for the discussion of the anion geometries are not always given in publications (P–P–P angles, for instance), so these values have been calculated by the author from the atomic parameters found in the literature.

Most of the drawings employ a polyhedral representation of the anions. This approach seems to us the most pedagogically suitable one in a book mainly devoted to the geometry of tetrahedral anions. All the polyhedral drawings were produced using the *STRUPLO* system.[45]

The chapter sections entitled "Present State of Chemistry" may appear deceptively short, given that a relatively large number of condensed phosphates have been well characterized. However the papers reporting these investigations are, in most cases, very short, describing as briefly as possible a more or less reproducible chemical preparation, followed very often by a good-quality crystal structure determination. Thus, in some respects, these sections will often resemble a simple annotated bibliography.

A consequence of the brevity of the articles reporting the characterization of new condensed phosphates is that the fundamental chemical properties and the behavior of most of these compounds are not yet known, even in the case of some most important synthesis reagents.

2

Oligophosphates

2.1. Introduction

Oligophosphates whose anions correspond to the general formula $P_nO_{(3n+1)}^{(n+2)-}$ have not yet been systematically investigated. A large number of diphosphates and triphosphates ($n = 2$ and 3) are presently well characterized, but examples with $n = 4$ and $n = 5$ are rather rare. This slow development, common to many areas of condensed phosphate chemistry, can probably be attributed to the lack of convenient starting materials for the preparation of these higher oligophosphates.

There is evidence from paper chromatography analysis for the existence of oligophosphates with $n > 5$. Griffith and Buxton,[46] for instance, reported careful and convincing experiments showing clearly the existence of oligophosphates up to $n = 8$ (the P_8O_{25} anion). However, such compounds have not yet been produced in a good crystalline state, and no structural data are available. Corbridge[1] discussed several reasons why crystallization of higher oligophosphates is difficult.

A problem that has not yet been solved is that of **trömelite**, a calcium phosphate characterized by Hill *et al.*[47] during their investigation of the P_2O_5–CaO system. Originally identified as $Ca_7(P_5O_{16})_2$, a formula corresponding to a possible pentaphosphate, this salt was further investigated by Van Wazer and Ohashi,[48,49] who confirmed the existence of a pentaphosphate anion by paper chromatography experiments. Later Wieker *et al.*[50] produced crystalline trömelite and by careful paper chromatography analysis assigned this compound the formula of a hexaphosphate, $Ca_4P_6O_{19}$. These authors reported the following triclinic unit cell dimensions for trömelite:

$$a = 9.40, b = 13.39, c = 7.07 \text{ Å}$$

$$\alpha = 109.5, \beta = 87.9, \gamma = 108.9°$$

Up to now, no structural data exist for this phosphate. In addition, starting from aqueous solutions of trömelite, the same authors claimed to have prepared of $[Co(NH_3)_6]_2$-$(P_6O_{19})_3 \cdot 20H_2O$.

2.2. Diphosphates

2.2.1. Introduction

Diphosphates, still frequently called pyrophosphates, constitute the largest family of condensed phosphates. Since the first characterization of such a salt by Berzelius in 1816, they have been the subject of hundreds of investigations, but nevertheless much remains to be done in this field, especially in regard to structural investigations.

In condensed phosphate chemistry, examples of acidic anions are generally very rare, being limited to one or two examples in the class of cyclophosphates and about the same number of long-chain polyphosphates. In contrast, acidic anions are common among the lower oligophosphates, the diphosphates and triphosphates. In diphosphate chemistry, the existence of $HP_2O_7^{3-}$, $H_2P_2O_7^{2-}$ and $H_3P_2O_7^{-}$ groups has been well established for a long time. Such groups exhibit a characteristic in crystalline materials, interconnected by very strong hydrogen bonds, they build infinite entities, chains, ribbons, or planes , commonly but perhaps improperly, called "macroanions." We will report in detail the various geometrical configurations of these "macroanions" for the case of diphosphates at the end of this chapter. In monophosphates, the $H_2PO_4^{-}$ and HPO_4^{2-} groups have the same tendency to form "macroanions," but, no systematic survey of their geometries has yet been performed.

2.2.2. Present State of Diphosphate Chemistry

2.2.2.1. Monovalent Cation Diphosphates

(i) $M_4P_2O_7 \cdot xH_2O$

$Li_4P_2O_7$. $Li_4P_2O_7$ appears as a congruent-melting compound (mp = 1158 K) in the $LiPO_3$–Li_3PO_4 phase-equilibrium diagram elaborated by Nakano et al.[51] This salt is dimorphous with a transition temperature of 903 K. These results are in good agreement with the findings previously reported by Tien and Hummel.[52] As far as we know, no crystal data exist for this salt.

$Na_4P_2O_7$. It may be of interest to note that very little is known about this very simple diphosphate, characterized by Berzelius as early as 1816 and by Clark in 1827. This compound was in fact the first condensed phosphate ever described in the chemical literature.

According to Lazarev et al.[53] $Na_4P_2O_7$ exhibits five reversible polymorphic transformations at 675.7, 785.0, 791.3, 815.7 and 827.4 K, respectively. The enthalpies of the phase transitions have been determined by differential scanning calorimetry (DSC) and the melting temperature was found to be 1261 K.

The crystal structure of an orthorhombic form of this salt has been reported by Leung and Calvo.[54]

$Na_4P_2O_7 \cdot 10H_2O$. The first X-ray investigations of this salt were reported by Sundara-Rao and Nampoothiri[55] and Corbridge,[56] and its crystal structure, first obtained by MacArthur and Beevers,[57] was later reinvestigated by Cruickshank[58] and by McDonald and Cruickshank.[59]

$Ag_4P_2O_7$. This diphosphate appears as a congruent-melting compound (mp = 916 K), with a phase transition at 647K, in the Ag_3PO_4–$AgPO_3$ phase-equilibrium diagram established by Osterheld and Moser.[60]

The growth of single crystals of $Ag_4P_2O_7$ by the Czochralski method was reported by Yamada and Koizumi.[61] In this study, the melting point was observed at 843 K and the phase transition at 623 K. Electrical and optical properties were described. Based on the present author's investigations, the crystal data reported by these authors seem questionable.

$K_4P_2O_7$ and $K_4P_2O_7 \cdot 3H_2O$. The existence of these two salts is well established. The very hygroscopic anhydrous salt has two crystalline forms. The low-temperature form, or β form, reported by Thilo and Dostal,[62] can be produced by heating K_2HPO_4 at 673 K. By heating this low-temperature form to 773 K Osterheld and Audrieth[63] obtained the α form. Dehydration of the trihydrate at 378 K also leads to the α form, whose optical properties were reported by Lehr *et al.*[64] Both dimorphs are stable but hygroscopic.

The trihydrate was characterized by Brun[65] and by Frazier *et al.*[66] during their investigation of the K_2O–$H_4P_2O_7$–H_2O system.

$(NH_4)_4P_2O_7$ and $(NH_4)_4P_2O_7 \cdot H_2O$. The chemical preparation of these two salts have been reported by Frazier *et al.*[67] The monohydrate crystallizes at room temperature from a concentrated aqueous solution kept at pH 6.5. In air, this salt dehydrates rapidly to the anhydrous salt. From optical measurements, this salt is monoclinic 2/m.

The anhydrous salt is obtained, at temperatures above 298 K, from moderately concentrated aqueous solutions kept at pH 6 or above by addition of ethanol. In air, it loses ammonia and alters to $(NH_4)_3P_2O_7$. On being heated rapidly to 378 K, it decomposes into a mixture of $(NH_4)H_2PO_4$ and orthorhombic $(NH_4)_2H_2P_2O_7$. Crystal data for the anhydrous salt were given by Frazier *et al.*[67]

$Tl_4P_2O_7$. Two crystalline forms of this salt were characterized by Dostal *et al.*[68] The transformation between the two forms occurs at 538 K and the melting point is 618 K. Unindexed powder diagrams were given by the authors.

(ii) $M_3HP_2O_7 \cdot xH_2O$

$Na_3HP_2O_7 \cdot 9H_2O$. After a previous crystallographic investigation by Corbridge,[56] crystal structure determination for this salt was performed by Emmerson and Corbridge[69] (see also Ingerson and Morey.[70])

$K_3HP_2O_7$, $K_3HP_2O_7 \cdot 1/2H_2O$ and $K_3HP_2O_7 \cdot 3H_2O$. Verdier *et al.*[71] investigated several chemical preparations of $K_3HP_2O_7$ and observed the existence of two crystalline forms for the anhydrous salt. Both forms are hygroscopic. The same authors characterized a hemihydrate, which is also hygroscopic.

The chemical preparation of the trihydrate and its main properties were described by Brun.[65] Its existence was later confirmed by Frazier *et al.*[66] and its atomic arrangement was determined by Dumas *et al.*[72]

$(NH_4)_3HP_2O_7$ and $(NH_4)_3HP_2O_7 \cdot H_2O$. These two salts were investigated by Frazier *et al.*[67] The monohydrate can be crystallized from an aqueous solution kept at a temperature below 328 K at pH 6. It is nonhygroscopic and relatively stable at room temperature. The anhydrous salt can be obtained by dehydration of the monohydrate, but large crystals are produced from an aqueous solution at temperatures above 328 K. Crystal and optical data were reported for these two salts.

(iii) $M_2H_2P_2O_7 \cdot xH_2O$

$Na_2H_2P_2O_7$ and $Na_2H_2P_2O_7 \cdot 6H_2O$. The unit-cell dimensions of the anhydrous salt were reported by Ingerson and Morey[70] and in ASTM Card 10-192. From these data, Averbuch-Pouchot and Durif[73] concluded that this salt is isotypic with the corresponding silver salt.

The unit-cell dimensions and space group of the hexahydrate were first reported by Corbridge,[56] and its crystal structure was determined by Collin and Willis,[74] who observed a strong structural analogy with $Na_2H_2P_2O_6 \cdot 6H_2O$. The authors prepared single crystals by adding glacial acetic acid to a solution of $Na_4P_2O_7 \cdot 10H_2O$ kept at 293 K. Crystals of $Na_2H_2P_2O_7 \cdot 6H_2O$ appeared when the solution was cooled to 283 K.

$Ag_2H_2P_2O_7$. This compound has been observed several times during the hydrolytic decyclization of various silver cyclophosphates, for instance, in the case of silver cyclo-triphosphate monohydrate the following reaction occurs:

$$2Ag_3P_3O_9 \cdot H_2O + H_2O \rightarrow 3Ag_2H_2P_2O_7$$

According to Lee,[75] it is also one of the silver phosphates observed during the thermal hydrolytic degradation of silver triphosphate dihydrate.

Averbuch-Pouchot and Durif[73] produced single crystals of $Ag_2H_2P_2O_7$ by the action of dilute nitric acid on $Ag_4P_2O_7$ and determined its crystal structure. Since its characterization this salt has been extensively used for the preparation of dihydrogendiphosphate derivatives through a metathesis reaction derived from Boullé's process[23] (see Chapter 4).

$K_2H_2P_2O_7$ and $K_2H_2P_2O_7 \cdot 1/2H_2O$. The chemical preparation of the anhydrous salt was been reported by Larbot[76] and its crystal structure determination later performed by Larbot *et al.*[77] The existence of these two salts was confirmed by Frazier *et al.*[66] during their investigation, at 298 K, of the $K_2O–H_4P_2O_7–H_2O$ system.

The unit cell dimensions and space group of the hemihydrate were first reported by Corbridge,[56] and its crystal structure, first determined by Emmerson and Corbridge,[78] was later reexamined by Dumas *et al.*[79]

$(NH_4)_2H_2P_2O_7$. Frazier *et al.*[67] described two crystalline forms of this salt, a monoclinic one stable at room temperature or below and an orthorhombic one stable above room temperature. Both forms are obtained by crystallization from an aqueous solution kept at pH 3.4. If the temperature of the solution is kept above room temperature the orthorhombic form is obtained; otherwise the monoclinic form is obtained. The orthorhombic form, stable up to about 388 K, converts rapidly at room temperature to the monoclinic form, which is stable and nonhygroscopic. Optical data are given for the two forms and crystal data reported for the monoclinic one. Crystal structure of the monoclinic form was recently determined by Averbuch-Pouchot and Durif.[80]

$Rb_2H_2P_2O_7$ or $Rb_2H_2P_2O_7 \cdot 1/2H_2O$ and $Cs_2H_2P_2O_7$. These two salts, prepared by the action of $H_4P_2O_7$ on the corresponding carbonates, were described by Larbot *et al.*[81] The cesium salt can also be obtained by heating CsH_2PO_4 at 468–478 K. Crystal data were reported by Larbot *et al.*[81] Averbuch-Pouchot and Durif[82] determined the crystal structure of a hemihydrate of the rubidium salt having the same unit-cell dimensions and space group as the previously reported anhydrous salt. Crystal structure of the cesium salt has been performed by Averbuch-Pouchot and Durif.[83]

$Tl_2H_2P_2O_7$ and $Tl_2H_2P_2O_7 \cdot 1/2H_2O$. These two thallium salts were characterized by Dostal *et al.*[68] among the intermediate products of the thermal dehydration of TlH_2PO_4. Unindexed X-ray powder diffraction patterns were given by these authors. The main crystallographic features of the anhydrous salt have been recently determined by Averbuch-Pouchot and Durif.[84]

(iv) $MH_3P_2O_7 \cdot xH_2O$

$NaH_3P_2O_7$. This salt was first characterized by Giran.[85] Another preparation was described by Partington and Wallsom.[86] Norbert and Dautel[87] measured the melting point (458–473K), described some chemical properties, and reported an unindexed X-ray powder diagram.

$KH_3P_2O_7 \cdot H_2O$. This diphosphate has been characterized by Frazier *et al.*[66] during their investigation of the $K_2O - H_4P_2O_7 - H_2O$ system.

$CsH_3P_2O_7 \cdot H_2O$. This salt prepared by action of $H_4P_2O_7$ on Cs_2CO_3,[81] transforms into the anhydrous salt at 343 K and into $CsH_5(PO_4)_2$ when heated rapidly to 353 K. Crystal structure was performed by Larbot *et al.*[88]

(v) Other Monovalent Cation Diphosphates

Some salts with a more complex stoichiometry have also been characterized:

$K_3H_5(P_2O_7)_2$ was characterized by by Frazier *et al.*[66] during the investigation of the $K_2O - H_4P_2O_7 - H_2O$ system.
$K_4H_8(P_2O_7)_3$ was characterized by Brun,[89] who obtained this salt by adding potassium hydroxide to a methanolic solution of $H_4P_2O_7$.
$K_3H(H_2P_2O_7)_2$ was characterized by Brun,[89,90] who reported its chemical preparation. Its crystal structure was later determined by Dumas.[91]

Table 2.2.1 reports the main crystallographic data for monovalent cation diphosphates.

(vi) Mixed-Monovalent Cation Diphosphates

$NaK_3(H_2P_2O_7)_2$. Crystal structure of this salt was determined by X-ray diffraction by Dumas and Lapasset[92] and was reexamined with neutron diffraction by Dumas *et al.*[93]

$NaK_3P_2O_7 \cdot 4H_2O$. The chemical preparation and crystal structure of $NaK_3P_2O_7 \cdot 4H_2O$ have been reported by Dumas *et al.*[94]

Solubility studies on the system $NH_3 - K_2O - H_3PO_4 - H_4P_2O_7 - H_2O$ at 298 K by Frazier *et al.*[95] produced three new ammonium–potassium diphosphates, corresponding approximately to the following formulas:

$$(NH_4)_3K(H_2P_2O_7)_2 \qquad N/K \sim 2.98$$

$$(NH_4)_{1.6}K_{2.4}(H_2P_2O_7)_2 \cdot H_2O \qquad N/K \sim 0.68$$

Table 2.2.1. Main Crystallographic Data for Monovalent Cation Diphosphates

Formula	a (Å) α (°)	b (Å) β (°)	c (Å) γ (°)	S.G.	Z	Reference(s)
$Na_4P_2O_7$	9.367	5.390	13.480	$P2_12_12_1$	4	54
$Na_4P_2O_7 \cdot 10H_2O$	7.01(2)	6.96(1) 112.0(2)	14.85(2)	$C2/c$	4	55–59
$Ag_4P_2O_7$	9.538	9.538	40.83	$P31?$	18	61
$(NH_4)_4P_2O_7$	11.77	6.51 104.7	13.63	$C2/c$	4	67
$Na_3HP_2O_7 \cdot 9H_2O$	8.60	31.46 113.8	6.12	$P2_1/a$	4	56
$K_3HP_2O_7 \cdot 3H_2O$	6.114(4)	10.35(6) 90.1(1)	16.97(1)	$P2_1/c$	4	72
$(NH_4)_3HP_2O_7$	6.63	19.74 111.0	7.09	$P2_1/c$	4	67
$(NH_4)_3HP_2O_7 \cdot H_2O$	9.13 115.4	6.33 77.2	9.96 109.2	$P\bar{1}$	2	67
$Na_2H_2P_2O_7$	27.49	12.35	6.856	$Fddd$	16	70
$Ag_2H_2P_2O_7$	27.78(2)	12.385(6)	7.026(4)	$Fddd$	16	73
$Na_2H_2P_2O_7 \cdot 6H_2O$	14.099(6)	6.959(4) 117.69(4)	13.455(8)	$C2/c$	4	74
$K_2H_2P_2O_7$	7.001(1) 101.26(3)	8.918(2) 106.20(2)	6.576(1) 107.73(3)	$P\bar{1}$	2	77
$K_2H_2P_2O_7 \cdot 1/2H_2O$	17.96(1)	6.958(5) 120.9(1)	14.24(1)	$C2/c$	8	78, 79
$(NH_4)_2H_2P_2O_7$	9.058(7)	11.199(8) 108.40(1)	7.764(6)	$P2_1/a$	4	67, 80
$Rb_2H_2P_2O_7 \cdot 1/2H_2O$	19.55(4)	10.534(3)	7.784(3)	$Pna2_1$	8	81, 82
$Cs_2H_2P_2O_7$	7.967(3)	9.055(3) 90.00(1)	11.404(3)	Cc or $C2/c$	4	81
$Tl_2H_2P_2O_7$	17.896(8)	7.094(2) 120.09(5)	14.698(6)	$C2/c$	4	84
$K_3H(H_2P_2O_7)_2$	17.19(1)	8.975(5) 128.59(5)	10.921(8)	$C2/c$	4	91
$CsH_3P_2O_7 \cdot H_2O$	8.142(3) 89.03(2)	6.883(3) 104.73(2)	7.765(3) 110.86(2)	$P\bar{1}$ or $P1$	2	81, 88

$$(NH_4)_3K_3(HP_2O_7)_2 \cdot 2H_2O \qquad N/K \sim 1.02$$

These three diphosphates were later, clearly characterized by Waerstad and Frazier,[96] who reported the unit cells and main crystallographic features for these three species. From the crystal data given for the second of these salts, Averbuch-Pouchot and Durif[82] showed it to be isotypic with $Rb_2H_2P_2O_7 \cdot 1/2H_2O$.

The main crystal data for mixed-monovalent cation diphosphates are given in Table 2.2.2.

2.2.2.2. Divalent Cation Diphosphates

A large number of divalent metal diphosphates, $M_2P_2O_7$ have been investigated during the past three decades. Most of these compounds are polymorphic. For smaller divalent

Table 2.2.2. Main Crystallographic Data for Mixed-Monovalent Cation Diphosphates

Formula	a (Å) α (°)	b(Å) β (°)	c(Å) γ (°)	S.G.	Z	Reference(s)
$NaK_3(H_2P_2O_7)_2$	6.912	7.251 100.25	11.015	$C2/c$	4	92, 93
$NaK_3P_2O_7 \cdot 4H_2O$	5.681	11.993 92.54	17.509	$P2_1/c$	4	94
$(NH_4)_3K(H_2P_2O_7)_2$	7.950(3) 90.89(4)	7.270(3) 117.61(3)	7.184(3) 81.86(4)	$P\bar{1}$ or $P1$	1	96
$(NH_4)_{1.6}K_{2.4}(H_2P_2O_7)_2 \cdot H_2O$	19.22(2)	7.64(1)	10.39(1)	$Pnma$	4	82, 96
$(NH_4)_3K_3(HP_2O_7)_2 \cdot 2H_2O$	19.41(2)	6.158(6) 107.5(1)	16.90(2)	Cc or $C2/c$	4	96

cations (e.g. Cu, Zn, Co), the high-temperature form is in many cases isotypic with thortveitite, $Sc_2Si_2O_7$, a disilicate previously investigated by Zachariasen.[97] Good surveys of anhydrous divalent cation diphosphates have been published by Nord and Kierkegaard[98] and Brown and Calvo.[99]

Because of their potential use as fertilizers, calcium diphosphates have received more attention than the other members of the divalent cation diphosphate family.

(i) Anhydrous Diphosphates

$Be_2P_2O_7$. According to Jaulmes,[100] $Be_2P_2O_7$ appears to be stable up to its melting point (about 1573 K). She observed two forms for $Be_2P_2O_7$ as well a third salt with a similar global formula but formulated as $Be(PO_3)_2 \cdot BeO$.

The existence of two forms of this salt was confirmed by de Sallier-Dupin.[101,102] The high-temperature form cannot be produced below 1173 K. Both authors reported only unindexed powder diagrams. Boullé and de Sallier-Dupin[103] investigated the thermal decomposition of $BeNH_4PO_4 \cdot H_2O$ and observed the formation of $Be_2P_2O_7$ at about 923 K.

Bleyer and Müller[104] reported the characterization of a nonahydrate, $Be_2P_2O_7 \cdot 9H_2O$ whose existence has not yet been confirmed.

$Mg_2P_2O_7$. This diphosphate appears as a congruent-melting compound (mp = 1655 K) in the P_2O_5–MgO phase-equilibrium diagram elaborated by Berak.[105] Two crystalline forms are known for this salt, and, according to Katnack and Hummel,[106] the $\alpha \rightarrow \beta$ transition occurs at 343 K and is reversible. The crystal structure of the α form was described simultaneously by Lukaszewicz[107] and Calvo.[108] Both authors discussed the mechanism of the transition. The crystal structure of the β form was first determined by Lukaszewicz[109] and was later refined by Calvo.[110]

$Mn_2P_2O_7$. A part of the P_2O_5–MnO phase-equilibrium diagram was determined by Konstant and Dimante.[111] $Mn_2P_2O_7$ appears in this diagram as a congruent melting compound (1474 K).

Stefanidis and Nord[112] investigated $Mn_2P_2O_7$ by neutron diffraction and reported a structure isotypic with thortveitite.

$Fe_2P_2O_7$. Royen and Korinth[113] obtained this diphosphate by reduction of $FePO_4$ with hydrogen. They reported a powder diagram. Stefanidis and Nord[114] produced single crystals from a mixture of Fe_2O_3, Fe, and $(NH_4)H_2PO_4$ in the molar proportions 1:1:3 kept under vacuum at 1170 K for two days and then slowly cooled. They performed crystal structure determination for $Fe_2P_2O_7$, assuming the non-centrosymmetric $P1$ space group. Later Hoggins et al.[115] described the same compound with the non-conventional $C1$ space group to compare the unit cell with that of the other monoclinic $M_2P_2O_7$ diphosphates. They concluded that this diphosphate is isostructural with the high-temperature form of $Mg_2P_2O_7$. In fact, it is a triclinic distortion of this form.

$Co_2P_2O_7$. $Co_2P_2O_7$ appears as a congruent melting compound (mp = 1513 K) in the P_2O_5–CoO phase diagram elaborated by Sarver.[116] Crystal structure of the α form was determined by Krishnamachari and Calvo.[117] This structure is almost identical to that of α-$Mg_2P_2O_7$.

A detailed procedure for growing large single crystals of $Co_2P_2O_7$ was described by Wanklyn et al.[118] Crystals up to 20 mm were obtained.

$Ni_2P_2O_7$. Four crystalline forms of $Ni_2P_2O_7$ are presently known. The crystal structure of α-$Ni_2P_2O_7$, the stable form at room temperature, was determined by Lukaszewicz.[119] This form was shown to be isotypic with the corresponding magnesium salt. Crystal structure of a high temperature form, β-$Ni_2P_2O_7$, isotypic with thortveitite, was determined at 853 K by Pietraszko and Lukaszewicz.[120] The $\alpha \rightarrow \beta$ transition temperature determined by X-ray experiments is 838 K. A γ form, sometimes denoted α', investigated by Lukaszewicz[119] is very probably isotypic with α-$Zn_2P_2O_7$. A fourth variety, δ-$Ni_2P_2O_7$, prepared and investigated by Masse et al.[121] is isotypic with $Er_2Si_2O_7$.

$Ni_2P_2O_7$ appears as a congruent-melting compound (mp = 1668 K) in the P_2O_5–NiO system investigated by Sarver.[116] Solid solutions corresponding to $Ni_{2-x}M_xP_2O_7$ for M = Mg, Mn, Co, Cu, and Zn were examined by Ericsson and Nord.[122]

$Cu_2P_2O_7$. Lukaszewicz and Nagler[123] were the first to measure unit-cell dimensions of α-$Cu_2P_2O_7$. Crystal structure of this form was later on determined by Robertson and Calvo,[124] and that of the β form was determined at 373 K by the same authors.[125]

A study of the phase-equilibrium relationships in the CuO–P_2O_5 system by Ball[126] confirmed the existence of $Cu_2P_2O_7$.

$Zn_2P_2O_7$. The ZnO–$Zn(PO_3)_2$ phase-equilibrium diagram has been determined by Katnack and Hummel.[106] In this diagram, $Zn_2P_2O_7$ appears as a congruent-melting compound (mp = 1290 K) with a reversible transition at 405 K. This transition was observed at 401.2 K by Lazarev et al.[127] and its enthalpy determined by DSC. Thermal expansion coefficients for both α- and β-$Zn_2P_2O_7$ have been determined by Brown and Hummel[128] and correlated with previous phase-equilibrium data.

Crystal structure of the low-temperature form (α) was determined by Robertson and Calvo,[129] and the atomic arrangement of the high-temperature form (β) was investigated by Calvo.[130] The existence of a third modification at higher temperature has been suspected but has not yet been clearly established.

$ZnH_2P_2O_7$. The chemical preparation of $ZnH_2P_2O_7$ and crystal data for this salt reported by Averbuch-Pouchot.[131]

$Ca_2P_2O_7$ and $CaH_2P_2O_7$. The anhydrous salt is trimorphous. The crystal structure of the α form was determined by Calvo.[132] The atomic arrangement is closely related to that of α-$Sr_2P_2O_7$ although the latter has orthorhombic symmetry.

The unit-cell dimensions and space group for β-$Ca_2P_2O_7$ were first reported by Corbridge,[56] and the crystal structure determined by Webb.[133] The latter author prepared β-$Ca_2P_2O_7$ using $CaHPO_4 \cdot 2H_2O$ as starting material.

$$2CaHPO_4 \cdot 2H_2O \xrightarrow{\text{393 K } -4H_2O} 2CaHPO_4 \xrightarrow{\text{723 K } -H_2O} \gamma\text{-}Ca_2P_2O_7 \xrightarrow{\text{393 K}} \beta\text{-}Ca_2P_2O_7$$

Crystals were obtained by heating a sample of $CaHPO_4 \cdot 2H_2O$ up to 1673 K and slowly cooling the melt.

The chemical preparation of $CaH_2P_2O_7$ was described by Brown et al.,[134] Hill et al.,[135] and Lehr et al.[136] From optical examination, it was described as orthorhombic *mmm*. This salt is not stable in aqueous solutions and precipitates as $Ca_3H_2(P_2O_7)_2 \cdot 4H_2O$ or as the triclinic form of $Ca_2P_2O_7 \cdot 2H_2O$ depending on the concentration. In dilute solutions, only the second salt precipitates.

$Ca_3H_2(P_2O_7)_2 \cdot H_2O$ and $Ca_3H_2(P_2O_7)_2 \cdot 4H_2O$. The tetrahydrate was first observed by Brown and co-workers,[137,138] during the hydrolysis of vitreous long-chain calcium polyphosphate. The chemical preparation of these two hydrates was reported by Brown et al.[134,139] Both hydrates dissolve very slowly in water and transform into the triclinic form of $Ca_2P_2O_7 \cdot 2H_2O$. According to optical measurements, the monohydrate is ortho-rhombic *mmm* and the tetrahydrate is triclinic.

$Sr_2P_2O_7$ and $SrH_2P_2O_7$. Several strontium diphosphates were characterized by Ropp et al.[14] $SrH_2P_2O_7$ is obtained by thermal condensation of $Sr(H_2PO_4)_2$ at 463–483 K:

$$Sr(H_2PO_4)_2 \rightarrow SrH_2P_2O_7 + H_2O$$

This reaction is extremely slow. Further heating at 593–603 K leads to the polyphosphate:

$$SrH_2P_2O_7 \rightarrow Sr(PO_3)_2 + H_2O$$

Two forms of $Sr_2P_2O_7$ and $Sr_2P_2O_7 \cdot 1/2H_2O$ were also described. According to the authors, β-$Sr_2P_2O_7$ can be obtained by heating α-$SrHPO_4$ at 673 K:

$$2\alpha\text{-}SrHPO_4 \rightarrow \beta\text{-}Sr_2P_2O_7 + H_2O$$

the hemihydrate by heating β-$SrHPO_4$ at 573 K, and β-$Sr_2P_2O_7$ by heating the hemihydrate at 848 K.

Kreidler and Hummel[141] investigated the phase relationships in the system SrO–P_2O_5. From their investigations, $Sr_2P_2O_7$ appears as a congruent-melting compound (mp = 1648

K). The salt is dimorphous with a transition temperature at 1048 K. This transformation and its reversibility were discussed by the authors.

Hoffman and Mooney[142] were the first to recognize β-$Sr_2P_2O_7$ as an isotype of the corresponding calcium salt.

The crystal structure of α-$Sr_2P_2O_7$ was determined almost simultaneously by Hagman et al.[143] and Grenier and Masse.[144,145]

$Sr_2P_2O_7$ and $Ba_2P_2O_7$ were prepared by Klement[146] by the reaction of SrO or BaO on molten $NaPO_3$.

$Ba_2P_2O_7$. Mc Cauley and Hummel[147] investigated a part of the BaO–P_2O_5 phase-equilibrium diagram and reported the existence of two crystalline forms of this diphosphate, melting congruently at 1703K. The low temperature α form transforms at about 1073 K into the δ form. The α form is isomorphous with α-$Sr_2P_2O_7$ and α-$Ca_2P_2O_7$.

The properties of $Ba_2P_2O_7$ as a phosphor material were investigated by Ranby et al.[148] and a mass-spectroscopic study of its thermal dissociation was performed by Lopatyin and Semenov.[149]

Table 2.2.3. Main Crystallographic Data for Divalent Cation Diphosphates

Formula	a (Å) α (°)	b (Å) β (°)	c (Å) γ (°)	S.G.	Z	Reference(s)
α–$Mg_2P_2O_7$	6.981(5) 	8.295(5) 113.0(1)	9.072(5)	$P2_1/c$	4	107, 108
β–$Mg_2P_2O_7$	6.494(7) 	8.28(1) 103.8(1)	4.522(5)	$C2/m$	2	109, 110
$Co_2P_2O_7$	13.248(6) 	8.345(3) 104.60(6)	9.004(3)	$B2_1/c$	8	117
α–$Zn_2P_2O_7$	20.07(2) 	8.259(6) 106.35(5)	9.099(8)	$I2/c$	12	129
β–$Zn_2P_2O_7$	6.61(1) 	8.29(1) 105.4(2)	4.51(1)	$C2/m$	2	130
α–$Ni_2P_2O_7$	13.093 	8.275 104.94	8.974	$B2_1/c$	8	119
β–$Ni_2P_2O_7$ (853 K)	6.501 	8.239 104.14	4.480	$C2/m$	2	120
β–$Cu_2P_2O_7$	6.827(8) 	8.118(10) 108.85(10)	4.576(6)	$C2/m$	2	125
α–$Cu_2P_2O_7$	6.876(5) 	8.113(5) 109.54(6)	9.162(5)	$C2/c$	4	123, 124
$Ni_2P_2O_7(IV)$	5.212(3) 	9.913(5) 97.46(10)	4.475(3)	$P2_1/a$	2	121
$ZnH_2P_2O_7$	9.660(9) 	12.65(1) 106.20(1)	9.099(6)	$P2_1/a$	8	131
$Mn_2P_2O_7$	6.633 	8.584 102.67	4.546	$C2/m$	2	112
$Fe_2P_2O_7$	6.649(2) 90.04(3)	8.484(2) 103.89(3)	4.488(1) 92.82(3)	$C\bar{1}$	2	115
$Fe_2P_2O_7$	5.517 98.73	5.255 98.33	4.488 103.81	$P1?$	1	114

Table 2.2.4. Main Crystallographic Data for Divalent Cation Diphosphates and their Hydrates

Formula	a (Å) α (°)	b (Å) β (°)	c (Å) γ (°)	S.G.	Z	Reference(s)
α–Ca$_2$P$_2$O$_7$	12.66(1)	8.542(8) 90.3(1)	5.315(5)	$P2_1/n$	4	132
β–Ca$_2$P$_2$O$_7$	6.684(6)	6.684(6)	24.14(1)	$P4_1$	8	133
β–Sr$_2$P$_2$O$_7$	6.920	6.920	24.79	$P4_1$	8	142
α–Sr$_2$P$_2$O$_7$	8.910(1)	5.403(1)	13.105(1)	$Pnma$	4	143–145
Cd$_2$P$_2$O$_7$	6.858 82.38	6.672 95.80	6.623 115.38	$P\bar{1}$	2	150
Pb$_2$P$_2$O$_7{}^a$	6.963(1) 96.78(1)	6.975(1) 91.16(1)	12.764(1) 89.68(1)	$P\bar{1}$	4	151, 153
Pb$_2$P$_2$O$_7{}^b$	12.852(7)	7.076(5) 95.4(4)	7.096(3)	$P2_1/n$	4	152
Zn$_2$P$_2$O$_7$·4H$_2$O	9.07	25.27	8.32	$Pnma$	8	56
Ca$_2$P$_2$O$_7$·2H$_2$O	7.365(4) 102.96(1)	8.287(4) 72.73(1)	6.691(4) 95.01(1)	$P\bar{1}$	2	177
Ca$_2$P$_2$O$_7$·4H$_2$O	6.01	25.0 109.54	6.86	$P2_1/c$	4	134
Mg$_2$P$_2$O$_7$·2H$_2$O	7.367(1)	13.906(3) 94.37(3)	6.277(1)	$P2_1/n$	4	167
Co$_2$P$_2$O$_7$·2H$_2$O	7.637(1)	13.997(2) 94.77(2)	6.334(1)	$P2_1/n$	4	174
Mn$_2$P$_2$O$_7$·2H$_2$O	7.570	14.325 95.20	6.461	$P2_1/n$	4	172

^{a}LT: low-temperature form.
^{b}HT: high-temperature form.

Cd$_2$P$_2$O$_7$. As established by Brown and Hummel[128] during their investigation of the CdO–P$_2$O$_5$ phase diagram, Cd$_2$P$_2$O$_7$ is a congruent-melting compound (mp = 1393 K). Crystal structure determination for this triclinic diphosphate was later performed by Calvo and Au.[150]

Pb$_2$P$_2$O$_7$. The crystal growth and some properties of Pb$_2$P$_2$O$_7$ were described by Brixner et al.[151] Bruckner and Worzala[152] observed a reversible phase transformation for this salt at 943 K and proposed an atomic arrangement for the high-temperature form. The crystal structure of the room-temperature form was performed by Mullica et al.[153]

Crystallographic data for divalent cation diphosphates are gathered in Tables 2.2.3 and 2.2.4.

(ii) Anhydrous Mixed Divalent Cation Diphosphates

A good number of MM′P$_2$O$_7$ diphosphates have been prepared and investigated. They belong to two structural types, one represented by a series of ten monoclinic compounds (form I) and the other by a group of five triclinic compounds (form II). According to Riou and Raveau[154] the monoclinic salts have an atomic arrangement similar to that of α-Ca$_2$P$_2$O$_7$. The diphosphates corresponding to form II are isotypic with KAgCr$_2$O$_7$. The

Table 2.2.6. Main Crystallographic Data for Mixed-Divalent Cation Diphosphates Belonging to Form II and for $MM'_3(P_2O_7)_2$ Compounds

Formula	a (Å) α (°)	b (Å) β (°)	c (Å) γ (°)	S.G.	Z	Reference(s)
$BaNiP_2O_7$	5.323(1) 101.22(1)	7.580(1) 84.19(1)	7.117(1) 89.32(1)	$P\bar{1}$	2	158, 160
$BaCdP_2O_7$	5.641(1) 78.55(2)	7.038(2) 89.83(2)	7.624(2) 86.79(2)	$P\bar{1}$	2	157, 160
$BaCoP_2O_7$	5.370(1) 102.06(1)	7.580(1) 85.31(1)	7.151(1) 88.99(1)	$P\bar{1}$	2	158, 160
$BaZnP_2O_7$	5.316(1) 102.68(2)	7.309(2) 92.13(2)	7.579(2) 94.08(2)	$P\bar{1}$	2	160
$BaCuP_2O_7$	7.353(2) 90.83(2)	7.578(2) 95.58(2)	5.231(1) 103.00(2)	$P\bar{1}$	2	159
$PbNi_3(P_2O_7)_2$	7.419(3)	9.455(4) 112.26(5)	7.676(3)	$P2_1/b$	2	162, 163
$PbCo_3(P_2O_7)_2$	7.495(4)	9.522(5) 111.94(6)	7.753(4)	$P2_1/b$	2	162

main crystallographic data for these compounds are reported in Tables 2.2.5 and 2.2.6. Crystal structure determinations for these two classes of diphosphates have been performed by various authors.

A 1:3 stoichiometry for the associated cations is also observed in this category of diphosphates, $PbNi_3(P_2O_7)_2$ and $PbCo_3(P_2O_7)_2$ being examples.

$CaCuP_2O_7$, $SrCoP_2O_7$ and $SrCuP_2O_7$. These three diphosphates belong to form I. The calcium-copper salt was synthesized and investigated by Riou and Goreaud,[155] and the strontium–cobalt salt by Riou and Raveau.[154] Crystal structures determinations were performed for both compounds. The atomic arrangement is closely related to that of α-$Ca_2P_2O_7$. Crystal data for $SrCoP_2O_7$ were reported by Boukhari *et al.*[156]

$SrCdP_2O_7$ and $CdBaP_2O_7$. These two diphosphates were described by Alaoui *et al.*[157] Crystals were obtained by slow cooling from the melt. The atomic arrangement in $CdSrP_2O_7$ resembles that in the high-temperature form of $Sr_2P_2O_7$, whereas $CdBaP_2O_7$ is isostructural with $BaCoP_2O_7$ and $BaCuP_2O_7$.

$BaMP_2O_7$ (M = Ni, Co, Cu, Cd, Zn, Mg). Riou *et al.*[158] described the syntheses of two mixed diphosphates, $BaNiP_2O_7$ and $BaCoP_2O_7$, prepared from powders at 1173 and 1373 K respectively. The two compounds are isotypic. The reported atomic arrangement of the cobalt salt showed the existence of a layered structure. Some discrepancies in the thermal coefficients indicate that additional work is needed to confirm the proposed atomic arrangement. Incommensurability phenomena appear for the nickel salt.

The chemical preparation and crystal structure of $BaCuP_2O_7$ were recently reported by Moqine *et al.*[159] The crystal structures of $BaCdP_2O_7$ and $BaZnP_2O_7$ were performed by Murashova *et al.*[160] In addition the authors reported crystal data for $BaCoP_2O_7$ and $BaNiP_2O_7$. Crystal structure determinations for $BaMgP_2O_7$ and $PbZnP_2O_7$ were performed

Table 2.2.5. Main Crystallographic Data for Mixed Divalent Cation Diphosphates Belonging to Form I

Formula	a (Å) α (°)	b (Å) β (°)	c (Å) γ (°)	S.G.	Z	Reference
$CaCuP_2O_7$	5.2104(4)	8.0574(5) 91.356(6)	12.344(1)	$P2_1/n$	4	155
$SrCoP_2O_7$	5.3165(4)	8.2574(5) 90.133(5)	12.6755(7)	$P2_1/n$	4	154
$SrMgP_2O_7$	5.309(7)	8.299(9) 90.6(1)	12.68(2)	$P2_1/n$	4	161
$SrCuP_2O_7$	5.37(1)	8.13(2) 90.6(1)	12.46(2)	$P2_1/n$	4	156
$SrZnP_2O_7$	5.299(6)	8.198(9) 90.0(1)	12.72(1)	$P2_1/n$	4	161
$SrCdP_2O_7$	5.414(1)	8.615(3) 90.01(3)	12.878(5)	$P2_1/n$	4	157
$BaMgP_2O_7$	5.483(1)	8.561(3) 91.32(2)	12.626(2)	$P2_1/n$	4	161
$PbZnP_2O_7$	5.113(1)	8.294(1) 90.02(1)	12.847(2)	$P2_1/n$	4	161
$PbCoP_2O_7$	5.306	8.251 90.03	12.74	$P2_1/n$	4	161
$PbCuP_2O_7$	5.384(1)	8.209(1) 90.43(2)	12.559(3)	$P2_1/n$	4	104

by Murashova *et al.*[161] These authors also reported crystal data for four isotypic compounds, $SrMgP_2O_7$, $SrZnP_2O_7$, $PbCuP_2O_7$, and $PbCoP_2O_7$.

$PbCoP_2O_7$. A detailed procedure for growing large single crystals of $PbCoP_2O_7$ was reported by Wanklyn *et al.*[118] Crystals up to 10 mm. were obtained.

$PbNi_3(P_2O_7)_2$ and $PbCo_3(P_2O_7)_2$. Crystals of $PbNi_3(P_2O_7)_2$ and $PbCo_3(P_2O_7)_2$ were grown by slow cooling of a melt of $Ni_2P_4O_{12}$ or $Co_2P_4O_{12}$ and PbO by Dindune *et al.*[162] These two monoclinic salts are isotypic. Indexed powder diagrams were reported. A detailed determination of the atomic arrangement of the nickel salt was later reported by Krasnikov *et al.*[163]

Other Systems. A study of the $Sr_2P_2O_7$–$Mg_2P_2O_7$ system was performed by Calvo.[164] On the basis of crystal data, the author discussed the change in symmetry as a function of the cation ratios. Terpstra *et al.*[165] investigated the $Ca_2P_2O_7$–$Mg_2P_2O_7$ system at 1273 K. The $Zn_2P_2O_7$–$Cd_2P_2O_7$ system was investigated by Brown and Hummel.[166] The phase-equilibrium diagram is a simple eutectic type system.

(iii) Hydrated Divalent Cation Diphosphates

$Mg_2P_2O_7 \cdot 2H_2O$, $Mg_2P_2O_7 \cdot 5H_2O$, and $Mg_2P_2O_7 \cdot 8H_2O$. The dihydrate was first characterized by Oka and Kawahara.[167] The authors gave a detailed procedure for the synthesis of this salt and a complete description of its atomic arrangement, identical to that of the corresponding manganese salt. $Mg_2P_2O_7 \cdot 5H_2O$ and $Mg_2P_2O_7 \cdot 8H_2O$ were characterized by Kokhanovskii *et al.*[168] during an investigation of the $K_4P_2O_7$–$Mg(NO_3)_2$–H_2O system.

$Ni_3(HP_2O_7)_2 \cdot 10H_2O$, $Ni(H_3P_2O_7)_2 \cdot 2H_2O$ and $NiH_2P_2O_7 \cdot nH_2O$. The existence of these salts was established by Lavrov and Bykanova[169] in their investigation of the solubility isotherm for the $Ni_2P_2O_7$–$H_4P_2O_7$–H_2O system at 273 K under equilibrium conditions.

$Zn_2P_2O_7 \cdot nH_2O$ (n = 2, 3, 4, and 5). The tri- and pentahydrates were clearly characterized as crystalline compounds by Quimby and McCune[170] during a study of the precipitation of zinc phosphates from solutions of sodium diphosphate. The thermal stability of $Zn_2P_2O_7 \cdot 5H_2O$ was investigated by Selivanova *et al.*[171] According to these authors dehydration occurs according to the following scheme:

$$An_2P_2O_7 \cdot 5H_2O \xrightarrow{358K} Zn_2P_2O_7 \cdot 3H_2O \xrightarrow{410K} Zn_2P_2O_7 \text{ (amorphous)}$$

The melting point of $Zn_2P_2O_7$ reported by these authors is 1253 K, which is slightly lower than that reported by Katnack and Hummel.[106] Optical and density data were reported. Crystal data for the tetrahydrate were reported by Corbridge.[56]

Detailed procedures for the chemical preparation for the di- and pentahydrates were given by Lehr *et al.*[64] Both salts are not water soluble and are orthorhombic *mmm*.

$Mn_2P_2O_7 \cdot 2H_2O$ and $Mn_2P_2O_7 \cdot 5H_2O$. Crystal structure determination for the dihydrate was performed by Schneider and Collin,[172] and the existence of $Mn_2P_2O_7 \cdot 5H_2O$ was reported by Goloshchapov and Martynenko.[173]

$Co_2P_2O_7 \cdot 2H_2O$. Effenberger and Pertlik[174] determined the crystal structure of this salt, showing it to be isotypic with the corresponding manganese salt.

$Ca_2P_2O_7 \cdot 2H_2O$. As observed by Lindsay and co-workers,[175,176] this hydrate has two crystalline forms, one monoclinic and the other one triclinic. The stable form is the triclinic one. In contact with water, the monoclinic form slowly transforms into the triclinic one. This hydrate was first observed during the hydrolysis of calcium polyphosphates in large quantities of water by Brown *et al.*,[137] and its chemical preparation was described by Brown *et al.*[134] This salt was also observed by Brown *et al.*[139] during their investigation of the $(K,NH_4)_2O$–$H_4P_2O_7$–H_2O system.

Naturally occuring crystals of this salt have been observed among compounds identified as for responsible for gouty arthritis. The crystal structure of the triclinic form was determined by Mandell[177] using crystals prepared according to the procedure described by Brown *et al.*[134]

$Ca_2P_2O_7 \cdot 4H_2O$. This diphosphate is dimorphous, one form is orthorhombic and the other one monoclinic. Both forms were investigated by Brown *et al.*,[134] who reported their chemical preparation. The orthorhombic form is not water soluble and transforms slowly into the triclinic form of the dihydrate. The monoclinic form, which is more stable, is also not soluble and, on standing in water, transforms slowly into the orthorhombic form and then to triclinic $Ca_2P_2O_7 \cdot 2H_2O$. Except for the measurement of the unit-cell dimensions of the monoclinic form, no structural data have been reported.

2.2.2.3. Divalent–Monovalent Cation Diphosphates

Numerous types of chemical formulas have been observed in this class of compounds. The most important group corresponds to the $M^{II}M^I_2P_2O_7$ salts and their hydrates. In order to avoid repetition and duplication of references, our classification of the compounds in this family is not based on their chemical formulas but rather on the nature of the divalent cation. The main crystallographic data available for compounds reviewed in this section are presented in Tables 2.2.7–2.2.9.

(i) Magnesium Salts

$MgNa_2P_2O_7$. This diphosphate was prepared by Klement[178] by the action of MgO on molten $NaPO_3$.

$Mg_9Na_{14}(P_2O_7)_8$. In the $Na_4P_2O_7$–$Mg_2P_2O_7$ phase-equilibrium diagram established by Majling and Hanic,[179] $Mg_9Na_{14}(P_2O_7)_8$ is the only intermediate compound. It melts congruently at 1105 K. Its crystal structure was later determined by Hanic and Zac.[180]

$MgK_2P_2O_7{\cdot}4H_2O$, $Mg_3K_2(P_2O_7)_2{\cdot}6H_2O$, and $Mg_3K_2(P_2O_7)_2{\cdot}11H_2O$. These three diphosphates were characterized by Kokhanovskii et al.[181] during their investigation of the $K_4P_2O_7$–$Mg(NO_3)_2$–H_2O system.

$MgNH_4HP_2O_7$. This diphosphate was identified during an investigation of the interaction of MgO and $NH_4H_2PO_4$ by Atstinya et al.[182] Its formation is observed at 493–513 K when the starting mixture corresponds to a ratio P/Mg of 2–8. Its thermal behavior was investigated by the authors. The final product is magnesium cyclotetraphosphate:

$$MgNH_4HP_2O_7 \xrightarrow{\text{513–655 K}} \text{amorphous mixture} \xrightarrow{\text{913 K}} Mg_2P_4O_{12}$$

Table 2.2.7. Main Crystallographic Data for Diphosphates Characterized during the investigations of the CaO–$(NH_4)_2O$–$H_4P_2O_7$–H_2O and CaO–K_2O–$H_4P_2O_7$–H_2O Systems[a]

Formula	a (Å) α (°)	b (Å) β (°)	c (Å) γ (°)	S.G.	Z
$Ca_3(NH_4)_2(P_2O_7)_2{\cdot}6H_2O$	7.67	11.51 92.47	11.00	$P2_1/n$	2
$Ca_5(NH_4)_2(P_2O_7)_3{\cdot}6H_2O$	11.88	11.88	9.83	Hexagonal	2
$CaNH_4HP_2O_7$	10.57	17.56 90.63	7.27	$P2_1/n$	8
$Ca(NH_4)_2H_4(P_2O_7)_2$	7.17	19.99 102.87	9.33	$C2/c$	4
$CaK_2P_2O_7$	9.79	5.69 104.03	12.97	$P2_1/n$	4
$Ca_5K_2(P_2O_7)_3{\cdot}6H_2O$	11.88	11.88	9.83	Hexagonal	2
$CaK_2H_4(P_2O_7)_2$	7.17	19.99 102.87	9.33	$C2/c$	4

[a]Refs. 64, 134, 136–139, 186 and 187

Table 2.2.8. Main Crystallographic Data for Anhydrous $M^{II}M^{I}_2P_2O_7$ Diphosphates

Formula	a (Å) α (°)	b (Å) β (°)	c (Å) γ (°)	S.G.	Z	Reference
$BaLi_2P_2O_7$	7.078(4)	12.164(6)	13.856(6)	*C mcm*	8	192
$PdLi_2P_2O_7$	12.856(1)	7.4955(5)	5.8116(4)	*Imma*	4	217
$ZnNa_2P_2O_7$	7.690(3)	7.690(3)	10.271(4)		4	197
$CoNa_2P_2O_7$	7.698(2)	7.698(2)	10.282(2)		4	197
$PbNa_2P_2O_7$	6.881(1) 96.95(2)	9.363(2) 108.22(2)	5.482(1) 105.63(2)	?	2	214
$PdNa_2P_2O_7$	14.693(3)	5.855(4) 114.11(2)	7.922(2)	*C2/c*	4	216
$ZnK_2P_2O_7$	7.860(3)	7.860(3)	11.324(4)		4	197
$CoK_2P_2O_7$	7.884(3)	7.884(3)	11.262(5)		4	197
$CuK_2P_2O_7$	8.054(4)	8.054(4)	10.970(5)		4	197
$CaK_2P_2O_7$	9.79(1)	5.69(1) 104.03(5)	12.97(1)	$P2_1/n$	4	139
$SrK_2P_2O_7$	9.168(2)	5.712(4) 105.79(8)	14.720(6)	$P2_1/c$	4	189
$CdK_2P_2O_7$	9.715(5)	5.528(3) 106.41(5)	12.72(1)	*C2, Cm* or *C2/m*	4	210
$PbK_2P_2O_7{}^a$	9.992(1)	5.887(1) 113.91(1)	6.978(1)	*C...*	2	214
$CaRb_2P_2O_7$	10.012(3)	5.784(3) 104.79(2)	13.070(3)	$P2_1/n$	4	188
$SrRb_2P_2O_7$	10.270(2)	5.867(2) 116.48(1)	14.413(6)	*C2/c*	4	191
$PbRb_2P_2O_7$	10.261(3)	5.901(1) 106.12(2)	7.185(1)	*C...*	2	214
$CaCs_2P_2O_7$	10.302(3)	5.946(3) 104.73(6)	13.182(6)	$P2_1/n$	4	188
$SrCs_2P_2O_7$	10.528(2)	6.081(2) 118.34(8)	14.766(6)	*C2/c*	4	191

^{a}HT: High-temperature form.

$Mg(NH_4)_2P_2O_7\cdot4H_2O$. Two crystalline forms of this salt were obtained by Frazier *et al.*[183] by adding MgO to aqueous solutions of ammonium diphosphate under various conditions. Both forms are not water-soluble.

$Mg(NH_4)_6(P_2O_7)_2\cdot6H_2O$ and $Mg(NH_4)_2H_4(P_2O_7)_2\cdot2H_2O$. The same authors also described these two insoluble diphosphates. The first one is isotypic with the corresponding Mn, and Fe(II) derivatives whereas the second is isotypic with the zinc compound.

(ii) Calcium–Sodium Salts

$CaNa_2P_2O_7$ and $CaNa_2P_2O_7\cdot4H_2O$. The anhydrous salt was prepared by Klement[178] by the action of CaO on molten $NaPO_3$. Three polymorphic forms of the tetrahydrate have been reported. Cheng *et al.*[184] described a detailed procedure for the preparation of the α form and a complete determination of its atomic arrangement. The atomic arrangement of the β form was also determined by Cheng *et al.*[185]

Table 2.2.9. Main Crystallographic Data for Various Divalent–Monovalent Cation Diphosphates

Formula	a (Å) α (°)	b (Å) β (°)	c (Å) γ (°)	S.G.	Z	Reference
$Zn_3Rb_2(P_2O_7)_2$	13.22(1)	7.224(6) 92.08(2)	7.196(5)	$P2_1$	2	201
$Co_3Rb_2(P_2O_7)_2$	13.25(1)	7.248(7) 92.00(5)	7.213(6)	$P2_1$	2	201
$Pb_3K_2(P_2O_7)_2$	17.758(3)	9.280(2)	8.596(2)	Orthorhombic	4	214
$Pb_3Rb_2(P_2O_7)_2$	17.918(4)	9.394(3)	8.766(2)	Orthorhombic	4	214
$Co_3K_2(P_2O_7)_2 \cdot 2H_2O$	9.229(2)	8.110(1) 99.31(4)	9.122(4)	$P2_1/a$	2	198
$CaNH_4NaP_2O_7 \cdot 3H_2O$	10.39(1)	16.55(1) 103.31(1)	5.677(3)	Cc	4	218
$CdNH_4NaP_2O_7 \cdot 3H_2O$	10.211(9)	16.56(1) 103.73(1)	5.632(3)	Cc	4	218
$Mg_9Na_{14}(P_2O_7)_8$	10.882(1) 112.49(1)	9.734(1) 99.63(1)	6.372(1) 107.40(1)	$P\bar{1}$	1/2	180
$\alpha CaNa_2P_2O_7 \cdot 4H_2O$	5.689(6)	8.586(8) 106.3(1)	10.565(9)	Pc	2	184
$\beta CaNa_2P_2O_7 \cdot 4H_2O$	10.38	16.98 104.4	5.75	Cc	4	185
$CdK_2P_2O_7 \cdot 4H_2O$	10.68(1)	11.37(1) 118.29(5)	10.24(1)	$P2_1$ or $P2_1/m$	4	210
$CuNa_6(P_2O_7)_2 \cdot 16H_2O$	6.842(1) 89.50(1)	8.759(2) 95.96(1)	12.727(3) 112.58(1)	$P?$	1	204

(iii) Calcium–Potassium and Calcium–Ammonium Salts

Calcium-potassium and calcium-ammonium diphosphates have been very carefully investigated as possible fertilizers. Thus 19 diphosphates were characterized in the CaO–K_2O–$H_4P_2O_7$–H_2O and CaO–$(NH_4)_2O$–$H_4P_2O_7$–H_2O systems by Brown, Lehr, Frazier, Smith, and coworkers.[64,134,136–139,186,187] We report below the main results obtained by these authors. We consider their studies of these two systems seem to be the most outstanding investigations of this type in the field of oligophosphate chemistry. For each characterized compound, a detailed optimized chemical preparation was reported and optical measurements were given; in many cases, unit-cell dimensions and space groups were determined. Unfortunately no further structural studies followed these remarkable investigations. The main crystal data obtained during the investigations of these two systems are reported in Table 2.2.7.

We now summarize very briefly some chemical features of the numerous compounds characterized during these investigations.

$CaK_2P_2O_7$ and $CaK_2P_2O_7 \cdot 4H_2O$. The anhydrous salt forms at 298 K at pHs above 6.5. This salt melts at 1453 K and is very stable and not water-soluble. The tetrahydrate is observed for pHs in the range 6–10.5 but is stable only at pHs above 8. Like the anhydrous salt, it is not water-soluble and dehydrates above 373 K.

$Ca_3K_2(P_2O_7)_2 \cdot 2H_2O$. This salt has been observed in the pH region 5–6.5. It dissolves slowly in water and does not seem to be isotypic with the corresponding ammonium salt.

$Ca_5K_2(P_2O_7)_3 \cdot 6H_2O$. This hexagonal salt appears upon addition of a calcium salt to dilute or basic solutions of $K_4P_2O_7$. It is very insoluble in water and is isotypic with the corresponding ammonium salt.

$CaKHP_2O_7$ and $CaKHP_2O_7 \cdot 2H_2O$. The anhydrous diphosphate forms in the system $CaO–K_2O–H_4P_2O_7–H_2O$ between pH 2.7 and 5 at 298 K. It also is relatively insoluble in water and, isotypic with $Ca(NH_4)HP_2O_7$. The dihydrate is slightly water-soluble and dehydrates slowly at 298 K to form $CaKHP_2O_7$.

$CaK_2H_4(P_2O_7)_2$, $CaK_4H_2(P_2O_7)_2$, $Ca_2KH_3(P_2O_7)_2 \cdot 3H_2O$, and $Ca_3KH(P_2O_7)_2 \cdot 4H_2O$. These salts were also characterized during these investigations. The first one is very soluble in water and is isotypic with the corresponding ammonium compound. The second one is moderately soluble and is a very stable phase. The third salt is formed in the pH range 2–2.7, dissolves slowly in H_2O, and is isotypic with the corresponding NH_4 salt. The last salt is relatively insoluble in H_2O.

For the $CaO–(NH_4)_2O–H_4P_2O_7–H_2O$ system the main results obtained by the same authors can be summarized in the list of the new salts they characterized, together with a short comment on their properties.

$Ca(NH_4)_2P_2O_7 \cdot H_2O$. This diphosphate is relatively insoluble in water, dissolving very slowly and incongruently with the formation of $Ca_3(NH_4)_2(P_2O_7)_2 \cdot 6H_2O$.

$Ca_3(NH_4)_2(P_2O_7)_2 \cdot 6H_2O$. This salt dissolves very slowly and incongruently in H_2O, producing a precipitate of orthorhombic $Ca_2P_2O_7 \cdot 4H_2O$. A second crystalline form was observed.

$Ca_5(NH_4)_2(P_2O_7)_3 \cdot 6H_2O$. This diphosphate is not water-soluble and is isotypic with the corresponding potassium salt.

$CaNH_4HP_2O_7$. This salt is also isotypic with the potassium derivative and dissolves very slowly in H_2O.

$Ca(NH_4)_2H_4(P_2O_7)_2$. This diphosphate, also isotypic with the potassium derivative, is observed in the $CaO–(NH_4)_2O–H_4P_2O_7–H_2O$ system below pH 2. It is stable in air and very water soluble. In water it is altered rapidly to $Ca_2KH_3(P_2O_7)_2 \cdot 3H_2O$.

$Ca(NH_4)_4H_2(P_2O_7)_2$. This salt can be produced by adding calcium carbonate to a saturated solution of $(NH_4)_4P_2O_7$ kept at a pH between 4 and 6. In contact with water, it transforms into $Ca_3(NH_4)_2(P_2O_7)_2 \cdot 6H_2O$.

$Ca_2NH_4H_3(P_2O_7)_2 \cdot H_2O$. The preparation of this insoluble salt is long, requiring several months. A detailed procedure for its synthesis was given by the authors.

$Ca_2NH_4H_3(P_2O_7)_2 \cdot 3H_2O$. This insoluble salt is isomorphous with the corresponding potassium derivative.

$Ca_3(NH_4)_4H_6(P_2O_7)_4 \cdot 3H_2O$. This salt is very soluble in water transforming slowly into $CaNH_4HP_2O_7$.

(iv) Other Calcium Salts

$CaRb_2P_2O_7$ and $CaCs_2P_2O_7$. Lyutsko *et al.*[188] prepared the rubidium and cesium salts and found them to be isotypic with the previously described potassium salt. Samples were simply prepared by heating a mixture of $CaH_2P_2O_7$ and M^INO_3 at 1073 K. IR spectra were presented by the authors.

(v) Strontium and Barium Salts

$SrK_2P_2O_7$, $SrRb_2P_2O_7$, and $SrCs_2P_2O_7$. Single crystals of the potassium salt (mp = 1348 K) were prepared by Trunov *et al.*[189] by slow cooling of a melt prepared from a stoichiometric mixture of K_2CO_3, $SrCO_3$, and $NH_4H_2PO_4$ first heated at 773 K for 30h before melting. A complete determination of the atomic arrangement was reported. This compound is isotypic with $CaK_2As_2O_7$, previously described by Faggiani and Calvo.[190]

The same authors[191] performed the crystal structure determination of $SrRb_2P_2O_7$ and $SrCs_2P_2O_7$. These last two compounds are isotypic but not isomorphous with the potassium salt.

$BaLi_2P_2O_7$. Liebertz and Stähr[192] described the chemical preparation of $BaLi_2P_2O_7$. By using the Czochralski method they prepared crystals up to 30 mm long. This salt melts congruently at about 1100 K.

(vi) Manganese Salts

$Mn(NH_4)_2P_2O_7 \cdot 2H_2O$. This salt was prepared by Frazier *et al.*[183] by the addition of $MnSO_4$ to an aqueous solution of NH_4 polyphosphate having a pH in the range 4–8. This diphosphate is not water-soluble. The corresponding Fe(II) salt is isotypic.

$MnNa_2P_2O_7 \cdot 4H_2O$, $MnK_2P_2O_7 \cdot 3H_2O$, $MnRb_2P_2O_7 \cdot 3H_2O$ and $MnHNH_4P_2O_7 \cdot H_2O$. These salts were characterized by Goloshchapov and Martynenko[193] during the study of various $MnCl_2–M_4P_2O_7–H_2O$ systems at 298 K. The rubidium salt was also reported by these authors.[194]

(vii) Iron Salts

$Fe(NH_4)_2P_2O_7 \cdot 2H_2O$ and $Fe(NH_4)_6(P_2O_7)_2 \cdot 6H_2O$. These two diphosphates were described by Frazier *et al.*[83] Both are prepared by adding $FeSO_4$ to ammonium diphosphate aqueous solutions. The first one is isotypic with the corresponding Mn, Zn, and Cu salts, and the second with the Zn and Mg derivatives.

(viii) Cobalt Salts

$CoNa_2P_2O_7$, $CoK_2P_2O_7$ and $CoK_2P_2O_7 \cdot 4H_2O$. Syntheses, main properties and reactivities of the two potassium salts were reported by Kokhanovskii and Prodan.[195] Kokhanovskii[196] examined the thermal behavior up to 1073 K. Dehydration occurs at 436 K and the melting point is observed at 905 K. The anhydrous salt has three forms.

$CoNa_2P_2O_7$ and $CoK_2P_2O_7$ were prepared by Gabelica-Robert.[197] According to this author, they are tetragonal and isomorphous with $ZnNa_2P_2O_7$, $ZnK_2P_2O_7$, and $CuK_2P_2O_7$.

$Co_3K_2(P_2O_7)_2 \cdot 2H_2O$. This salt was prepared as single crystals by Lightfoot et al.[198] under hydrothermal conditions at 973 K from a starting mixture of KH_2PO_4 and $CoCl_2$ with an external pressure of 3 kbar. Polycrystalline samples can be obtained under milder conditions (493 K and 30 bar). The authors reported a complete determination of the atomic arrangement.

$CoK_2P_2O_7 \cdot 4H_2O$, $Co_3K_2(P_2O_7)_2 \cdot 10H_2O$, $Co_3K_2(P_2O_7)_2 \cdot 4H_2O$. These three compounds were characterized by Kokhanovskii and Prodan[199] during an investigation of the $K_4P_2O_7$–$Co(NO_3)_2$–H_2O system at 293 K. According to these authors this system is very similar to the $K_4P_2O_7$–$Mg(NO_3)_2$–H_2O system.

$Co(NH_4)_2P_2O_7 \cdot 4H_2O$ and $Co(NH_4)_2P_2O_7 \cdot H_2O$. The chemical preparation and thermal transformations of these two hydrates were investigated by Zemtsova.[200]

$Co_3Rb_2(P_2O_7)_2$. The chemical preparation of this diphosphate was described by Averbuch-Pouchot.[201] It is isotypic with the corresponding zinc salt, characterized by the same author.

(ix) Nickel Salts

$Ni_3Na_2(P_2O_7)_2 \cdot 6H_2O$ and $Ni_3M(P_2O_7)_2 \cdot 10H_2O$, M = K, Rb,Cs, and NH_4. The thermal behavior of this series of compounds was investigated by Bykanova and Lavrov.[202] In the case of the ammonium salt, the thermal decomposition scheme is

$$2Ni_3(NH_4)_2(P_2O_7)_2 \xrightarrow{773-1073 \text{ K}} Ni_2P_4O_{12} + 2Ni_2P_2O_7 + 4NH_3 + 2H_2O$$

for the other alkali metals

$$Ni_3M_2(P_2O_7)_2 \rightarrow Ni_3(PO_4)_2 + 2MPO_3$$

under the same conditions.

(x) Copper Salts

$CuLi_2P_2O_7$ and $CuNa_2P_2O_7$. These two salts were characterized by Sokolova et al.[203] The lithium salt decomposes at 1047 K whereas the sodium salt melts congruently at 922 K. The authors presented IR spectra and DSC heating curves as well as the results of specific conductance measurements.

$CuNa_6(P_2O_7)_2 \cdot 16H_2O$. This triclinic salt was prepared by Yuen and Collin[204] who gave a complete description of its atomic arrangement.

$CuK_2P_2O_7$. This diphosphate was prepared by Gabelica-Robert.[197] According to the author, it is tetragonal and isomorphous with $CoNa_2P_2O_7$, $CoK_2P_2O_7$, $ZnNa_2P_2O_7$, and $ZnK_2P_2O_7$.

$Cu_3(NH_4)_2(P_2O_7)_2 \cdot 3H_2O$. This salt was characterized by Spintse *et al.*[205] during their investigation of the $CuO-NH_4H_2PO_4$ system between 298 and 1273 K. The authors reported a complete scheme of the reactions in this system in various temperature ranges as a function of the initial P/Cu ratio.

(xi) Zinc Salts

$ZnNa_2P_2O_7$ and $ZnK_2P_2O_7$. These two diphosphates were prepared by Gabelica-Robert.[197] According to this author these compounds are tetragonal and isomorphous. $ZnNa_2P_2O_7$ was also investigated by Krivoviazov *et al.*[206] during a study of the $Zn(PO_3)_2-Na_2O$ system. It is a congruent-melting compound (mp = 1054 K). It was later investigated by Lazarev *et al.*[207] who reported its melting point to be 1052 K and measured its enthalpy of melting. They did not observe the polymorphic transformation previously reported for this diphosphate.

$ZnK_2P_2O_7$ was observed by Krivoviazov *et al.*[206] as a incongruent-melting compound decomposing at 951 K during their investigation of the $Zn(PO_3)_2-K_2O$ and KPO_3-ZnO systems.

Klement[178] described the preparation of the sodium derivative by the action of ZnO on molten $NaPO_3$.

$Zn_3K_2(P_2O_7)_2 \cdot 3H_2O$, $ZnK_2P_2O_7 \cdot 2H_2O$, $ZnK_6(P_2O_7)_2 \cdot 10H_2O$. These three crystalline zinc–potassium diphosphates were characterized by Morozova and Selivanova[208] during a study of the $K_4P_2O_7-Zn_2P_2O_7-H_2O$ system at 298 K.

$Zn(NH_4)_2P_2O_7 \cdot H_2O$. This salt prepared by Frazier *et al.*[183] precipitates when ZnO is dissolved in a dilute ammonium diphosphate aqueous solution kept at pH 8.

$Zn(NH_4)_6(P_2O_7)_2 \cdot 6H_2O$. According to the same authors[183] this salt can be prepared by drying a wet mixture of ZnO and $(NH_4)_3HP_2O_7$ at 298 K. The salt is water-soluble and isotypic with the Mg and Fe derivatives.

$Zn_3(NH_4)_2(P_2O_7)_2 \cdot 2H_2O$. This non-water-soluble diphosphate was prepared by the same authors[183] by adding ZnO to a concentrated aqueous solution of ammonium diphosphate kept at pH 4–6.

$Zn_3(NH_4)_2H_4(P_2O_7)_2 \cdot 2H_2O$. This salt was obtained by addition of ethanol to an acidic solution of zinc–ammonium diphosphate. It dissolves incongruently in water and is isomorphous with the magnesium derivative.[183]

$Zn_3Rb_2(P_2O_7)_2$. The chemical preparation of this diphosphate and a complete description of its atomic arrangement were reported by Averbuch-Pouchot.[201] The corresponding cobalt salt is isotypic.

(xii) Cadmium Salts

$CdNa_2P_2O_7 \cdot 4H_2O$. This salt was characterized as a crystalline material by Selivanova *et al.*[209] during a study of the $Cd_2P_2O_7-Na_4P_2O_7-H_2O$ system at 298 K. Some properties of this salt were reported by the authors.

$CdK_2P_2O_7$ and $CdK_2P_2O_7·4H_2O$. The chemical preparations and crystal data for $CdK_2P_2O_7$ and its tetrahydrate were reported by Averbuch-Pouchot.[210] In spite of a strong similarity in their unit-cell dimensions, $CaK_2P_2O_7$ and $CdK_2P_2O_7$ are not isotypic.

$CdK_6(P_2O_7)_2·6H_2O$. A crystalline cadmium–potassium diphosphate, $CdK_6(P_2O_7)_2·6H_2O$, was characterized by Selivanova and Kudryavtsev.[211] Some properties of this salt were described. $CdK_6(P_2O_7)_2·6H_2O$ is converted at 398 K to the anhydrous salt, which melts at 979 K.

$Cd(NH_4)_2P_2O_7·H_2O$ and $Cd(NH_4)_2P_2O_7·3H_2O$. Solubility of cadmium diphosphate in aqueous solutions of ammonium diphosphate was investigated at 298 K by Selivanova and Krudryavtsev.[212] They observed the formation of two mixed diphosphates, $Cd(NH_4)_2P_2O_7·H_2O$ and $Cd(NH_4)_2P_2O_7·3H_2O$, when the concentration of $Cd_2P_2O_7$ was increased. Detailed procedures for the chemical preparation of these two salts were reported. When heated at 523 K, the monohydrate is transformed into long-chain cadmium polyphosphate:

$$Cd(NH_4)_2P_2O_7·H_2O \rightarrow Cd(PO_3)_2 + 2NH_3 + 2H_2O$$

The monohydrate was previously described by Rosenheim.[213]

(xiii) Lead Salts

$PbNa_2P_2O_7$, $PbK_2P_2O_7$, $Pb_3K_2(P_2O_7)_2$, $PbRb_2P_2O_7$, $Pb_3Rb_2(P_2O_7)_3$. The pseudo-binary system $K_4P_2O_7–Pb_2P_2O_7$ was investigated by Rulmont et al.[214] From X-ray diffraction, vibrational spectroscopy, and differential thermal analysis (DT.A) experiments the authors characterized two new compounds: $PbK_2P_2O_7$ and $Pb_3K_2(P_2O_7)_2$. The first one is dimorphous, with a reversible transition temperature at 844 K.(mp = 920 K). They reported unit-cell dimensions for the high-temperature form of $PbK_2P_2O_7$ and for $Pb_3K_2(P_2O_7)_2$. In addition, the two corresponding rubidium salts, $PbRb_2P_2O_7$ and $Pb_3Rb_2(P_2O_7)_3$, were characterized and found to be isotypic with the potassium compounds. The existence of $PbNa_2P_2O_7$ was also reported. A triclinic unit cell was proposed for this last salt, whose melting point has been measured as 1089 K.

(xiv) Palladium Salts

$PdLi_2P_2O_7$ and $PdNa_2P_2O_7$. These double diphosphates were obtained by Sokolova et al.[215] via solid-phase reactions between $PdCl_2$ and LiH_2PO_4 or NaH_2PO_4. They melt at 1153 and 1113 K, respectively. The corresponding potassium salt could not be obtained by the authors. Crystal structures for these two salts were reported some years later by Laligant.[216,217]

(xv) Triple Salts

$CaNH_4NaP_2O_7·3H_2O$. The chemical preparation for this diphosphate was reported by Averbuch-Pouchot and Guitel.[218] The corresponding cadmium salt is isotypic, and its atomic arrangement was described by these authors.

$CdNH_4NaP_2O_7 \cdot 3H_2O$. The chemical preparation and crystal structure of this diphosphate were reported by Averbuch-Pouchot and Guitel.[218] The corresponding calcium salt is isotypic.

The main crystallographic data for divalent–monovalent-cation diphosphates are reported in Tables 2.2.8. and 2.2.9.

2.2.2.4. Trivalent Cation Diphosphates

The chemical literature dealing with trivalent cation diphosphates is relatively extensive but the atomic arrangements of very few of these compounds have been determined.

$MnHP_2O_7$. According to Selevich and Lyutsko[219] three forms of $MnHP_2O_7$ exist. These authors reported the experimental conditions for the preparation of these three forms by interaction of phosphoric acid with manganese(II) salts in the presence of an oxidant, in the temperature range 373–573 K. The thermal behavior of $MnHP_2O_7$ can be represented as follows:

$$6MnHP_2O_7 \xrightarrow{\ 823\ K\ -O_2\ } \text{intermediate phase} \xrightarrow{\ 1073\ K\ -\frac{1}{2}\,O_2\ } 3Mn_2P_4O_{13}$$

In the intermediate phase the only crystalline products are $Mn_2P_2O_7$ and $Mn(PO_3)_3$ but this phase includes also an amorphous part composed of various manganese oligophosphates. The authors did not report any crystal data, but from X-ray powder diagrams, identified the β form as an isotype of $FeHP_2O_7$, previously described by d'Yvoire.[220]

One form of $MnHP_2O_7$ was synthesized by a flux method at 473 K by Durif and Averbuch-Pouchot.[221] The main feature of the atomic arrangement reported by the authors is the existence of Mn_2O_{10} clusters built by two edge-sharing MnO_6 octahedra.

$YbHP_2O_7$. During an investigation between 373 and 673 K, of the $Yb_2O_3-P_2O_5-H_2O$ system Chudinova et al.[222] determined the region of crystallization for $YbHP_2O_7$.

$CrHP_2O_7 \cdot 10H_2O$. This salt was characterized by Medvedev et al.[223] during an investigation of the $Cr(NO_3)_2-M_4P_2O_7-H_2O$ and $Cr_4(P_2O_7)_3-H_4P_2O_7-H_2O$ systems. It was reported to crystallize as hexagonal platelets, and its thermal behavior was described.

$LaHP_2O_7 \cdot 3H_2O$. This hydrate was observed by Tananaev et al.[224] during their investigation of the solubility isotherm at 298 K in the system $La_4(P_2O_7)_3-H_4P_2O_7-H_2O$.

$BiHP_2O_7$ and $BiHP_2O_7 \cdot 2H_2O$. During a study, at 273K, of the system $Bi_4(P_2O_7)_3-H_4P_2O_7-H_2O$ Tezikova et al.[225] characterized an acidic diphosphate, $BiHP_2O_7 \cdot 2H_2O$. This salt crystallizes as large needles or prisms. The thermal dehydration and decomposition of $BiHP_2O_7 \cdot 2H_2O$ was described. Carefully controlled dehydration of this salt leads to the formation of crystalline bismuth tetraphosphate, $Bi_2P_4O_{13}$.

$$BiHP_2O_7 \cdot 2H_2O \rightarrow BiHP_2O_7 + 2H_2O \rightarrow Bi_2P_4O_{13} + H_2O$$

Chudinova[226] reported the X-ray electron spectrum of $BiHP_2O_7$.

$Cr_4(P_2O_7)_3$, $Fe_4(P_2O_7)_3$, $V_4(P_2O_7)_3$, and $Cr_4(P_2O_7)_3$. The existence of $Cr_4(P_2O_7)_3$ and $Fe_4(P_2O_7)_3$ was reported by Rémy and Boullé[227] and by d'Yvoire.[220] From X-ray diffraction studies they concluded that these compounds are isotypic. The solubility of $Fe_4(P_2O_7)_3$ was investigated by Leitsin and Grekov.[228]

The anhydrous diphosphate $Cr_4(P_2O_7)_3$ was obtained as brown crystals by Schlesinger *et al.*[229] This salt was produced by thermal decomposition of chromium polyphosphate at 1573 K; it is isotypic with the corresponding vanadium salt.

$V_4(P_2O_7)_3$ crystals were grown by thermal transformation of an amorphous intermediate synthesized from V_2O_5 and aqueous H_3PO_3 and H_3PO_4 solution by Schlesinger *et al.*[229] Unit-cell dimensions were reported. $V_4(P_2O_7)_3$ is isotypic with the corresponding chromium and iron salts. Its crystal structure was determined by Palkina *et al.*[230] This atomic arrangement is mainly characterized by the existence of V_2O_9 groups formed by a pair of VO_6 octahedra sharing a face.

$Cr_4(P_2O_7)_3 \cdot 40H_2O$. This salt was characterized by Medvedev *et al.*[223] during an investigation of the $Cr(NO_3)_2$–$M_4P_2O_7$–H_2O (M = Li, Na, K, Rb, Cs, NH_4, Tl, Ag) and $Cr_4(P_2O_7)_3$–$H_4P_2O_7$–H_2O systems. It crystallizes as hexagonal platelets. Its thermal behavior was reported.

$La_4(P_2O_7)_3$ and $La_4(P_2O_7)_3 \cdot 12H_2O$. The existence of $La_4(P_2O_7)_3 \cdot 12H_2O$ has been reported by Tananaev and Vasil'eva[231] during a study of the $La(NO_3)_3$–$M_4P_2O_7$–H_2O systems (M = Li, Na, K) at 298 K. This compound was also observed by Kuznetsov and Vasil'eva[232] as the initial compound formed during their study of the same systems. By calcination at high temperature, they obtained three crystalline forms of $La_4(P_2O_7)_3$.

$La_4(P_2O_7)_3 \cdot La(OH)_3 \cdot 12H_2O$ and $La_4(P_2O_7)_3 \cdot 1/2La_2O_3$. Tananaev and Vasil'eva[233] characterized $La_4(P_2O_7)_3 \cdot La(OH)_3 \cdot 12H_2O$ during a study of the $La(NO_3)_3$–$M_4P_2O_7$–H_2O systems (M = Rb, Cs). Some properties of this salt were examined, and its thermal behavior was later investigated by Kuznetsov and Vasil'eva.[234] According to these authors, it transforms into $La_4(P_2O_7)_3 \cdot 1/2La_2O_3$ at 643 K.

$Gd_4(P_2O_7)_3$. The preparation of anhydrous gadolinium diphosphate, $Gd_4(P_2O_7)_3$, was reported by Kizilyalli,[235] but the crystallographic features of this salt as reported by this author need revision.

$Gd_4(P_2O_7)_3 \cdot 13H_2O$ and $Gd_4(P_2O_7)_3 \cdot Gd(OH)_3 \cdot 9H_2O$. Tananaev *et al.*[236] characterized $Gd_4(P_2O_7)_3 \cdot 13H_2O$ and $Gd_4(P_2O_7)_3 \cdot Gd(OH)_3 \cdot 9H_2O$ during an investigation of the reaction of $GdCl_3$ with alkali metal diphosphates. The thermal stability of these salts was studied. In the case of $Gd_4(P_2O_7)_3 \cdot Gd(OH)_3 \cdot 9H_2O$, it is observed that $Gd_4(P_2O_7)_3 \cdot 1/2Gd_2O_3$ is formed at 558 K.

Tananaev and Petushko[237] described a method for the preparation of $Gd_4(P_2O_7)_3 \cdot 13H_2O$ by the action of $Li_4P_2O_7$ on $GdCl_3$ and reported some properties of this diphosphate. The same authors obtained $Gd_4(P_2O_7)_3 \cdot Gd(OH)_3 \cdot 9H_2O$ by the action of $GdCl_3$ on $Cs_4P_2O_7$.

$Al_8H_{12}(P_2O_7)_9$. d'Yvoire[220] described a method for the preparation of this salt.

Table 2.2.10. Main Crystallographic Data for Trivalent Cation Diphosphates

Formula	a (Å) α (°)	b (Å) β (°)	c (Å) γ (°)	S.G.	Z	Reference
$V_4(P_2O_7)_3$	7.443(1)	9.560(2)	21.347(4)	4	$Pmcn$	229, 230
$Cr_4(P_2O_7)_3$	7.26(2)	9.38(1)	21.00(4)	4	$Pmcn$	229
$MnHP_2O_7$	7.951(4)	12.645(8) 100.92(5)	4.922(2)	4	$P2_1/n$	221
$Cr(HP_2O_7)(NH_3)_3(H_2O)\cdot2H_2O$	7.825(2)	10.107(3) 103.92(5)	15.322(5)	4	$P2_1/c$	238
$Cr(HP_2O_7)(NH_3)_3\cdot2H_2O$	8.695(2)	10.327(3) 97.81(5)	11.913(4)	4	$P2_1/c$	238
$Cr(HP_2O_7)(NH_3)_4\cdot2H_2O$	7.491	13.840 93.39	11.554	4	$P2_1/c$	239

Triamminechromium Diphosphate Dihydrates. Two forms of triamminechromium diphosphate dihydrate were synthesized and investigated by Haromy *et al.*[238] The first one is a monomer, $Cr(HP_2O_7)(NH_3)_3H_2O\cdot2H_2O$, while the second form is a dimer of $Cr(HP_2O_7)(NH_3)_3\cdot2H_2O$.

$CrHP_2O_7(NH_3)_4\cdot2H_2O$. Haromy *et al.*[239] described the crystal structure of this salt and identified it as an isotype of the corresponding cobalt complex.[240]

2.2.2.5. Monovalent–Trivalent Cation Diphosphates

Four classes of monovalent–trivalent cation diphosphates are presently well characterized. They correspond to the following general formulas:

$$M^IM^{III}P_2O_7\cdot xH_2O, \ M^IM^{III}(H_2P_2O_7)_2, \ M^I_2M^{III}H_3(P_2O_7)_2, \text{ and } M^I_6M^{III}_2(P_2O_7)_3$$

A large number of anhydrous $M^IM^{III}P_2O_7$ diphosphates have been clearly characterized, and what is rather surprising is that they belong to only three structure types. The first structural type is common to $LiFeP_2O_7$ and $LiInP_2O_7$. The two other structural types are usually designated as $NaFeP_2O_7$ (I) low temperature and (II) high temperature and have numerous representatives. Some of them are dimorphous. All the K, NH_4, Rb, Cs, and Tl derivatives crystallize with the $NaFeP_2O_7$ (I) structure while all the Na and Ag derivatives, with the exception of the dimorphous $NaFeP_2O_7$, crystallize with the $NaFeP_2O_7$ (II) structure.

Numerous structure determinations of compounds belonging to the last two structural types have been performed.

Tables 2.2.11–2.2.15 report the main crystallographic data for monovalent–trivalent cation diphosphates.

(i) Anhydrous $M^IM^{III}P_2O_7$ Diphosphates

$LiFeP_2O_7$ and $LiInP_2O_7$. Genkina *et al.*[241] reported a detailed procedure for the preparation of single crystals of $FeLiP_2O_7$ and an accurate crystal structure determination for this

Table 2.2.11. Main Crystallographic Data for Lithium-, Sodium-, and Silver-Trivalent Cation Diphosphates[a]

Formula	a (Å) α (°)	b (Å) β (°)	c (Å) γ (°)	S.G.	Z	Reference(s)
LiFeP$_2$O$_7$	6.938(2) 109.38(2)	8.079(2)	4.825(1)	P2$_1$	2	241
LiInP$_2$O$_7$	7.084(2) 110.75(2)	8.436(2)	4.908(3)	P2$_1$	2	243
NaAlP$_2$O$_7$	7.201(5) 111.75(8)	7.700(5)	9.317(5)	P2$_1$/c	4	244, 246, 247
NaVP$_2$O$_7$	7.324(5) 111.96(4)	7.930(4)	9.586(6)	P2$_1$/c	4	245
NaCrP$_2$O$_7$	7.294(2) 111.72(2)	7.838(2)	9.484(2)	P2$_1$/c	4	246–247
NaFeP$_2$O$_7$ (I)	7.11(1) 109.23(5)	10.00(5)	8.08(6)	P2$_1$/c	4	244
NaFeP$_2$O$_7$ (II)	7.324(1) 111.86(1)	7.905(1)	9.574(1)	P2$_1$/c	4	248, 249
NaGaP$_2$O$_7$	7.283(5) 111.98(8)	7.821(5)	9.462(5)	P2$_1$/c	4	246, 247
NaMoP$_2$O$_7$	7.4195(3) 111.87(1)	8.1084(4)	9.7598(4)	P2$_1$/c	4	250
NaInP$_2$O$_7$	7.486(8) 112.21(1)	8.182(7)	9.823(9)	P2$_1$/c	4	247
AgCrP$_2$O$_7$	7.302(5) 111.80(1)	7.932(5)	9.483(5)	P2$_1$/c	4	247
AgGaP$_2$O$_7$	7.299(8) 112.18(1)	7.921(10)	9.458(10)	P2$_1$/c	4	247
AgFeP$_2$O$_7$	7.335(4) 111.86(1)	7.998(4)	9.575(5)	P2$_1$/c	4	247
AgInP$_2$O$_7$	7.490(10) 112.12(1)	8.267(13)	9.833(11)	P2$_1$/c	4	247

[a]With the exception of the dimorphous NaFeP$_2$O$_7$, all the NaM^{III}P$_2$O$_7$ and AgMP$_2$O$_7$ compounds in this table belong to the NaFeP$_2$O$_7$ (II) form.

compound. This salt melts congruently at about 1263 K. Grunze and Grunze[242] also described a chemical preparation of this diphosphate.

LiInP$_2$O$_7$ was prepared by Tranqui *et al.*[243] The authors reported the determination of the atomic arrangement, identical to that previously determined for the corresponding iron salt.

NaAlP$_2$O$_7$. This salt has been characterized by Gamondes et al.[244] Chemical preparation and unit-cell dimensions were reported by the authors, showing it to be isotypic with NaFeP$_2$O$_7$ (II). Its chemical preparation was also reported by Grunze and Grunze.[242]

NaVP$_2$O$_7$. This salt was prepared by Wang *et al.*[245] by heating a mixture of Na$_4$V$_2$O$_7$, V, VO$_2$, and P$_2$O$_5$ at 1170 K in a sealed silica tube. Yellowish green crystals were obtained. The authors reported a complete structural determination for this salt.

Table 2.2.12. Main Crystallographic Data for Potassium–Trivalent Cation Diphosphates[a]

Formula	a (Å) α (°)	b (Å) β (°)	c (Å) γ (°)	S.G.	Z	Reference(s)
$KAl(H_2P_2O_7)_2$	4.888(3) 97.81(5)	7.453(4) 95.96(5)	7.764(5) 73.86(5)	$P\bar{1}$	2	247
$KAlP_2O_7$	7.308(8)	9.662(6) 106.69(7)	8.025(4)	$P2_1/c$	4	244, 247, 253
$KCrP_2O_7$	7.337(5)	9.892(9) 106.53(1)	8.151(6)	$P2_1/c$	4	247
$KFeP_2O_7$ (I)	7.3523(7)	9.9875(6) 106.50(1)	8.187(1)	$P2_1/c$	4	244, 247, 255
$KGaP_2O_7$	7.311(8)	9.816(8) 106.51(1)	8.131(14)	$P2_1/c$	4	247
$KMoP_2O_7$	7.375(1)	10.348(1) 106.88(1)	8.351(2)	$P2_1/c$	4	257
$KInP_2O_7$	7.408(6)	10.38(2) 106.30(1)	8.383(8)	$P2_1/c$	4	247
$KTmP_2O_7$	7.534(4)	10.855(6) 106.80(1)	8.561(4)	$P2_1/c$	4	247
$KYbP_2O_7$	7.535(4)	10.819(7) 106.67(1)	8.542(6)	$P2_1/c$	4	247
$KLuP_2O_7$	7.545(4)	10.776(4) 106.65(1)	8.537(4)	$P2_1/c$	4	247

[a]All the KMP_2O_7 compounds in this table belong to the $NaFeP_2O_7$ (I) form.

$NaCrP_2O_7$ and $NaGaP_2O_7$. Crystal structure of $NaCrP_2O_7$ was determined by Bohathy *et al.*[246] The unit-cell parameters for three isotypic diphosphates, $NaAlP_2O_7$, $NaGaP_2O_7$, and $NaFeP_2O_7$ are also reported by the authors. $NaCrP_2O_7$ and $NaGaP_2O_7$ were also investigated by Gabelica-Robert.[247]

$NaFeP_2O_7$. Gamondes *et al.*[244] described the chemical preparation and measured unit-cell dimensions for the two forms of this salt. The crystal structure of $NaFeP_2O_7$ (II) was determined by Gabelica-Robert *et al.*[248] and later reexamined by Moya-Pizarro *et al.*[249] These authors report a detailed magnetic and Mössbauer resonance study for this salt. Wanklyn *et al.*[118] published a detailed procedure for obtaining large single crystals of $NaFeP_2O_7$ by a flux method. They reported the preparation of single crystals up to 2 mm. Grunze and Grunze[242] also reported the chemical preparation of the two forms.

$NaMoP_2O_7$ and $NaInP_2O_7$. The molybdenum salt, $NaMoP_2O_7$, was prepared for the first time by Leclaire et al.[250] A complete description of the atomic framework was given by these authors. $NaInP_2O_7$ was characterized by Avaliani *et al.*[251] during an investigation of the P_2O_5–In_2O_3–Na_2O–H_2O system between 423 and 773 K. These authors also observed he existence of $NaIn(H_2P_2O_7)_2$.

$NaLnP_2O_7$. Anisimova *et al.*[252] prepared 14 compounds of the general formula $NaLnP_2O_7$. The unit cell dimensions reported by the authors, obtained from powder data, need to be confirmed by single-crystal investigations.

Table 2.2.13. Main Crystallographic Data for Rubidium–Trivalent Cation Diphosphates

Formula	a (Å) α (°)	b (Å) β (°)	c (Å) γ (°)	S.G.	Z	Reference(s)
$RbAlP_2O_7$	7.448(4)	9.616(6) 105.97(1)	8.108(5)	$P2_1/c$	4	247
$RbVP_2O_7$	7.511(4)	10.035(2) 105.74(2)	8.254(2)	$P2_1/c$	4	261
$RbCrP_2O_7$	7.478(4)	9.891(5) 105.98(1)	8.227(4)	$P2_1/c$	4	247
$RbFeP_2O_7$	7.495(4)	9.972(6) 105.76(1)	8.278(4)	$P2_1/c$	4	247, 265
$RbGaP_2O_7$	7.453(6)	9.802(6) 105.87(1)	8.229(4)	$P2_1/c$	4	247
$RbInP_2O_7$	7.539(4)	10.402(6) 105.42(1)	8.482(5)	$P2_1/c$	4	247
$RbYP_2O_7$	7.705(3)	10.947(5) 105.36(1)	8.662(3)	$P2_1/c$	4	247, 263
$RbDyP_2O_7$	7.700(3)	11.001(6) 105.30(1)	8.676(4)	$P2_1/c$	4	247, 263
$RbHoP2O7$	7.695(4)	10.936(4) 105.25(4)	8.657(4)	$P2_1/c$	4	263
$RbErP_2O_7$	7.681(4)	10.881(4) 105.37(4)	8.631(3)	$P2_1/c$	4	263
$RbTmP_2O_7$	7.655(5)	10.850(5) 105.54(4)	8.608(5)	$P2_1/c$	4	263
$RbYbP_2O_7$	7.641(5)	10.838(4) 105.44(4)	8.601(4)	$P2_1/c$	4	263
$RbLuP_2O_7$	7.643(4)	10.785(5) 105.42(4)	8.606(5)	$P2_1/c$	4	263

[a]All the $RbM^{III}P_2O_7$ compounds in this table belong to the $NaFeP_2O_7$ (I) structure type.

$KAlP_2O_7$ and $KMnP_2O_7$. Gamondes *et al.*[244] described the chemical preparation of the aluminum salt and measured its unit-cell dimensions. The crystal structure was determined by Ng and Calvo,[253] who also reported its chemical preparation and the conditions for crystal growth. Grunze and Grunze[242] also examined the conditions for formation of this salt, which is isotypic with $NaFeP_2O_7$ (I). The isotypic manganese salt was characterized by Guzeeva *et al.*[254] during an investigation of the K_2O–Mn_2O_3–P_2O_5–H_2O system between 413 and 673 K.

$KFeP_2O_7$. Gamondes *et al.*[244] described the chemical preparation and determined the unit-cell dimensions of $KFeP_2O_7$. The crystal structure determination of the low temperature form of $KFeP_2O_7$ was performed by Riou *et al.*[255] Grunze and Grunze[242] determined the conditions for formation of this salt.

$KInP_2O_7$. During a study of the P_2O_5–In_2O_3–K_2O–H_2O system Avaliani *et al.*[256] identified $KInP_2O_7$ and $KIn(H_2P_2O_7)_2$ as isotypes of the corresponding gallium salts.

Table 2.2.14. Main Crystallographic Data for Cesium–Trivalent Cation Diphosphates

Formula	a (Å) α (°)	b (Å) β (°)	c (Å) γ (°)	S.G.	Z	Reference(s)
$CsVP_2O_7$	7.701(3)	9.997(2) 104.82(4)	8.341(4)	$P2_1/c$	4	265
$CsCrP_2O_7$	7.714(3)	9.920(5) 105.00(4)	8.359(3)	$P2_1/c$	4	267
$CsFeP_2O_7$	7.693(6)	9.959(9) 104.86(1)	8.391(5)	$P2_1/c$	4	247, 265
$Cs_2GaH_3(P_2O_7)_2$	5.076(1)	7.9755(1) 96.96(2)	16.898(2)	$P2_1/c$	2	279
$CsMoP_2O_7$	7.724(2)	10.30(1) 104.77(2)	8.488(2)	$P2_1/c$	4	268
$CsInP_2O_7$	7.739(4)	10.322(8) 104.74(1)	8.586(6)	$P2_1/c$	4	247
$CsYP_2O_7$	7.897(4)	10.904(6) 104.21(1)	8.758(4)	$P2_1/c$	4	247, 263
$CsGdP_2O_7$	7.920(3)	11.072(5) 104.10(1)	8.824(3)	$P2_1/c$	4	247, 263
$CsTbP_2O_7$	7.906(5)	11.047(4) 104.28(4)	8.814(5)	$P2_1/c$	4	263
$CsDyP_2O_7$	7.907(5)	10.959(3) 104.11(3)	8.809(4)	$P2_1/c$	4	263
$CsHoP_2O_7$	7.886(4)	10.886(4) 104.18(4)	8.751(4)	$P2_1/c$	4	263
$CsErP_2O_7$	7.883(5)	10.847(4) 104.20(4)	8.738(4)	$P2_1/c$	4	263
$CsTmP_2O_7$	7.869(4)	10.805(5) 104.18(3)	8.717(5)	$P2_1/c$	4	263
$CsYbP_2O_7$	7.853(6)	10.750(5) 104.23(4)	8.692(6)	$P2_1/c$	4	263, 269
$CsLuP_2O_7$	7.844(6)	10.694(4) 104.24(4)	8.658(5)	$P2_1/c$	4	263

$KMoP_2O_7$. The chemical preparation and crystal structure of $KMoP_2O_7$ were described by Leclaire *et al.*[257]

$NH_4AlP_2O_7$ and $(NH_4)_2Al(OH)P_2O_7 \cdot 2H_2O$. These two diphosphates prepared by Frazier *et al.*[183] are not water-soluble. The first one is isomorphous with the members of the $(Al,Fe)(NH_4,K)_2P_2O_7$ series. Both are stable compounds.

$NH_4GaP_2O_7$. This salt was characterized during an investigation of the Ga_2O_3–P_2O_5–NH_3 system performed by Chudinova *et al.*[258] This diphosphate is the only phase crystallizing, at 623 K after two days of heating, when the starting mixture corresponds to a molar ration P:N:Ga = 20:40:1. If the heating is prolonged this diphosphate transforms into a triphosphate, $NH_4GaHP_3O_{10}$.

Table 2.2.15. Main Crystallographic Data for Thallium–Trivalent Cation Diphosphates[a]

Formula	a (Å) α (°)	b (Å) β (°)	c (Å) γ (°)	S.G.	Z	Reference
$TlAlP_2O_7$	7.468(2)	9.639(3) 105.98(1)	8.106(4)	$P2_1/c$	4	247
$TlCrP_2O_7$	7.483(2)	9.891(2) 105.93(1)	8.208(2)	$P2_1/c$	4	247
$TlFeP_2O_7$	7.513(4)	9.987(5) 105.82(1)	8.259(4)	$P2_1/c$	4	247
$TlGaP_2O_7$	7.469(4)	9.822(5) 105.82(1)	8.212(4)	$P2_1/c$	4	247
$TlInP_2O_7$	7.545(3)	10.384(4) 105.57(1)	8.455(4)	$P2_1/c$	4	247
$TlYP_2O_7$	7.703(4)	10.932(4) 105.59(1)	8.636(5)	$P2_1/c$	4	247

[a]All the $TlM^{III}P_2O_7$ compounds in this table belong to the $NaFeP_2O_7$ (I) structure type.

$NH_4FeP_2O_7$. The chemical preparation of $NH_4FeP_2O_7$ was reported by Grunze and Grunze.[242] A detailed procedure for the chemical preparation of this salt and its main chemical and optical properties are given in Ref.[183]

$AgFeP_2O_7$. This diphosphate has been prepared and investigated by Belkouch et al.[259] Their crystallographic investigation confirmed the previous observation of Gabelica-Robert.[247]

$RbGaP_2O_7$ and $RbGa(H_2P_2O_7)_2$. During a study of the $Ga_2O_3–Rb_2O–P_2O_5–H_2O$ system between 423 and 773 K Chudinova et al.[260] characterized $RbGaP_2O_7$ and two crystalline forms of $RbGa(H_2P_2O_7)_2$, one of them being isotypic with the corresponding gallium–potassium salt.

$RbVP_2O_7$. A crystal structure determination for $RbVP_2O_7$ was performed by Florke.[261]

$RbMnP_2O_7$. This salt was characterized by Guzeeva and Tananaev[262] during an investigation of the $MnO_2–Rb_2O–P_2O_5–H_2O$ system between 413 and 673 K. This diphosphate melts congruently at 933 K.

$RbInP_2O_7$. During a study of the $P_2O_5–In_2O_3–Rb_2O–H_2O$ system, Avaliani et al.[256] identified $RbInP_2O_7$.

$RbLnP_2O_7$ (Ln = Ho, Er, Tm, Yb, Lu). Syntheses and unit-cell dimensions of these five compounds were reported by Akrim et al.[263] In addition, the authors performed, using powder data, a refinement of the atomic arrangement of $RbYP_2O_7$ previously described by Gabelica-Robert.[247]

$CsFeP_2O_7$ and $RbFeP_2O_7$. Crystal structure refinements from powder data were reported by Millet and Mentzen[264] for these two salts. Dvoncova and Lii[265] synthesized

$CsFeP_2O_7$ and $RbFeP_2O_7$, measured their magnetic susceptibilities, and determined the crystal structure of the cesium salt. The two compounds are isotypic with $KFeP_2O_7$.

$CsVP_2O_7$, $CsCrP_2O_7$ and $CsMoP_2O_7$. The chemical preparation and crystal structure determination of the vanadium salt was reported by Wang and Lii.[266] Linde and Gorbunova[267] reported the crystal structure of $CsCrP_2O_7$ while Lii and Haushalter[268] prepared $CsMoP_2O_7$ and reported an accurate determination of its atomic arrangement.

$CsLnP_2O_7$ (Ln = Gd, Tb, Dy, Ho, Er, Tm, Yb, Lu, Y). Syntheses and unit-cell measurements for these nine isotypic compounds were reported by Akrim et al.[263] All are of the $NaFeP_2O_7$ (I) form. Among them the Y and Gd derivatives were previously reported by Gabelica-Robert and Tarte[248] and the Yb derivative by Jansen et al.[269]

Gabelica-Robert[247] reported the chemical preparations, IR and crystallographic data for a number of $M^IM^{III}P_2O_7$ compounds. Unit-cell parameters measured by the authors are reported in "Crystal Data." The authors reported that a third form of these salts was observed for KMP_2O_7 (M = Er, Y, Ho, Dy), but the orthorhombic unit-cell dimensions assigned to these compounds from powder data need to be confirmed by single-crystal investigations.

(ii) $M^IM^{III}P_2O_7 \cdot xH_2O$ Diphosphates

Chromium Derivatives. Various chromium derivatives, $LiCrP_2O_7 \cdot 7H_2O$, $NaCrP_2O_7 \cdot 10H_2O$, $AgCrP_2O_7 \cdot 5H_2O$, $KCrP_2O_7 \cdot 7H_2O$, $NH_4CrP_2O_7 \cdot 8H_2O$, $RbCrP_2O_7 \cdot 15/2H_2O$, $TlCrP_2O_7 \cdot 5H_2O$ and $CsCrP_2O_7 \cdot 6H_2O$ were characterized by Medvedev et al.[270] during investigations of the $Cr(NO_3)_3-M_4P_2O_7-H_2O$ and $Cr_4(P_2O_7)_3-H_4P_2O_7-H_2O$ systems. All these hydrates crystallize as needles, and their thermal behavior was reported.

$MLaP_2O_7 \cdot 4H_2O$ (M = Li, Na, and K). These three diphosphates were characterized by Tananaev and Vasil'eva[271] and by Kuznetsov and Vasil'eva[272] during their investigations of the $La(NO_3)_3-M_4P_2O_7-H_2O$ systems. The authors also reported the existence of several crystalline forms for the corresponding anhydrous salts.

$KGdP_2O_7$, $KGdP_2O_7 \cdot 3H_2O$ and $CsGdP_2O_7$, $CsGdP_2O_7 \cdot 9/2H_2O$. The two hydrates were characterized by Tananaev et al.[273] during an investigation of various $M_4P_2O_7-GdCl_3-H_2O$ systems. Some properties of these two salts were reported. Anhydrous $KGdP_2O_7$ is obtained at 423 K and has probably three crystalline modifications. Anhydrous $CsGdP_2O_7$ is said to have two crystalline forms. The existence of $CsGdP_2O_7 \cdot 9/2H_2O$ was later confirmed by Tananaev and Petushkova[274] during another investigation of the $GdCl_3-Cs_4P_2O_7-H_2O$ system. Unfortunately, no crystal data are available for these various diphosphates.

(iii) $M^IM^{III}(H_2P_2O_7)_2$ Diphosphates

This section is dominated by a careful and systematic investigation of the $M_2O-N_2O_3-P_2O_5-H_2O$ systems for M = Li, Na, NH_4, K, Rb, and Cs and N = Al, and Fe performed by Grunze and Grunze.[242] The authors report the existence of 33 new condensed phosphates, including 6 diphosphates $MN(H_2P_2O_7)_2$ characterized for M = Li, Na, and K. Some of them are dimorphous. This investigation was conducted between 443 and 723 K using MH_2PO_4, $Al(OH)_3$, $Fe(NO_3)_3 \cdot 9H_2O$, and H_3PO_4 (85%) as starting materials. The authors reported in detail the various ratios $M:N:P_2O_5$ of the starting mixtures and the thermal treatments used for the syntheses. Unindexed powder X-ray diffraction data were given.

Several new diphosphates belonging to this category were later characterized, and the thermal behavior of some of them was carefully investigated. The preparation and thermal transformation of $KAl(H_2P_2O_7)_2$ was investigated by Grunze *et al.*[275] This salt crystallized as a polycrystalline powder when a mixture of $Al(OH)_3$, KH_2PO_4, and H_3PO_4 in the molar ratio 1:5:15 was heated at 343 K for 6 to 8 days. According to the authors at 663 K this salt is almost completely transformed into potassium-aluminum cyclooctaphosphate:

$$2KAl(H_2P_2O_7)_2 \xrightarrow{663K} K_2Al_2P_8O_{24} + 4H_2O$$

At higher temperature, this cyclophosphate decomposes into $KAlP_2O_7$ and $Al_4(P_4O_{12})_3$. In the same study, a similar scheme of thermal behavior was observed for $KGa(H_2P_2O_7)_2$.

$KMn(H_2P_2O_7)_2$ was characterized by Guzeeva *et al.*[254] during an investigation of the $K_2O-Mn_2O_3-P_2O_5-H_2O$ system between 413 and 673 K. This salt is said to be isotypic with the corresponding gallium salt. According to the authors, the thermal decomposition scheme is

$$4KMn(H_2P_2O_7)_2 \xrightarrow{T > 603K} 2Mn(PO_3)_3 + 2KMnP_3O_9 + 2KPO_3 + P_2O_5 + 8H_2O + 1/2O_2$$

$$4KMn(H_2P_2O_7)_2 \xrightarrow{T > 873K} Mn_2(P_4O_{12}) + 2KMnP_3O_9 + 2KPO_3 + 2P_2O_5 + 8H_2O + O_2$$

$KIn(H_2P_2O_7)_2$ was identified by Avaliani *et al.*[256] during a study of the $In_2O_3-K_2O-P_2O_5-H_2O$ system and recognized as an isotype of the corresponding gallium salt.

The existence of $NaIn(H_2P_2O_7)_2$ was reported by Avaliani *et al.*[251] during an investigation of the $In_2O_3-Na_2O-P_2O_5-H_2O$ system between 423 and 773 K. According to these authors, the thermal transformations of $NaIn(H_2P_2O_7)_2$ can be represented as follow:

$$NaIn(H_2P_2O_7)_2 \xrightarrow{588-678 \text{ K}} NaInP_4O_{12} \xrightarrow{973-1023 \text{ K}} In(PO_3)_3 + melt$$

They identified the intermediate compound, $NaInP_4O_{12}$, as a cyclotetraphosphate.

$MnRb(H_2P_2O_7)_2$ was characterized by Guzeeva and Tananaev[262] during their investigation of the $MnO_2-Rb_2O-P_2O_5-H_2O$ system between 413 and 673 K. According to these authors, this compound is isotypic with the corresponding potassium salt. The thermal decomposition of this diphosphate was studied by the authors, who observed the formation of manganese(II)–rubidium cyclotriphosphate, $MnRbP_3O_9$, during this decomposition.

The crystal structure determination for $AlK(H_2P_2O_7)_2$ was performed by Yakubovich *et al.*[276]

During a study of the $In_2O_3-K_2O-P_2O_5-H_2O$ system, Avaliani *et al.*[251] identified $KInP_2O_7$ and $KIn(H_2P_2O_7)_2$ as isotypes of the corresponding gallium salts.

$RbGaP_2O_7$ and $RbGa(H_2P_2O_7)_2$. During a study of the $Ga_2O_3-Rb_2O-P_2O_5-H_2O$ system between 423 and 773 K, Chudinova *et al.*[260] characterized $RbGaP_2O_7$ and two forms of $RbGa(H_2P_2O_7)_2$, one of them being isotypic with the corresponding gallium–potassium salt.

(iv) $M^I_2M^{III}H_3(P_2O_7)_2$ Diphosphates

Several $M^I_2M^{III}H_3(P_2O_7)_2$ diphosphates have been characterized during the investigations of various $M^I_2O-M^{III}_2O_3-P_2O_5-H_2O$ or $M^I_2O-M^{IV}O_2-P_2O_5-H_2O$ systems.

$Rb_2MnH_3(P_2O_7)_2$. This salt was characterized during the study of the $MnO_2-Rb_2O-P_2O_5-H_2O$ system between 413 and 673 K by Guzeeva and Tananaev.[262]

$Cs_2MnH_3(P_2O_7)_2$. The same authors[277] characterized this salt during an investigation of the $MnO_2-Cs_2O-P_2O_5-H_2O$ system between 423 and 673 K. Its thermal behavior was investigated. When it was heated two types of decompositions occurred:

$$Cs_2MnH_3(P_2O_7)_2 \xrightarrow{563-603 \text{ K}} CsMnHP_3O_{10} + CsPO_3 + H_2O$$

and on further heating:

$$CsMnH_3P_3O_{10} \xrightarrow{793-873 \text{ K}} CsMnP_2O_7 + 1/2\ P_2O_7 + 1/2\ H_2O$$

$Cs_2GaH_3(P_2O_7)_2$. This salt was first prepared by Grunze et al.[278] and its crystal structure determined shortly thereafter.[279] It is stable up to about 613 K. The scheme for the decomposition of this salt is similar to that for the corresponding manganese–cesium diphosphate reported above.

(v) $M^I_6M^{III}_2(P_2O_7)_3$ Diphosphates

$Li_6Fe_2(P_2O_7)_3$, $K_6Fe_2(P_2O_7)_3$, and $Cs_6Fe_2(P_2O_7)_3$. These three diphosphates were characterized by Berul and Sizova during their investigations of the $LiPO_3-Fe_2O_3$,[280] $KPO_3-Fe_2O_3$,[281] and $CsPO_3-Fe_2O_3$[282] systems. The lithium and cesium salts are congruent-melting compounds.

$Na_7Fe_3(P_2O_7)_4$. This diphosphate has been recently characterized. It exhibits a reversible transformation at 438–463 K. The monoclinic high-temperature β form is isostructural with $Na_7Fe_3(As_2O_7)_4$ and was investigated by Masquelier et al.[283] A crystal structure determination for the α form was performed by the same authors.[284]

2.2.2.6. Trivalent-Divalent Cation Diphosphates

$Fe(II)Fe(III)_2(P_2O_7)_2$. This mixed-valence iron diphosphate was described by Ijjaali et al.[285] It was prepared by heating at 1173 K mixtures of $FePO_4 + Fe_2P_4O_{12}$ or $Fe_2P_2O_7 + Fe_4(P_2O_7)_3$, to which an excess of $FeCl_2$ was added to improve crystal growth. The authors reported an accurate determination of the atomic arrangement, mainly characterized by the existence of $[Fe_3O_{12}]$ clusters composed of a central $Fe(II)O_6$ trigonal prism sharing its two triangular faces with two adjacent $Fe(III)O_6$ octahedra. This salt is antiferromagnetic with a Néel temperature of 18 K. Recently, a neutron diffraction study was performed by Venturini et al.[286]

$Fe_7(P_2O_7)_4$. Malaman et al.[287] prepared this mixed Fe^{II}–Fe^{III} diphosphate by the solid-state reaction at 1173 K between $Fe_2P_2O_7$ and $Fe_4(P_2O_7)_3$. The authors performed the structure determination and discussed the phase relationships in the system $Fe_2P_2O_7$–$FePO_4$–$Fe_4(P_2O_7)_3$–$Fe_2P_4O_{12}$. The Mössbauer spectrum and magnetic properties of $Fe_7(P_2O_7)_4$ were also reported.

$Fe_2Pb(P_2O_7)_2$. Large single crystals, up to 6 mm, of $Fe_2Pb(P_2O_7)_2$ were obtained by Wanklyn et al.[118] The authors reported a detailed procedure for the growth of such crystals by the use of a flux method.

$Ti_2Ba(P_2O_7)_2$, $Mo_2Ba(P_2O_7)_2$, and $V_2Ba(P_2O_7)_2$. Wang and Hwu[288] prepared the titanium derivative by heating a mixture of P_2O_5, Ti and Ba_2TiO_4 in a molar ratio 4:2:1.at 1173 K for 72 h in an evacuated silica tube. This calcination is followed by a slow cooling (5°/h) to room temperature. The crystals obtained are gemlike navy blue. A crystal structure determination for this compound was reported by the authors.

The corresponding molybdenum and vanadium compounds were prepared and described by Leclaire et al.[289] and Benhamada et al.[290] respectively. In both cases, the atomic arrangements are refined.

Crystallographic Data. Table 2.2.16. gives the main crystal data for trivalent–divalent cation diphosphates.

2.2.2.7. Tetravalent–Cation Diphosphates

As early as 1935, Levi and Peyronel[28] investigated MP_2O_7 diphosphates for M = Si, Ti, Sn, Zr, and Hf and performed a crystal structure determination for the zirconium salt. This structural study was the first structural investigation of a condensed phosphate.

SiP_2O_7. Several forms of SiP_2O_7 have been characterized. A very careful investigation of the SiO_2–P_2O_5 system was performed by Liebau and co-workers.[291,292] These authors reported the existence of seven forms of SiP_2O_7 and discussed the structural relations between them.

Table 2.2.16. Main Crystallographic Data for Trivalent–Divalent Cation Diphosphates

Formula	a (Å) α (°)	b (Å) β (°)	c (Å) γ (°)	S.G.	Z	Reference
$Ti_2Ba(P_2O_7)_2$	10.680(3)	10.564(4) 102.88(3)	9.834(4)	$C2/c$	4	288
$Mo_2Ba(P_2O_7)_2$	10.814(1)	10.641(1) 103.63(1)	9.821(1)	$C2/c$	4	289
$V_2Ba(P_2O_7)_2$	10.621(1)	10.468(1) 103.07(1)	9.706(1)	$C2/c$	4	290
$Fe^{III}_2Fe^{II}(P_2O_7)_2$	8.950(1)	12.235(2)	10.174(1)	$Pnma$	4	285
$(Fe^{II},Fe^{III})_7(P_2O_7)_4$	8.4327(7)	9.695(1)	23.663(4)	$C222_1$	4	287

The crystal structure of the monoclinic variety, AIII, was later determined by Bissert and Liebau.[293] The AIV form crystal structure was described by Liebau and Hesse[294] and Hesse.[295] A structural investigation of the cubic form was also performed by Tillmanns *et al.*[296]

TiP_2O_7. The mechanism of the formation of TiP_2O_7 by heat treatment of $Ti(HPO_4)_2 \cdot H_2O$ was investigated by Pechkovskii *et al.*[297]

GeP_2O_7. The P_2O_5–GeO_2 phase diagram was constructed by Mal'shikov *et al.*[298] GeP_2O_7 is observed as a congruent-melting compound (mp = 1513 K). Two forms of this salt were are reported by the authors: a monoclinic one stable between 823 K and 1323 K and a cubic one stable from 1323 K to the melting point. In addition, they observed that below 823 K an hydrated form corresponding to the formula $GeP_2O_7 \cdot 1.6H_2O$ can be obtained. This third form is hexagonal. Avduevskaya and Tananaev[299] confirmed the existence of two forms for the anhydrous salt.

ZrP_2O_7 and $(ZrO)_2P_2O_7$. The crystal structure determination of $Zr_2P_2O_7$, performed as early as 1935 by Levi and Peyronel,[28] was the first structural investigation of a condensed phosphate.

The ZrO_2–P_2O_5 phase-equilibrium diagram was established by Mal'shikov and Bondar.[300] The most stable compound in this system is $(ZrO)_2P_2O_7$, melting congruently at 2343 K. This compound is polymorphic with a transition temperature of 1373 K. ZrP_2O_7 decomposes at 1823 K and transforms into $(ZrO)_2P_2O_7$ with vaporization of P_2O_5. No crystal data have been reported for $(ZrO)_2P_2O_7$.

Solid electrolytes based on cubic ZrP_2O_7 were investigated by Sacks *et al.*[301]

HfP_2O_7 and $(HfO)_2P_2O_7$. $(HfO)_2P_2O_7$ melting congruently at 2423 K is the most stable compound observed in the HfO_2–P_2O_5 system investigated by Mal'shikov and Bondar.[302] Unlike the corresponding zirconium salt, $(HfO)_2P_2O_7$ does not seem to be polymorphic but is isotypic with one form of $(ZrO)_2P_2O_7$. The other diphosphate observed in this system, HfP_2O_7 decomposes at 1923 K, transforming into $(HfO)_2P_2O_7$.

CeP_2O_7 and $CeP_2O_7 \cdot 9H_2O$. The chemical preparation of various cerium condensed phosphates, including CeP_2O_7, was discussed by Tsuhako *et al.*[303] Merkusheva *et al.*[304] characterized the nonahydrate and investigated its stability.

PtP_2O_7. PtP_2O_7 has been synthesized by reaction between platinum metal and P_4O_{10} vapor in a dry oxygen stream at temperatures above 823 K. It dissociates into platinum and phosphorus oxide between 1073 and 1233 K. A crystal structure determination was performed by Wellmann and Liebau.[305] These authors also presented a good survey of tetravalent diphosphate crystal chemistry.

MoP_2O_7. The synthesis of MoP_2O_7 and a detailed determination of its atomic arrangement were reported by Leclaire *et al.*[306] This salt is isotypic with the cubic form of ZrP_2O_7.

ReP_2O_7. Banks and Sacks[307] prepared ReP_2O_7 by heating a mixture of ReO_2 and $(NH_4)_2HPO_4$ at 873 K in a sealed platinum tube. On the basis of their crystal data, this compound is isotypic with the cubic form of ZrP_2O_7.

MP_2O_7 (M = U, Th). Douglass and Staritzky[308] described the chemical preparation and reported crystal data for the orthorhombic form of UP_2O_7. The thermal expansion of this compound was investigated by Kirchner *et al.*[309] The thorium derivative was described by Douglass and Staritzky[308] and Burdese and Borlera.[310,311,312] The latter authors also investigated the ThP_2O_7–UP_2O_7 system.

Other Studies of MP_2O_7 Salts. The Infrared and Raman spectra of SiP_2O_7, ZrP_2O_7, TiP_2O_7, and SnP_2O_7 were performed by Steger and Leukroth.[313] Vollenkle *et al.*[314] investigated MP_2O_7 compounds, where M = Si, Ge, Sn, Pb, Ti, Zr, Hf, Ce, and U. From crystallographic measurements on powder specimens or single crystals, they reported for these salts cubic unit cells with $22.4 < a < 25.8$ Å. Lecomte *et al.*[315–317] investigated ZrP_2O_7, GeP_2O_7, and three forms of SiP_2O_7 by IR spectroscopy.

Crystallographic Data. Tables 2.2.17 and 2.2.18 gather the main crystal data for tetravalent cation diphosphates.

2.2.2.8. Higher-Valency Cation Diphosphates

(ii) Pentavalent–Divalent Cation Diphosphates

$Nb_2Mg(P_2O_7)_2$ and $Nb_2Co(P_2O_7)_2$. The chemical preparation and crystal structures of these two isotypic salts were reported by Averbuch-Pouchot and Durif.[318] A certain amount of disorder is observed between Nb and the divalent cation. The authors discussed the analogy between this atomic arrangement and that of the cubic $M^{IV}P_2O_7$ series.

$Ta_2Cd(P_2O_7)_2$. This salt was prepared by Averbuch-Pouchot and Durif[319] using a flux method. The authors reported the crystal structure of this compound and discussed its structural analogies with $Nb_2Co(P_2O_7)_2$ and $Nb_2Mg(P_2O_7)_2$.

Table 2.2.17. Unit-Cell Parameters for the Cubic $M_2P_2O_7$ Diphosphates[a]

Formula	a (Å)	Reference(s)
ReP_2O_7	7.94(2)	307
MoP_2O_7	7.944(1)	306
SiP_2O_7(AI)	22.428(6)	291, 292, 314
GeP_2O_7	22.854	298, 314
TiP_2O_7	23.592(3)	314
ZrP_2O_7	24.720(3)	314
HfP_2O_7	24.630(6)	314
CeP_2O_7	25.74(1)	314
SnP_2O_7	23.826(6)	314
PbP_2O_7	24.105(3)	314
UP_2O_7	25.881(6)	314

[a] The space group for all the compounds in this table is $Pa3$ with $Z = 4$ for ReP_2O_7 and MoP_2O_7 and $Z = 108$ for all the other compounds.

Table 2.2.18. Crystal Data for Various Tetravalent Cation Diphosphates

Formula	a (Å) α (°)	b (Å) β (°)	c (Å) γ (°)	S.G.	Z	Reference(s)
GeP_2O_7	4.798	6.48 90.4	15.04	$P2_1/c$	4	298
$GeP_2O_7 \cdot 1.6H_2O$	7.994	7.994	24.87	$R3$	3	298
PtP_2O_7	7.095(2)	7.883(2) 111.37(3)	9.302(3)	$P2_1/n$	4	305
SiP_2O_7 (AII)	22.4	22.4	14.9	Tetragonal.	72	291, 292
SiP_2O_7 (AIII)	4.73	6.33 90.1	14.71	$P2_1/c$	4	291–293
SiP_2O_7 (AIV)	4.73	12.02 91.3	7.62	$P2_1/n$	4	291–292, 294–295
SiP_2O_7 (BI)	8.18	8.18	11.85	$P6_3...$	6	291, 292
UP_2O_7	11.526(5)	12.810(5)	7.045(5)	$Pna...$	8	308

(ii) Pentavalent–Trivalent Cation Diphosphates

$Sb^{III}Sb^V(P_2O_7)_2$. Two forms of $Sb^{III}Sb^V(P_2O_7)_2$ were described by Verbaere *et al.*[320] The authors reported detailed procedures for the chemical preparation of these two phases and accurate determinations of the two atomic arrangements. Both structures are closely related to that of the zirconium diphosphate, ZrP_2O_7.

Crystallographic Data. Table 2.2.19 contains crystal data for the various mixed V–II and V–III diphosphates.

2.2.2.9. Titanyl, Vanadyl, Molybdenyl, Niobyl and Uranyl Diphosphates

A great number of diphosphates include as associated cations various inorganic radicals such as titanyl, vanadyl, niobyl, molybdenyl and uranyl. The main crystal data for these various diphosphates are given in Tables 2.2.20 and 2.2.21.

Table 2.2.19. Main Crystallographic Data for Higher Valency Cation Diphosphates

Formula	a (Å) α (°)	b (Å) β (°)	c (Å) γ (°)	S.G.	Z	Reference
$Ta_2Cd(P_2O_7)_3$	13.094(10)	8.365(8)	15.796(12)	$Pnam$	2	319
$Nb_2Mg(P_2O_7)_3$	15.36(1)	7.930(5) 90.51(1)	6.487(5)	$P2_1/n$	2	318
$Nb_2Co(P_2O_7)_3$	15.32(1)	7.890(5) 90.76(1)	6.490(5)	$P2_1/n$	2	318
$Sb_2(P_2O_7)_2$ (α)	8.088(1)	16.015(3) 90.17(2)	8.135(5)	$P2_1/c$	4	320
$Sb_2(P_2O_7)_2$ (β)	8.018(1)	16.134(3)	8.029(5)	$Pna2_1$	4	320

(i) Vanadyl and Titanyl Diphosphates

$K_2(VO)P_2O_7$ Gorbunova *et al.*[321] described the chemical preparation of $K_2(VO)P_2O_7$. The product obtained appears as mica-like platelets of a greenish-blue color. An accurate structure determination was reported by the authors.

$Rb_2(VO)P_2O_7$ and $Cs_2(VO)P_2O_7$. These two vanadyl diphosphates were prepared by Lii and Wang[322] by heating a mixture of M_3VO_4, VO_2, V, and P_2O_5 in evacuated silica ampoules at 1123 K. The authors reported crystal structure determinations for the two salts.

$K_2(VO)_3(P_2O_7)_2$. The chemical preparation of this compound was described by Leclaire *et al.*[323] These authors reported a detailed description of the atomic arrangement and discussed structural relationships between this compound and other pentavalent and tetravalent phosphates.

$Cs_2(VO)_3(P_2O_7)_2$. Lii *et al.*[324] synthesized this diphosphate and performed a complete determination of its atomic arrangement. From the unit-cell dimensions this salt seems to be isotypic with the corresponding potassium salt,[271] but it was described in a centrosymmetric space group.

$K(V_2O)(VO)(P_2O_7)_2$. This compound containing vanadium atoms in both the 3+ and 4+ states has been recently synthesized by Lee and Lii.[325] It can be obtained as a single-phase

Table 2.2.20. Main Crystallographic Data for Titanyl, Vanadyl, and Niobyl Diphosphates

Formula	a (Å) α (°)	b (Å) β (°)	c (Å) γ (°)	S.G.	Z	Reference
$K_2(VO)P_2O_7$	8.277(3)	8.277(3)	5.420(2)	$P4bm$	2	321
$Rb_2(VO)P_2O_7$	7.101(10)	9.174(2)	12.801(4)	$P2_12_12_1$	4	322
$Cs_2(VO)P_2O_7$	13.280(3)	7.247(2)	9.518(2)	$Pnma$	4	322
$K_2(VO)_3(P_2O_7)_2$	17.407(1)	11.3438(7)	7.2964(15)	$Pna2_1$	4	323
$Cs_2(VO)_3(P_2O_7)_2$	17.613(5)	7.328(2)	11.600(4)	$Pnma$	4	324
$(V_2O)(VO)K(P_2O_7)_2$	5.201(1)	12.661(2) 94.11(2)	9.476(2)	$P2_1/m$	2	325
$Cs_2(TiO)P_2O_7$	7.275(2)	9.452(1)	13.569(2)	$P2_12_12_1$	4	326
$Na_{0.3}MoP_2O_7$	4.8813(6) 91.400(5)	7.0110(5) 92.466(8)	8.2563(4) 106.551(9)	$P1$	2	327
$K_{0.17}MoP_2O_7$	21.278(2)	21.278(2)	4.921(6)	$I4_1/a$	16	328
$K(MoO)OP_2O_7$	5.0862(4)	11.720(1) 90.91(1)	11.486(1)	$P2_1/n$	4	330
$(NH_4)_2(MoO_2)P_2O_7$	13.984(9)	8.297(6) 99.11(1)	15.81(1)	$C2/c$	8	332
$Cs(MoO_2)HP_2O_7$	9.670(5)	14.23(1) 100.10(7)	6.265(5)	$P2_1/a$	4	333
$Cs(NbO)P_2O_7$	4.925(1)	8.887(2) 95.42(2)	18.365(2)	$P2_1/n$	4	332
$K(NbO)P_2O_7$	5.171(1)	11.843(3) 90.98(2)	11.727(3)	$P2_1/n$	4	329

material by heating a mixture of $K_4P_2O_7$, VO_2, V_2O_3, and P_2O_5 in a 1:8:2:7 molar ratio in a silica tube at 1023 K for one day after an intermediate grinding at 1093 K for another day. These also authors reported a procedure for crystal growth and an accurate determination of the atomic arrangement. Magnetic susceptibility data confirm the valency distribution of the vanadium atoms.

$Cs_2(TiO)P_2O_7$. The crystal structure of this salt, which is isotypic with $Rb_2VOP_2O_7$ was reported by Protas *et al.*[326]

(ii) Molybdenyl and Niobyl Diphosphates

$Na_{(0.3)}(Mo)P_2O_7$. The chemical preparation of this mixed valency molybdenum diphosphate was described by Leclaire *et al.*[327] The crystal structure as determined by these authors is closely related to that of $NaMoP_2O_7$.

$K_{(0.17)}(Mo)P_2O_7$. This tetragonal mixed-valency molybdenum diphosphate was prepared by Leclaire *et al.*[328] A complete determination of the crystal structure was performed by these authors.

$K(NbO)P_2O_7$ and $K(MoO)P_2O_7$. Linde *et al.*[329] described the preparation of crystalline $K(NbO)P_2O_7$ by a flux method and reported a detailed determination of its atomic arrangement. The molybdenum salt was later prepared by Gueho *et al.*[330] and identified as an isotype of the niobium derivative. A detailed determination of the atomic arrangement was also performed.

$Cs(NbO)P_2O_7$. Nikolaev *et al.*[331] reported the preparation of single crystals of a cesium–niobyl diphosphate, $Cs(NbO)P_2O_7$ and made an accurate determination of its structural arrangement.

$(NH_4)_2(MoO_2)P_2O_7$ and $Cs(MoO_2)HP_2O_7$. These two compounds were prepared by Averbuch-Pouchot[332,333] using flux methods at a relatively low temperature (573 K). In both cases, the atomic arrangements were accurately determined by the author.

(iii) Uranyl Diphosphates

$(UO_2)_2P_2O_7$ and $(UO_2)_2P_2O_7 \cdot 9H_2O$. The thermal stability of $(UO_2)_2P_2O_7$ was investigated by Barten and Cordfunke.[334,335] The nonahydrate was characterized during an investigation of the $Na_4P_2O_7$–$UO_2(NO_3)_2$–H_2O and $K_4P_2O_7$–$UO_2(NO_3)_2$–H_2O systems by Lavrov *et al.*[336]

Table 2.2.21. Main Crystallographic Data for Uranyl Diphosphates

Formula	a (Å) α (°)	b (Å) β (°)	c (Å) γ (°)	S.G.	Z	Reference
$(UO_2)Na_2P_2O_7$	13.259(2)	8.127(2)	6.973(1)	$Pna2_1$	4	337
$(UO_2)Rb_2P_2O_7$	12.500(3)	12.792(2)	5.783(3)	$Pmmn$	4	340
$(UO_2)Cs_2P_2O_7$	12.670(2)	12.807(2)	6.1522(8)	$Pmmn$	4	340

$(UO_2)Na_2P_2O_7$, $(UO_2)Na_2P_2O_7 \cdot 5H_2O$ and $(UO_2)_3Na_2(P_2O_7)_2 \cdot 14H_2O$. Crystals of $(UO_2)Na_2P_2O_7$ were prepared by Linde et al.[337] by melting a mixture of $(UO_2)Na_2P_2O_7$ and $NaPO_3$ and slowly cooling the melt from 1113 to 973 K. The same authors determined the crystal structure of this salt.

The $Na_4P_2O_7$–$(UO_2)_2P_2O_7$ system was investigated by Lavrov et al.[338] These authors reported a part of the $Na_4P_2O_7$–$(UO_2)_2P_2O_7$ phase-equilibrium diagram, showing clearly that $(UO_2)Na_2P_2O_7$ melts congruently at 1173 K.

The $Na_4P_2O_7$–$UO_2(NO_3)_2$–H_2O system was also investigated by Lavrov et al.[339] They reported the existence of two hydrates, $(UO_2)_3Na_2(P_2O_7)_2 \cdot 14H_2O$ and $(UO_2)Na_2P_2O_7 \cdot 5H_2O$.

$(UO_2)_3K_2(P_2O_7)_2 \cdot 14H_2O$ and $(UO_2)K_2P_2O_7 \cdot nH_2O$ ($5 < n < 7$) have been characterized during an investigation of the $K_4P_2O_7$–$UO_2(NO_3)_2$–H_2O system.[339]

$(UO_2)Rb_2P_2O_7$ and $(UO_2)Cs_2P_2O_7$. These two isotypic diphosphates were obtained and clearly characterized by Linde et al.[340] Crystals of the rubidium salt were obtained by melting a mixture of $Rb_4P_2O_7$ and $(UO_2)Rb_2P_2O_7$ (1:1 stoichiometry) at 1493 K and slowly cooling it to 1270 K. In the case of the cesium salt, crystals were obtained by melting a 1:2 mixture of $(UO_2)Cs_2P_2O_7$ and $CsCl$ at 1460 K and slowly cooling the melt to 950 K. Crystal structure determinations were performed by the authors.

2.2.2.10. Organic Cation Diphosphates

A relatively small number of organic cation diphosphates have been today well characterized. The main crystallographic data for organic cation diphosphates are given in Table 2.2.22.

$[C(NH_2)_3]_3HP_2O_7$, $[C(NH_2)_3]_4P_2O_7 \cdot H_2O_2 \cdot 3/2H_2O$, $[C(NH_2)_3]_4P_2O_7 \cdot H_2O$. Three guan-id-inium derivatives, trisguanidinium hydrogen diphosphate, tetraguanidinium diphosphate monohydrate, and guanidinium diphosphate monoperhydrate sesquihydrate, were prepared by Adams and Ramdas,[341–343] who reported in all cases determinations of the atomic arrangements.

$[NH_3–(CH_2)_2–NH_3]_3(HP_2O_7)_2 \cdot 2H_2O$ and $[NH_3–(CH_2)_2–NH_3]_2P_2O_7$. These two ethy-lenediammonium derivatives were synthesized by Kamoun et al.[344,345] by using a metathesis reaction involving $Ag_4P_2O_7$ as starting material. The two atomic arrangements were determined. Infrared and Raman spectra were also reported.

$[NH_3–(CH_2)_2–NH_3]H_2P_2O_7$ and $[NH_3–(CH_2)_2–OH]_2H_2P_2O_7$. Ethanolammonium dihydrogenodiphosphate[346] and ethylenediammonium dihydrogenodiphosphate[347] were prepared by Averbuch-Pouchot and Durif by using a metathesis reaction involving $Ag_2H_2P_2O_7$ as starting material. The authors reported in both cases the determination of the atomic arrangements.

$Cu_2(EDA)_3(HP_2O_7)_2 \cdot 3H_2O$. This complex atomic arrangement recently investigated by Gharbi et al.[348] must, in fact, be considered as an adduct between ethylenediamine and a copper–ethylenediammonium diphosphate and should be formulated as follows:

$$Cu_2[(NH_3(CH_2)_2NH_3](HP_2O_7)_2 \cdot [NH_2(CH_2)_2NH_2]_2 \cdot 3H_2O$$

Table 2.2.22. Crystal Data for Organic Cation Diphosphates

Formula	a (Å) α (°)	b (Å) β (°)	c (Å) γ (°)	S.G.	Z	Reference
$(EDA)_3(HP_2O_7)_2 \cdot 2H_2O$	11.860(2)	6.463(1) 112.36(1)	14.997(3)	$P2_1/n$	2	344
$(Gua)_3HP_2O_7$	9.29(2)	17.01(2)	9.47(2)	$P2_12_12_1$	4	341
$(Gua)_4P_2O_7 \cdot H_2O$	7.05(2)	11.69(2) 111.7(1)	12.01(2)	$P2_1$	2	342
$(Gua)_4P_2O_7 \cdot H_2O_2 \cdot 3/2H_2O$	17.84(3)	14.15(2) 116.8(2)	17.72(3)	$P2_1/a$	4	343
$(EDA)H_2P_2O_7$	9.965(9) 101.56(4)	7.666(5) 102.58(3)	6.117(7) 90.25(6)	$P\bar{1}$		347
$(EDA)_2P_2O_7$	8.724(1)	13.511(2) 96.25(1)	10.039(1)	$C2/c$	4	345
$(EthA)_2H_2P_2O_7$	9.003(1) 107.82(3)	8.773(1) 93.16(3)	8.151(1) 94.14(3)	$P\bar{1}$	2	346
$Cu_2(EDA)_3(HP_2O_7)_2 \cdot 3H_2O^b$	18.808(8)	9.631(2) 109.63(5)	14.019(8)	$C2/c$	4	348

[a] Abbreviations: EDA, ethylenediammonium, $NH_3-(CH_2)_2-NH_3$; Gua, guanidinium, $C(NH_2)_3$; EthA, ethanolammonium, $NH_3-(CH_2)_2-OH$.
[b] In fact this compound includes both ethylenediammonium and ethylenediamine groups.

2.2.2.11. Adducts Between Diphosphates and Telluric Acid

Like all other alkali phosphates, condensed or not, diphosphates have the property of forming adducts with telluric acid. A relatively small number of such adducts are presently known.

The first example, $Te(OH)_6 \cdot K_3HP_2O_7 \cdot H_2O$, was reported by Averbuch-Pouchot and Durif,[349] who determined the crystal structure. Three new examples of such adducts, $Te(OH)_6 \cdot 2(NH_4)_2H_2P_2O_7$, $Te(OH)_6 \cdot 2Rb_2H_2P_2O_7$, and $Te(OH)_6 \cdot Cs_2H_2P_2O_7$ were recently described by the same authors. The first two[350–351] are isotypic, and the atomic arrangements have been determined using the ammonium salt. Crystal data for the cesium salt are given in Ref.[352] The main crystal data for these adducts are gathered in Table 2.2.23.

Table 2.2.23. Crystal Data for Adducts between Telluric Acid and Diphosphates

Formula	a (Å) α (°)	b (Å) β (°)	c (Å) γ (°)	S.G.	Z	Reference(s)
$Te(OH)_6 \cdot K_3HP_2O_7 \cdot H_2O$	15.98(8) 109.49(6)	7.226(5) 84.34(7)	6.253(5) 101.83(7)	$P\bar{1}$	2	349
$Te(OH)_6 \cdot 2(NH_4)_2H_2P_2O_7$	7.651(2)	21.79(1) 113.85(3)	6.689(2)	$P2_1/n$	4	350, 351
$Te(OH)_6 \cdot 2Rb_2H_2P_2O_7$	7.652(2)	21.76(1) 114.12(5)	6.674(2)	$P2_1/n$	4	350 351
$Te(OH)_6 \cdot Cs_2H_2P_2O_7$	20.49(1)	8.362(9) 106.54(5)	16.63(1)	$C2/c$	8	352

2.2.2.12. Miscellaneous Diphosphates

Some other types of diphosphates that are difficult to classify have been reported. We list them below.

$Na_2MZr(P_2O_7)_2$, M = Ni and Co. Gali et al.[353] produced crystals under hydrothermal conditions (473–573 K and 1–10 GPa) to which they assigned the above formula. They performed crystal structure determinations for these two isotypic salts using the triclinic noncentrosymmetric $P1$ space group. Later Marsh[354] contested these results and, using the same X-ray diffraction data collection, refined the two structures in the centrosymmetric $P\bar{1}$ space group and suggested a very different chemical formula—$Co(Ni)NaHP_2O_7$—with, nevertheless possibly a very low content of zirconium.

$K_2P_2O_5F_2$. This difluorodiphosphate, one of the few examples of a substituted P_2O_7 group was prepared by Durand et al.[355] who also determined its crystal structure. Its unit-cell dimensions are given below.

$$a = 12.614, b = 7.585, c = 7.195 \text{ Å}, \beta = 90.91°$$

Space group is $C2/c$, and $Z = 4$.

$Ag_{16}I_{12}P_2O_7$. Diphosphate groups have also been observed in some less conventional compounds such as $Ag_{16}I_{12}P_2O_7$—in fact, $Ag_{42}O_7 \cdot 12AgI$—described by Garrett et al.[356] Its unit-cell dimensions are given below.

$$a = 12.05, c = 7.504 \text{ Å}$$

The space group is $P6/mcc$, and $Z = 1$. The structure is a hexagonal close-packed arrangement of iodine atoms, forming large channels containing disordered P_2O_7 groups.

$NH_4NO_3 \cdot (NH_4)_3HP_2O_7 \cdot H_2O$. This ammonium diphosphate nitrate, was characterized by Frazier et al.[67] The authors reported several procedures for its preparation. This salt is stable at relative humidities less than 50%. At higher humidity, it decomposes into ammonium nitrate and $(NH_4)_3HP_2O_7 \cdot H_2O$. Near 348 K it undergoes a reversible phase transition, leading to a cubic modification, and above 373 K it slowly decomposes to $NH_4H_2PO_4$ and NH_4NO_3.

$Cd_2SiP_4O_{14}$. A diphosphate group has also been observed in this cadmium silicate–phosphate. Its preparation and crystal structure were reported by Trojan et al.[357] Diphosphate groups also appear in some mixed-anion phosphates. These compounds will be discussed in Chapter 5.

2.2.3. Chemical Preparation of Diphosphates

In contrast to the situation for most of the other classes of condensed phosphates, several starting materials are commonly available for the synthesis of diphosphates. Thus, sodium, potassium, and ammonium diphosphates as well as diphosphoric acid are commercially available chemicals. Nevertheless, there are some preparative routes that are more or less specific to this category of phosphates, and we describe these in this section.

2.2.3.1. Dehydration–Condensation of Mono- or Dihydrogen Monophosphates

A good number of diphosphates can be prepared as polycrystalline samples by dehydration-condensation of monohydrogen monophosphates. Thus, $Na_4P_2O_7$, for instance, can be easily prepared by heating Na_2HPO_4:

$$2Na_2HPO_4 \rightarrow Na_4P_2O_7 + H_2O$$

Divalent cation diphosphates can also be prepared by the same process; $Ca_2P_2O_7$, for instance, can be obtained as follows:

$$2CaHPO_4 \cdot 2H_2O \xrightarrow{\ 393\ K\ -4H_2O\ } 2CaHPO_4 \xrightarrow{\ 723\ K\ -H_2O\ } \gamma\text{-}Ca_2P_2O_7 \xrightarrow{\ 723\ K\ -H_2O\ } \beta\text{-}Ca_2P_2O_7$$

Some dihydrogen diphosphates can also be synthesized by a very similar process but using dihydrogen monophosphate as starting material:

$$Sr(H_2PO_4)_2 \rightarrow SrH_2P_2O_7 + H_2O$$

In this particular case the reaction must be conducted at 463–483 K and is extremely slow. One also must not forget that, in reactions of this last type, prolonged heating or a slight increase of the reaction temperature leads to a more condensed phosphate:

$$SrH_2P_2O_7 \rightarrow Sr(PO_3)_2 + H_2O$$

This last step occurs, in this case, at temperature as low as 593–603 K.

2.2.3.2. Use of Diphosphoric Acid

Many alkali diphosphates have been obtained by neutralizing diphosphoric acid with appropriate amounts of the corresponding alkali carbonates. When this process is used, the reaction tank must be kept at low temperature to avoid the decondensation of the acid.

2.2.3.3. Thermal Methods

A good number of diphosphates can be prepared by now classical thermal methods. For instance, most of divalent cation diphosphates can be obtained as polycrystalline specimens according to the following reaction:

$$2MCO_3 + 2(NH_4)_2HPO_4 \rightarrow M_2P_2O_7 + 4NH_3 + 2CO_2 + 3H_2O$$

The temperature to be used and the time of heating are mainly dependent on the nature of the divalent cation involved in the reaction.

2.2.3.4. Exchange Reactions

Since the recent characterization of $Ag_2H_2P_2O_7$ and the description of a convenient procedure for its preparation, this salt has been successfully used for the preparation of dihydrogen diphosphates through an exchange process derived from Boullé's method. Schematically, the reaction is for instance:

$$Ag_2H_2P_2O_7 + 2CsCl \rightarrow Cs_2H_2P_2O_7 + 2AgCl$$

This process is also used for the preparation of organic derivatives as described below.

2.2.3.5. Flux Methods

The various steps in the condensation of monophosphoric acid have been carefully investigated as a function of temperature, and it is well known that the first step corresponding to the formation of diphosphoric acid, $H_4P_2O_7$, occurs at a relatively low temperature (T < 473 K). This fact has often been used as the basis for the preparation of diphosphates by flux methods, employing monophosphoric acid as starting reagent. For example, single crystals of $MnHP_2O_7$ can be prepared by heating a mixture of H_3PO_4 and $MnCO_3$ containing a large excess of phosphoric acid for 18 h at 473 K. When using such a method, one must take into account that the state of polymerization of phosphoric acid at a given temperature is highly dependent of the nature of the added cations.

The chemical preparation of $ZnNa_2P_2O_7$, $MgNa_2P_2O_7$ and $CaNa_2P_2O_7$ by the action of the divalent oxides on molten $NaPO_3$ has been reported.

2.2.3.6. Preparation of the Organic Derivatives

Most of the organic cation diphosphates have been prepared using two main processes:

(a) The direct action of diphosphoric acid on the corresponding amine, according to the following reaction scheme:

$$2R-NH_2 + H_4P_2O_7 \rightarrow (R-NH_3)_2H_2P_2O_7$$

(b) The use of an exchange reaction similar to that described above. In this case, the amine chloride is used as starting material, and the reaction is:

$$2R-NH_3 \cdot Cl + Ag_2H_2P_2O_7 \rightarrow (R-NH_3)_2H_2P_2O_7 + 2AgCl$$

In both cases, the reaction must be run at room temperature or below in order to avoid hydrolysis of the P_2O_7 group.

2.2.4. Some Atomic Arrangements of Diphosphates

Diphosphate crystal chemistry is presently the most investigated part of condensed phosphate chemistry. Most of the major types of structures exhibited by compounds in this family are now well established. Therefore, in this section we will not describe as great a number of atomic arrangements as we will do for less common condensed phosphates. In the next section, we will describe in some detail the geometrical aspects of the infinite networks built by the association of acidic diphosphate groups interconnected by hydrogen bonds. These infinite anionic entities, sometimes called "macroanions," have, up to date, never been reviewed. Thus, in order to limit the number of drawings, most of the examples selected in the present section correspond to compounds that are relevant to our later discussion of these "macroanions."

2.2.4.1. Diammonium-Dihydrogenodiphosphate

$$(NH_4)_2H_2P_2O_7,^{80} \text{ monoclinic, } P2_1/n, Z = 4$$

$$a = 9.058(7), b = 11.199(8), c = 7.764(6) \text{ Å}, \beta = 108.40(1)°$$

Figure 2.2.1. Projection, along the *c* direction, of the atomic arrangement in $(NH_4)_2H_2P_2O_7$. The larger open circles represent the nitrogen atoms, and the smaller ones the hydrogen atoms of the phosphoric entities. The hydrogen atoms of the ammonium groups have been omitted.

The atomic arrangement in $(NH_4)_2H_2P_2O_7$ is a typical layer organization. As can be seen in Fig. 2.2.1, which shows a projection along the **c** axis, the planes y = 0 and 1/2 contain the $[H_2P_2O_7]^{2-}$ groups and alternate with planes (y = 1/4 and 3/4) containing the ammonium groups. These two types of planes are slightly corrugated. Within a plane containing the phosphoric entities, the $[H_2P_2O_7]^{2-}$ groups, linked by strong hydrogen bonds, form infinite ribbons parallel to the *a* direction. Figure 2.2.2 represents such a linear organization in projection along the *b* axis.

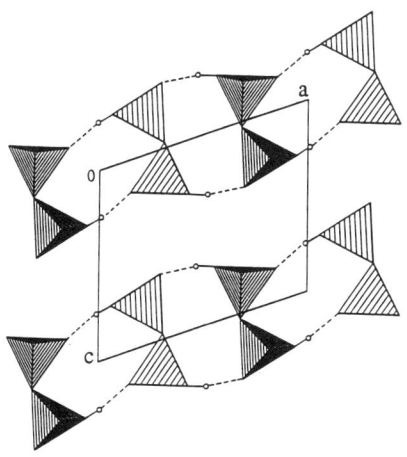

Figure 2.2.2. Projection, along the *b* axis, of the $[H_2P_2O_7]^{2n-}$ ribbons located in the plane y = 1/2. Hydrogen bonds are represented by solid and dashed lines.

In this arrangement, the NH_4 groups are rather regular tetrahedra with N–H distances ranging from 0.76 to 0.94 Å and H–N–H angles ranging between 99 and 115°. The oxygen coordination of the two crystallographically independent ammonium groups is different. Within a range of 3.50 Å, the first one has nine neighbors, and the second one seven. Inside these NO_n polyhedra, the N–O distances range from 2.818 to 3.408 Å.

In an infinite $[H_2P_2O_7]^{2n-}$ ribbon, each phosphoric group is connected to its two adjacent neighbors by strong hydrogen bonds since the corresponding O–O distances, 2.541 and 2.514 Å, are of of the same order of magnitude as the O–O distances inside the PO_4 tetrahedra. In addition, the two H···O distances observed in this ribbon, 1.77 and 1.68 Å, are the shortest in the hydrogen-bond network of this arrangement.

2.2.4.2. Silver Dihydrogenodiphosphate

$$Ag_2H_2P_2O_7,[73] \text{ orthorhombic, } Fddd, Z = 16$$

$$a = 27.779(20), b = 12.385(6), c = 7.026(4) \text{ Å}$$

As the central oxygen atom of the $H_2P_2O_7$ group is located on a twofold axis, this anion has a binary internal symmetry and so is composed of only one independent PO_4 tetrahedron. This type of internal symmetry is not very among diphosphate groups. These phosphoric groups, interconnected by strong hydrogen bonds, assemble themselves so as to build very corrugated ribbons spreading along the *b* direction. Figure 2.2.3 shows in

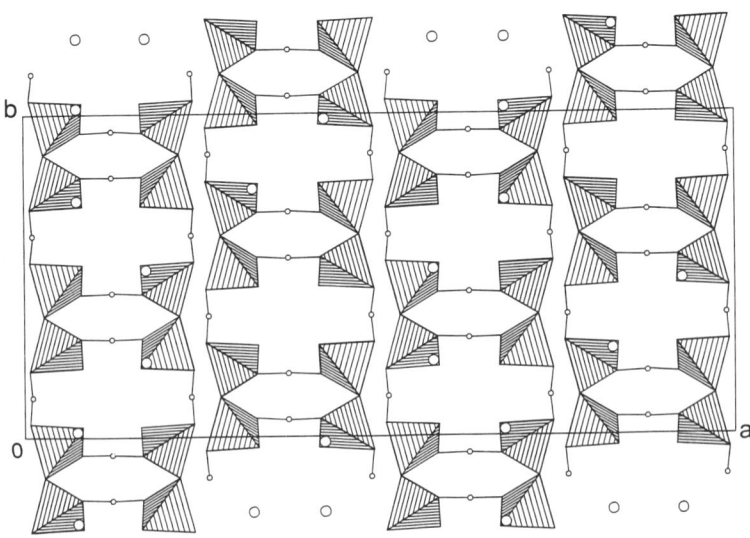

Figure 2.2.3. Projection, along the *c* direction, of the atomic arrangement in $Ag_2H_2P_2O_7$. The smaller circles represent the hydrogen atoms, and the larger ones the silver atoms.

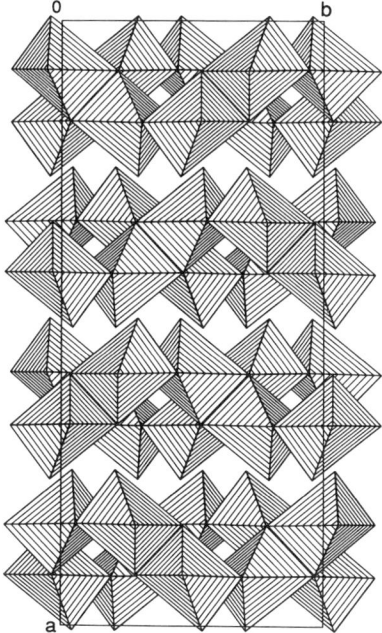

Figure 2.2.4. Projection, along the *c* axis, of the layers of AgO_6 octahedra in Ag_2H_2P_2O_7.

projection along the *c* direction, the organization of the four phosphoric ribbons crossing the unit cell.

The silver atom located in general position has sixfold coordination; as usual, only the external oxygen atoms of the phosphoric anion and not the central oxygen atom are involved in the cation polyhedron. These AgO_6 octahedra, through edge and corner sharing, build thick bidimensional layers parallel to the (*b*,*c*) plane. The general aspect of this layer arrangement is depicted in Figure 2.2.4., which shows a projection of this AgO_6 framework along the *c* axis, and the internal organization of the polyhedra inside one layer is illustrated in Fig. 2.2.5, representing a projection along the *a* axis. Within an AgO_6 octahedron, the Ag–O distances range from 2.439 to 2.781 Å, and the O–Ag–O angles from 76.1 to 113.2°.

Figure 2.2.3 gives a clear description of the hydrogen bonds interconnecting the phosphoric groups so as to build the $[H_2P_2O_7]_n$ infinite ribbons, in which each $H_2P_2O_7$ group is bonded to its three adjacent neighbors by very short hydrogen bonds. The O–O distances involved in these hydrogen bonds are all shorter than 2.50 Å. As is very commonly the case with short H bonds, they are symmetrical. Here the symmetry is a binary one, the two hydrogen atoms of the $H_2P_2O_7$ group being located on the twofold axes parallel to the *c* direction. The O–H distances in these symmetrical hydrogen bonds are in both cases 1.23Å, and the O–H–O angles are 169 and 177°. The corresponding sodium salt is isotypic.

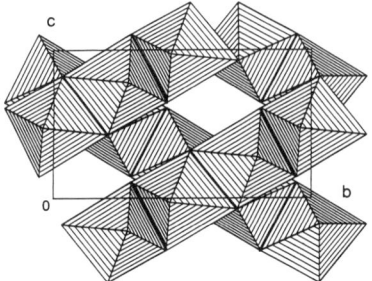

Figure 2.2.5. Projection along the *a* axis of the layers of AgO_6 octahedra in $Ag_2H_2P_2O_7$. Projection restricted to $-0.02 < x < 0.27$.

2.2.4.3. Rubidium Dihydrogenodiphosphate Hemihydrate

$$Rb_2H_2P_2O_7 \cdot 1/2H_2O,[82] \text{ orthorhombic, } Pnam, Z = 8$$

$$a = 19.568(10), b = 10.545(7), c = 7.773(5) \text{ Å}$$

Due to the organization of the strong hydrogen-bond network connecting the $H_2P_2O_7$ groups, this atomic arrangement must be described as a layer structure. Thick layers composed by the $(H_2P_2O_7)_n$ macroanion lie perpendicular to the *a* direction, approximately centered by the planes x = 1/4 and 3/4, while all the other atoms are located in the mirror planes situated at z = 1/4 and 3/4. Figure 2.2.6. gives a projection of this structure along the *b* direction.

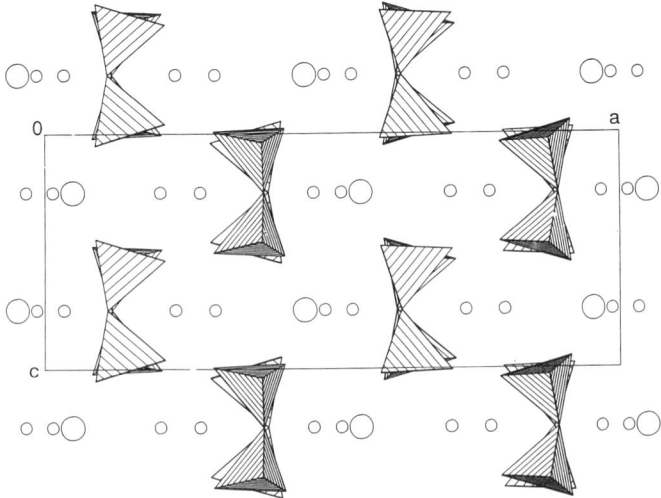

Figure 2.2.6. Projection of the atomic arrangement in $Rb_2H_2P_2O_7 \cdot 1/2H_2O$ along the *b* direction. The large open circles represent water molecules, and the smaller ones the rubidium atoms.

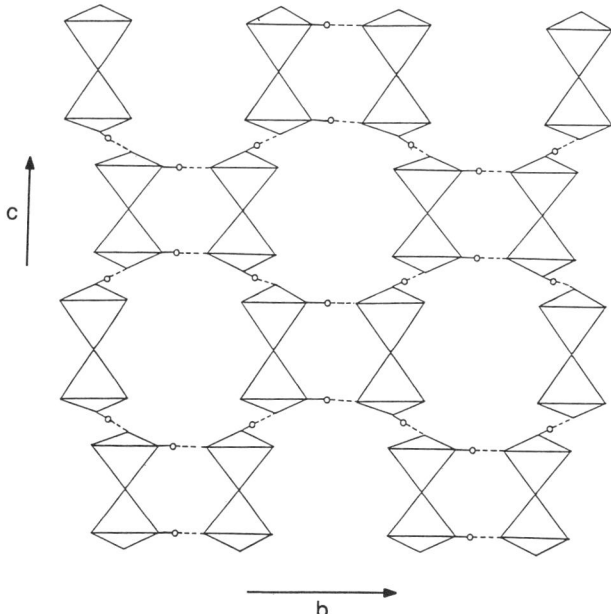

Figure 2.2.7. Schematic representation of the two-dimensional macroanion in projection along the *a* direction.

The behavior of the two independent $H_2P_2O_7$ groups is rather different. Both have internal mirror symmetry, their bonding oxygen atoms being located in the mirror planes. One of these two phosphoric groups has a quite usual configuration with, as can be expected, three types of P–O bond–distances, inside the PO$_4$ tetrahedrona, a long one corresponding to the P–O–P bridge (1.613 Å), an intermediate one (1.560 Å) corresponding to the P–OH bond, and two short ones (1.490 and 1.501 Å) corresponding to the normal P–O bonds. The average values for the P–O distances and the O–P–O angles are 1.541 Å and 109.2° respectively. In the second $H_2P_2O_7$ group, the situation is different; the central oxygen atom is disordered and must be described as two fragments both located on the mirror plane with occupancy rates of 0.215 and 0.285. This type of disorder is commonly observed in diphosphate groups having a linear P–O–P bond, but rarely for other types of internal symmetries. The hydrogen bonds interconnecting these $H_2P_2O_7$ entities correspond to O–O distances (2.519 and 2.508 Å) shorter than the average O–O distance inside the PO$_4$ tetrahedra. In Fig. 2.2.7, a schematic representation of the two-dimensional macroanion as observed in this arrangement is presented.

The four independent rubidium atoms, all located in the mirror planes, differ in their coordination to oxygen; two of them have 8 neighbors, one has 10 and one has 12. The hydrogen atoms of the water molecule were not located with certainty.

2.2.4.4. Cesium Dihydrogenodiphosphate

$Cs_2H_2P_2O_7$,[83] monoclinic, $C2/c$, $Z = 4$

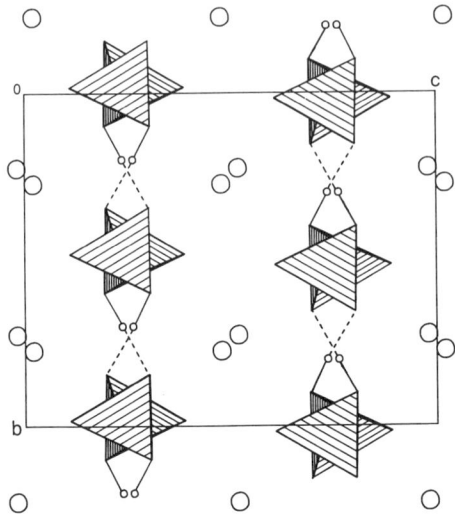

Figure 2.2.8. Projection of the atomic arrangement in $Cs_2H_2P_2O_7$ along the *a* direction. The smaller circles represent the hydrogen atoms, and the larger ones the cesium atoms. Hydrogen bonds connecting the phosphoric entities are represented by solid and dashed lines.

$$a = 7.977(3), b = 9.064(6), c = 11.406(6) \text{ Å}, \beta = 90.29°$$

The atomic arrangement in $Cs_2H_2P_2O_7$ is another typical layer organization. Figure.2.2.8 shows a projection of this structure along the *a* direction. The phosphoric entities are located in layers perpendicular to the *c* axis at z = 1/4 and 3/4. These layers alternate with planes of cesium atoms at z = 0 and 1/2. Within a phosphoric layer, each $[H_2P_2O_7]^{2-}$ group is connected to its four adjacent neighbors by strong hydrogen bonds, corresponding to an O· · ·O distance of 2.534 Å, shorter than the average O–O distance in the PO_4 tetrahedron. One of these anionic layers is depicted in projection along the *c* direction in Fig. 2.2.9.

The central oxygen atom of the $[H_2P_2O_7]^{2-}$ group is located on a twofold axis, imparting a binary internal symmetry to the phosphoric entity, built so by only one crystallographically independent PO_4 tetrahedron. Within the PO_4 tetrahedron, three types of P–O distances are observed, the longest one (1.618 Å) corresponding to the bonding oxygen atom, the intermediate one (1.566 Å) to the P–OH bond, and the shortest two (1.483 and 1.503 Å) to the P–O bonds. The average values for the P–O distances and O–P–O angles, 1.542 Å and 109.2°, are those generally observed for this type of tetrahedron.

The cesium atoms have a ninefold coordination, with Cs–O distances ranging from 3.076 to 3.471 Å; it should be noticed that the bonding oxygen atom takes part in this coordination.

2.2.4.5. Nickel Diphosphate Form(IV)

$$Ni_2P_2O_7,^{[121]} \text{ monoclinic, } P2_1/a, Z = 2$$

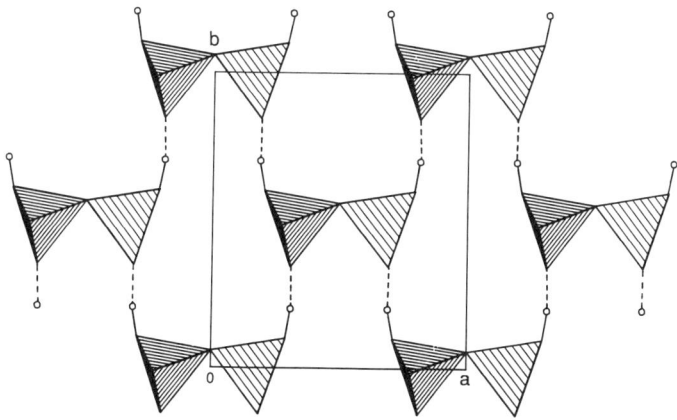

Figure 2.2.9. Projection, along the *c* direction, of a layer of phosphoric anions. The hydrogen atoms are represented by small open circles, and the hydrogen bonds by solid and dashed lines.

$$a = 5.212(3), \ b = 9.913(5), \ c = 4.475(3) \ \text{Å}, \ \beta = 97.46(10)°$$

$Ni_2P_2O_7$ is isotypic with the high temperature form of $Er_2Si_2O_7$. The most interesting feature of its very simple atomic arrangement is the existence of a P_2O_7 group with a linear P–O–P bond. The central oxygen atom of the phosphoric group is located on the inversion center at 1/2, 0, 1/2. The projection along the *c* axis in Fig. 2.2.10 shows the respective locations of the nickel atoms and the P_2O_7 groups. The nickel atom has sixfold oxygen coordination, forming a distorted octahedron with Ni–O distances ranging from 2.027 to 2.204 Å. Each NiO_6 octahedron shares three of its edges with its three adjacent neighbors, thus building, corrugated layers perpendicular to the *c* direction. Figure 2.2.11, representing a projection along the *a* axis, shows clearly this layer organization. Inside a layer, as shown

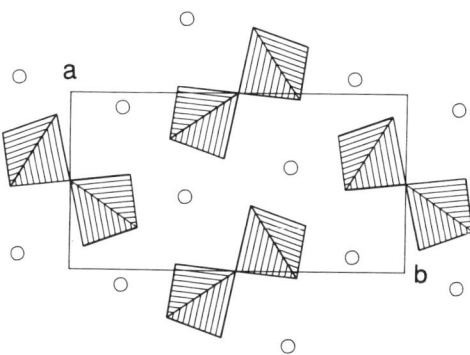

Figure 2.2.10. Projection of the structure of $Ni_2P_2O_7$ along the *c* direction. The open circles represent the nickel atoms.

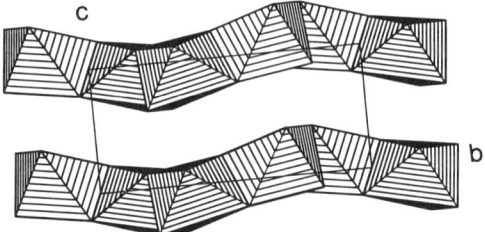

Figure 2.2.11. Projection, along the *a* direction, showing a profile view of the NiO_6 layer organization in $Ni_2P_2O_7$.

in Fig. 2.2.12, the stacking of the NiO_6 octahedra creates large elongated hexagonal voids in which the phosphoric groups are located.

2.2.4.6. β-Calcium Diphosphate

$$\beta\text{-}Ca_2P_2O_7,^{133} \text{ tetragonal, } P4_1, Z = 8$$

$$a = 6.684(6), c = 24.14(1)\text{Å}$$

As shown in Fig. 2.2.13, which is a projection along the *a* direction, the atomic arrangement of β-$Ca_2P_2O_7$ can be described as a succession of layers perpendicular to the *c* direction. The organization of calcium atoms and P_2O_7 groups in one of these layers is given in Fig. 2.2.14, a view of an isolated layer in projection along the *c* direction. The phosphoric groups form arrays parallel to the (1 1 0) direction alternating with parallel arrays of calcium atoms.

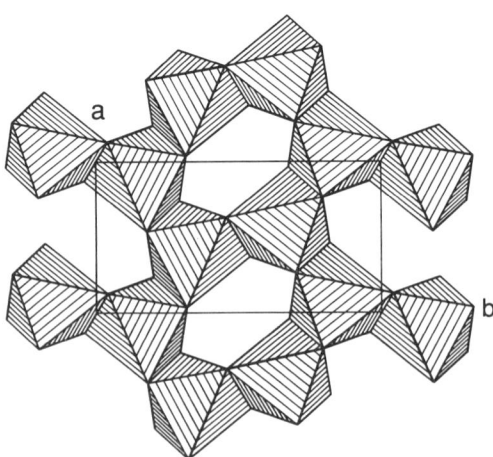

Figure 2.2.12. Projection, along the *c* direction, of a layer of NiO_6 octahedra showing the elongated hexagonal channels in which are inserted the phosphoric groups.

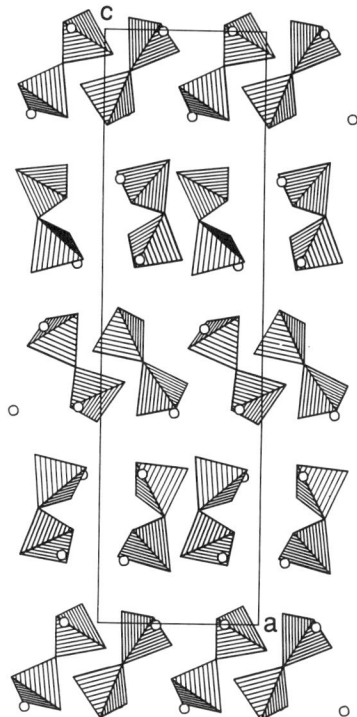

Figure 2.2.13. Projection, along the *a* axis, of the atomic arrangement in β-$Ca_2P_2O_7$. The open circles represent the calcium atoms.

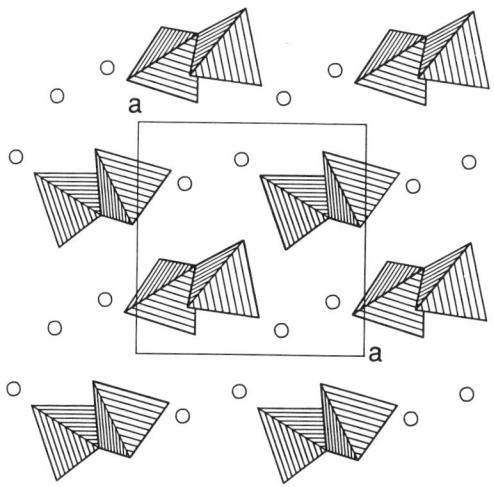

Figure 2.2.14. Projection, along the *c* direction ($0.17 < z < 0.45$), of an isolated layer in β-$Ca_2P_2O_7$, showing the alternating arrays of calcium atoms and phosphoric groups.

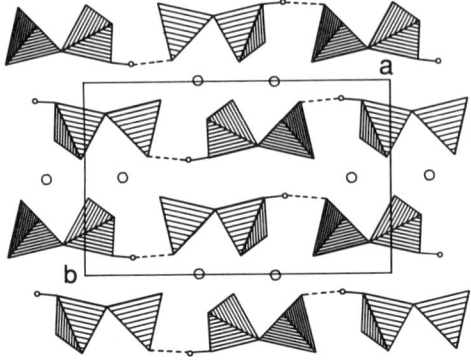

Figure 2.2.15. Projection, along the *c* direction, of the atomic arrangement in $MnHP_2O_7$. The larger open circles represent the manganese atoms, and the smaller ones the hydrogen atoms.

Two independent P_2O_7 groups coexist inside the structure; both have no internal symmetry. Within a range of 3 Å, the four independent calcium atoms have various coordinations. Two of them have seven oxygen neighbors, one has eight, and the last one has nine. The Ca–O distances range from 2.318 to 2.927 Å in these various polyhedra.

2.2.4.7. Manganese(III) Monohydrogenodiphosphate

$$MnHP_2O_7,^{221} \text{ monoclinic, } P2_1/n, Z = 4$$

$$a = 7.951(4), b = 12.645(8), c = 4.922(2) \text{ Å}, \beta = 100.92(5)°$$

As shown in Fig. 2.2.15, representing a projection along the *c* direction, infinite chains built by $P_2O_7H^{3-}$ groups connected by strong hydrogen bonds (O–O = 2.577 Å) lie along the *b* direction in planes at approximately x = 1/4 and 3/4. Between these phosphoric layers are located the manganese atoms in planes at x = 0 and 1/2. These manganese atoms have sixfold coordination with five Mn–O distances ranging from 1.876 to 2.113 Å and a longer one of 2.433 Å, forming a distorted octahedron (78.71° < O–Mn–O < 94.5°). By sharing edges, the MnO_6 octahedra assemble in pairs yielding centrosymmetric Mn_2O_{10} clusters with an Mn–Mn distance of 3.417 Å.

Within this arrangement the Mn–P distances are rather short, since ranging from 3.171 to 3.273 Å, and this accounts for the compactness of this structure (17.3 Å³ per oxygen atom), probably one of the most compact ever observed in a condensed phosphate.

2.2.4.8. Molybdenyl Cesium Monohydrogendiphosphate

$$(MoO_2)CsHP_2O_7,^{333} \text{ monoclinic, } P2_1/a, Z = 4$$

$$a = 9.670(5), b = 14.231(10), c = 6.265(5) \text{ Å}, \beta = 100.10(7)°$$

A projection of the structure of $(MoO_2)CsHP_2O_7$ along the *b* axis is shown in Fig. 2.2.16. Infinite ribbons composed of P_2O_7 and MoO_6 octahedra lie along the c direction. Within such a ribbon, a P_2O_7 group is connected to three MoO_6 octahedra. A projection

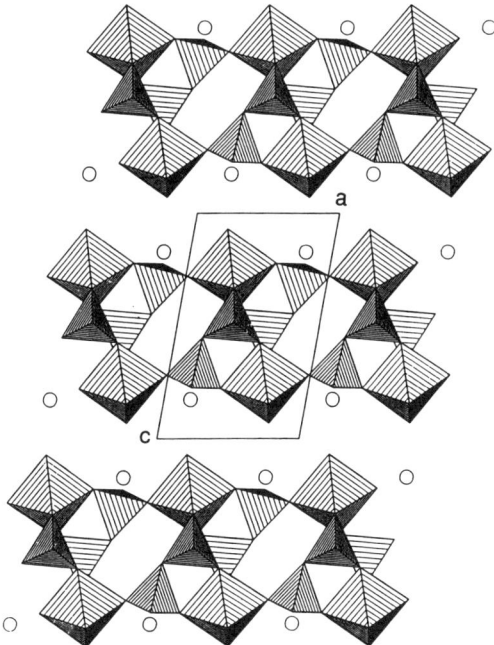

Figure 2.2.16. Projection of the structure of $(MoO_2)CsHP_2O_7$ along the b axis $(0.70 < y < 1.30)$. The MoO_6 groups are represented by hatched octahedra, and the open circles represent the cesium atoms.

along the c axis, (Fig. 2.2.17), that is, to say, along the ribbon direction, allows a better understanding of this structure. Perpendicular to the b axis, the layers containing the ribbons alternate with corrugated layers of cesium atoms. This projection also illustrates the connections established by strong hydrogen bonds between adjacent HP_2O_7 groups giving rise to infinite phosphoric chains extending along the a direction. The O–O distance of the hydrogen bond, 2.516 Å, is shorter than some O–O distances inside the PO_4 tetrahedra.

The MoO_6 octahedron is rather distorted, as usual for the molybdenyl group, with two short Mo–O distances (1.688 and 1.701 Å) and four larger ones ranging between 1.997 and 2.190 Å.

Within a range of 3.50 Å, the cesium atom has sixfold coordination with Cs–O distances ranging from 3.102 to 3.182 Å.

2.2.4.9. *Ethylenediammonium Dihydrogenodiphosphate*

$$[NH_3-(CH_2)_2-NH_3]H_2P_2O_7,^{347} \text{ triclinic, } P\bar{1}, Z = 2$$

$$a = 9.965(9), b = 7.666(5), c = 6.117(7) \text{ Å}$$

$$\alpha = 101.56(4), \beta = 102.58(3), \gamma = 90.25(6)°.$$

Figure 2.2.17. Projection of the structure of $(MoO_2)CsHP_2O_7$ along the *c* axis. The larger open circles represent the cesium atoms, and the smaller ones the hydrogen atoms. Hydrogen bonds are indicated by solid and dashed lines in the lower part of the figure.

The $[H_2P_2O_7]^{2-}$ anions in this structure have no internal symmetry. They are interconnected by strong hydrogen bonds (O–O = 2.513 and 2.588 Å) and form infinite ribbons extending parallel to the *c* direction. Figure 2.2.18, a projection along the *b* direction, depicts the geometry of one of these ribbons, while Fig. 2.2.19 shows the complete atomic arrangement in projection along the axis of the ribbon and serves to illustrate clearly the layer organization of the structure, essentially composed of layers of organic groups alternating with layers containing the phosphoric ribbons.

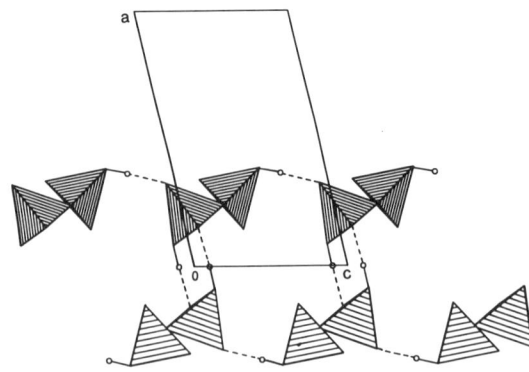

Figure 2.2.18. Projection, along the *b* direction, of an isolated infinite $[H_2P_2O_7]_n$ ribbon. The small open circles denote the hydrogen atoms. The hydrogen bonds are represented by solid and dashed lines.

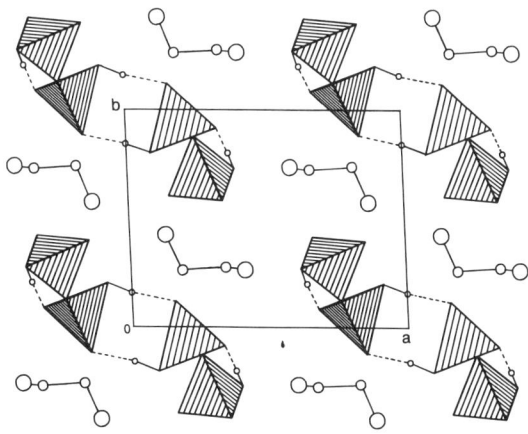

Figure 2.2.19. Projection along the *c* direction of the atomic arrangement in $[NH_3-(CH_2)_2-NH_3]H_2P_2O_7$. The phosphoric entities are depicted in polyhedral representation. The open circles represent, in order of decreasing size, nitrogen, carbon, and hydrogen atoms. The H atoms of the organic entities are omitted. Hydrogen bonds, in the phosphoric ribbons, are represented by solid and dashed lines.

2.2.4.10. Bis(ethanolammonium) Dihydrogenodiphosphate

$$(NH_3-C_2H_4-OH)_2H_2P_2O_7,[346]\ triclinic,\ P\bar{1},\ Z = 2$$

$$a = 9.003(1),\ b = 8.772(1),\ c = 8.151(1)\ \text{Å}$$

$$\alpha = 107.82(3),\ \beta = 93.16(3),\ \gamma = 94.14(3)°$$

The main feature of the atomic arrangement of $(NH_3-C_2H_4-OH)_2H_2P_2O_7$ is the existence of infinite ribbons composed of $[H_2P_2O_7]^{2-}$ groups. In such a ribbon, each $[H_2P_2O_7]^{2-}$ group is connected to its two adjacent neighbors by four hydrogen bonds. Figure 2.2.20 shows the projection along *b* and *c* of this ribbon. This chain crosses the center of the unit cell parallel to the *a* direction with a period of two $[H_2P_2O_7]^{2-}$ units. These rows are interconnected in a three-dimensional way through hydrogen bonds established between the NH_3 and OH groups of the ethanolammonium cations and some oxygen atoms of the phosphoric arrays.

Two crystallographically independent ethanolammonium groups coexist in this atomic arrangement. It is worth noting that two hydrogen bonds, $N-H\cdots O$, interconnect these organic groups. Such bonding between organic cations has rarely been observed in the numerous organic cation phosphates previously investigated.

The two hydrogen atoms belonging to a $[H_2P_2O_7]^{2-}$ group are involved in hydrogen bonds with oxygen atoms of the two adjacent $[H_2P_2O_7]^{2-}$ groups, giving rise to infinite ribbons extending along the *a* axis. This type of hydrogen bond is very strong since the corresponding O–O distances, 2.553 and 2.598 Å, are of the same order of magnitude as the O–O distances inside the PO_4 tetrahedra. In addition, the $H \rightarrow O$ acceptor distances, 1.78 and 1.70 Å, are the shortest in the present atomic arrangement. The hydrogen bonds involving the hydrogen atoms of the NH_3 groups and OH radicals of the ethanolammonium

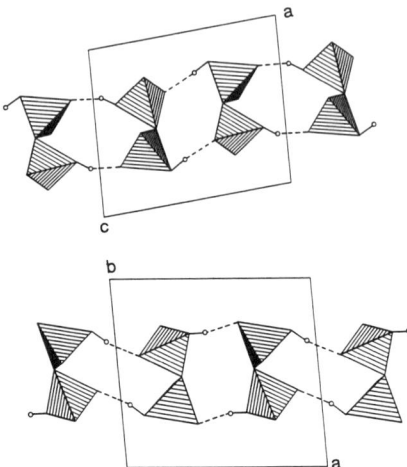

Figure 2.2.20. Projection, along the **b** axis, (upper part) and along the **c** axis (lower part), of the $[H_2P_2O_7]^{2-}$ ribbons in $(NH_3-C_2H_4-OH)_2H_2P_2O_7$. The organic entities have been omitted.

groups are responsible for the cohesion between the phosphoric chains. This second type of hydrogen bond is weaker than the first type since the corresponding N–O or O–O distances range between 2.742 and 3.085 Å.

The entire structure is shown in projection along **b** in Fig. 2.2.21.

2.2.4.11. Ammonium Dihydrogenodiphosphate–Telluric Acid Adduct

$$Te(OH)_6 \cdot 2(NH_4)_2H_2P_2O_7,^{350-351} \text{monoclinic, } P2_1/n, Z = 4$$

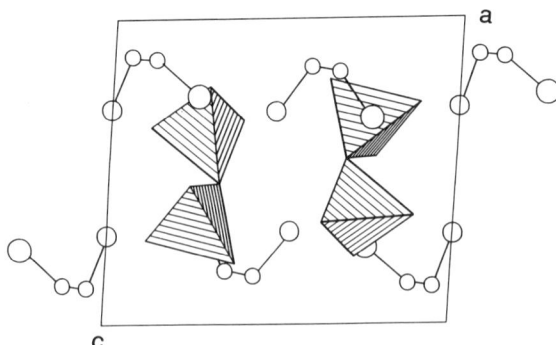

Figure 2.2.21. Projection, along the **b** direction, of the atomic arrangement in $(NH_3-C_2H_4-OH)_2H_2P_2O_7$. The largest open circles represent the oxygen atoms of the hydroxyl groups, the intermediate the nitrogen atoms, and the smallest ones the carbon atoms. The hydrogen atoms have been omitted.

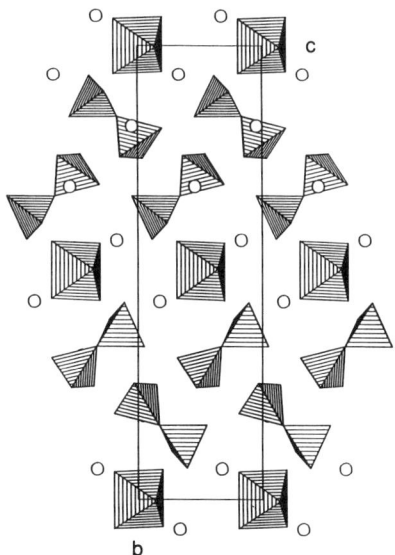

Figure 2.2.22. Projection, along the *a* direction, of the atomic arrangement in Te(OH)$_6$·2(NH$_4$)$_2$H$_2$P$_2$O$_7$. Hatched octahedra represent the Te(OH)$_6$ groups, and open circles the NH$_4$ groups. The H atoms have been omitted for clarity.

$$a = 7.651(2), \ b = 31.790(10), \ c = 6.689(2) \ \text{Å}, \ \beta = 113.85(3)°$$

As can be seen in Fig. 2.2.22 the atomic arrangement in Te(OH)$_6$·2(NH$_4$)$_2$H$_2$P$_2$O$_7$ can be described as a succession of thick layers containing the phosphoric entities and layers of Te(OH)$_6$ groups alternating perpendicular to the *b* direction. The phosphoric layers contain infinite chains composed of H$_2$P$_2$O$_7^{2-}$ groups interconnected by strong hydrogen bonds. Figure 2.2.23 represents the organization of these chains within a phosphoric layer. The H bonds connecting the phosphoric groups within a chain are, as usual, rather strong since corresponding to O–O distances of 2.523 and 2.594 Å.

The Te(OH)$_6$ groups, located in the planes $y = 0$ and 1/2 are, as usual, composed of an almost regular centrosymmetric TeO$_6$ octahedron, with Te–O distances ranging from 1.909 to 1.919 Å. The Te–O–H angles vary from 106 to 112°.

The two crystallographically independent NH$_4$ groups are almost regular tetrahedra with N–H distances ranging from 0.78 to 1.03 Å. Within a range of 3.50 Å one of them has sixfold oxygen coordination, and the other one tenfold. Within these two polyhedra, the N–O distances vary from 2.780 to 3.402 Å.

The corresponding rubidium salt is isotypic.

2.2.5. The Diphosphate Group

Composed of only two corner-sharing PO$_4$ tetrahedra, the diphosphate group is the simplest condensed anion. There is a large amount of literature dealing with the geometry of the P$_2$O$_7$ group. Among these investigations, probably the most numerous are the

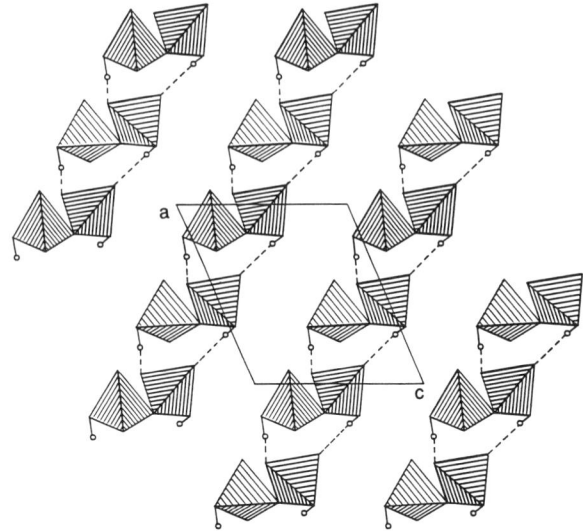

Figure 2.2.23. Projection, along the *b* direction ($0.05 < y < 0.45$), of a layer of phosphoric groups in $Te(OH)_6 \cdot 2(NH_4)_2H_2P_2O_7$, showing the organization of the $[H_2P_2O_7{}^{2-}]_n$ chains. The ammonium groups lining the layer represented have been omitted. The hydrogen bonds connecting the phosphoric groups are represented by solid and dashed lines.

spectroscopic studies, mainly employing IR absorption spectroscopy, aimed at establishing correlations between the P–O–P angles and the positions of some characteristic absorption lines. These correlations seem well established now, and from IR spectroscopy one can deduce, with rather surprising accuracy, many geometrical features of a P_2O_7 group. Our purpose is not to discuss these techniques and their results but to ascertain, through examination of some tens of accurate X-ray structural investigations, what are the usual main geometrical features of a P_2O_7 group.

In the early papers dealing with structural investigations of diphosphates, many aspects of the geometrical configuration of this anion were discussed, with the P–O–P angle and the relative orientation of the two PO_4 tetrahedra being the main aspects considered. A terminology has been developed in order to distinguish between "eclipsed" (cis) or "staggered" (trans) conformations. In fact, except for some very rare examples, involving very restrictive symmetry elements, a P_2O_7 group is never purely staggered or eclipsed. So, in our opinion, such a group can be clearly described by the P–P distance, the P–O–P angle, and its internal symmetry when one or several of its components are located on symmetry elements.

P_2O_7 groups may adopt various internal symmetries: m, 2, $\bar{1}$ etc. We will classify the examples given in the present survey in terms of these internal symmetries. The $\bar{1}$ internal symmetry, corresponding to a central oxygen atom located on an inversion center and leading to a P–O–P angle of 180° has been the subject of much controversy. One must clearly say that in all the studies leading to the conclusion of the existence of such a symmetry in a P_2O_7 group, the central oxygen atom has a high thermal factor, indicating, either a disordered location for this oxygen atom or an error in the symmetry assignment or in the choice of the unit-cell. Most P_2O_7 groups have, in fact, no internal symmetry but 2 and m symmetries are nevertheless frequently observed.

One must distinguish between two types of mirror symmetries. Most commonly this symmetry element is perpendicular to the P–P direction and consequently includes the central oxygen atom. In some rarer cases, five components of the P_2O_7 group are located in the mirror plane, the two phosphorus atoms, the central oxygen atom, and two external oxygen atoms. This latter configuration has, thus far, been observed only twice, in $Ta_2Cd(P_2O_7)_3$ and in $Nb_2OCa(P_4O_{13})(P_2O_7)$, a mixed-anion phosphate.

In the case of twofold internal symmetry, the central oxygen atom is located on a binary axis. This internal symmetry, not very common, has been observed recently for the diphosphate groups of $Cs_2H_2P_2O_7$, $Ag_2H_2P_2O_7$, and ethylenediammonium dihydrogeno-diphosphate.

We report in Table 2.2.24 the main geometrical features of a PO_4H entity in as $H_2P_2O_7$ group, as observed in $(NH_4)_2H_2P_2O_7$.

The values observed for P–O distances and O–P–O angles within a PO_4 tetrahedron belonging to a diphosphate group never depart significantly from those reported in Table 2.2.24. Two main types of P–O distances must be distinguished—the P–O(L) distances, corresponding to the P–O–P bonds, and the P–O(E) distances, corresponding to the external oxygen atoms. In fact, when the diphosphate group is an acidic one, as in the example given in Table 2.2.24, a third type of P–O distance, corresponding to the P–OH bonds, must be considered.

One has evidently

$$P–O(L) > P–OH > P–O(E)$$

In spite of an apparent large distortion inside this polyhedron, the average P–O distance inside a tetrahedron never departs significantly from 1.540 Å, an average value calculated from several dozens accurate crystal structures. Similarly, the O–P–O average angle is always very close by 109.1°, a value calculated in the same manner. In addition, for acidic diphosphate groups, the P–O–H angles are always within the range 110–120°.

Table 2.2.24. Main Geometrical Features of a PO_4H Entity in a $H_2P_2O_7$ Group as Observed in $(NH_4)_2H_2P_2O_7$ by Averbuch-Pouchot and Durif[a,b]

P	O(L)	O(E1)	OH	O(E2)
O(L)	1.613(2)	2.520(3)	2.442(3)	2.527(3)
O(E1)	108.0(1)	1.500(2)	2.497(3)	2.546(3)
OH	100.7(1)	109.4(1)	1.559(2)	2.547(3)
O(E2)	108.5(1)	116.2(1)	112.8(1)	1.499(2)

[a]Ref. 80.

[b]The four P–O values (underlined) are given along the diagonal, the six O–O distances above the diagonal, and the six O–P–O angles below the diagonal. O(L) is the central oxygen atom of the $H_2P_2O_7$ anion, and the other three oxygens are the external oxygen atoms, OH representing the H-bearing oxygen Estimated standard deviations are given in parentheses. Distances are in angstrom, and angles are in decimal degrees.

Table 2.2.25. Some Geometrical Features of P_2O_7 Groups with
No Internal Symmetry

Formula[a]	P–P distance (Å)	P–O–P angle (°)	Reference
$Na_4P_2O_7$	2.936	127.5	54
$(NH_4)_2H_2P_2O_7$	2.980	133.7	80
$Co_2P_2O_7$	2.998	142.6	117
β-$Ca_2P_2O_7$[b]	2.955	130.5	133
	2.991	137.8	
$NaFeP_2O_7(II)$	2.937	133.0	244
$KFeP_2O_7$	2.843	124.3	244
$Nb_2Co(P_2O_7)_3$ [c]	2.954	138.3	318
$Nb_2Mg(P_2O_7)_3$ [c]	2.961	140.0	318
$Ta_2Cd(P_2O_7)_3$ [d]	2.962	139.8	319
$CsMoO_2HP_2O_7$	2.862	126.6	333
$(NH_4)_2MoO_2P_2O_7$	2.911	131.3	332
$(EDA)_2H_2P_2O_7$	2.893	128.1	
$(EthA)_2H_2P_2O_7$	2.960	132.9	346
$Cu_2(EDA)_3(HP_2O_7)_2 \cdot 3H_2O$[e]	2.902	128.6	347
$Te(OH)_6 \cdot Cs_2H_2P_2O_7$	2.907	129.5	352
$Te(OH)_6 \cdot K_3HP_2O_7 \cdot H_2O$	2.971	132.6	349
$Te(OH)_6 \cdot 2(NH_4)_2H_2P_2O_7$	2.903	128.5	350

[a]Abbreviations: EDA, ethylenediammonium, NH_3–$(CH_2)_2$–NH_3; EthA, ethanolammonium, NH_3–$(CH_2)_2$–OH.
[b]Two independent P_2O_7 groups coexist in this arrangement.
[c]This arrangement includes also a centrosymmetric P_2O_7 group.
[d]This arrangement includes also a P_2O_7 group with mirror symmetry.
[e]This compound contains, in fact, both ethylenediammonium and ethylenediamine groups.

We present in Tables 2.2.25 and 2.2.26 some numerical values obtained for the P–P distances and the P–O–P angles in recent, accurate structure determinations. We have classified these data according to the internal symmetries of the diphosphate groups.

The P–P distances , ranging from 2.843 to 3.116 Å, are within the range that we will find, with a very small number of exceptions, for all the other types of condensed phosphoric anions that we will examine in this book. With the exception of the three reported examples of centrosymmetric groups in which all the P–P distances are longer than 3 Å, it seems that no correlation can be established between the internal symmetry of a diphosphate group and the P–P distance.

For *the P–O–P angles*, if one excludes the centrosymmetric diphosphate groups in which this angle is 180°, the values commonly observed vary between 127 and 146°, a range of values that we will find again and again in our examination of most condensed phosphoric anions.

To date, and as is the case for most classes of condensed phosphates, it has not been possible to establish any meaningful correlation between the geometrical configuration of the diphosphate anions and the associated cations.

2.2.6. The Infinite Networks of Acidic Diphosphate Groups

At the beginning of this chapter, we had briefly discussed the association of the various acidic diphosphate anions via strong hydrogen bonds. In all atomic arrangements of

Table 2.2.26. Some Geometrical Features of P_2O_7 Groups with Various Internal Symmetries

Formula	P–P distance (Å)	P–O–P angle (°)	Reference
P_2O_7 groups with twofold symmetry			
$(EDA)_2P_2O_7$[a]	2.968	141.9	345
$Ag_2H_2P_2O_7$	2.910	131.2	73
$Cs_2H_2P_2O_7$	2.963	132.7	81
P_2O_7 groups with mirror symmetry			
$Rb_2H_2P_2O_7 \cdot 1/2H_2O$[b]	2.964	133.5	81, 82
	3.009	137.7[c]	
$Fe_3(P_2O_7)_2$[b]	2.870	128.1	285
	3.012	142.2	
$Ta_2Cd(P_2O_7)_3$[d]	2.936	135.8	319
$CaNb_2O(P_4O_{13})(P_2O_7)$	3.005	145.6	499
Centrosymmetric P_2O_7 groups			
$Ni_2P_2O_7$	3.116	180.0	121
$Nb_2Co(P_2O_7)_3$	3.064	180.0	318
$Nb_2Mg(P_2O_7)_3$	3.054	180.0	318

[a]EDA, ethylenediammonium, $NH_3–(CH_2)_2–NH_3$

[b]Two crystallographic independent P_2O_7 groups coexist in the arrangement.

[c]The central oxygen atom is disordered.

[d]Another independent P_2O_7 group with no internal symmetry coexists in this arrangement.

phosphates that include acidic anions, these anions are observed to have a strong tendency to assemble via strong hydrogen bonds and build infinite networks. In many cases the O–O distances corresponding to these H bonds are shorter than the O–O distances observed in the individual PO_4 tetrahedra. Such infinite networks have sometimes been called "macroanions."

This phenomenon is quite general in the field of monophosphates, the simplest phosphate family, but is generally ignored or very briefly considered in the structural descriptions given in articles describing their atomic arrangements. Thus, even for this very simple class of phosphates, no systematic survey of these network geometries exists.

Our purpose is now to describe briefly the main types of geometries that have been observed to date for these associations in acidic diphosphates. Four types of geometrical configurations are known :

- Infinite chains (the most common)
- Infinite ribbons (generally double chains)
- Infinite bidimensional layers.
- Infinite tridimensional network

It should be noted that no finite association—clusters or rings, for instance—has been observed in acidic diphosphates.

In *infinite chains* each diphosphate group is bonded to its two adjacent neighbors. In fact, two types of chains can be distinguished. In the first type, the H-bonds are established so that the P–O–P bond is parallel to the chain axis. An example of such an association is found in the atomic arrangement of $MnHP_2O_7$ (See Fig. 2.2.15). In the second type of chain arrangement, the P–O–P bond is perpendicular to the chain direction. Two examples of this type of chains, as observed in $(NH_4)_2H_2P_2O_7$ and in ethanolammonium dihydrogenodiphosphate can be seen in Fig. 2.2.2 and 2.2.20. Infinite chains represent the most common type of organization.

Ribbons, generally built up by a double chain, are also frequently observed. Figure 2.2.3 depicts such a ribbon as found in $Ag_2H_2P_2O_7$ while Fig. 2.2.18 gives a representation of a similar organization in $(EDA)H_2P_2O_7$. In this type of association, each diphosphate group is connected to three adjacent groups.

Infinite bidimensional layers have been observed in, for example, $Rb_2H_2P_2O_7 \cdot 1/2H_2O$ and in $Cs_2H_2P_2O_7$. These two arrangements are shown in Fig. 2.2.7 and 2.2.9 respectively. In the first example each diphosphate group is connected to three others and, in the second example to four others.

Infinite three-dimensional networks are rare. The only example of such a network that we can cite is that observed very recently in $Cs_2H_2P_2O_7 \cdot Te(OH)_6$. This structure has unfortunately not been reported here. In this last example each diphosphate group is connected to four of its neighbors.

A systematic survey of these types of networks in all acidic oligophosphates is presently in preparation.

2.3. Triphosphates

2.3.1. Introduction

As we mentioned earlier the existence of oligophosphates for $n > 2$ was formerly disputed, even after the elaboration of the $NaPO_3$–$Na_4P_2O_7$ and KPO_3–$K_4P_2O_7$ phase-equilibrium diagrams by Morey and Ingerson,[358–360] which showed clearly the existence of $Na_5P_3O_{10}$ and $K_5P_3O_{10}$. These two diagrams are presented in Fig. 2.3.1. and 2.3.2.

The development of paper chromatography and of structural X-ray diffraction analysis put an end to this controversy. In 1958, Davies and Corbridge[32] performed the first structural investigation of a triphosphate when they solved the crystal structure of one form of the anhydrous sodium salt $Na_5P_3O_{10}$. Since then a large amount of chemical literature has been devoted to triphosphates, but the number of accurately determined atomic arrangements is still relatively small.

With relatively few exceptions, triphosphates are difficult to crystallize in aqueous media. This fact probably explains why most of the well-characterized triphosphates were discovered during investigations of various systems by flux methods at relatively high temperatures.

The existence of acidic triphosphate anions, $[HP_3O_{10}]^{4-}$ and $[H_2P_3O_{10}]^{3-}$, leads to the formation of infinite anionic networks similar to those we described in the previous section for acidic diphosphate groups.

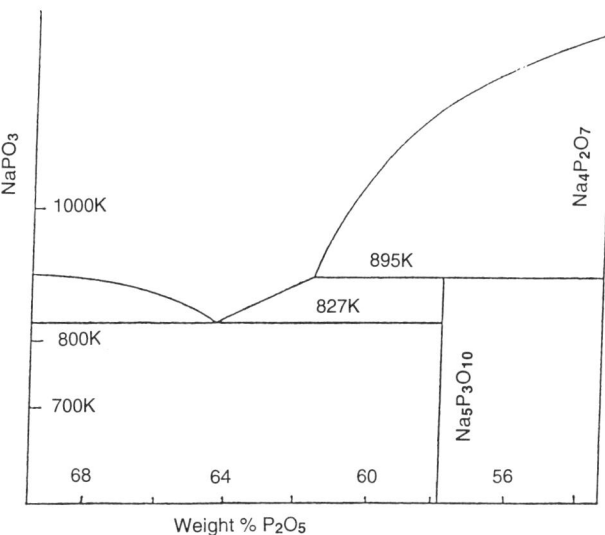

Figure 2.3.1. The NaPO₃–Na₄P₂O₇ phase-equilibrium diagram.

2.3.2. Present State of Triphosphate Chemistry

2.3.2.1. Alkali and Monovalent Cation Triphosphates

$Li_5P_3O_{10} \cdot 5H_2O$. Crystalline $Li_5P_3O_{10} \cdot 5H_2O$ was obtained in good yield by Sotnikova-Yuzhik *et al.*[361] during an investigation of the alkaline hydrolytic decyclization of lithium

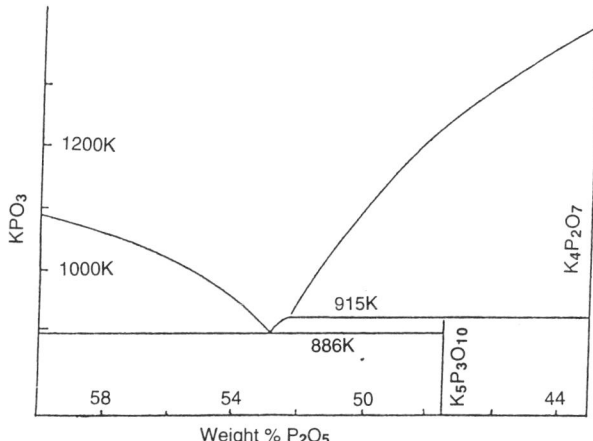

Figure 2.3.2. The KPO₃–K₄P₂O₇ phase-equilibrium diagram.

cyclotriphosphate trihydrate in an aqueous solution of LiOH at 313–353 K. The decyclization reaction is

$$Li_3P_3O_9 + 2LiOH \rightarrow Li_5P_3O_{10} + H_2O$$

$Na_5P_3O_{10}$ and $Na_5P_3O_{10} \cdot 6H_2O$. Sodium triphosphate, $Na_5P_3O_{10}$, occurs in two anhydrous forms, commonly denoted (I) and (II), and as the hexahydrate. All three forms have important application in detergent industry and so their physical properties and chemical preparation have been extensively studied and reported in hundreds of publications, mostly of technical nature, which we cannot review here. Dymon and King[362] were the first to propose a method for obtaining single crystals of both anhydrous forms and to start crystal chemistry investigations on these forms. Corbridge[363] and Davies and Corbridge[32] performed the crystal structure determinations for the two anhydrous forms, thereby providing the first structural characterization of a triphosphate anion. The hexahydrate has been the subject of a very large number of studies. Bonneman and Bassiere[364] performed the first X-ray investigation of this salt, whose crystal structure was first determined by Dyroff[365] and was later reexamined by Wiench et al.[366]

The transition from form (II) to phase (I) occurs at about 688 K and occurs more easily than the reverse process. Form (I) dissolves in water more readily than form (II). Corbridge[363] postulated that this difference is due to the presence in form (I) of a fourfold-coordinated sodium atom, which represents a relatively unstable arrangement.

$Ag_5P_3O_{10}$ and $Ag_5P_3O_{10} \cdot 2H_2O$. The ASTM data file contains two unindexed powder diagrams (14-264 and 11-642) for two forms of $Ag_5P_3O_{10}$, denoted I and II by analogy with the two varieties of the anhydrous sodium salt, but no crystal data are given. Lee[367] investigated the chemical preparation and properties of silver triphosphates. He described the preparation of silver triphosphate dihydrate, $Ag_5P_3O_{10} \cdot 2H_2O$, and investigated its thermal behavior. On heating, this hydrate decomposes into anhydrous silver salt forms I and II and subsequently into crystalline silver diphosphate together with a glass containing mono-, di-, and cyclotetraphosphates. A detailed procedure for the production of $Ag_5P_3O_{10}$ (form I) was given.

According to Lamotte and Merlin[368] the anhydrous salt cannot be obtained. Using radiochromatography with phosphorus-32 they investigated the degradation into diphosphate during the dehydration of the dihydrate.

Ammonium triphosphates. Six ammonium triphosphates were characterized by Frazier et al.[67]: $(NH_4)_3H_2P_3O_{10}$, $(NH_4)_3H_2P_3O_{10} \cdot 2H_2O$, $(NH_4)_9H(P_3O_{10})_2 \cdot 2H_2O$, $(NH_4)_4H P_3O_{10}$, $(NH_4)_5P_3O_{10} \cdot H_2O$ and $(NH_4)_5P_3O_{10} \cdot 2H_2O$. The last one appears as a metastable phase evolving rapidly to the monohydrate.

A study of the NH_3–$H_5P_3O_{10}$–H_2O system at 273 and 298 K, made some years later by Farr et al.[369] confirmed clearly the existence of these salts and provided accurate data on their hydrolysis under various conditions (see also Waerstad and McClellan[370] for accurate crystal data on $(NH_4)_5P_3O_{10} \cdot 2H_2O$ and $(NH_4)_5P_3O_{10} \cdot H_2O$).

Crystalline ammonium triphosphate diperoxohydrate monohydrate, $(NH_4)_5P_3O_{10} \cdot 2H_2O_2 \cdot H_2O$, was isolated by Sotnikova-Yuzhik and Prodan.[371]

Potassium Triphosphates. At least four crystalline potassium triphosphates have been described. $K_5P_3O_{10}$ was characterized by Morey[360] as an incongruent-melting compound decomposing at 915 K during an investigation of the $K_4P_2O_7$–KPO_3 phase-equilibrium diagram given in Fig. 2.3.2. Various physico-chemical investigations have been performed on $K_2H_3P_3O_{10}\cdot 2H_2O$, $K_3H_2P_3O_{10}\cdot H_2O$ and $K_5P_3O_{10}\cdot 4H_2O$ by Shashkova and co-workers.[372–377] In the case of $K_3H_2P_3O_{10}\cdot H_2O$ Lyutsko and Johansson[378] described a detailed process of crystal growth and an accurate structure determination.

$Rb_5P_3O_{10}$ and $Cs_5P_3O_{10}$. These two triphosphates were first observed by Krivoviazov *et al.*[379] during their investigation of the $RbPO_3$–$Rb_4P_2O_7$ and $CsPO_3$–$Cs_4P_2O_7$ systems. They appear as incongruent-melting compounds decomposing at 889 K(Rb) and 791 K(Cs).

Recently, Sotnikova-Yuzhik[380] produced the rubidium triphosphate by thermal treatment of a phosphate mixture corresponding to the ratio Rb:P = 5:3 and itself prepared by the action of ion-exchange resins on an aqueous solution of sodium triphosphate. The compound is hygroscopic, and at least two hydrates exist. The decomposition temperature was given as 878K in this last study.

Other Alkali and Monovalent Cation Triphosphates. Some other systems have also been investigated:

(i) The crystalline phases deposited in the $Na_5P_3O_{10}$–$K_5P_3O_{10}$–H_2O system were analyzed by Bulavkina *et al.*,[381] who established the existence of a dihydrate, $Na_xK_{(5-x)}P_3O_{10}\cdot 2H_2O$ for $2 < x < 3$. This system was also investigated by Griffith and Buxton.[382]

(ii) Some mixed potassium–ammonium triphosphates were prepared by Shashkova *et al.*[377] by the action of gaseous ammonia on $K_2H_3P_3O_{10}\cdot 2H_2O$. They reported the existence of $(NH_4)_3K_2P_3O_{10}\cdot 2H_2O$, $(NH_4)_2K_2HP_3O_{10}\cdot nH_2O$ and $(NH_4)_3K_2P_3O_{10}\cdot nH_2O$ ($2 < n < 3$).

(iii) The $Na_5P_3O_{10}$–$(NH_4)_5P_3O_{10}$–H_2O system was investigated by Prodan *et al.*[383] $(NH_4)_4NaP_3O_{10}\cdot 4H_2O$, one of the intermediate compounds, was investigated in more detail. Its thermal behavior was studied by Prodan *et al.*,[384] who reported the following scheme of transformation:

$$(NH_4)_4NaP_3O_{10}\cdot 4H_2O \xrightarrow{308K} (NH_4)_4NaP_3O_{10}\cdot H_2O \xrightarrow{323K} (NH_4)_3NaHP_3O_{10}\cdot H_2O$$

Later, Averbuch-Pouchot and Durif[385] produced single crystals of $(NH_4)_4NaP_3O_{10}\cdot 4H_2O$ and reported a detailed determination of its atomic arrangement.

Crystallographic Data. The main crystallographic data for alkali and monovalent cation triphosphates are summarized in Table 2.3.1.

2.3.2.2. Divalent and Divalent–Monovalent Cation Triphosphates

$Zn_2HP_3O_{10}\cdot 6H_2O$ and $Zn_5(P_3O_{10})_2\cdot 17H_2O$. These two salts were characterized by Averbuch-Pouchot and co-workers.[386–388] Both compounds were produced by the hydrolytic decyclization of P_3O_9 ring anions during investigations to produce mixed cyclotriphosphates. The authors reported accurate determinations of the atomic arrangements for the

Table 2.3.1. Crystal Data for Monovalent Cation Triphosphates

Formula	a (Å) α (°)	b (Å) β (°)	c (Å) γ (°)	S.G.	Z	Reference(s)
$Na_5P_3O_{10}$(I)	9.61(3)	5.34(2) 112.0(5)	19.73(5)	$C2/c$	4	32
$Na_5P_3O_{10}$(II)	16.00(5)	5.24(2) 93.0(5)	11.25(3)	$C2/c$	4	363
$Na_5P_3O_{10}\cdot6H_2O$	10.370(2) 92.24(7)	9.848(4) 94.55(9)	7.615(3) 90.87(6)	$P\bar{1}$	2	365, 366
$(NH_4)_3H_2P_3O_{10}\cdot2H_2O$	7.59	10.90 126.1	8.78	$P2/c$	2	369
$(NH_4)_3H_2P_3O_{10}$	7.31	8.08 101.2	21.73	$P2_1/c$	4	369
$(NH_4)_4HP_3O_{10}$	14.24	7.47 109.5	12.59	$C2/c$	4	369
$(NH_4)_9H(P_3O_{10})_2\cdot2H_2O$	22.71	6.98 112.2	18.86	$C2/c$	4	369
$(NH_4)_5P_3O_{10}\cdot H_2O$	10.949(6)	10.377(3) 91.06(4)	6.426(3)	$P2/n$ or Pn	2	370
$(NH_4)_5P_3O_{10}\cdot2H_2O$	16.643(5)	6.719(5) 98.61(3)	13.643(6)	$P2_1/n$	4	370
$K_3H_2P_3O_{10}\cdot H_2O$	7.588(2)	11.163(4)	26.697(8)	$Pbca$	8	378
$Na_5K_5(P_3O_{10})_2\cdot4H_2O$	10.754(2)	9.560(3) 67.7	10.232(4)	$P2_1/c$	?	381, 382
$(NH_4)_4NaP_3O_{10}\cdot4H_2O$	11.813(8)	7.292(3) 105.19(5)	20.20(8)	$C2/c$	4	385

two salts. In the first compound, 4 of the 6 water molecules are of a zeolitic nature; in the second one, 7 of the 17 are also of this nature.

$Pb_2HP_3O_{10}$. This salt was synthesized by Worzala and Jost[389] by firing $Pb_2P_4O_{12}.2H_2O$ at 423 K. The crystals that the authors obtained were formed topotactically, always twinned, and contained H_3PO_4. The authors reported a model for the atomic arrangement of this triphosphate. The reaction for its formation can be written as

$$Pb_2P_4O_{12}\cdot2H_2O \rightarrow Pb_2HP_3O_{10} + H_3PO_4$$

$Be_2NH_4P_3O_{10}$. An anhydrous beryllium–ammonium triphosphate, $Be_2(NH_4)P_3O_{10}$, was first described by Bagieu-Beucher *et al.*[390] and later investigated by Averbuch-Pouchot *et al.*,[391] who reported a complete determination of its atomic arrangement.

$Zn_2LiP_3O_{10}\cdot8H_2O$. Lyakhov *et al.*[392] produced single crystals of this triphosphate by slow evaporation of a solution prepared by mixing 0.1M aqueous solutions of $Li_5P_3O_{10}$ and $Zn(NO_3)_2$ in the ratio 1:1. The crystal structure determined by the authors is largely identical to that reported for $Zn_2NaP_3O_{10}\cdot9H_2O$

$Zn_2NaP_3O_{10}\cdot9H_2O$. Several authors, including Schwarz,[10] Stange,[18] Huber,[393] and Bonneman-Bemia[24] reported the existence of a sodium–zinc triphosphate identified as

$Zn_2NaP_3O_{10}\cdot19/2H_2O$. In more recent work: Quimby and McCune[170] and Corbridge and Tromans,[394] the formula reported was that of a nonahydrate. Finally, Averbuch-Pouchot and Guitel[395] succeeded in producing single crystals of this compound by hydrolytic decyclization of a cyclotriphosphate and determined its atomic arrangement, confirming that this salt is a nonahydrate and that four of the nine water molecules are of a zeolitic nature. The thermal behavior of this salt was investigated up to 1173 K by Prodan and Zonov.[396]

$M^{II}Na_3P_3O_{10}\cdot12H_2O$ (M = Zn, Ni, Mg, Co, Mn, Cu, and Cd). A large series of triphosphates corresponding to the above general formula have been well investigated. Their existence was reported at the end of last century by Schwarz[10] and Stange[18] and later confirmed by Bonneman-Bémia.[24,397] The first crystallographic investigation of this series was performed by Rakotomahanina et al.[398] The crystal structure of the manganese salt was solved by Herceg.[399] The copper salt was reexamined by Jouini and Durif,[400] and its crystal structure determined by Jouini *et al.*[401] A very accurate crystal structure determination of the cadmium salt was performed by Lyustko and Johansson.[402] Dehydration of the magnesium salt at various temperatures was studied by Dewald.[403]

$Zn_2Ag_{0.62}H_{0.38}P_3O_{10}\cdot9H_2O$. A crystal structure determination for a silver–zinc triphosphate corresponding to the above formula was performed by Averbuch-Pouchot and Guitel.[404] This salt was produced by hydrolytic decyclization of various cyclotriphosphates during experiments involving $Ag_3P_3O_9\cdot H_2O$ as starting material.

Mixed Potassium–Divalent Cation Triphosphates. Various mixed potassium–divalent cation triphosphates have been reported in the chemical literature. $Zn_3K_4(P_3O_{10})_2$ and $ZnK_3P_3O_{10}$, both incongruent-melting compounds, were identified by Krivoviazov *et al.*[206] during investigations of the $Zn(PO_3)_2$–K_2O and KPO_3–ZnO systems. The first one decomposes at 876 K, the second at 931 K. $Co_2KP_3O_{10}\cdot7H_2O$, $Ni_2KP_3O_{10}\cdot9H_2O$, $Cu_2KP_3O_{10}5H_2O$ were described by Prodan and Ol'Shevskaya,[405] and $Zn_2KP_3O_{10}\cdot7H_2O$ was described by Ol'Shevskaya and Prodan.[406]

$Ca(NH_4)_3P_3O_{10}\cdot2H_2O$. The reaction of ammonium triphosphate with calcium chloride in aqueous solutions was investigated by Lyakhov *et al.*[407] They observed the formation of three sparingly water-soluble compounds: $Ca(NH_4)_3P_3O_{10}\cdot2H_2O$, $Ca(NH_4)_3P_3O_{10}\cdot2H_2O$, and $Ca_5(P_3O_{10})_2\cdot10H_2O$. The two mixed compounds were obtained in a crystalline state while the calcium triphosphate is amorphous. The authors reported an accurate structure determination for $Ca(NH_4)_3P_3O_{10}\cdot2H_2O$.

Other Divalent Cation–Ammonium Derivatives. The thermal behavior of $Mg_2NH_4P_3O_{10}\cdot7H_2O$, $Mn_2NH_4P_3O_{10}\cdot6H_2O$, $Zn_2NH_4P_3O_{10}\cdot7H_2O$, $Zn_3(NH_4)_4(P_3O_{10})_2\cdot9H_2O$ and $Ca(NH_4)_3P_3O_{10}\cdot2H_2O$ were investigated by Ol'shevskaya *et al.*[408] in the temperature range 293–1173 K. At temperatures higher than 773 K the Mg and Mn salts transform into the corresponding $M_2P_4O_{12}$ cyclotetraphosphates, whereas the Ca salt gives the long-chain polyphosphate $Ca(PO_3)_2$. The two zinc salts differ in their thermal behavior:

$$Zn_2NH_4P_3O_{10}\cdot7H_2O \rightarrow Zn_2P_4O_{12}$$

$$Zn_3(NH_4)_4(P_3O_{10})_2\cdot9H_2O \rightarrow Zn(PO_3)_2$$

The following salts of general formula $M^{II}_2NH_4P_3O_{10} \cdot nH_2O$ have been characterized: $Co_2NH_4P_3O_{10} \cdot 8H_2O$ and $Ni_2NH_4P_3O_{10} \cdot 8H_2O$ by Prodan and Ol'Shevskaya,[405] $Mg_2NH_4P_3O_{10} \cdot 6H_2O$ by Galkova and Prodan,[409] $Zn_2NH_4P_3O_{10} \cdot 7H_2O$ by Prodan and Galkova,[410] $Mn_2NH_4P_3O_{10} \cdot 5H_2O$ by Galkova and Prodan,[411] $Mn_2NH_4P_3O_{10} \cdot 6H_2O$ by Prodan et al.,[412] and $Zn_2NH_4P_3O_{10}$ by Konstant et al.[413] With the exception of the last one, all these salts were prepared by the action of various divalent salts on $(NH_4)_5P_3O_{10}$ in aqueous solutions. Recently, the atomic arrangement of $Zn_2NH_4P_3O_{10} \cdot 7H_2O$ was determined by Lyakhov et al.[414]

$Sr_2KP_3O_{10} \cdot 7H_2O$ and $Sr_2NH_4P_3O_{10} \cdot 5H_2O$ were also prepared by Ol'Shevskaya and Prodan.[415]

Table 2.3.2. Crystal Data for Divalent and Divalent–Monovalent
Cation Triphosphates

Formula	a (Å) α (°)	b (Å) β (°)	c (Å) γ (°)	S.G.	Z	Reference(s)
$Zn_5(P_3O_{10})_2 \cdot 17H_2O$	10.766(8) 111.39(5)	10.316(8) 115.08(5)	8.525(5) 70.19(5)	$P\bar{1}$	1	387, 388
$Zn_2HP_3O_{10} \cdot 6H_2O$	10.71 4(8) 114.51(1)	10.658(8) 103.21(1)	8.391(5) 74.31(1)	$P\bar{1}$	2	386
$Pb_2HP_3O_{10}$	6.93(1)	14.34(1) 135.1(1)	5.97(1)	Cm		389
$Be_2NH_4P_3O_{10}$	12.202(8)	8.645(3) 117.41(5)	8. 949(3)	$C2/c$	4	390, 391
$Zn_2LiP_3O_{10} \cdot 8H_2O$	10.305(1) 101.79(1)	10.505(1) 113.42(1)	8.671(1) 94.24(1)	$P\bar{1}$	2	392
$Zn_2NaP_3O_{10} \cdot 9H_2O$	10.454(5) 101.14(2)	10.675(5) 109.85(2)	8.629(4) 99.03(2)	$P\bar{1}$	2	395
$NiNa_3P_3O_{10} \cdot 12H_2O$	15.01(1)	9.208(3) 90.00(5)	14.71(1)	$P2_1/n$	4	398
$ZnNa_3P_3O_{10} \cdot 12H_2O$	15.03(1)	9.241(3) 90.00(5)	14.70(1)	$P2_1/n$	4	398
$MgNa_3P_3O_{10} \cdot 12H_2O$	15.05(1)	9.245(3) 90.00(5)	14.72(1)	$P2_1/n$	4	399
$CoNa_3P_3O_{10} \cdot 12H_2O$	15.06(1)	9.238(3) 90.00(5)	14.70(1)	$P2_1/n$	4	398
$MnNa_3P_3O_{10} \cdot 12H_2O$	15.13(1)	9.320(3) 90.00(5)	14.76(1)	$P2_1/n$	4	398
$CdNa_3P_3O_{10} \cdot 12H_2O$	14.835(12)	9.397(10) 90.00(6)	15.244(9)	$P2_1/n$	4	398, 402
$CuNa_3P_3O_{10} \cdot 12H_2O$	15.052(8)	9.234(3) 90.03(5)	14.767(8)	$P2_1/n$	4	400
$Zn_2Ag_{(1-x)}H_xP_3O_{10} \cdot 9H_2O$	10.473(4) 101.08(1)	10.683(5) 109.81(1)	8.629(3) 98.87(1)	$P\bar{1}$	2	404
$Zn_2NH_4P_3O_{10} \cdot 7H_2O$	10.529(3) 102.93(2)	10.737 (3) 111.90(2)	8.497(3) 96.01(2)	$P\bar{1}$	2	414
$Ca(NH_4)_3P_3O_{10} \cdot 2H_2O$	6.962(2)	18.824(2) 102.76(2)	10.529(2)	$P2_1/n$	4	407

Crystallographic Data. The main crystallographic data for the divalent and divalent–monovalent cation triphosphates investigated to date are presented in Table 2.3.2.

2.3.2.3. Trivalent Cation Triphosphates

Most of the well-characterized trivalent cation triphosphates are of the general formula $MH_2P_3O_{10} \cdot nH_2O$.

D'Yvoire[220] was the first to identify two forms of $AlH_2P_3O_{10}$ and to analyze their thermal behavior through X-ray analysis and paper chromatography. According to this author the two corresponding iron salts, which he also characterized, are isotypic with the aluminum ones. Soon after, Rémy and Boullé[227] described three forms of $CrH_2P_3O_{10}$, hydrated to different degrees

During the past 15 years, a good number of trivalent cation triphosphates have been reported in the chemical literature. Most of them have been obtained by various flux methods.

$AlH_2P_3O_{10}$. This triphosphate is dimorphous. The crystal structure of form II was determined by Lyutsko *et al.*[416]

$MnH_2P_3O_{10} \cdot 2H_2O$ and $MnH_2P_3O_{10}$. The formation of $MnH_2P_3O_{10} \cdot 2H_2O$ and MnH_2-P_3O_{10} in various systems involving P_2O_5 and MnO or Mn_2O_3 was described by Guzeeva *et al.*[254] and Selevich and Lyustko.[417] The latter authors investigated the interaction of phosphoric acid with manganese(II) salts in the presence of an oxidant (nitric acid) in the temperature range 373–573 K. $MnH_2P_3O_{10} \cdot 2H_2O$ is formed when the ratio P/Mn is in the range 4–8 and for temperatures ranging from 483 to 513 K. The crystals are thin rectangular prisms with an intense violet color and usually grow in tufts. The thermal behavior of this salt can be represented by the following scheme:

$$MnH_2P_3O_{10} \cdot 2H_2O \xrightarrow{\;363-523\ K\;} MnH_2P_3O_{10} + 2H_2O \xrightarrow{\;653-713\ K\;} Mn(PO_3)_3 + H_2O$$

The anhydrous salt, $MnH_2P_3O_{10}$, is produced between 513 and 523 K when the initial P/Mn ratio is 15. The thermal behavior is identical to that of the dihydrate.

$GaH_2P_3O_{10}$ and $GaH_2P_3O_{10} \cdot 2H_2O$. The existence of $GaH_2P_3O_{10}$ was established by Chudinova *et al.*,[418] and the formation of the dihydrate by thermal condensation of $Ga(H_2PO_4)_3$ was investigated by Mel'Nikov *et al.*[419] The following scheme was reported by the latter authors:

$$Ga(H_2PO_4)_3 \xrightarrow{\;413K\;} GaH_2P_3O_{10} \cdot 2H_2O \underset{reversible}{\overset{413-503K}{\rightleftharpoons}} GaH_2P_3O_{10} \rightarrow Ga(PO_3)_3(C)$$

$ScH_2P_3O_{10}$. Kanepe and Konstant[420] described this salt, which was obtained by the same process, starting from $Sc(H_2PO_4)_3$. The corresponding ytterbium salt, $YbH_2P_3O_{10}$ was prepared by Palkina *et al.*[421] who performed its structure determination, but unfortunately with a poor accuracy.

Table 2.3.3. Crystal Data for Trivalent Cation Triphosphates

Formula	a (Å) α (°)	b (Å) β (°)	c (Å) γ (°)	S.G.	Z	Reference
$AlH_2P_3O_{10}$	7.234(1)	8.522(1) 101.95(2)	11.618(2)	$P2_1/n$	4	416
$VH_2P_3O_{10}$	7.378(3)	8.786(3) 102.82(3)	11.702(4)	$P2_1/n$	4	422
$FeH_2P_3O_{10}$	7.381(2)	8.808(4) 112.68(2)	12.399(3)	$P2_1/c$	4	378
$FeH_2P_3O_{10} \cdot H_2O$	12.076(9)	8.443(7) 112.10(1)	9.352(7)	$C2/c$	4	423
$YbH_2P_3O_{10}$	5.617(2)	6.666(2) 97.32(2)	10.011(3)	$A2$	2	421

$VH_2P_3O_{10}$. A crystal structure determination for this salt was performed by Lyakhov *et al.*,[422] showing it to be isotypic with the corresponding aluminum salt.

Structural Studies. Detailed and accurate crystal structure determinations have been performed for a number of trivalent cation triphosphates. Most of these structural studies describe reproducible procedures for crystal growth, providing in many cases, a more detailed and clearer description of the preparation of these salts than the previous chemical literature on these compounds. In this domain, Averbuch-Pouchot and Guitel[423] described the atomic arrangement in $FeH_2P_3O_{10} \cdot H_2O$ while Genkina *et al.*[424] and Lyutsko and Johansson[378] reported that in $AlH_2P_3O_{10}$ and in $FeH_2P_3O_{10}$, respectively (see also Lyutsko *et al.*[425]). Intercalation compounds of the "host–guest" type, based on the layer organization of $CrH_2P_3O_{10} \cdot 2H_2O$ and $AlH_2P_3O_{10} \cdot 2H_2O$ have been obtained with hydrazine and hydroxylamine derivatives by Lyutsko *et al.*[426,427]Table 2.3.3 contains the main crystallographic data for the five trivalent cation triphosphates that have been well characterized to date.

Indium, Yttrium and Rare-Earth Triphosphates. All these compounds correspond to the general formula $M_5(P_3O_{10})_3 \cdot nH_2O$. In most of cases, the lanthanide compounds are poorly crystallized and their hydration states are uncertain. The following compounds have been reported:

- $In_5(P_3O_{10})_3 \cdot 21H_2O$ (Rodicheva *et al.*[428])
- $Y_5(P_3O_{10})_3 \cdot 16H_2O$ (Giesbrecht and Melardi[429])
- $La_5(P_3O_{10})_3 \cdot 16H_2O$ (Rodicheva *et al.*[430])
- $Pr_5(P_3O_{10})_3 \cdot 22$–$25H_2O$ (Petushkova *et al.*[431] and Kuznetsov *et al.*[432])
- $Nd_5(P_3O_{10})_3 \cdot 22H_2O$ (Rodicheva *et al.*[433])
- $Sm_5(P_3O_{10})_3 \cdot 22H_2O$ and $Gd_5(P_3O_{10})_3 \cdot 22H_2O$ (Petushkova *et al.*[431]). These two salts are said to be isotypic with the corresponding praseodymium compound.[432]
- $Er_5(P_3O_{10})_3 \cdot 20H_2O$ (Rodicheva *et al.*,[434] Rodicheva and Romanova,[435] Petushkova *et al.*[431,436]).
- $Yb_5(P_3O_{10})_3 \cdot 20H_2O$ (Petushkova *et al.*[431,436]).

According to Rodicheva *et al.*,[432] the thermal decomposition of these salts leads to a mixture of poly- and monophosphates. In the case of lanthanum, for instance:

$$La_5(P_3O_{10})_3 \cdot 16H_2O \xrightarrow{1223K} 2La(PO_3)_3 + 3LaPO_4 + 16H_2O$$

No crystallographic investigation has been performed on these salts.

2.3.2.4. Trivalent–Monovalent Cation Triphosphates

Most of the well-characterized compounds in the category of trivalent–monovalent cation triphosphates have the general formula $M^{III}M^{I}HP_3O_{10}$, and a good number of them are dimorphous. Table 2.3.4. contains the main crystallographic data for the trivalent–monovalent cation triphosphates that have been investigated to date.

$AlNH_4HP_3O_{10}$ and $AlKHP_3O_{10}$. D'Yvoire[220] was the first to describe the chemical preparation of $AlNH_4HP_3O_{10}$ by firing $Al(NH_4)_2P_3O_{10} \cdot H_2O$ at 473 K.

$$Al(NH_4)_2P_3O_{10} \cdot H_2O \rightarrow AlNH_4HP_3O_{10} + NH_3 + H_2O$$

Averbuch-Pouchot *et al.*[437] described a crystal growth method and reported a detailed structural investigation of this salt. These authors also reported the chemical preparation and crystal data for the isotypic potassium salt, $AlKHP_3O_{10}$. The thermal behavior of the ammonium salt was recently investigated by Lyutsko and Pap.[438]

$M^{III}Cs_2P_3O_{10}$ (M = Al, Ga, Cr, Fe). These four isotypic triphosphates were characterized by Lyutsko *et al.*[439] All of them are stable up to 1000 K. The melting points reported by the authors are 1053 (Al), 1008 (Ga), 1310 (Cr), and 1253 K (Fe).

$ScNa_2P_3O_{10}$ and $ScNaHP_3O_{10}$. These two triphosphates were recently characterized by Avaliani[440] during an investigation of the P_2O_5–Na_2O–Sc_2O_3–H_2O system between 423 and 773 K. They are observed between 523 and 588 K, the first one when the initial starting P_2O_5:Na_2O:Sc_2O_3 ratio is 15:10:1 and the second one when the initial ratio is 15:5:1. According to the author, the second compound contains a small quantity of zeolitic water. No crystal data were reported.

$VCsHP_3O_{10}$. This triphosphate was prepared by Klinkert and Jansen,[441] who reported an accurate description of the atomic arrangement, showing this salt to be isotypic with $FeHNH_4P_3O_{10}$.

$GaNH_4HP_3O_{10}$. This diphosphate was observed by Chudinova *et al.*[258] as one of the compounds formed during the preparation of $Ga(PO_3)_3$ by reaction of Ga_2O_3 and $(NH_4)_2HPO_4$ at 623 K. The following reaction scheme, corresponding to a progressive increase in condensation, was reported

$$Ga_2O_3 + (NH_4)_2HPO_4 \rightarrow GaNH_4P_2O_7 \rightarrow$$
$$GaNH_4HP_3O_{10} \rightarrow Ga_2(NH_4)_2P_8O_{24} \rightarrow Ga(PO_3)_3 \,(C)$$

The two crystalline forms of $GaNH_4HP_3O_{10}$ were observed during this reaction.

$VNH_4HP_3O_{10}$. This salt was described by Teterevkov and Mikhailovskaya[442] and later reexamined by Krasnikov et al.[443]

$CrHNH_4P_3O_{10}$. Vaivada et al.[444] characterized this triphosphate and carried out a detailed structural investigation. This salt is dimorphous. The thermal behavior of its α form was examined by Lyustko and Pap.[438] According to these authors, this form is always hydrated, and its behavior may be represented as follows:

$$\alpha\text{-}CrNH_4HP_3O_{10}\cdot H_2O \text{ ----> } \alpha\text{-}CrNH_4HP_3O_{10}$$

$CrKHP_3O_{10}$, $CrRbHP_3O_{10}$ and $CrCsHP_3O_{10}$. These three triphosphates were prepared by Grunze and Chudinova.[445] The first two are dimorphous.

$AlMHP_3O_{10}$ and $FeMHP_3O_{10}$ (M = K, NH_4, Rb, and Cs). Grunze and Grunze[242] characterized these eight triphosphates. The atomic arrangement of the iron–ammonium salt was determined by Krasnikov et al.[443]

$GaKHP_3O_{10}$. The existence of $GaKHP_3O_{10}$ was first reported by Chudinova et al.,[446] and its main properties were reexamined in a more detailed way by Chudinova et al.[447] This triphosphate decomposes at 823 K according to the following scheme:

$$GaKHP_3O_{10} \rightarrow GaKP_2O_7 + 1/2P_2O_5 + 1/2H_2O$$

$GaCsHP_3O_{10}$. According to Chudinova et al.,[448] $GaCsHP_3O_{10}$ is the most stable phase in the Ga_2O_3–Cs_2O–P_2O_5–H_2O system. The authors observed three crystalline forms. On prolonged heating above 623 K this salt transforms into a mixed cyclododecaphosphate $Ga_3Cs_3P_{12}O_{36}$.

$MnKHP_3O_{10}$. This salt was obtained by Guzeeva et al.[254] According to these authors, it decomposes at 823–873 K into a mixed-divalent manganese–potassium cyclotriphosphate, $MnKP_3O_9$.

$$2MnKHP_3O_{10} \rightarrow 2Mn(II)KP_3O_9 + H_2O + 1/2O_2$$

$MnRbHP_3O_{10}$ by Guzeeva and Tananaev[262] characterized this triphosphate during their investigation of the P_2O_5–MnO_2–Cs_2O–H_2O system between 413 and 623 K. The study was run at a constant P/Mn ratio (15) and at three different Rb/Mn ratios (n = 5, 7.5, and 10). For n =5, $MnRbHP_3O_{10}$ is the most stable compound and is produced over the whole temperature range. When n increases, the region in which crystallization of $MnRbHP_3O_{10}$ occurs is displaced to higher temperatures (473–573 K) and disappears when n = 10. Two forms of $MnRbHP_3O_{10}$ are observed. Form I is isotypic with $FeNH_4HP_3O_{10}$, and form II with $FeCsHP_3O_{10}$. The thermal behavior of these two forms is different. Form I decomposes at 873 K:

$$MnRbHP_3O_{10}(I) \rightarrow MnRbP_2O_7 + MnRbP_3O_9 + \text{amorphous phase}$$

whereas the decomposition of form II begins at 833 K and follows a different path:

$$MnRbHP_3O_{10}(II) \rightarrow MnRbP_2O_7 + Mn_2Rb_2P_8O_{24} + MnRbP_3O_9$$

$MnCsHP_3O_{10}$. This triphosphate characterized by Guzeeva and Tananaev,[277] is the most stable compound appearing in the $Cs_2O-P_2O_5-MnO_2-H_2O$ system. Investigation of this system for a constant ratio P/Mn ratio of 15 and for various initial Cs/Mn ratios (Cs/Mn = n) in the temperature range 423–673 K showed that $MnCsHP_3O_{10}$ crystallizes over a relatively narrow temperature range (423–473 K) when $n = 5$. For higher values ($n = 7.5$ and 10), the $MnCsHP_3O_{10}$ crystallization region is broadened, and this compound is formed over practically the whole investigated temperature range.

$MnCsHP_3O_{10}$ is stable up to 923 K and then decomposes according to two different schemes

$$2MnCsHP_3O_{10} \rightarrow 2MnCsP_2O_7 + P_2O_5 + H_2O$$

$$6MnCsHP_3O_{10} \rightarrow 3MnCsP_2O_7 + Mn_3Cs_3(P_6O_{18})_2 + 3H_2O$$

The first scheme of decomposition was observed for rapid heating, and the second one for slow heating rates. It should be noted that $Mn_3Cs_3(P_6O_{18})_2$, obtained in the second process and said to be a double cyclohexaphosphate, is very probably a cyclododecaphosphate isotypic with $V_3Cs_3P_{12}O_{36}$.

$MnCs_2P_3O_{10}$. This compound prepared by Lyustsko *et al.*,[449] is said to be isotypic with the corresponding Al, Ga, Cr, and Fe salts previously investigated by Lyutsko *et al.*[439] but its behavior is different. In moist air, it can absorb up to one water molecule without any change in structure.

$InNaHP_3O_{10}$, $InRbHP_3O_{10}$ and $InCsHP_3O_{10}$. These three triphosphates were obtained by Avaliani *et al.*[450] during an investigation of the $P_2O_5-In_2O_3-M_2O$ systems. The sodium salt is obtained between 423 and 493 K for initial mixtures corresponding to a $P_2O_5:Na_2O:In_2O_3$ ratio of 15:5:1; the rubidium triphosphate between 423 and 573 K for mixtures with a $P_2O_5:Rb_2O:In_2O_3$ initial ratio of 15:(2.5–5):1 and the cesium compound for an initial $P_2O_5:Cs_2O:In_2O_3$ ratio of 15:5:1 (423–573 K) or 15:7.5:1 (423–683 K). These three triphosphates are said to be isotypic with the corresponding gallium salts.

$LnMHP_3O_{10}$. Vinogradova and Chudinova[451] reported the existence of nine compounds of general formula $LnMHP_3O_{10}$ with Ln = Nd, Er, Yb, Tb, and Dy and M = K, Rb, and Cs. $NdNH_4HP_3O_{10}$ and $CeNH_4HP_3O_{10}$ were characterized, respectively, by Chudinova *et al.*[452] and by Vaivada and Konstant.[453] $EuKHP_3O_{10}$ characterized by Chudinova *et al.* [454] was also investigated by Hilmer *et al.*[455,456] It decomposes at 833 K according to the following scheme:

$$2EuKHP_3O_{10} \rightarrow EuK(PO_3)_4 + EuPO_4 + KPO_3$$

$BiHNH_4P_3O_{10}$. A method of crystal growth and a complete description of the atomic arrangement of $BiNH_4HP_3O_{10}$ were reported by Averbuch-Pouchot and Bagieu-Beucher.[457]

Other Trivalent–Monovalent Cation Triphosphates. A large number of mixed triphosphates of various formulas have also been reported in the chemical literature, but

Table 2.3.4. Crystal Data for Trivalent–Monovalent Cation Triphosphates

Formula	a (Å) α (°)	b (Å) β (°)	c (Å) γ (°)	S.G.	Z	Reference
$AlNH_4HP_3O_{10}$	11.643(8)	4.918(4) 119.27(5)	8.705(5)	$P2/a$	2	437
$AlKHP_3O_{10}$	11.619(8)	4.905(5) 120.84(5)	8.614(5)	$P2/a$	2	437
$FeNH_4HP_3O_{10}$	12.077(2)	8.448(1) 112.21(1)	9.357(1)	$C2/c$	4	443
$VCsHP_3O_{10}$	12.087(5)	8.777(3) 110.76(3)	8.944(3)	$C2/c$	4	441
$VNH_4HP_3O_{10}$	12.12	8.42 111.6	9.32	$C2/c$	4	442
β-$CrNH_4HP_3O_{10}$	12.051(5)	8.436(3) 112.22(3)	9.204(4)	$C2/c$	4	444
α-$CrNH_4HP_3O_{10}$	11.806(10)	4.993(5) 120.3(1)	8.790(8)	$P2/a$	2	444
$CrRbHP_3O_{10}$	11.687(7)	4.943(2) 118.96(5)	8.798(8)	$P2/a$	2	445
$CrCsHP_3O_{10}$	11.742(2)	4.975(3) 117.10(2)	8.86(1)	$P2/a$	2	445
$BiNH_4HP_3O_{10}$	7.023(5) 106.20(4)	7.697(6) 105.78(4)	8.664(7) 82.82(5)	$P\bar{1}$	2	457

most of them have never been investigated from a structural point of view. We simply present a list of these salts.

- $AlNa_2P_3O_{10}\cdot4H_2O$ (Lyustko *et al.*[458])
- $Fe_2Na_9(P_3O_{10})_3$ (Berul and Voskresenskaya[459])
- $Fe_2K_9(P_3O_{10})_3$ (Berul and Sizova[282])
- $Fe_4(NH_4)_3(P_3O_{10})_3\cdot13H_2O$, $Fe_5(P_3O_{10})_3\cdot20H_2O$ and $Fe(NH_4)_7(P_3O_{10})_2\cdot3H_2O$ (Galkova and Prodan[460])
- $PrNa_2P_3O_{10}\cdot3/2H_2O$ (Petusshkova *et al.*[461])
- $La_3Na(P_3O_{10})_2\cdot12H_2O$ and $LaNa_2(P_3O_{10})\cdot5H_2O$ (Rodicheva *et al.*[462])
- $Er_3M(P_3O_{10})_2\cdot12H_2O$ and $Er_4M_3(P_3O_{10})_3\cdot14H_2O$, M = Cs, NH_4 (Rodicheva
- and Romanova[463])
- $Nd_3NH_4(P_3O_{10})_2\cdot12H_2O$ and $Nd_4(NH_4)_3(P_3O_{10})_3\cdot14H_2O$ (Rodicheva *et*
- *al.*[464])
- $Cr_3M(P_3O_{10})_2\cdot24H_2O$, M = H, Li, Na, Ag, K, Rb, Cs, NH_4, and Tl (Tananaev *et al.*[465])
- $VK_{2-x}H_xP_3O_{10}$ (x ~ 0.5) (Lavrov *et al.*[41])

2.3.2.5. Triphosphates of Higher Valency Cation

Giesbrecht and Vicentini[466] prepared $Th_5(P_3O_{10})_4\cdot29/2H_2O$, $(UO_2)_5Na_5(P_3O_{10})_3\cdot19H_2O$ and $(UO_2)Na_3P_3O_{10}\cdot8H_2O$. $ThKP_3O_{10}$ has been described by Ruzic-Toros *et al.*,[467] but the atomic arrangement proposed by these authors needs to be revised.

2.3.3. Chemical Preparation of Triphosphates

2.3.3.1. $Na_5P_3O_{10}$

The most common starting material for the preparation of triphosphates is the sodium salt, which has been for a long time commercially available in a very good state of purity. The first preparation of pure crystalline $Na_5P_3O_{10}$ was performed by Schwarz[10] by melting together $Na_4P_2O_7$ and $NaPO_3$ in the proper ratio:

$$Na_4P_2O_7 + NaPO_3 \rightarrow Na_5P_3O_{10}$$

Several other processes involving commonly available sodium monophosphates can also be used.

2.3.3.2. Ion-Exchange Resins

Most alkali triphosphates can be prepared from the sodium salt by the use of ion-exchange resins. Various systems involving $(NH_4)_5P_3O_{10}$ or $K_5P_3O_{10}$ as components have been investigated by using alkali triphosphates prepared by this process.

2.3.3.3. Flux Methods

A good number of triphosphates have been characterized during investigations carried out with a large excess of phosphoric acid at relatively low temperatures. In most cases, the optimization of the initial component ratios and of the temperature leads to a reproducible process for their preparation.

2.3.3.4. Classical Methods

Several mixed-cation triphosphates have been successfully prepared as well-crystallized samples by using classical methods of aqueous chemistry. The $M^{II}Na_3P_3O_{10} \cdot 12H_2O$ (M = Mg, Ni, Co, Cu,...) series of triphosphates, for instance, can be simply prepared by the following reaction:

$$Na_5P_3O_{10} + MSO_4 \rightarrow MNa_3P_3O_{10} + Na_2SO_4$$

2.3.3.5. Decyclization of Cyclotriphosphate Ring Anions

In our review of triphosphates given in Section 2.3.2., we reported three examples of the production of P_3O_{10} anions by hydrolytic decyclization of cyclotriphosphate anions[362,387-389,396] according to the following scheme:

$$H_3P_3O_9 + H_2O \rightarrow H_5P_3O_{10}$$

In spite of its apparent simplicity and elegance, this procedure for the synthesis of triphosphates, although reproducible in the examples reported above, is generally unreliable and will very probably remain only of academic interest.

2.3.4. Atomic Arrangements of Some Triphosphates

2.3.4.1. Pentasodium Triphosphate (Form I)

$$Na_5P_3O_{10},^{32} \text{ monoclinic, } C2/c, Z = 4$$

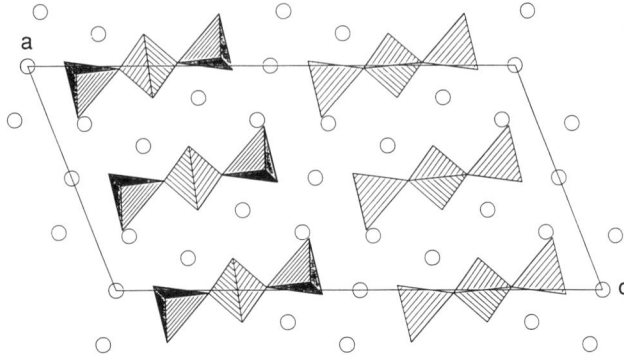

Figure 2.3.3. Projection, along the *b* direction, of the atomic arrangement in $Na_5P_3O_{10}$ (I). The open circles represent the sodium atoms.

$$a = 9.61(3), b = 5.34(2), c = 19.73(5) \text{ Å}, \beta = 112.0(5)°$$

Figure 2.3.3 shows the atomic arrangement of form I of $Na_5P_3O_{10}$ in projection along the *b* axis. The triphosphate anion has twofold internal symmetry, its central phosphorus atom being located on a twofold axis. The chain axis is approximately parallel to $(\bar{2}\ 0\ 1)$. Two of the three crystallographically independent sodium atoms have sixfold oxygen coordination, in a distorted octahedral arrangement, and the third one has fourfold oxygen coordination. One of the six-coordinated sodium atoms is located on an inversion center. The Na–O distances range between 2.26 and 2.41 Å in the NaO_6 polyhedra and between 2.22 and 2.59 Å in the NaO_4 polyedron. Corbridge[32,363] explained the easier solubility of form I by the existence of fourfold coordination for one of the sodium atoms.

2.3.4.2. Pentasodium Triphosphate (Form II)

$$Na_5P_3O_{10},[363] \text{ monoclinic, } C2/c, Z = 4$$

$$a = 16.00(5), b = 5.24(2), c = 11.25(3) \text{ Å}, \beta = 93.0(5)°$$

As in form I, the central phosphorus atom of the triphosphate anion is located on a twofold axis. Figure 2.3.4 shows a projection of this structure along the *b* direction. The three crystallographically independent sodium atoms are surrounded by oxygen atoms in distorted octahedral arrangements, which link together to form a complex three-dimensional network. A relatively large variation of the Na–O distance is observed within the octahedra, this distance ranging from 2.25 to 2.76 Å.

2.3.4.3. Potassium Dihydrogen Triphosphate Monohydrate

$$K_3H_2P_3O_{10} \cdot H_2O,[378] \text{ orthorhombic, } Pbca, Z = 8$$

$$a = 7.588(2), b = 11.163(4), c = 26.697(8) \text{ Å}$$

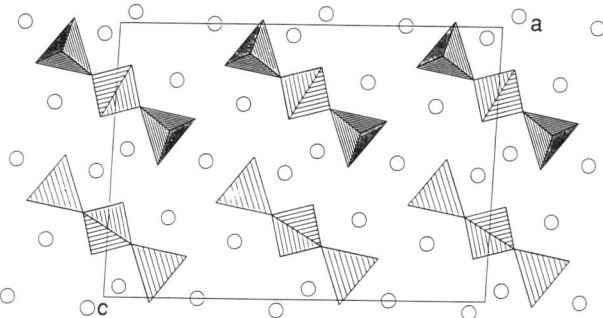

Figure 2.3.4. Projection, along the *b* direction, of the atomic arrangement in $Na_5P_3O_{10}$ (II). The open circles represent the sodium atoms.

The atomic arrangement in $K_3H_2P_3O_{10} \cdot H_2O$ is another typical layer organization. Perpendicular to the *b* direction, planes containing the potassium atoms ($y = 0$ and $1/2$) alternate with layers ($y = 1/4$ and $3/4$) composed of the phosphoric entities and the water molecules. Within the $H_2P_3O_{10}$–H_2O layers the phosphoric groups are connected by strong hydrogen bonds (O–O = 2.581 and 2.544 Å). Each $H_2P_3O_{10}$ group is connected by such bonds to its four adjacent neighbors. This organization of the phosphoric groups creates voids in which are located the water molecules, themselves connected by hydrogen bonds to the adjacent phosphoric groups. Figure 2.3.5 shows such a layer in projection along the *b* axis. The phosphoric group has no internal symmetry.

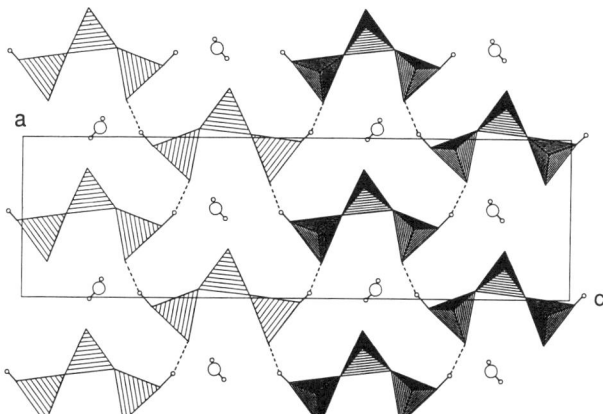

Figure 2.3.5. Representation, in projection along the *b* axis, of an $H_2P_3O_{10}$-H_2O layer in $K_3H_2P_3O_{10} \cdot H_2O$. The hydrogen bonds connecting the phosphoric groups and the water molecules are denoted by solid and dashed lines. The larger open circles represent the oxygen atoms of the water molecules, and the smaller ones the hydrogen atoms of these molecules.

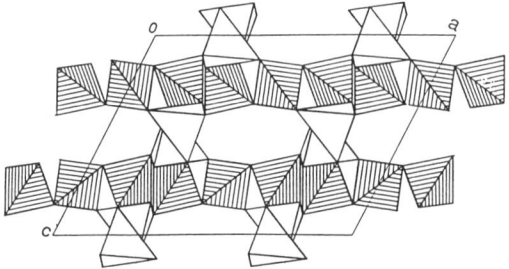

Figure 2.3.6. Projection of the $Be_2NH_4P_3O_{10}$ atomic arrangement along the **b** axis. The hatched tetrahedra correspond to the PO_4 groups, and the plain ones to the BeO_4 tetrahedra. In this projection, the ammonium groups are superimposed on the central phosphorus atoms of the P_3O_{10} anions.

Among the three independent potassium atoms, two have sevenfold coordination and the third one has sixfold coordination. Within these various polyhedra, the K–O distances range between 2.749 and 3.093 Å. One of the K–P distances reported, 3.092 Å, seems abnormally short for this type of compound and is probably a misprint.

2.3.4.4. Beryllium–Ammonium–Triphosphate

$$Be_2NH_4P_3O_{10},^{390,391} \text{ monoclinic, } C2/c, Z = 4$$

$$a = 12.202(8),\ b = 8.645(3),\ c = 8.949(3) \text{ Å},\ \beta = 117.41(5)°$$

The atomic arrangement in $Be_2NH_4P_3O_{10}$ can be easily described as a three-dimensional network of tetrahedra, built by P_3O_{10} groups and BeO_4 tetrahedra, in which the ammonium groups are inserted in channels parallel to the **c** axis. Figure 2.3.6 shows a projection of the structure along the **b** direction. P_3O_{10} groups, all centered around the planes $z = 1/4$ and $3/4$ are interconnected by BeO_4 tetrahedra located in the planes $z = 0$ and $1/2$. As central phosphorus atom of the P_3O_{10} anion is located on a twofold axis, this anion adopts a binary symmetry rather common in this class of anions. The other representation, a projection along the **c** axis (Fig. 2.3.7) shows how the organization of this three-dimensional network creates large channels parallel to the **c** direction, in which the ammonium groups are located.

Each BeO_4 tetrahedron shares its four corners with the four adjacent PO_4 tetrahedra, with Be–P distances ranging from 2.811 to 2.884 Å. Within a BeO_4 tetrahedron, the Be–O distances range from 1.603 to 1.662 Å. It should be noted that the P–P distance in the phosphoric group (2.821 Å) is the shortest ever observed in a condensed phosphate anion and that the P–O–P angle (112.6°) is also abnormally small in this class of compounds.

The ammonium group located on a twofold axis has sixfold coordination, with N–O distances ranging from 2.935 to 3.099 Å.

2.3.4.5. Iron–Ammonium Hydrogen Triphosphate

$$FeNH_4HP_3O_{10},^{443} \text{ monoclinic, } C2/c, Z = 4$$

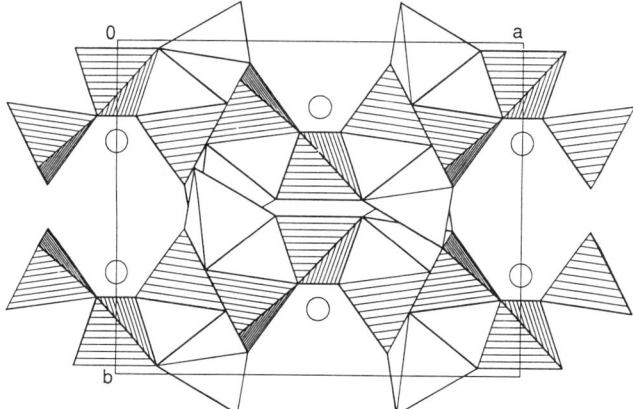

Figure 2.3.7. Projection of the $Be_2NH_4P_3O_{10}$ atomic arrangement along the *c* direction. The conventions used for the polyhedral representation are the same as those used for the *b* projection given in Fig. 2.3.6. The open circles figurate the ammonium groups.

$$a = 12.077(2), b = 8.448(1), c = 9.357(1) \text{ Å}, \beta = 112.21(1)°$$

As can be seen in Fig. 2.3.8, the atomic arrangement in $FeNH_4HP_3O_{10}$ can be considered as successive layers alternating along the *c* direction. The planes $z = 0$ and 1/2 contain the iron atoms and are separated by thick layers composed of the phosphoric entities and the ammonium groups. The central phosphorus atom is located on a binary axis, and the phosphoric group has twofold internal symmetry. The hydrogen atoms located on the inversion center at 1/4,1/4,1/2 connect the phosphoric groups by strong centrosymmetric

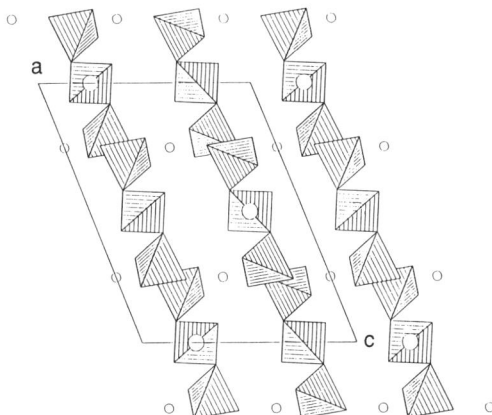

Figure 2.3.8. Projection of the atomic arrangement in $FeNH_4HP_3O_{10}$ along the *b* axis. The hydrogen bonds are represented by solid lines. Ammonium groups have been omitted. In this projection, the hydrogen and iron atoms are superimposed.

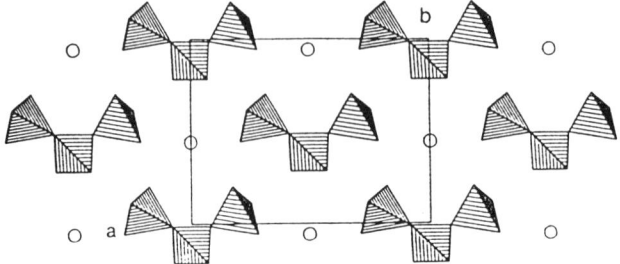

Figure 2.3.9. Projection, along the c direction, of a layer of P_3O_{10} anions and NH_4 groups centered by the plane $z = 1/4$. The open circles represent the ammonium groups.

H bonds (O–O = 2.436 Å) resulting in the formation of infinite chains extending across the (a,c) planes. The projection along the c direction in Fig. 2.3.9. shows in the organization of the NH_4 and P_3O_{10} groups within a layer centered by the plane $z = 1/4$.

The iron atoms, located on inversion centers, have a very regular octahedral coordination with Fe–O distances ranging from 1.976 to 2.002 Å, whereas the nitrogen atoms located on twofold axes, have ten oxygen neighbors with N–O distances ranging from 2.850 to 3.435 Å.

$VCsHP_3O_{10}$, $VNH_4HP_3O_{10}$, and the β form of $CrNH_4HP_3O_{10}$ are isotypic.

2.3.5. The Triphosphate Anion

To date, the atomic arrangements of no more than 20 triphosphates have been determined with reasonable accuracy. We summarize in Table 2.3.5 the main features describing the geometry of P_3O_{10} groups as observed in some accurately determined crystal structures. In this table, we present only the data that we consider essential for the description of a P_3O_{10} group: the P–P distances and the P–O–P and P–P–P angles.

Table 2.3.5. Main Geometrical Features Measured in Triphosphate Groups

Formula	distance P–P (Å)	angle P–P–P (°)	angle P–O–P (°)	Symmetry	Reference(s)
$Na_5P_3O_{10}$ (I)	2.87	_a	121.8	Twofold	363
$Na_5P_3O_{10}$ (II)	2.87	_a	121.5	Twofold	32
$Na_5P_3O_{10}\cdot6H_2O$	2.881	151.3	124.4	None	365, 366
	2.887		123.4		
$(NH_4)_4NaP_3O_{10}\cdot4H_2O$	2.944	103.9	130.6	Twofold	384
$K_3H_2P_3O_{10}\cdot H_2O$	2.978	98.1	139.0	None	375
	2.991		138.4		
$Be_2NH_4P_3O_{10}$	2.821	105.4	112.6	Twofold	390, 391
$FeH_2P_3O_{10}$	2.932	124.0	133.5	None	378
	2.933		134.1		
$FeNH_4HP_3O_{10}$	2.897	112.2	130.5	Twofold	443
$BiNH_4HP_3O_{10}$	2.878	84.5	126.3	None	457
	2.978		137.6		

^{a}Not reported.

The values for the P–P distances and P–O–P angles reported in this table are within the limits generally observed in condensed phosphate crystal chemistry, with the exception of those reported for $Be_2NH_4P_3O_{10}$, where the P–P distance (2.821 Å) and the P–O–P angle (112.6°) are among the smallest ever measured in a condensed phosphate anion. The situation is similar in the isotypic $Be_2RbP_3O_{10}$, in which P–P = 2.818 Å and P–O–P = 105.3°.

On the other hand, the wide variation in the P–P–P angles, ranging from 84.5 to 151.3°, may seem unusual, but a similar range of values is commonly observed in the crystal chemistry of long-chain polyphosphates and even in large ring anions, such as cyclo-hexaphosphates, for instance.

If one examines the internal symmetry of the P_3O_{10} groups, one observes that about 50% of them exhibit a twofold symmetry, the central phosphorus atom being located on a binary axis. The remaining ones have no internal symmetry or a pseudo-binary one.

2.4. Tetraphosphates

2.4.1. Introduction

The first evidence for the existence of a tetraphosphate was given by Osterheld and Langguth[468] and Langguth *et al.*[469] The $Pb_2P_2O_7-Pb(PO_3)_2$ phase-equilibrium diagram elaborated by these authors shows clearly the existence of $Pb_3P_4O_{13}$ as an incongruent-melting compound decomposing at 973 K. This diagram is reported in Fig. 2.4.1.

For a long time, the only suggested method for the preparation of water-soluble tetraphosphates was the alkaline hydrolysis of cyclotetraphosphates based on the following scheme:

$$H_4P_4O_{12} + H_2O \rightarrow H_6P_4O_{13}$$

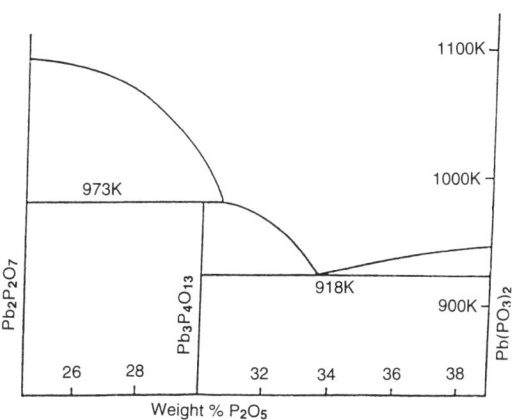

Figure 2.4.1. The $Pb_2P_2O_7$-$Pb(PO_3)_2$ phase-equilibrium diagram.

This procedure was first used by Thilo and Ratz[11] during an attempt to prepare $Na_6P_4O_{13}$, but they did not succeed in crystallizing this salt. Investigations in which this process was used include those by Westman and Scott,[25] Quimby,[470] and Watters et al.[471] This preparative method is tedious, time-consuming and most of the time, deceiving. The only successful use of this process seems to have been the preparation of $[Co(NH_3)_6]_2P_4O_{13}$·$5H_2O$ by Schulz and Jansen.[472]

Griffith[473] was the first to describe a reproducible method for the preparation of convenient amounts of a water-soluble tetraphosphate, $(NH_4)_6P_4O_{13}$·$6H_2O$, through a very simple reaction using $Pb_3P_4O_{13}$ as starting material:

$$Pb_3P_4O_{13} + 3(NH_4)_2S \rightarrow (NH_4)_6P_4O_{13}·6H_2O + 3PbS$$

In spite of the possibilities offered by this process, all the recent developments in the chemistry of tetraphosphates have come from the use of various flux methods. The understanding of tetraphosphates is still relatively poor and it was, only in 1976 that the first structural characterization of a P_4O_{13} anion was performed, during investigations by Averbuch-Pouchot and Durif[474] and Durif et al.[38] of $(NH_4)_2SiP_4O_{13}$. To date, only 11 tetraphosphates have been characterized by X-ray diffraction. The main crystallographic data for these compounds are presented in Table 2.4.1.

2.4.2. Present State of Tetraphosphate Chemistry

2.4.2.1. Alkali and Monovalent Cation Tetraphosphates

Sodium tetraphosphate has not yet been prepared in a crystalline state. Thilo and Ratz[11] described a preparation of this salt, but the final product obtained by these authors was oily and its hydration state uncertain. For their preparation, the authors used the hydrolytic decyclization of sodium cyclotetraphosphate by a solution of sodium hydroxide according to the following scheme:

$$Na_4P_4O_{12} + 2NaOH(H_2O) \rightarrow Na_6P_4O_{13}·xH_2O$$

The reaction was conducted at 313 K and was completed, based on paper chromatography experiments, after about 100 h. The resulting tetraphosphate was then precipitated by acetone as an oily syrup. According to the authors, this salt decomposes rapidly:

$$Na_6P_4O_{13}·xH_2O \rightarrow 2\ Na_3HP_2O_7·H_2O$$

Griffith[473] did not succeed in reproducing the preparation of this salt and reported that all attempts to produce a crystalline potassium tetraphosphate proved to be fruitless.

The reaction scheme used by Griffith[473] to produce $(NH_4)_6P_4O_{13}$·$6H_2O$ was presented in Section 2.4.1. This salt decomposes when heated at relatively low temperatures. At 323 K total decomposition is achieved within 8 h. The same reaction scheme was used by Waerstad and McClellan[475] to synthesize $(NH_4)_6P_4O_{13}$·$2H_2O$ and $(NH_4)_4H_2P_4O_{13}$.

A detailed study of the NH_3–$H_6P_4O_{13}$–H_2O system at 273 K has been performed by Farr et al.[476] The rate of hydrolysis of $(NH_4)_6P_4O_{13}$ in dilute solutions at 298 K was investigated.

2.4.2.2. Divalent Cation Tetraphosphates

Careful chemical investigations by Kreidler and Hummel[141] and Mc Keag and Steward[477] provided evidences for the existence of $Sr_3P_4O_{13}$. Two forms of $Ba_3P_4O_{13}$ were reported by McCauley and Hummel[478] and in Ref. 479. Single crystals of the two forms were prepared by Millet *et al.*[480] who report the unit-cell dimensions. A crystal structure of the low-temperature form was proposed by Gatehouse *et al.*[481] but needs to be revised.

The existence of the lead salt, $Pb_3P_4O_{13}$, and of its potential applications have already been reported in Section 2.4.1. Single crystals of this salt were prepared by Averbuch-Pouchot and Durif.[482,483] Their structural investigation confirmed the geometrical nature of the anion suggested by the chemical formula. Some other investigations of this salt were performed by Schulz[484] and Argyle and Hummel.[485]

2.4.2.3. Trivalent Cation Tetraphosphates

Several tetraphosphates of trivalent cations have been reported, but only three of them $[Co(NH_3)_6]_2P_4O_{13}\cdot5H_2O$, $Bi_2P_4O_{13}$, and $Cr_2P_4O_{13}$, have been investigated from a structural point of view.

The chemical preparation and crystal structure of $[Co(NH_3)_6]_2P_4O_{13}\cdot5H_2O$ were recently reported by Schulz and Jansen.[472] This compound is the only example of a crystalline tetraphosphate prepared by controlled hydrolysis of a P_4O_{12} ring anion.

$Cr_2P_4O_{13}$ was characterized for the first time by Lii *et al.*[486] during attempts to synthesize a chromium analog of $Mo_3P_5SiO_{19}$ at 1303 K in a sealed silica tube containing Cr_2O_3 and P_2O_5 in a molar ratio of 3:5. The authors reported a complete description of the atomic arrangement.

$Bi_2P_4O_{13}$ was first characterized by Schulz[484] as a congruently melting compound (mp = 1083 K). Hilmer *et al.*[487] confirmed this melting point during a general investigation of condensed bismuth phosphates. Later, Bagieu-Beucher and Averbuch-Pouchot[488] synthesized single crystals of this salt and reported a detailed determination of its atomic arrangement.

$Y_2P_4O_{13}$ and $Gd_2P_4O_{13}$ were fully characterized by Agrawal and Hummel,[489] while $La_2P_4O_{13}$ was identified by Park and Kreidler[490] and said to be isotypic with these two salts. $In_2P_4O_{13}\cdot10H_2O$ is said to have been observed by Rodicheva *et al.*[493] during an investigation of the $InCl_3$–$Li_6P_4O_{13}$–H_2O system.

2.4.2.4. Trivalent–Monovalent Cation Tetraphosphates

Rodicheva *et al.*[491–494] investigated several $InCl_3$–$M_6P_4O_{13}$–H_2O systems (M = Li, Na, NH_4, Cs) at 273K or 298 K(Na) and reported the existence of some new mixed salts, several of them not having well-defined hydration state: $InNa_3P_4O_{13}\cdot xH_2O$ ($7 < x < 11$), $In_3Li_3(P_4O_{13})_2\cdot16H_2O$, $InLi_3P_4O_{13}\cdot11H_2O$, $In_3(NH_4)_3(P_4O_{13})_2\cdot15H_2O$, $In(NH_4)_3P_4O_{13}\cdot3H_2O$, $In_3Cs_3(P_4O_{13})_2\cdot xH_2O$ with $x \sim 12$, and $InCs_3P_4O_{13}\cdot xH_2O$ with $x \sim 5$. No structural characterization has been performed for these compounds, and so the nature of the phosphoric anion is still uncertain.

2.4.2.5. Tetravalent–Monovalent Cation Tetraphosphates

The chemical preparation and crystal data for $(NH_4)_2SiP_4O_{13}$ were reported by Averbuch-Pouchot and Durif.[474] The atomic arrangement of this salt determined by Durif *et al.*[38] provided the first structural proof of the existence of the P_4O_{13} anion. This

Table 2.4.1. Crystal Data for Tetraphosphates

Formula	a (Å) α (°)	b (Å) β (°)	c (Å) γ (°)	S.G.	Z	Reference(s)
$(NH_4)_6P_4O_{13} \cdot 2H_2O$	13.426(5) 93.80(3)	11.874(5) 84.78(2)	6.524(2) 106.97(6)	$P\bar{1}$ or $P1$	2	475, 476
$(NH_4)_4H_2P_4O_{13}$	13.359(9)	13.244(6)	8.214(7)	$P22_12_1$	4	475, 476
$Ba_3P_4O_{13}(LT)^a$	7.240(1) 104.02(2)	8.011(1) 109.51(2)	5.689(1) 83.62(2)	$P\bar{1}$	1	480, 481
$Pb_3P_4O_{13}$	7.826(5) 104.36(5)	7.338(5) 101.77(5)	10.206(8) 94.31(5)	$P\bar{1}$	2	482, 483
$Cr_2P_4O_{13}$	8.097(2)	8.787(3) 105.54(2)	13.098(4)	$P2_1/c$	4	486
$Bi_2P_4O_{13}$	11.977(4)	6.878(2) 106.50(2)	13.285(4)	$C2/c$	4	488
$Si(NH_4)_2P_4O_{13}$	15.14(1) 97.86(1)	7.684(5) 96.74(1)	4.861(5) 83.89(1)	$P\bar{1}$	2	38, 474
$Ge(NH_4)_2P_4O_{13}$	15.08(1) 98.18(1)	7.763(5) 96.74(1)	4.914(5) 83.89(1)	$P\bar{1}$	2	495
$(NbO)_2P_4O_{13}$	6.586(1) 106.37(1)	8.400(1) 90.35(1)	10.842(1) 89.97(1)	$P\bar{1}$	2	496
$(MoO)_2P_4O_{13}$	8.288(2)	10.690(3) 106.71(3)	19.529(5)	Pb	6	497
$[Co(NH_3)_6]_2P_4O_{13} \cdot 5H_2O$	15.478(8)	12.636(8) 112.23(4)	14.511(8)	$P2_1/c$	4	472

^{a}LT: Low-temperature form.

compound is very simply prepared by heating a mixture of 20 g of $(NH_4)_2HPO_4$ and 1 g of silica wool at 623 K for 12 h. The corresponding germanium salt, $Ge(NH_4)_2P_4O_{13}$, was characterized by Averbuch-Pouchot[495] as an isotype of the first one.

2.4.2.6. Pentavalent Cation Tetraphosphates

The existence of a niobyl tetraphosphate, $(NbO)_2P_4O_{13}$, was established by Nikolaev et al.[496] during an investigation of the $Nb_2O_5-P_2O_5-H_2O$ system at 703–733 K. Unfortunately, further work is necessary to confirm the atomic arrangement proposed by the authors. A molybdenyl tetraphosphate, $(MoO)_2P_4O_{13}$ was described by Minacheva et al.,[497] but in this case also, the structural aspects of the work need reinvestigation.

Lavrov et al.[498] reported a detailed procedure for the preparation of crystalline $(VO)_2P_4O_{13}$ by reaction at 493–573 K of a mixture of H_3PO_4 and V_2O_5 with an initial P/V ratio of 6–10. This tetraphosphate is stable in air up to 873 K but decomposes at 943 K:

$$(VO)_2P_4O_{13} \rightarrow 2VO(PO_3)_2 + 1/2O_2$$

2.4.3. Chemical Preparations of Tetraphosphates

Although Griffith[473] described a clear and reproducible procedure for the production of $(NH_4)_6P_4O_{13} \cdot 6H_2O$, this compound has never been used as a starting material for the investigation of tetraphosphate chemistry. With the exception of $[Co(NH_3)_6]_2P_4O_{13} \cdot 5H_2O$ produced by hydrolytic decyclization of a P_4O_{12} ring, all tetraphosphates investigated to date have been characterized during the elaboration of phase-equilibrium diagrams, as in the case of $Pb_3P_4O_{13}$, or during investigations of various systems by flux methods. In most cases, the true nature of the anion was, in fact, established by structural characterization of the tetraphosphates.

2.4.4. Some Atomic Arrangements of Tetraphosphates

2.4.4.1. Lead Tetraphosphate

$$Pb_3P_4O_{13},^{482,483} \text{ triclinic, } P\bar{1}, Z = 2$$

$$a = 7.826(5), b = 7.338(5), c = 10.206(8) \text{ Å}$$

$$\alpha = 104.36(5), \beta = 101.77(5), \gamma = 94.31(5)°$$

Figure 2.4.2 shows a projection of the atomic arrangement of $Pb_3P_4O_{13}$ along the b direction. Here the P_4O_{13} group has no internal symmetry and is thus is composed of by independent PO_4 tetrahedra. Layers of phosphoric groups centered in $x = 0$ alternate with the lead atoms, all located in planes corresponding approximately to $x = \pm 1/4$ and $x = 1/2$.

Within a range of 3 Å, two of the three independent lead atoms have sevenfold coordination and the third one has eightfold coordination. Within these polyhedra, the Pb–O distances range between 2.442 and 2.950 Å.

2.4.4.2. Chromium Tetraphosphate

$$Cr_2P_4O_{13},^{486} \text{ monoclinic, } P2_1/c, Z = 4$$

$$a = 8.097(2), b = 8.787(3), c = 13.098 \text{ Å}, \beta = 105.54(2)°$$

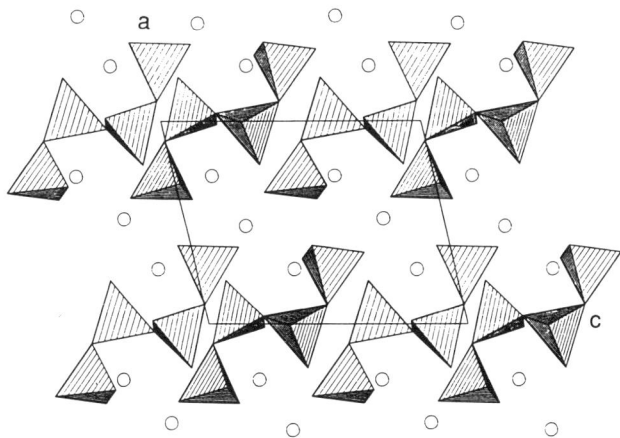

Figure 2.4.2. Projection of the $Pb_3P_4O_{13}$ structure along the b direction. The open circles represent the lead atoms.

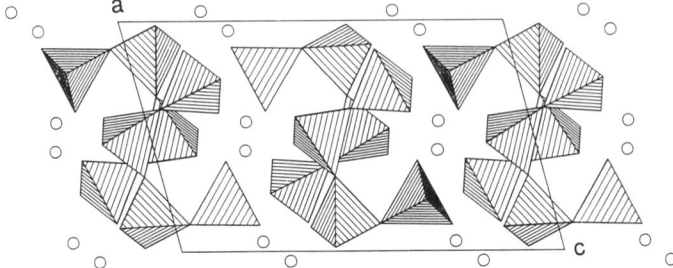

Figure 2.4.3. Projection, along the *b* axis, of the atomic arrangement in $Cr_2P_4O_{13}$. The small open circles represent the chromium atoms.

Figure 2.4.3 gives a view of the atomic arrangement in $Cr_2P_4O_{13}$ in projection along the *b* axis and shows the layer organization of phosphoric anions centered in planes $z = 0$ and 1/2 alternating with chromium atoms in planes corresponding approximately to $z = \pm 1/4$. Here, as in $Pb_3P_4O_{13}$, the phosphoric group has no internal symmetry, but, unlike all the previously investigated groups of the same kind, has a very accentuated horseshoe configuration, represented in Fig. 2.4.4 in projection along the *b* direction. By edge-sharing the CrO_6 octahedra assemble in pairs, forming Cr_2O_{10} units. Each Cr_2O_{10} unit shares its ten oxygen atoms with seven different P_4O_{13} groups. This stacking of P_4O_{13} groups and Cr_2O_{10} units produces hexagonal and tetragonal channels running parallel to the *a* direction. Within a Cr_2O_{10} group, the Cr–Cr distance is 3.150 Å, and the Cr–O distances range from 2.028 to 1.928 Å in one of the CrO_6 octahedra and from 1.896 to 2.048 Å in the other one.

2.4.4.3. Bismuth Tetraphosphate

$$Bi_2P_4O_{13},^{488} \text{ monoclinic, } C2/c, Z = 4$$

$$a = 11.977(4), b = 6.878(2), c = 13.285(4) \text{ Å}, \beta = 106.50(2)°$$

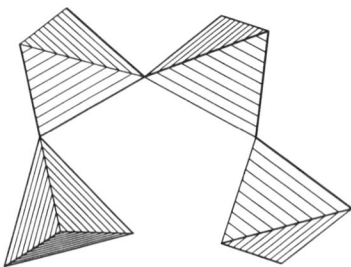

Figure 2.4.4. An isolated P_4O_{13} anion as observed in $Cr_2P_4O_{13}$. Projection made along the *b* direction.

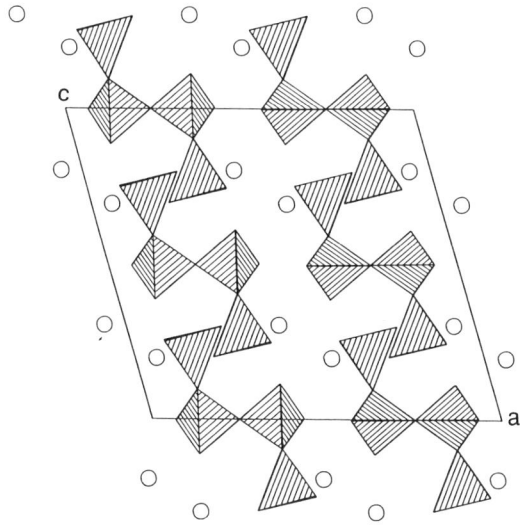

Figure 2.4.5. Projection, along the **b** direction, of the atomic arrangement in $Bi_2P_4O_{13}$. The open circles represent the bismuth atoms.

The atomic arrangement of $Bi_2P_4O_{13}$ in projection along the **b** direction is shown in Fig. 2.4.5. Layers of edge sharing BiO_8 polyhedra are centered around the planes $z = 0$ and $1/2$. These polyhedra are distorted dodecahedra, an usual coordination for Bi atoms. Figure 2.4.6 represents such a layer in projection along the **c** direction.

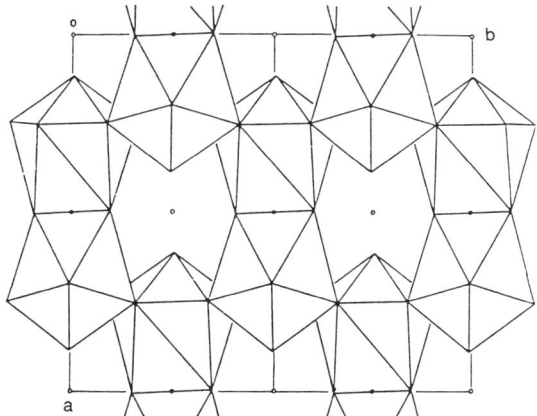

Figure 2.4.6. Projection, along the **c** axis, of a layer of BiO_8 polyhedra in $Bi_2P_4O_{13}$. The small open circles are the inversion centers.

The phosphoric anions are all located around the planes $z = 0.25$ and 0.75, thus alternating with the bismuth layers. The central oxygen atom of the phosphoric anion is located on a twofold axis, inducing a binary internal symmetry for the P_4O_{13} group.

2.4.4.4. Silicium–Diammonium Tetraphosphate

$$Si(NH_4)_2P_4O_{13},^{38,474}, \text{ triclinic, } P\bar{1}, Z = 2$$

$$a = 15.14(1), b = 7.684(5), c = 4.861(5) \text{ Å}$$

$$\alpha = 97.86(1), \beta = 96.74(1), \gamma = 83.89(1)°$$

The atomic arrangement in $Si(NH_4)_2P_4O_{13}$ is a typical layer organization. Layers composed of a two-dimensional network of SiO_6 octahedra and P_4O_{13} groups located around the planes $x = 1/4$ and 3/4 alternate with layers containing the ammonium groups and located in the planes $x = 0$ and 1/2. The projection of this structure along the c axis (Fig. 2.2.7.) shows clearly the layer aspect of the arrangement.

The projection along the a axis in Fig. 2.4.8 shows the details of the SiP_4O_{13} layer located around the plane $x = 1/4$. Each SiO_6 octahedron shares its six oxygen atoms with four adjacent P_4O_{13} groups. This network creates large rectangular voids in the center of the unit cell. The SiO_6 octahedron is rather regular, with Si–O distances ranging from 1.762 to 1.788 Å and O–Si–O angles ranging from 86.5 to 93.2°. The two independent ammonium

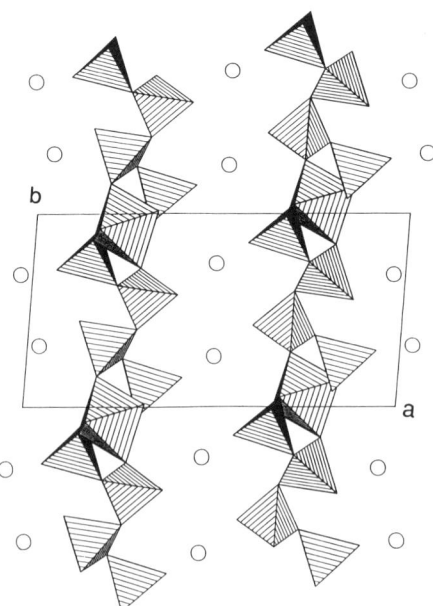

Figure 2.4.7. Projection, along the c direction, of the atomic arrangement in $(SiNH_4)_2P_4O_{13}$. Hatched octahedra represent the SiO_6 groups, and the open circles represent the ammonium groups.

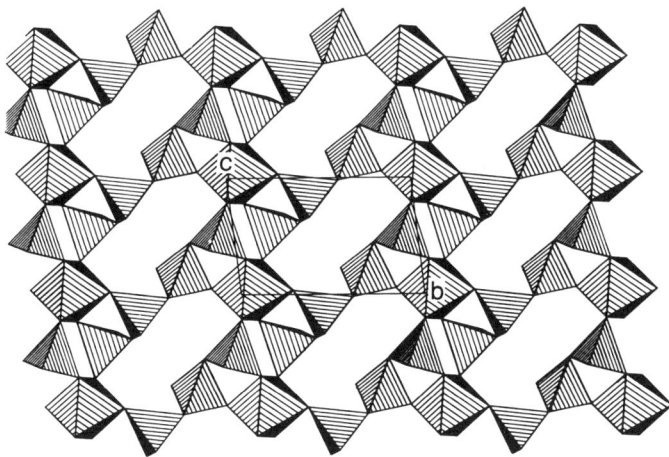

Figure 2.4.8. Projection, along the *a* direction, of the $Si(NH_4)_2P_4O_{13}$ layer located around the plane $x = 1/4$. Projection made for $0.05 < x < 0.45$.

groups, which establish the cohesion between the SiP_4O_{13} planes, are different in their coordination. Within a range of 3.20 Å, the first one has fivefold coordination, and the second one sixfold.

The corresponding germanium salt, $Ge(NH_4)_2P_4O_{13}$, is isotypic.[495]

2.4.5. The Tetraphosphate Group

To date, only six atomic arrangements of tetraphosphates have been accurately determined. Thus, any discussion aiming at comparing the geometries of P_4O_{13} anions or establishing geometrical correlations seems premature, and we simply report in Table 2.4.2

Table 2.4.2. Geometry of the Tetraphosphate Anions

Formula	distance P–P (Å)	angle P–P–P (°)	angle P–O–P (°)	Symmetry	Reference(s)
$Pb_3P_4O_{13}$	2.895	84.6	131.3	None	482, 483
	3.034	85.5	143.6		
	2.898		128.7		
$Cr_2P_4O_{13}$	2.979	108.6	136.8	None	486
	2.997	92.6	142.7		
	2.932		135.0		
$Bi_2P_4O_{13}$	2.974	95.8	140.9	Twofold	488
	2.941		134.4		
$Si(NH_4)_2P_4O_{13}$	2.912	95.6	131.8	None	38, 474
	2.931	108.6	133.7		
	2.856		125.8		
$[Co(NH_3)_6]_2P_4O_{13}\cdot5H_2O$	2.958	121.0	134.4	None	472
	2.919	111.7	129.6		
	2.993		139.8		
$Nb_2OCa(P_4O_{13})(P_2O_7)$	2.848	96.3	126.3	Twofold	499
	2.927		134.7		

the main geometrical features of the known P_4O_{13} groups. A P_4O_{13} anion has also been observed in a mixed-anion phosphate, $Nb_2OCa(P_2O_7)(P_4O_{13})$, recently characterized by Averbuch-Pouchot,[499] and its geometrical parameters are included in the same Table 2.4.2.

Among the six examples reported in Table 2.4.2, it may be noted that two P_4O_{13} groups have twofold internal symmetry, while the other four have no internal symmetry. Most of these groups are zigzag chains of tetrahedra, quite comparable to a fragment of a long-chain polyphosphate, whereas the P_4O_{13} group observed in the chromium salt[486] departs very significantly from this conformation, with its very accentuated horseshoe geometry (Fig. 2.4.4).

The remarks made in Section 2.3.5 concerning the P–P distances and the P–P–P and P–O–P angles in the triphosphate group remains valid for the tetraphosphate group.

2.5. Pentaphosphates

2.5.1. Present State of Pentaphosphate Chemistry

The first reported evidence for the existence of a crystalline pentaphosphate was that obtained by Majling and Hanic[507] during an investigation of the $MgO–Na_2O–P_2O_5$ system. These authors characterized, in this system, a compound of the formula $Mg_2Na_3P_5O_{16}$, which they suspected to be a pentaphosphate. The structural characterization of this salt was performed by Smolin et al.,[39] providing the first geometrical data for a P_5O_{16} anion. Crystals up to 2–3 mm long were prepared by slow cooling of a melt of composition 35 wt. % $Mg_2P_2O_7$, 65 wt. % $NaPO_3$.

Recently, Klinkert and Jansen[501] identified two isotypic compounds, $V_2CsP_5O_{16}$ and $Fe_2CsP_5O_{16}$, as pentaphosphates. Crystals were produced by slow cooling (10 K/h) of melts of Cs_2CO_3, H_3PO_4, and M_2O_3 (M = V, Fe), prepared at 823–873 K for the vanadium salt and at 873–923 K in the case of iron. The vanadium compound can also be obtained by firing pure $VCsHP_3O_{10}$ at 823 K according to the following scheme:

$$2VCsHP_3O_{10} \rightarrow V_2CsP_5O_{16} + CsPO_3 + H_2O$$

This very rare type of anion has also been observed in a mixed-anion phosphate, $Ta_2Rb_2H(PO_4)_2(P_5O_{16})$, a mono–pentaphosphate described by Sadikov et al.[502] and discussed in Chapter 5 which is devoted to mixed-anion condensed phosphates.

Table 2.5.1 gives crystal data for the three pentaphosphates that have been investigated to date.

Table 2.5.1. Main Crystallographic Data for Pentaphosphates

Formula	a (Å) α (°)	b (Å) β (°)	c (Å) γ (°)	S.G.	Z	Reference(s)
$Mg_2Na_3P_5O_{16}$	18.617(5)	6.844(3) 90.15(3)	5.174(3)	$P2/a$	2	39, 500
$V_2CsP_5O_{16}$	7.538(1)	9.410(1) 111.16(9)	10.173(1)	Pn	2	501
$Fe_2CsP_5O_{16}$	7.530(2)	9.378(4) 111.02(1)	10.209(2)	Pn	2	501

As one can guess from the rare occurrence of pentaphosphates, no general procedure exists for their preparation. All the pentaphosphates investigated to date have been characterized either during the elaboration of phase-equilibrium diagrams or during the investigation of various systems by flux methods, the true nature of their anion being recognized during structural studies.

2.5.2. Some Atomic Arrangements of Pentaphosphates

2.5.2.1. Magnesium–Sodium Pentaphosphate

$$\text{Mg}_2\text{Na}_3\text{P}_5\text{O}_{16},\text{[39,500]} \text{ monoclinic, } P2/a, Z = 2$$

$$a = 18.617(5),\ b = 6.844(3),\ c = 5.174\ (3)\ \text{Å},\ \beta = 90.15(3)°$$

A projection of the structure of $\text{Mg}_2\text{Na}_3\text{P}_5\text{O}_{16}$ along the b direction is represented in Fig. 2.5.1. This drawing shows clearly the layer organization of this arrangement, in which layers of P_5O_{16} groups alternate with layers of the associated cations. As the central phosphorus atom of the phosphoric anion is located on a binary axis, this group adopts twofold internal symmetry.

The magnesium atom, located on a general position, has a slightly distorted octahedral oxygen coordination. This MgO_6 octahedron shares its six oxygen atoms with five adjacent phosphoric groups. Within the MgO_6 octahedron, the Mg–O distances range from 2.006 to 2.197 Å. One of the two crystallographically independent sodium atoms is situated on the twofold axis and has sixfold coordination, with Na–O distances ranging from 2.365 to 3.053 Å; the second one, in general position, has only five neighbors, with Na–O distances ranging from 2.322 to 2.448 Å.

Another way to describe this arrangement is to consider it as composed of a three-dimensional network of P_5O_{16} groups and MgO_6 octahedra. Such a network, represented in Fig. 2.5.2 in projection along the c axis, creates two kinds of channels parallel

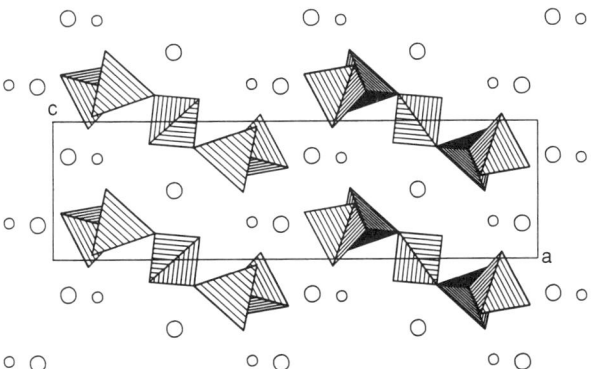

Figure 2.5.1. Projection, along the b direction, of the atomic arrangement in $\text{Mg}_2\text{Na}_3\text{P}_5\text{O}_{16}$, showing the layer organization of this arrangement. The large open circles represent sodium atoms, and the smaller ones the magnesium atoms.

Figure 2.5.2. Projection, along the c direction, of the atomic arrangement in $Mg_2Na_3P_5O_{16}$. The MgO_6 groups are represented by hatched octahedra, and the sodium atoms by open circles.

to the c axis, in which the sodium atoms are located. This second representation also shows clearly the horseshoe configuration of the phosphoric group and its twofold symmetry.

2.5.2.2. Vanadium–Cesium Pentaphosphate

$$V_2CsP_5O_{16},^{501} \text{ monoclinic, } Pn, Z = 2$$

$$a = 7.538(1), b = 9.410(1), c = 10.173(1) \text{ Å}, \beta = 111.16(9)°$$

In $V_2CsP_5O_{16}$, the P_5O_{16} group has no internal symmetry but still has the horseshoe configuration already observed in $Mg_2Na_3P_5O_{16}$. Figure 2.5.3 shows a projection along

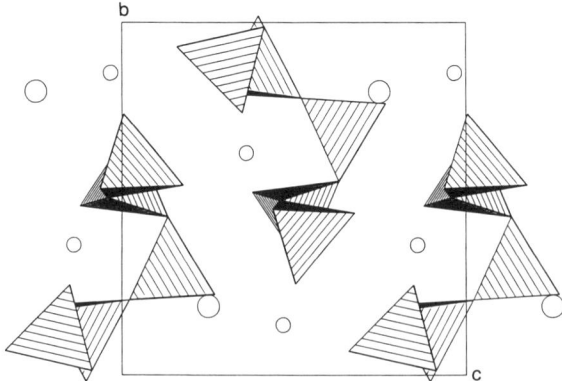

Figure 2.5.3. Projection of the structure of $V_2CsP_5O_{16}$ along the a direction. The large circles represent the cesium atoms, and the smaller ones the vanadium atoms.

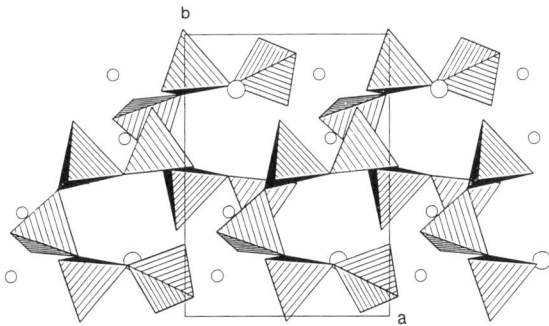

Figure 2.5.4. Projection of the structure of $V_2CsP_5O_{16}$ along the c direction. The large circles represent the cesium atoms, and the smaller ones the vanadium atoms.

the a direction of the respective locations of the phosphoric groups and associated cations in the unit cell, while Fig. 2.5.4 depicts the same arrangement in projection along the c axis. The stacking of the P_5O_{16} groups, whose mean planes are almost perpendicular to the c direction, is probably easier to discern in the second drawing.

The two independent vanadium atoms have sixfold oxygen coordination, in a slightly distorted octahedral arrangement. Within the VO_6 groups, the V–O distances range from 1.914 to 2.057 Å. The cesium atom also has sixfold coordination, with Cs–O distances ranging between 3.108 and 3.310 Å.

The corresponding iron compound is isotypic.

2.5.3. The Pentaphosphate Anion

The three P_5O_{16} anions that are presently well known all have a more or less accentuated horseshoe conformation. One of them has no internal symmetry, one has twofold symmetry, and the last one has mirror symmetry. Table 2.5.2 summarizes the main geometrical features of these three P_5O_{16} groups. Here, as in the case of the tetraphosphates, the limited number of examples obviously precludes any meaningful discussion or comparison of the geometry of the P_5O_{16} groups.

Table 2.5.2. Geometry of the Pentaphosphate Anions

Formula	distance P–P (Å)	angle P–P–P (°)	angle P–O–P (°)	Symmetry	Reference(s)
$Mg_2Na_3P_5O_{16}$	2.949	120.0	136.7	Twofold	39, 500
	2.898	133.9	127.3		
$Ta_2Rb_2H(PO_4)_2(P_5O_{16})$	2.864	106.17	128.74	Mirror	502
	2.879	122.16	129.70		
$V_2CsP_5O_{16}$	2.875	115.40	128.0	None	501
	2.908	87.27	129.6		
	2.817	113.13	122.4		
	2.857		126.3		

3

Long-Chain Polyphosphates

3.1. Introduction

The highly polymerized type of anion found in what we denote as long-chain polyphosphates, has a formula approximating $(PO_3)_n^{n-}$. Sometimes, these anions are considered as having terminal OH groups and thus a formula $(H_2P_nO^{3n+1})^{n-}$. Long-chain polyphosphates have been the subject of a great number of physical investigations, mainly devoted to the estimation of their degree of polymerization. From these studies, which we do not report here, one arrives at the following two conclusions:

(a) The degree of polymerization is very high. Molecular weights estimated by various methods, such as sedimentation, dialysis, end-group titration, viscosity, and pH titration, can vary from 500,000 to several million.

(b) For a given compound, the values so obtained are mainly dependent on the thermal history of the specimen and on its stoichiometry. A very small departure from the theoretical O/P ratio of 3 can induce large variations in the degree of polymerization.

Although many authors have claimed to have characterized hydrated long-chain polyphosphates, no structural evidence has been obtained to date for the existence of such compounds. These claims remain to be confirmed by structural investigations.

Acidic long-chain polyphosphates are rare. The only reported examples are $(UO_2)H(PO_3)_3$, $Ba_2H_3(PO_3)_7$, and of a series of $M^{III}H(PO_3)_4$ polyphosphates for $M^{III} = Bi$ and Ln.

Some chemical features are common to all long-chain polyphosphates:

- They are relatively stable compounds. Under normal conditions of temperature and humidity they can be kept for many years in a perfect state of crystallinity.
- They are not water-soluble as may be inferred from their estimated molecular weights.
- They all produce glasses when heated to their melting points.

From a structural point of view, the great flexibility of this type of anion gives rise to a wide number of configurations for the phosphoric chains. Some of them are almost straight linkages of tetrahedra, as observed for instance in zinc polyphosphate $Zn(PO_3)_2$ (Fig. 3.1). Some others have zigzag arrangements of various amplitudes. A typical example of such an anion is observed in $BeK(PO_3)_3$ (Fig. 3.2). Less often, the phosphoric chains spiral around 2_1 or 3_1 helical axes, as is observed in $CoK(PO_3)_3$ (see Section 3.4).

These almost purely aesthetic considerations are not sufficient for even a rudimentary classification of long-chain anions. As we mentioned above, these chains are very long. In a crystalline polyphosphate, characterized, like all crystalline materials, by the periodic

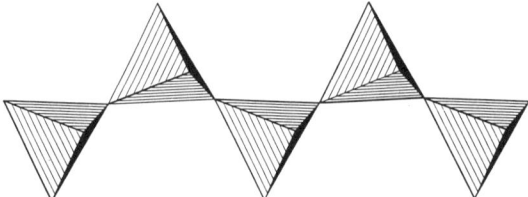

Figure 3.1. A phosphoric chain as observed in $Zn(PO_3)_2$.

repetition of a motif, it is evident that the chains themselves must be built up by the repetition of a fundamental unit composed of a group of PO_4 tetrahedra. This repeating unit is often called the "chain period," and one can distinguish or classify chains according to the number of tetrahedra in this basic unit. Attempts have been made to classify long-chain polyphosphates on the basis of this criterion. For a time, when the number of well-investigated chains was small, it was also assumed that the chain direction corresponded to the direction of the shortest unit-cell dimension. As more structural determinations were performed, not only was this last assumption, shown to be incorrect, but it also became apparent that there is no correlation between chain periods and chemical formulas, thus invalidating all attempts to elaborate a classification based upon this concept. It is still, nevertheless, a useful concept with which to illustrate the great flexibility of these macroanions.

For a time, also, it appeared that the longest chain periods were observed in polyphosphates having unusual chemical formulas. Thus, long periods of 10 tetrahedra and 14 tetrahedra were observed in $Ba_2K(PO_3)_5$ and $Ba_2Li_3(PO_3)_7$, respectively, in fact corresponding to uncommon formulas in this field. This possible relation was rapidly invalidated by the discovery of a period of 16 tetrahedra in $YNH_4(PO_3)_4$, a compound with a very simple and common chemical formula.

Attempts were also made to evaluate the degree of stretching of a given chain by comparing the number of tetrahedra in the repeating unit to the length of this unit. Liebau[503] had applied this concept to silicates and was able to find correlations between this stretching

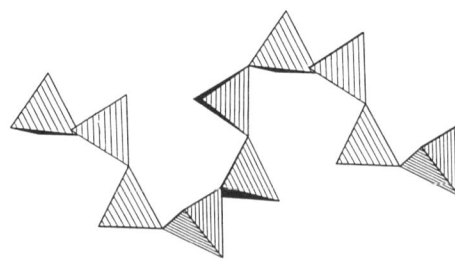

Figure 3.2. A corrugated phosphoric chain as observed in $BeK(PO_3)_3$.

factor and the nature of the associated cations. In the field of phosphates, and probably because of the more restricted number of examples, such attempts were inconclusive.

Thus, in the present state of knowledge in the field of long-chain polyphosphates, no correlation has been clearly established between the nature of the associated cations and the geometry of the corresponding chain.

To date, organic derivatives have not been observed in this family of phosphates.

3.2. Present State of Long-Chain Polyphosphate Chemistry

3.2.1. Alkali and Monovalent Cation Long-Chain Polyphosphates

(i) M^IPO_3 polyphosphates

The most common way to prepare M^IPO_3 polyphosphates is by the thermal dehydration of the corresponding dihydrogenomonophosphate according to the following scheme:

$$MH_2PO_4 \rightarrow MPO_3 + H_2O$$

This process was used as early as 1833 by Graham, who first described $NaPO_3$. Over the years, the preparation of $NaPO_3$ and KPO_3, in most cases by this route gave rise to a very abundant but often confusing chemical literature. A large number of more or less water-soluble varieties of $NaPO_3$ were described. In most cases, distinction between these various forms was based upon such properties as their solubility, behavior toward various chemical reagents, state of crystallinity, and fibrous or non fibrous nature. Various names were used to denote these forms, known for a long time as Graham, Maddrell, and Kurrol salts. In these early investigations, the polymeric nature of the anion was evidently not suspected, and only recently were most of these forms recognized as mixtures of various sodium polyphosphates with different chain lengths, the thermal history of the sample being the main parameter controlling the distribution of the chain lengths. Thus, the application of modern techniques of structural analysis reduced considerably the number of well-established chemical species in this field; for instance, today the existence of only two crystalline forms of long-chain sodium polyphosphate is clearly established. Thus, in the following survey we will not present an exhaustive review of the tremendous amount of chemical literature dealing with some of these polyphosphates, but instead we will examine only what is now firmly established through structural investigations.

$LiPO_3$. The normal way to prepare $LiPO_3$ is by the thermal dehydration of LiH_2PO_4 according to the scheme reported above. The nature of the intermediate compounds observed during this process has been the subject of some controversy. Beglov[504] compared the various proposed schemes and performed a thermodynamic analysis of this reaction. Recently, Schülke and Kayser[505] published a detailed study on the various possible paths of this reaction.

The existence of two crystalline forms of $LiPO_3$ based on X-ray diffraction patterns was first reported by Thilo and Grunze.[506] According to these authors, the low-temperature form is isotypic with $LiAsO_3$, whereas the high-temperature form is similar to Li_2SiO_3. These two forms were found to exist below and above 523 K. Grenier and Durif[507] were the first to determine the unit-cell dimensions of $LiPO_3$ from a twinned crystal. Untwinned crystals were later obtained during a study of the P_2O_5–CdO–Li_2O system by flux methods, and the crystal structure of this salt was then determined by Guitel and Tordjman.[508] Benkhoucha and Wunderlich[509] confirmed the dimorphism of this salt and identified the high-temperature form as the monoclinic salt investigated by Guitel and Tordjman.[508] No crystal data exist for the low-temperature form.

A theoretical study of the structure and stability of the $LiPO_3$ "molecule" was reported by Zakzhevskii et al.[510]

$NaPO_3$. Three crystalline forms of $NaPO_3$, denoted A, B and C, have been characterized to date.

Corbridge[511] measured unit cells and determined space groups for the A and B forms of $NaPO_3$. Later, Jost[512,513] determined crystal structures for these two forms, and a refinement of the crystal structure of the A form was performed by McAdam et al.[514] Recently, Immirzi and Porzio[515] reported the characterization of a new form of $NaPO_3$, obtained by prolonged grinding of the B form and denoted as the C form by the authors. They performed a crystal structure study of this new form based on unit-cell, space-group, and crystal data obtained from X-ray powder diffraction patterns and proposed two structural models, refined with two slightly different unit cells. Unfortunately, the work is confused and needs to be confirmed with more reliable data.

Two other salts have the global formula $NaPO_3$—a cyclotriphosphate, $Na_3P_3O_9$, and a cyclotetraphosphate, $Na_4P_4O_{12}$.

KPO_3. Andress and Fischer[516] identified KPO_3 as a long-chain polyphosphate and reported unit-cell dimensions of the room- temperature form, which were later confirmed by Corbridge.[511]

According to Jost,[517] KPO_3 undergoes phase transformations at 548, 733, and 923 K. These transformations are all reversible and do not exhibit any hysteresis. The phase stable up to 548 K is denoted KPO_3-(T), and its crystal structure was established by Jost.[517] The terminology used by Jost is as follows:

$$KPO_3\text{-(T)} \overset{548K}{\rightleftharpoons} KPO_3\text{-(Z)} \overset{733K}{\rightleftharpoons} KPO_3\text{-(H)} \overset{923K}{\rightleftharpoons} KPO_3\text{-(HT)}$$

The mechanisms of these transformations were investigated by Jost and Schulze,[518,519] and the crystal structure of the high-temperature form was described by the same authors.[519] Bekturov et al.[520] investigated the thermal behavior of KPO_3 and confirmed that this compound undergoes a phase transition at 733 K.

The effect of various factors on the kinetics of the thermal transformation of potassium cyclotriphosphate into potassium polyphosphate was reported by Dombrovskii and Koval.[521] The low-temperature heat capacity and entropy of KPO_3 were accurately measured by Egan and Wakefield.[522]

The same stoichiometric formula, KPO_3, is showed by a cyclotriphosphate, $K_3P_3O_9$.

$AgPO_3$. A crystal structure determination for $AgPO_3$ was performed by Jost.[523] $AgPO_3$ is isotypic with the A form of $NaPO_3$.

NH_4PO_3. Many attempts have been made to produce long-chain ammonium polyphospates. The thermal dehydration of ammonium monophosphates leads to a mixture of amorphous polyphosphates with a low ammonia content. Knorre,[20] Terem and Akalan,[524] and Margulis et al.[525] described some experiments of this type. Another approach involving the treatment of divalent cation polyphosphates with ammonium sulfide, was reported by Tammann[16,17] and Kiehl and Hill.[526] The products so obtained were later proved to be impure by Thilo and Grunze.[506] The formation of long-chain ammonium polyphosphates by the action of ammonia on polyphosphoric acids (80–90% P_2O_5) was investigated by Kopilevich and Shchegrov.[527]

Crystalline ammonium polyphosphates with a chain length greater than 50 were characterized by Shen et al.[528] They identified five forms of NH_4PO_3 and were able to produce crystals for three of them. Forms II, III, IV, and V were produced by various thermal treatments of form I; the latter form was obtained by heating an equimolar mixture of $NH_4H_2PO_4$ and urea under anhydrous ammonia at 453 K for 16 h. A detailed description of the various thermal processes used to transform form I into the other four forms was reported by the authors. The unit-cell dimensions of form IV are similar to those of KPO_3. These compounds are not water-soluble.

Two other compounds have the stoichiometric formula $(NH_4)PO_3$—a cyclotriphosphate, $(NH_4)_3P_3O_9$, and a cyclotetraphosphate, $(NH_4)_4P_4O_{12}$.

$RbPO_3$, $CsPO_3$, and $TlPO_3$. These three polyphosphates are isotypic. Corbridge[511] was the first to report the unit-cell dimensions and space groups for the first two compounds and showed them to be isotypic. He determined the crystal structure of the rubidium salt,[31] which was later reexamined by Cruickshank.[529]

The behavior of rubidium and cesium polyphosphates is quite similar to that described above for the potassium salt. During investigations of the thermal behavior of rubidium and cesium cyclohexaphosphate hexahydrates, Chudinova et al.[530] reported the existence of three new forms of rubidium polyphosphate and of one high-temperature form for the cesium salt. The authors reported unit-cell dimensions obtained from powder diagrams. For the high-temperature form of the cesium polyphosphate, they suggested an orthorhombic unit cell on the basis of a powder diagram. This form seems to be isotypic with $RbPO_3$ (HT). Beglov[504] did a careful thermodynamic analysis of the reaction leading to rubidium and cesium polyphosphates by thermal dehydration of the corresponding dihydrogenomonophosphates.

The existence of thallium polyphosphate was recognized by Dostal et al.[531] Thallium cyclotetraphosphate, $Tl_4P_4O_{12}$, the stable low-temperature form, converts into $TlPO_3$ at about 690 K. El-Horr[532] prepared single crystals of $TlPO_3$ and found this salt to be isotypic with the room temperature form of $RbPO_3$ and $CsPO_3$. The volatilization and thermodynamic properties of $TlPO_3$ were investigated by Alikhanyan et al.[533] $TlPO_3$ also exists as a cyclotetraphosphate, $Tl_4P_4O_{12}$.

Crystallographic data. The main crystallographic data for monovalent cation long-chain polyphosphates are given in Table 3.1.

Melting Temperatures. An examination by Majling and Hanic[542] of the melting temperatures reported for M^IPO_3 long-chain polyphosphates in a number of phase-equilibrium diagrams involving them as components shows a wide variation in these values. We report below the temperature ranges:

$LiPO_3$	911–939 K	$NaPO_3$	893–903 K
$AgPO_3$	754–763 K	KPO_3	1076–1086 K
$RbPO_3$	1053–1090 K	$CsPO_3$	993–1013 K
$TlPO_3$	707–736 K		

According to Majling and Hanic, the spread of these values can be attributed mostly to variations in sample purity rather than to the methods of temperature measurement.

Table 3.1. Main Crystallographic Data for Monovalent Cation Long-Chain Polyphosphates

Formula[a]	a (Å) α (°)	b (Å) β (°)	c (Å) γ (°)	S.G.	Z	Reference(s)
$LiPO_3$	16.453(2)	5.405(1) 98.99(2)	13.086(2)	Pn	20	507, 508
$NaPO_3$(A)	12.12(4)	6.20(2) 92.0(5)	6.99(3)	$P2_1/n$	8	511, 512
$NaPO_3$ (B)	11.37	6.01 85.7	7.63	$P2_1/n$	8	511, 513
$NaPO_3$ (C)	13.177(5)	13.177(5)	5.940(3)	$I4_1/a$	?	515
$AgPO_3$	11.86(4)	6.06(2) 93.5(5)	7.31(3)	$P2_1/n$	8	523
KPO_3(T)	14.02(3)	4.54(1) 101.5(2)	10.28(2)	$P2_1/n$	8	511, 517
KPO_3(HT)	12.94(3)	4.54(2)	5.92(2)	$Bbmm$	4	519
NH_4PO_3(II)	4.256	6.475	12.04	$P2_12_12_1$	4	528
NH_4PO_3 (IV)	14.5	4.62 100.0	11.0	$P2_1/n$	8	528
NH_4PO_3 (V)	4.346	6.135	13.646	?	4	528
$RbPO_3$	12.123(2)	4.228(2) 96.32(30)	6.479(2)	$P2_1/n$	4	31, 529
$RbPO_3$(HT)	13.338(9)	4.577(2)	6.147(3)	?	?	530
$RbPO_3$ (II)	14.47(1)	4.628(9) 82.5(4)	11.16(2)	?	?	530
$RbPO_3$(H)	13.07(4)	4.670(9)	6.064(8)	$Bbmm$	4	530
$CsPO_3$	12.71	4.32 83.0	6.83	$P2_1/n$	4	511
$CsPO_3$ (HT)	13.85(2)	4.64(20	6.508(9)	$Bbmm$	4	530
$TlPO_3$	12.270(7)	4.263(2) 96.72(3)	6.328(4)	$P2_1/n$	4	532

[a]In parentheses are given the designations used by the original authors to denote the various polymorphs of a given salt. HT: high-temperature form.

(ii) Mixed-Monovalent Cation Long-Chain Polyphosphates

Some mixed-monovalent cation polyphosphates have also been described. Most of them were characterized during investigations of phase-equilibrium diagrams.

$LiPO_3$–MPO_3 (M = Na, Ag) systems. The $LiPO_3$–$NaPO_3$ phase-equilibrium diagram established by Mardirosova and Bukhalova[534] shows the existence of solid solutions with

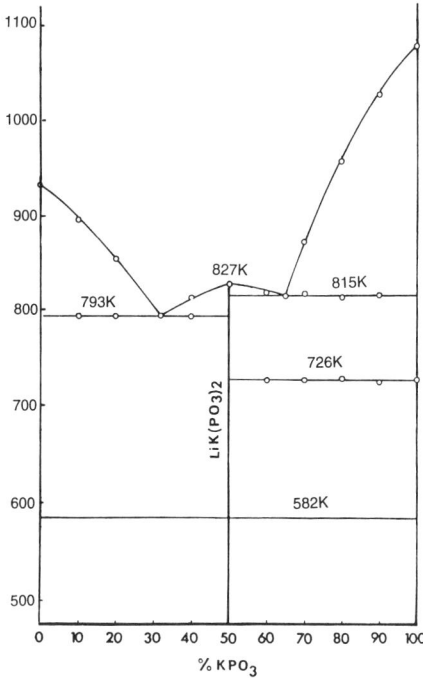

Figure 3.3. The LiPO₃–KPO₃ phase-equilibrium diagram. The $\alpha \rightarrow \beta$ transformation of LiK(PO₃)₂ at 582 K was not observed during the elaboration of the diagram. The transformation of KPO₃ is observed at 726 K.

a minimum on the solid-solutions curve at 50% LiPO₃ (749 K). The LiPO₃–AgPO₃ system investigated by Cavéro-Ghersi[535] is of the eutectic type (30% LiPO₃, 683 K).

LiK(PO₃)₂. The LiPO₃–KPO₃ phase-equilibrium diagram was first elaborated by Mardirosova and Bukhalova[534] and was later reinvestigated by El-Horr *et al.*[536] LiK(PO₃)₂, the only compound appearing in this system, has two crystalline forms with a transition temperature of 582 K and melts congruently at 827 K. The LiPO₃–KPO₃ phase-equilibrium diagram is presented in Fig. 3.3. Crystal structure determinations for both forms were performed by El-Horr and co-workers.[537,538]

Li₂NH₄(PO₃)₃, Li₂Rb(PO₃)₃, and Li₂Cs(PO₃)₃. The chemical preparation by a flux method and the crystal structure of the lithium–ammonium polyphosphate were reported by Averbuch-Pouchot *et al.*[539] Later, El-Horr and Bagieu-Beucher[540] prepared Li₂Rb(PO₃)₃ and Li₂Cs(PO₃)₃ and found them to be isotypic with this salt. Li₂Rb(PO₃)₃ and Li₂Cs(PO₃)₃ do not appear in the corresponding phase diagrams,[535] which are presented in Figs. 3.4. and 3.5.

LiCs(PO₃)₂ and LiRb(PO₃)₂. The LiPO₃–CsPO₃ phase-equilibrium diagram was first determined by Mardirosova and Bukhalova.[534] The authors established the existence of

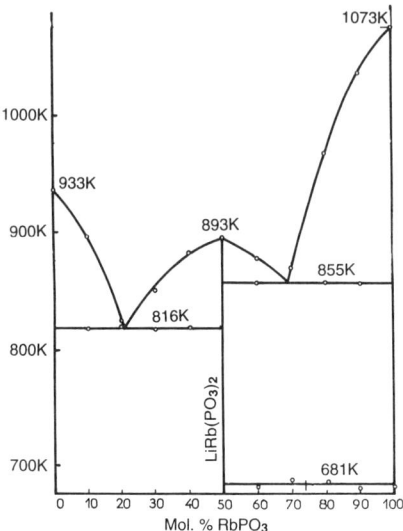

Figure 3.4. The LiPO₃–RbPO₃ phase-equilibrium diagram. The transformation of RbPO₃ appears at 681 K on the right-hand side of the diagram.

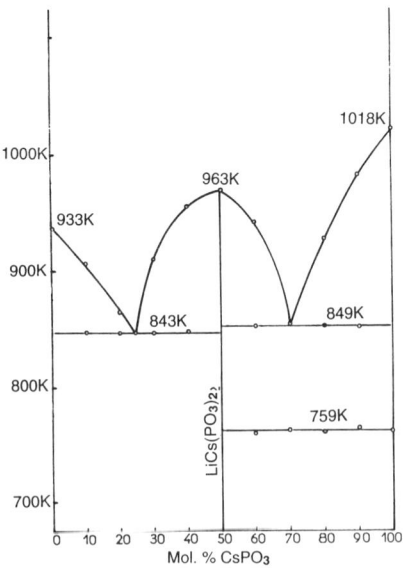

Figure 3.5. The LiPO₃–CsPO₃ phase-equilibrium diagram. The transformation of CsPO₃ appears at 759 K on the right-hand side of the diagram.

LiCs(PO$_3$)$_2$ but did not report any crystal data for this salt. This compound was characterized by Cavéro-Ghersi[535] during a new determination of the LiPO$_3$–CsPO$_3$ phase-equilibrium diagram. It melts congruently at 963 K and its atomic arrangement was later determined by El-Horr and Bagieu-Beucher.[541] LiRb(PO$_3$)$_2$ was characterized as an isotype of LiCs(PO$_3$)$_2$ by Cavéro-Ghersi during his determination of the LiPO$_3$–RbPO$_3$ phase-equilibrium diagram.[535] It melts congruently at 893 K.

LiTl(PO$_3$)$_2$ and Li$_2$Tl(PO$_3$)$_3$. During the elaboration of the LiPO$_3$–TlPO$_3$ phase-equilibrium diagram Cavéro-Ghersi[535] identified two compounds, LiTl(PO$_3$)$_2$ and Li$_2$Tl(PO$_3$)$_3$, decomposing at 723 and 645 K respectively. The only well-characterized compound in this system is Li$_2$Tl(PO$_3$)$_3$, identified by El-Horr and Bagieu-Beucher[540] as isotypic with Li$_2$NH$_4$(PO$_3$)$_3$. The phase-equilibrium diagram for this system is reproduced in Fig. 3.6.

NaAg(PO$_3$)$_2$. In most of the NaPO$_3$–M^{I}PO$_3$ phase-equilibrium diagrams that have been investigated to date, the only intermediate compounds are cyclophosphates, mainly cyclotriphosphates. These will be described in Chapter 4. The only exception seems to be the NaPO$_3$–AgPO$_3$ system investigated by Majling and Hanic[542] in which there exists a 1:1 intermediate compound, NaAg(PO$_3$)$_2$, melting congruently at 803 K. In a recent structural investigation of the M^{I}Ag(PO$_3$)$_2$ compounds, Averbuch-Pouchot[543] attempted to produce single crystals of NaAg(PO$_3$)$_2$ but always obtained crystals corresponding to various compounds of the solid solution between the two polyphosphates.

KAg(PO$_3$)$_2$, RbAg(PO$_3$)$_2$, CsAg(PO$_3$)$_2$ and TlAg(PO$_3$)$_2$. Savenkova *et al.*[544] elaborated the KPO$_3$–AgPO$_3$ phase-equilibrium diagram, (Fig. 3.8) and characterized KAg(PO$_3$)$_2$, the only compound in this system, as a long-chain polyphosphate melting incongruently at 703 K. Later, Averbuch-Pouchot[543] produced single crystals of

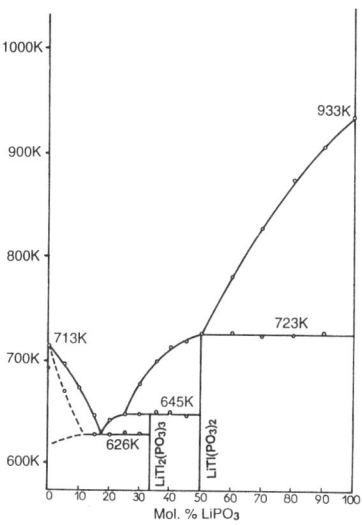

Figure 3.6. The LiPO$_3$–TlPO$_3$ phase-equilibrium diagram.

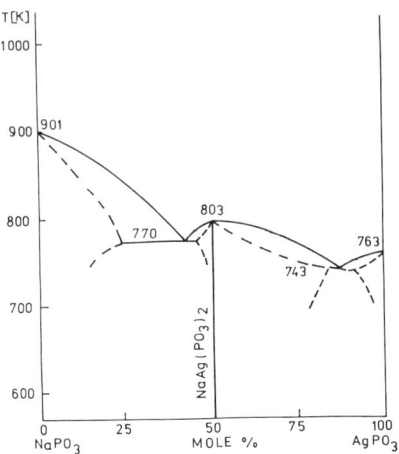

Figure 3.7. The NaPO₃–AgPO₃ system.

KAg(PO₃)₂ and determined the crystal structure of this salt, confirming its degree of polymerization.

The RbPO₃–AgPO₃ phase-equilibrium diagram (Fig. 3.9) established by Savenkova *et al.*[545] reveals the existence of RbAg(PO₃)₂, decomposing at 708 K. Using IR spectroscopy and paper chromatography, the authors established that this compound is a long-chain polyphosphate. Averbuch-Pouchot[543] showed it to be isotypic with KAg(PO₃)₂.

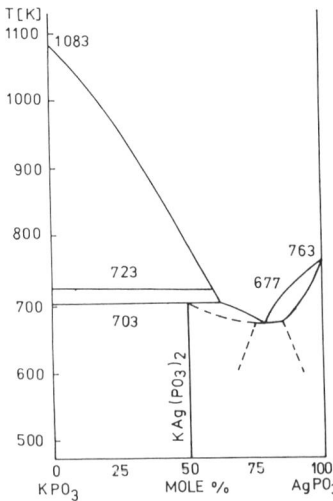

Figure 3.8. The KPO₃–AgPO₃ phase-equilibrium diagram. The $\alpha \to \beta$ transformation of KPO₃ appears at 723 K on the left-hand side of this diagram.

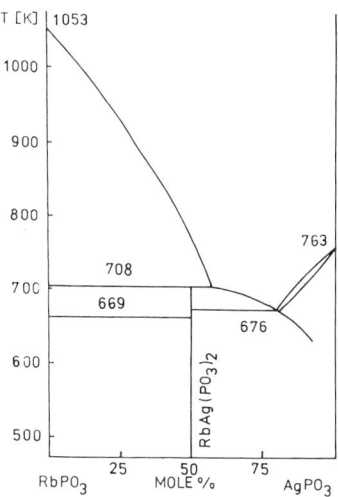

Figure 3.9. The RbPO₃–AgPO₃ phase-equilibrium diagram. The $\alpha \to \beta$ transformation of RbPO₃ appears at 669 K on the left-hand side of this diagram.

CsAg(PO₃)₂ and TlAg(PO₃)₂ were prepared by Averbuch-Pouchot[543] and identified as isotypes of the potassium–silver salt.

Crystallographic Data. The main crystallographic data for mixed-monovalent cation long-chain polyphosphates are given in Table 3.2.

Table 3.2. Main Crystallographic Data for Mixed-Monovalent Cation Polyphosphates

Formula	a (Å) α (°)	b (Å) β (°)	c (Å) γ (°)	S.G.	Z	Reference(s)
α-LiK(PO₃)₂	13.740(8)	13.776(8) 90.14(4)	11.832(7)	$C2/c$	16	536, 538
β-LiK(PO₃)₂	15.31(2)	5.563(7) 107.96(4)	13.72(1)	$P2_1/a$	8	536, 537
KAg(PO₃)₂	7.490(6)	13.175(10) 94.32(6)	6.037(5)	$P2_1/n$	4	543
RbAg(PO₃)₂	7.512(6)	13.355(10) 95.76(5)	6.183(5)	$P2_1/n$	4	543
TlAg(PO₃)₂	7.436(6)	13.331(10) 96.13(5)	6.161(5)	$P2_1/n$	4	543
CsAg(PO₃)₂	7.398(6)	13.602(10) 97.42	6.434(5)	$P2_1/n$	4	543
Li₂NH₄(PO₃)₃	12.199(5)	13.047(5)	10.537(5)	Pbca	8	539
Li₂Rb(PO₃)₃	12.177(5)	12.964(5)	10.616(4)	Pbca	8	540
Li₂Cs(PO₃)₃	12.284(5)	13.011(7)	10.735(5)	Pbca	8	540
Li₂Tl(PO₃)₃	12.039(7)	12.898(7)	10.556(7)	Pbca	8	540
LiRb(PO₃)₂	19.362(9)	19.025(7)	13.231(7)	Fdd2	32	535, 541
LiCs(PO₃)₂	19.440(5)	19.027(5)	13.222(5)	Fdd2	32	535, 541

(iii) Miscellaneous Investigations Involving Monovalent Cation Polyphosphates

A large number of investigations involving MPO_3 compounds have been performed. These include studies of the chemical interactions of these compounds with other phosphates and with nonphosphate compounds and measurements of the thermal stability, surface tension and various other properties. of these compounds. We review some of these studies below .

The surface tension, density, viscosity, and conductivity of molten $LiPO_3$, $NaPO_3$, KPO_3 and $CsPO_3$ were measured by Sokolova[546] and by Sokolova et al.[547] The thermal stability of alkali metal polyphosphates was studied by Steblevskii et al.[548] The authors used the Knudsen effusion method and analyzed the volatilization products by mass spectrometry. The stabilities decrease in the order Li > Na > K > Rb > Cs.

The reactions of solid potassium Kurrol salt with solutions of various metal salts were investigated by Ohashi and Yamagishi.[549] Particular attention was given to the interaction with nickel sulfate aqueous solutions. A reversible cation-exchange reaction occurs in this system.

The reaction between germanium dioxide and molten lithium, sodium, and potassium polyphosphates was investigated by Slobodyanik et al.[550] In all cases, they obtained the rhombohedral $MGe_2(PO_4)_3$ compounds.

The reaction of molten $NaPO_3$ and KPO_3 with NaCl at 1073K has been investigated by Markina and Voskresenskaya.[551] They observed the formation of gaseous products— HCl, Cl_2, and $POCl_3$, the latter in a very small quantity. Various parameters influencing the reaction, such as atmosphere, concentration of NaCl, and duration of the experiments, were examined. $NaPO_3$ is more reactive than KPO_3.

The kinetics of the reaction between molten $LiPO_3$, $NaPO_3$, and KPO_3 and the corresponding sulfates were investigated between 973 and 1023 K by Kochergin et al.[552]

The $NaF-NaPO_3$ and $KF-KPO_3$ phase-equilibrium diagrams were elaborated by Bukhalova and Mardirosova.[553] These authors reported the existence of four intermediate compounds: $NaF \cdot NaPO_3$, $2NaF \cdot NaPO_3$, $KF \cdot KPO_3$ and $2KF \cdot KPO_3$. The same authors[554] investigated the $LiCl-LiPO_3$, $NaCl-NaPO_3$, $KCl-KPO_3$, and $CsCl-CsPO_3$ systems. In the lithium system a double compound, $LiPO_3 \cdot 2LiCl$, is formed in the solid state at 777 K and exhibits a polymorphic transformation at 638 K. The $NaCl-NaPO_3$ system is of the eutectic type, with a compound, $NaPO_3 \cdot 2NaCl$, forming at 778 K in the solidus. For the $KF-KPO_3$ system the authors also obtained a eutectic diagram but showed the existence of an intermediate compound, $KPO_3 \cdot 2KCl$, forming at 673 K in the solidus. In a previous investigation by Amadori[555] this system was also found to be of the eutectic type. For the cesium system, the results are similar to those obtained in the case of potassium, with the probable formation of $CsPO_3 \cdot 2CsCl$ at 673 K.

Equilibrium chemical transformations in $NaPO_3-NaCl$ melts were investigated by Kovarskaya and Rodionov.[556] With increasing NaCl content in the mixture, the sodium polyphosphate gradually depolymerizes to tri-, di-, and monophosphate and the composition of the equilibrium melt is only dependent on the initial composition. The time necessary to reach the equilibrium is shortened when the temperature is increased.

The reaction of metal oxides with molten $NaPO_3$–$NaCl$ mixtures was studied by Kovarskaya et al.[557] Detailed results were given in the case of Fe_2O_3–$NaPO_3$–$NaCl$ mixtures.

3.2.2. Divalent Cation Long-Chain Polyphosphates

(i) $M^{II}(PO_3)_2$ Polyphosphates

$Be(PO_3)_2$. At present, at least four crystalline forms of $Be(PO_3)_2$ are known. All of them are long-chain polyphosphates. Bleyer and Müller[104] conducted the first investigations in this field. By firing BeO in a large excess of "metaphosphoric acid," they obtained a crystalline compound corresponding to the formula $Be(PO_3)_2$. Thilo and Grunze[558] concluded from their paper chromatography experiments, that this salt is very probably a long-chain polyphosphate. Later, Jaulmes,[100,559] in a general study of beryllium phosphates, showed the existence of two different crystalline forms of $Be(PO_3)_2$; the first form denoted (I), is observed at temperatures lower than 673 K, and the second one (II) at higher temperatures.

Bagieu-Beucher and Durif[560] reported the chemical preparation of $Be(PO_3)_2$ (II) single crystals and determined the unit-cell dimensions and space group of this compound from a twinned crystal. Averbuch-Pouchot et al.[561] described a procedure for obtaining untwinned crystals and performed the determination of the atomic arrangement.

A third form of $Be(PO_3)_2$ was characterized by Schultz[562] and denoted (III). During this study, it was observed that this form has itself two modifications, with a transition temperature at 369 K. The crystal structure of the high-temperature modification was determined at 392 K by Schultz and Liebau[563,564]; its atomic arrangement is closely related to that of the keatite form of silica.

$M^{II}(PO_3)_2$ (M^{II} = Ni, Co, Mg, Cu, Fe, Mn, Zn). Bagieu-Beucher et al.[565] reported the chemical preparation and crystal data for seven isotypic compounds corresponding to the high-pressure form of the $M^{II}_2P_4O_{12}$ cyclotetraphosphates. This high-pressure form had been for a long time considered a cyclotetraphosphate. A crystal structure determination showing that compounds of this type are in fact long-chain polyphosphates was performed by Averbuch-Pouchot et al.[566] using the zinc salt, which can be produced as single crystals by a flux method under atmospheric pressure. Another form of $Zn(PO_3)_2$ has been characterized by Katnack and Hummel.[106] This form was also produced by Schultz[562] by heating $Zn(PO_3)_2$ glass at 1073 K, for several days. He obtained in this way monoclinic crystals and reported crystal data. During this investigation the author observed by DTA a strong endothermic peak at 1048 K, which he attributed to the transformation:

$$Zn_2P_4O_{12} \rightarrow 2Zn(PO_3)_2$$

This transformation is not reversible.

The surface tension of molten $Zn(PO_3)_2$ was investigated by Krivovyazov et al.[567] According to these authors, the transformation of zinc cyclotetraphosphate into polyphosphate occurs at 913 K.

$Ca(PO_3)_2$. The CaO–P_2O_5 diagram established by Hill et al.[47,568] shows the existence of two crystalline forms of $Ca(PO_3)_2$. In this diagram, $Ca(PO_3)_2$ appears as a congruent-melting compound (mp = 1250 K). This system was also investigated in an earlier more general study by Trömel[569] dealing with the $CaO-P_2O_5-SiO_2$ system, which was extended by Trömel et al.[570] Additional information can be obtained from a study of the phase relationships in the system $CaO-Al_2O_3-P_2O_5$ by Stone et al.[571]

Thilo and Grunze[572] investigated the thermal behavior of $Ca(H_2PO_4)_2 \cdot H_2O$ and reported the formation of one cyclotriphosphate, one cyclotetraphosphate and two forms of long-chain polyphosphates during the course of the thermal transformation of this salt. Morin[573] investigated this same reaction, and reported the existence of a third crystalline form of $Ca(PO_3)_2$. In discussing the properties of condensed calcium phosphates, Ohashi and Van Wazer[49] distinguished four types of $Ca(PO_3)_2$.

Despite these numerous studies the only valuable structural data obtained to date concern the β form, which is the stable form at room temperature. Corbridge[511] was the first to determine the unit-cell dimensions and space group of this form, which is isotypic with the corresponding strontium and lead salts. Its atomic arrangement was refined by Schneider et al.[574] and later, reexamined by Rothammel et al.[575]

Thermodynamic properties of β-$Ca(PO_3)_2$ were measured by Egan and Wakefield[576] over the range 373 to 1370 K and equations for the thermal properties of this material were derived.

The hydrolytic degradation of glassy $Ca(PO_3)_2$ was investigated by Brown et al.[577] The authors reported crystal data for the various crystalline intermediate compounds, mainly calcium diphosphates. The same kind of study was also performed by Huffman and Fleming.[578] These types of investigations were mainly run in order to explore the possible use of condensed calcium phosphates as fertilizers.

$Cd(PO_3)_2$. Brown and Hummel[579] were the first to report the existence of two crystalline forms for this polyphosphate. The transformation of α-$Cd(PO_3)_2$ to β-$Cd(PO_3)_2$ occurs at 1008 K and is reversible. $Cd(PO_3)_2$ melts at 1143 K.

Thilo and Grunze[558] investigated the reorganization of $Cd(H_2PO_4)_2 \cdot 2H_2O$ upon thermal dehydration and observed the formation of $Cd(PO_3)_2$ and $Cd_2P_4O_{12}$.

The chemical preparation of and crystal data for the low-temperature form of $Cd(PO_3)_2$ were reported by Beucher and Tordjman.[580] This salt is isotypic with $Hg(PO_3)_2$. Its crystal structure determination was first performed by Tordjman et al.[581] and later refined by Bagieu-Beucher et al.[582]

Single crystals of the high-temperature (β) form, were obtained by Laügt et al.[583] by annealing a $Cd(PO_3)_2$ glass for some days at 1113 K. This form is stable for several months at room-temperature but transforms rapidly to the low-temperature form upon heating at 673 K. Its crystal structure was later determined by Bagieu-Beucher et al.[584]

$Sr(PO_3)_2$. Kreidler and Hummel[141] investigated the phase relationships in the $SrO-P_2O_5$ system and characterized two crystalline forms of $Sr(PO_3)_2$. They observed the transition temperature at 1078 K and the congruent melting temperature at 1255 K.

During earlier studies of the thermal behavior of $Sr(H_2PO_4)_2$ by Ropp and co-workers[140,585] three forms of $Sr(PO_3)_2$ were identified. According to these authors, the following behavior is observed:

The low-temperature form is possibly a cyclotriphosphate.[140,585]

$$Sr(H_2PO_4)_2 \xrightarrow{463-483 \text{ K}} SrH_2P_2O_7 + H_2O$$

$$SrH_2P_2O_7 \xrightarrow{593-603 \text{ K}} \gamma\text{-Sr(PO}_3)_2 \xrightarrow{673 \text{ K}} \beta\text{-Sr(PO}_3)_2 \xrightarrow{1123 \text{ K}} \alpha\text{-Sr(PO}_3)_2$$

Thilo and Grunze[572] also investigated the thermal condensation of $Sr(H_2PO_4)_2$ and found its course very similar to that observed in the case of calcium.

$Sr(PO_3)_2$ produced by calcination of $Sr_3(P_3O_9)_2 \cdot 7H_2O$ at temperatures higher than 873 K was identified by Durif et al.[586] as the β form, isotypic with calcium and lead polyphosphates.

$Ba(PO_3)_2$. Thilo and Grunze[572] identified $Ba(PO_3)_2$ during the study of the thermal dehydration condensation of $Ba(H_2PO_4)_2$. This polyphosphate is polymorphic. Grenier and Martin[587] reported the existence of three forms of $Ba(PO_3)_2$—the normal β form already described, a high-temperature form called "α," and a third form (γ) accidentally produced during their investigations. According to these authors,

$$\beta\text{-Ba(PO}_3)_2 \xrightarrow{1058 \text{ K}} \alpha\text{-Ba(PO}_3)_2$$

$$\gamma\text{-Ba(PO}_3)_2 \xrightarrow{978 \text{ K}} \beta\text{-Ba(PO}_3)_2$$

Crystal data for the γ form were reported.

The crystal structure of the orthorhombic β form of $Ba(PO_3)_2$ was determined by Grenier et al.,[588] and that of the γ form by Coing-Boyat et al.[589]

The preparation of barium polyphosphate by thermal dehydration of $Ba(H_2PO_4)_2$ was investigated by Kuz'menkov et al.[590] using paper chromatography, thermography, dilatometry, and X-ray diffraction, they showed the formation of intermediate compounds between the monophosphate and the polyphosphate during the condensation.

Lopatin and Semenov[591] investigated the thermal dissociation of $Ba(PO_3)_2$ by mass spectroscopy. Surface tension measurements for $Ca(PO_3)_2$, $Sr(PO_3)_2$, and $Ba(PO_3)_2$ were reported by Sokolova et al.[547]

$Pb(PO_3)$. $Pb(PO_3)_2$ was identified as a long-chain polyphosphate by Andress and Fischer[516] during a study of the thermal behavior of $Pb(H_2PO_4)_2$. Its atomic arrangement was determined by Jost.[592] It is isotypic with the β forms of calcium and strontium polyphosphate.

$Hg(PO_3)_2$. $Hg(PO_3)_2$ was described for the first time and identified as a long-chain polyphosphate by Thilo and Grunze[558] who measured the melting point of this salt at 853 K and suggested that it was possibly isomorphous with the low-temperature form of cadmium polyphosphate, $Cd(PO_3)_2$. The crystallographic investigations of $Hg(PO_3)_2$ by Beucher[593] and Beucher and Tordjman[580] confirmed this assumption. Polycrystalline $Hg(PO_3)_2$ is difficult to synthesize by normal thermal methods, but crystals can be prepared by firing a mixture of $(NH_4)_2HPO_4$ and HgO or $HgCO_3$ with P/Hg = 10 at 723 K for 24 h.

or a mixture of H_3PO_4 (85%) and HgO or $HgCO_3$ with P/Hg = 10 at 623 K for 15 h. The crystals are colorless rectangular or diamondlike thick plates.[580,593]

$Pd(PO_3)_2$. Palkina *et al.*[594] synthesized single crystals of $Pd(PO_3)_2$ by heating $PdCl_2$ and H_3PO_4 at 553 K and determined the crystal structure of this salt.

Crystallographic Data. The main crystal data for divalent cation long-chain polyphosphates are given in Tables 3.3 and 3.4.

Melting Temperatures. We discussed above of an examination made by Majling and Hanic[542] of the melting temperatures of monovalent cation long-chain polyphosphates. A similar survey reported by the same authors for some $M(PO_3)_2$ long-chain polyphosphates shows, as in the case of the monovalent cations, a wide variation in these values. We report below some of the temperature ranges:

$Ca(PO_3)_2$	1237–1251 K	$Zn(PO_3)_2$	1126–1153 K
$Sr(PO_3)_2$	1255–1277 K	$Cd(PO_3)_2$	1150–1169 K
$Ba(PO_3)_2$	1123–1153 K	$Pb(PO_3)_2$	937–953 K

Again, these authors attributed, this spread in the values mostly to the variations in sample purity rather than to the methods of temperature measurements.

Table 3.3. Crystallographic Data for Divalent Cation Polyphosphates[a]

Formula	a (Å) α (°)	b (Å) β (°)	c (Å) γ (°)	S.G.	Z	Reference(s)
$Be(PO_3)_2$ II	6.966(4)	12.875(8) 106.73(2)	4.844(3)	$P2_1/n$	4	560, 561
$Be(PO_3)_2$ III $(HT)^b$	9.968(4)	10.080(4)	8.692(2)	$C222_1$	8	562, 564
$Be(PO_3)_2$ III $(LT)^c$	14.063(2)	8.629(2) 90.80(2)	7.091(1)	$P2_1$	8	562, 564
$Ni(PO_3)_2$	9.609	8.743 108.26	4.980	$C2/c$	2	565
$Mg(PO_3)_2$	9.661	8.835 108.02	4.977	$C2/c$	2	565
$Cu(PO_3)_2$	9.820	8.899 109.69	4.491	$C2/c$	2	565
$Co(PO_3)_2$	9.730	8.872 108.15	4.978	$C2/c$	2	565
$Zn(PO_3)_2$	9.734(2)	8.889(2) 108.49(5)	4.963(1)	$C2/c$	2	565, 566
$Fe(PO_3)_2$	9.777	8.994 107.22	4.968	$C2/c$	2	565
$Mn(PO_3)_2$	9.943	9.144 107.23	4.968	$C2/c$	2	565
$Zn(PO_3)_2$	7.644(1)	7.618(1) 92.23(3)	16.355(2)	$C2/c$	4	562

[a]See also Table 3.4.
[b]HT: High-temperature form.
[c]LT: Low-temperature form.

Table 3.4. Crystallographic Data for Divalent Cation Polyphosphates[a]

Formula	a (Å) α (°)	b (Å) β (°)	c (Å) γ (°)	S.G.	Z	Reference(s)
$Ca(PO_3)_2$	6.9963(2)	7.7144(2) 90.394(5)	16.960(9)	$P2_1/c$	8	511, 575
$Sr(PO_3)_2$	7.204(4)	7.936(4) 90.64(5)	17.40(5)	$P2_1/c$	8	586
$Pb(PO_3)_2$	7.318(4)	7.997(4) 91.37(5)	17.42(2)	$P2_1/c$	8	592
$Cd(PO_3)_2(LT)^b$	9.607(3)	13.70(1)	7.037(3)	$Pbca$	8	580–582
$Cd(PO_3)_2(HT)^c$	7.428(2)	7.360(2)	8.577(2)	$P2_12_12_1$	4	583, 584
$Hg(PO_3)_2$	9.720(4)	13.75(1)	7.134(3)	$Pbca$	8	580
$Pd(PO_3)_2$	4.233(3)	4.571(1)	12.472(3)	$Pcnm$	2	594
β-$Ba(PO_3)_2$	4.510(2)	13.44(2)	8.360(5)	$P2_12_12_1$	4	588
γ-$Ba(PO_3)_2$	9.692(8)	6.894(4) 94.77(3)	7.512(5)	$P2_1/n$	4	587, 589

[a]See also Table 3.3.
[b]LT: Low-temperature form.
[c]HT: High-temperature form.

(ii) Mixed-Divalent Cation Long-Chain Polyphosphates

Several mixed-divalent cation long-chain polyphosphates have been characterized, most of them during various investigations of $M(PO_3)_2$–$M'(PO_3)_2$ phase-equilibrium diagrams.

$Ba_2Cu(PO_3)_6$ and $Ba_2Zn(PO_3)_6$. $Pb(PO_3)_2$–$Cu_2P_4O_{12}$, $Sr(PO_3)_2$–$Cu_2P_4O_{12}$, and $Cu_2P_4O_{12}$–$Ba(PO_3)_2$ phase-equilibrium diagrams were elaborated by Laügt.[595] The first two are of the eutectic type [921 K, 20% $Pb(PO_3)_2$ and 1056 K, 40% $Sr(PO_3)_2$], but the third one shows the existence of $Ba_2Cu(PO_3)_6$, a congruent-melting (1100 K) long-chain polyphosphate. This phase-equilibrium diagram is presented in Fig. 3.10. The atomic arrangement of $Ba_2Cu(PO_3)_6$ was later determined by Laügt and Guitel.[596]

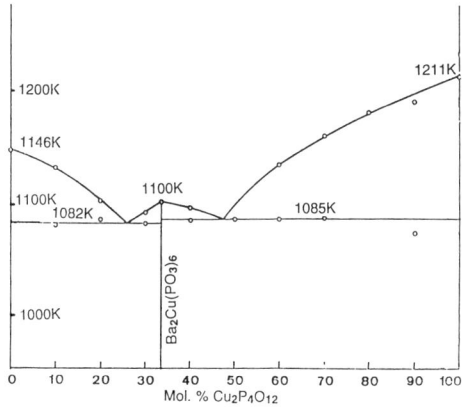

Figure 3.10. The $Ba(PO_3)_2$–$Cu_2P_4O_{12}$ phase-equilibrium diagram.

The $Zn(PO_3)_2$–$Ba(PO_3)_2$ phase-equilibrium diagram was first determined by Mardirosova *et al.*[597] As elaborated by these authors, it shows the existence of only one compound, $BaZn(PO_3)_4$, said to be a long-chain polyphosphate. A revised version of this diagram by Bagieu-Beucher and El-Horr[598] shows that the only compound in this system is, in fact, $Ba_2Zn(PO_3)_6$, a long-chain polyphosphate isotypic with $Ba_2Cu(PO_3)_6$. $Ba_2Zn_3P_{10}O_{30}$, a cyclodecaphosphate discovered during this investigation does not appear in the phase-equilibrium diagram.

$BaMn(PO_3)_4$, $BaCd(PO_3)_4$, $BaCa(PO_3)_4$, and $BaHg(PO_3)_4$. The $Ba(PO_3)_2$–$Cd(PO_3)_2$ and $Ba(PO_3)_2$–$Ca(PO_3)_2$ phase-equilibrium diagrams were elaborated by Bukhalova *et al.*[599] $BaCd(PO_3)_4$ and $BaCa(PO_3)_4$, both congruent-melting long-chain polyphosphates (1089 K for Cd and 1153 K for Ca), were observed in these systems. The existence of these two isotypic salts was confirmed by Averbuch-Pouchot[600] who reported their main crystallographic features as well as those of two other isotypic compounds, $BaMn(PO_3)_4$ and $BaHg(PO_3)_4$. The atomic arrangement in this series of compounds was determined by Averbuch-Pouchot *et al.*[601] using the cadmium salt. An IR spectroscopic investigation of $BaCa(PO_3)_4$ and $BaCd(PO_3)_4$ was reported by Tokman and Poletaev.[602]

Crystallographic Data. The main crystallographic features of mixed-divalent cation polyphosphates are given in Table 3.5.

(iii) Other $M^{II}(PO_3)_2$–$M^{II\prime}(PO_3)_2$ systems

Several other $M^{II}(PO_3)_2$–$M^{II\prime}(PO_3)_2$ systems have been investigated. Some of them show the existence of a continuous series of solid solutions; these include $Zn(PO_3)_2$–$Mg(PO_3)_2$ investigated by Sarver and Hummel,[603] $Cd(PO_3)_2$–$Ca(PO_3)_2$ investigated by Tokman and Bukhalova[604] and $Ca(PO_3)_2$–$Sr(PO_3)_2$, investigated by Bukhalova and Tokman.[605] Fluorescence properties were investigated in the case of the $Zn(PO_3)_2$–$Mg(PO_3)_2$ system.

Table 3.5. Crystallographic Data for Mixed-Divalent Cation Long-Chain Polyphosphates

Formula	a (Å) α (°)	b (Å) β (°)	c (Å) γ (°)	S.G.	Z	Reference(s)
$Ba_2Cu(PO_3)_6$	21.382(3)	7.286(1) 97.96(2)	9.520(1)	$P2_1/a$	4	595, 596
$Ba_2Zn(PO_3)_6$	21.52(1)	7.278(2) 97.83(3)	9.534(2)	$P2_1/a$	4	598
$BaMn(PO_3)_4$	14.69(1)	9.147(6) 90.57(1)	7.201(4)	$P2_1/n$	4	600
$BaCd(PO_3)_4$	14.94(1)	9.192(7) 90.79(1)	7.219(5)	$P2_1/n$	4	600, 601
$BaCa(PO_3)_4$	15.24(1)	9.173(7) 90.96(1)	7.231(5)	$P2_1/n$	4	600
$BaHg(PO_3)_4$	15.05(1)	9.236(8) 90.62(1)	7.239(6)	$P2_1/n$	4	600

The $Sr(PO_3)_2$–$Ba(PO_3)_2$ system investigated by Tokman and Bukhalova[604] is of the eutectic type [1075 K, 30% $Sr(PO_3)_2$].

The $Cd(PO_3)_2$–$Sr(PO_3)_2$ phase-equilibrium diagram elaborated by Tokman and Bukhalova[606] shows the existence of $CdSr(PO_3)_4$, assumed to be a long-chain polyphosphate. In the absence of structural data, the degree of condensation of this salt needs to be verified.

Phase equilibria in some portions of the $Zn(PO_3)_2$–$Mg(PO_3)_2$–$Cd(PO_3)_2$ system were determined by Brown and Hummel.[579] The manganese-activated luminescence was examined by the authors. In addition to this work, the $Zn(PO_3)_2$–$Cd(PO_3)_2$ system was investigated by Kuzmenkov *et al.*[607]

3.2.3. Divalent-Monovalent Cation Long-Chain Polyphosphates

The vast majority of divalent-monovalent cation polyphosphates have been characterized during the investigations of the $M^{II}(PO_3)_2$–$M^{I}PO_3$ phase-equilibrium diagrams. A careful and critical survey of these numerous diagrams has been published by Majling and Hanic,[542] so we will not reproduce all of them here. Instead, we will present a limited selection of the most representative ones. Most of the divalent–monovalent cation polyphosphates have been prepared as single crystals by various flux methods and their atomic arrangements determined with precision, but in most cases very little is known about their fundamental chemical and physical properties.

3.2.3.1. Divalent Cation–Lithium Polyphosphates

The divalent cation-lithium polyphosphates that have been investigated to date correspond to four different types of chemical formulas: $M^{II}Li(PO_3)_3$, $M^{II}Li_2(PO_3)_4$, $M^{II}_2Li(PO_3)_5$, and $M^{II}_2Li_3(PO_3)_7$. These four groups are considered individually below. The main crystallographic data for the divalent cation-lithium polyphosphates are gathered in Table 3.6.

(i) $M^{II}Li(PO_3)_3$ Polyphosphates (M^{II} = Cu, Fe, Zn, Mg, Ni, Co, Mn)

The seven $M^{II}Li(PO_3)_3$ polyphosphates corresponding to M^{II} = Zn, Cu, Mg, Fe, Ni, Co, and Mn are isomorphous. With the exception of the iron salt, all of them have been characterized during the elaboration of the $LiPO_3$–$M^{II}_2P_4O_{12}$ phase-equilibrium diagrams. The atomic arrangement in this class of compounds was first determined with the copper salt and reexamined with the iron salt.

$CuLi(PO_3)_3$. This compound appears as an incongruent-melting compound decomposing at 915 K in the $LiPO_3$–$Cu_2P_4O_{12}$ phase-equilibrium diagram elaborated by Laügt[608] (Fig. 3.11). $CuLi(PO_3)_3$ is the only compound observed in this system, and its low temperature form is a cyclohexaphosphate, $Cu_2Li_2P_6O_{18}$. The transformation

$$Cu_2Li_2P_6O_{18} \rightarrow 2CuLi(PO_3)_3$$

is not reversible and does not appear in the phase-equilibrium diagram. The crystal structure of $CuLi(PO_3)_3$ was determined by Laügt *et al.*[609]

$FeLi(PO_3)_3$. This compound was produced by hydrothermal synthesis at 523–723 K and 1500 atm during investigations of the system LiF–FeO–P_2O_5–H_2O by Genkina *et al.*[610,611]

Table 3.6. Main Crystallographic Data for Divalent Cation–Lithium Polyphosphates

Formula	a (Å) α (°)	b (Å) β (°)	c (Å) γ (°)	S.G.	Z	Reference(s)
CuLi(PO$_3$)$_3$	8.197(3)	8.613(3)	8.703(3)	$P2_12_12_1$	4	608, 609
MgLi(PO$_3$)$_3$	8.350(2)	8.527(2)	8.602(2)	$P2_12_12_1$	4	614
ZnLi(PO$_3$)$_3$	8.320(2)	8.517(2)	8.623(2)	$P2_12_12_1$	4	614
FeLi(PO$_3$)$_3$	8.362(2)	8.573(1)	8.690(2)	$P2_12_12_1$	4	614
NiLi(PO$_3$)$_3$	8.473(3)	8.550(3)	8.295(3)	$P2_12_12_1$	4	615
MnLi(PO$_3$)$_3$	8.673(3)	8.755(3)	8.447(3)	$P2_12_12_1$	4	612
CoLi(PO$_3$)$_3$	8.624(3)	8.349(3)	8.541(3)	$P2_1\underline{2}_12_1$	4	616
PbLi(PO$_3$)$_3$	7.245(3) 100.76(5)	7.409(3) 97.96(5)	6.795(3) 83.74(5)	$P\overline{1}$	2	620
CaLi(PO$_3$)$_3$	6.963(1) 99.37(5)	7.381(3) 98.27(5)	6.719(2) 83.79(5)	$P\overline{1}$	2	617
SrLi(PO$_3$)$_3$	7.163(4) 100.07(5)	7.360(4) 98.49(5)	6.767(4) 83.59(5)	$P\overline{1}$	2	618
MnLi$_2$(PO$_3$)$_4$	9.248(5)	10.09(1)	9.403(5)	$Pmcn$	4	612
CdLi$_2$(PO$_3$)$_4$	9.495(3)	10.15(1)	9.375(3)	$Pnam$	4	621, 622
HgLi$_2$(PO$_3$)$_4$	9.445(3)	9.983(3) 91.89(3)	9.518(3)	$P2_1/c$	4	622, 623
Pb$_2$Li(PO$_3$)$_5$	12.289(10)	9.689(8) 91.01(5)	5.523(5)	$P2/n$	2	624
Ba$_2$Li$_3$(PO$_3$)$_7$	18.014(8)	8.535(3) 104.48(2)	11.584(5)	$P2_1/a$	4	626, 627

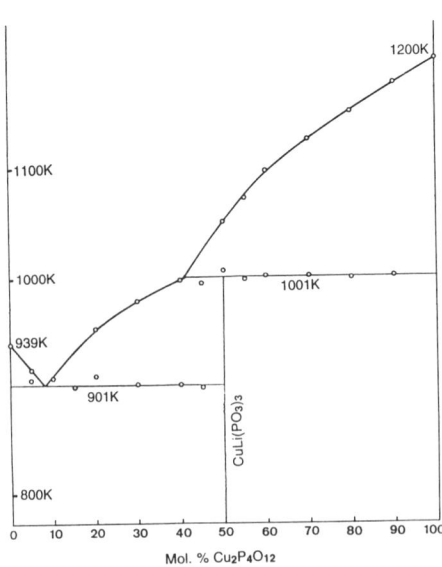

Figure 3.11. The LiPO$_3$–Cu$_2$P$_4$O$_{12}$ phase-equilibrium diagram.

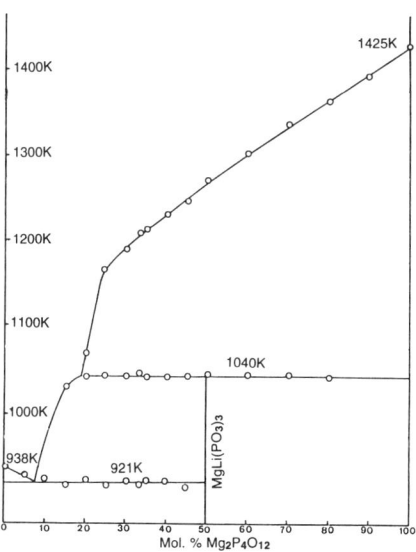

Figure 3.12. The $LiPO_3$–$Mg_2P_4O_{12}$ phase-equilibrium diagram.

The authors reported a complete determination of the atomic arrangement, confirming that this salt is isotypic with the corresponding copper compound.

$MnLi(PO_3)_3$. The chemical preparation of $MnLi(PO_3)_3$ and crystal data for this salt were reported by Averbuch-Pouchot and Durif.[612] This compound appears to be isotypic with $CuLi(PO_3)_3$. This polyphosphate is not observed in the $LiPO_3$–$Mn_2P_4O_{12}$ phase-equilibrium diagram elaborated by Bukhalova *et al.*[613]

$ZnLi(PO_3)_3$, $MgLi(PO_3)_3$, $NiLi(PO_3)_3$, and $CoLi(PO_3)_3$. These four compounds were characterized during investigations of the corresponding $LiPO_3$–$M^{II}_2P_4O_{12}$ phase-equilibrium diagrams by Averbuch-Pouchot and Rakotomahanina-Rolaisoa[614] (M^{II} = Zn and Mg), De Pontcharra and Durif[615] (M^{II} = Ni), and Rakotomahanina-Rolaisoa[616] (M^{II} = Co). The $LiPO_3$–$Mg_2P_4O_{12}$ phase-equilibrium diagram is presented in Fig. 3.12. The four $M^{II}Li(PO_3)_3$ polyphosphates observed in these diagrams are incongruent-melting compounds decomposing at 951 K (Zn), 1040 K (Mg), 1087 K (Ni) and 1053 K (Co).

(ii) $M^{II}Li(PO_3)_3$ Polyphosphates (M^{II} = Ca, Sr, and Pb)

$CaLi(PO_3)_3$. This polyphosphate is the only compound observed in the $LiPO_3$–$Ca(PO_3)_2$ phase-equilibrium diagram elaborated by Henry and Durif[617] (Fig. 3.13). $CaLi(PO_3)_3$ is an incongruent-melting compound decomposing at 1040 K and is isotypic with $PbLi(PO_3)_3$.

$SrLi(PO_3)_3$. The $LiPO_3$–$Sr(PO_3)_2$ phase-equilibrium diagram was elaborated by Martin and Durif.[618] The only compound observed in this system is $SrLi(PO_3)_3$, melting incon-

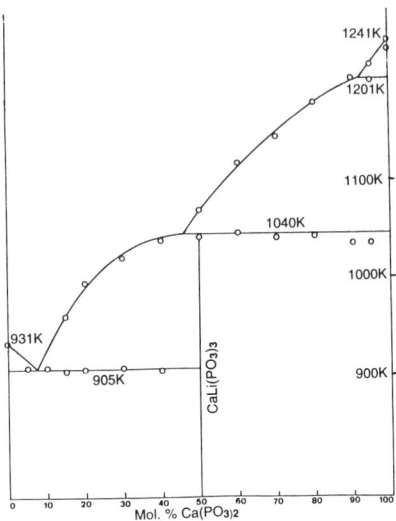

Figure 3.13. The LiPO₃–Ca(PO₃)₂ phase-equilibrium diagram. The $\alpha \rightarrow \beta$ transformation of Ca(PO₃)₂ is observed at 1201 K on the right-hand side of the diagram.

gruently at 1025 K. Crystallographic data were reported for this triclinic salt, which is isotypic with PbLi(PO₃)₃.

PbLi(PO₃)₃. PbLi(PO₃)₃ is not observed in the LiPO₃–Pb(PO₃)₂ phase-equilibrium diagram established by Grenier and Mahama[619] but has been characterized during attempts to crystallize Pb₂Li(PO₃)₅. Its atomic arrangement was determined by Guitel and Brunel-Laügt.[620]

(iii) MᴵᴵLi₂(PO₃)₄ Polyphosphates (Mᴵᴵ = Mn, Cd, Hg)

MnLi₂(PO₃)₄. Bukhalova *et al.*[613] constructed the LiPO₃–Mn₂P₄O₁₂ phase-equilibrium diagram and found that MnLi₂(PO₃)₄ is the only compound observed in this system. The crystal data reported by these authors for this salt are erroneous. They were revised by Averbuch-Pouchot and Durif[612] showing this salt to be isotypic with CdLi₂(PO₃)₄.

CdLi₂(PO₃)₄. This polyphosphate is the only compound observed in the LiPO₃–Cd(PO₃)₂ phase-equilibrium diagram elaborated by Averbuch-Pouchot and Durif[621] (Fig. 3.14). It melts congruently at 983 K. Its atomic arrangement was determined by Averbuch-Pouchot *et al.*[622]

HgLi₂(PO₃)₄. The LiPO₃–Hg(PO₃)₂ phase-equilibrium diagram elaborated by Raholison and Averbuch-Pouchot[623] shows the existence of HgLi₂(PO₃)₄, a congruent-melting compound (mp = 828 K). The crystal data reported by these authors indicated that this salt is a monoclinic distortion of CdLi₂(PO₃)₄. Its atomic arrangement was determined by Averbuch-Pouchot *et al.*[622]

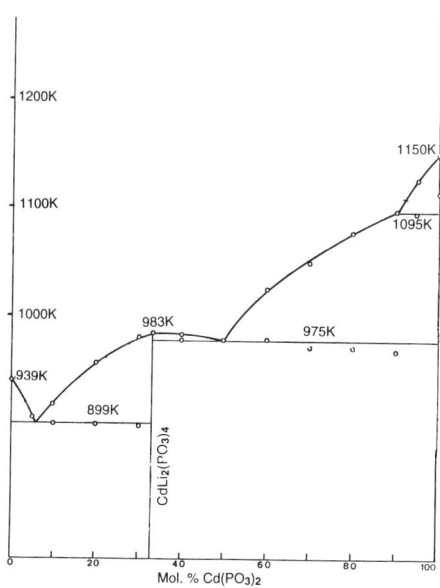

Figure 3.14. The $LiPO_3$–$Cd(PO_3)_2$ phase-equilibrium diagram. The transformation of $Cd(PO_3)_2$ appears at 1095 K in the right-hand side of the diagram.

(iv) $M^{II}_2Li(PO_3)_5$ and $M^{II}_2Li_3(PO_3)_7$ Polyphosphates

$Pb_2Li(PO_3)_5$. This polyphosphate, the only compound observed by Grenier and Mahama[619] during their study of the $LiPO_3$–$Pb(PO_3)_2$ phase-equilibrium diagram (Fig. 3.15) melts congruently at 893 K. Its crystal structure was determined later by El-Horr and Bagieu-Beucher.[624]

$Ba_2Li_3(PO_3)_7$. Owing to a lack of accuracy in the determination of the $LiPO_3$–$Ba(PO_3)_2$ phase-equilibrium diagram by Martin,[625] the only compound reported in this study was originally thought to correspond to $BaLi(PO_3)_3$. A reexamination of this diagram by El-Horr[626] showed that the correct formula is $Ba_2Li_3(PO_3)_7$. The crystal structure determination performed by El-Horr *et al.*[627] confirmed this formula.

3.2.3.2. Divalent Cation–Sodium Polyphosphates

In this category of compounds the only polyphosphates characterized with a good degree of certitude correspond to two types of chemical formulas: $M^{II}Na(PO_3)_3$ and $M^{II}Na_2(PO_3)_4$. These two groups are considered individually below. The main crystallographic data for the divalent cation–sodium polyphosphates are gathered in Table 3.7.

(i) $M^{II}Na(PO_3)_3$ (M^{II} = Mg, Co, Zn, Cd)

These four compounds were characterized during the elaboration of the $NaPO_3$–$Mg_2P_4O_{12}$, $NaPO_3$–$Zn(PO_3)_2$, $NaPO_3$–$Co(PO_3)_2$, and $NaPO_3$–$Cd(PO_3)_2$ phase-equilibrium diagrams performed by Thonnerieux *et al.*,[628] Krivovyasov *et al.*,[629] Rakotomahanina-Ro-

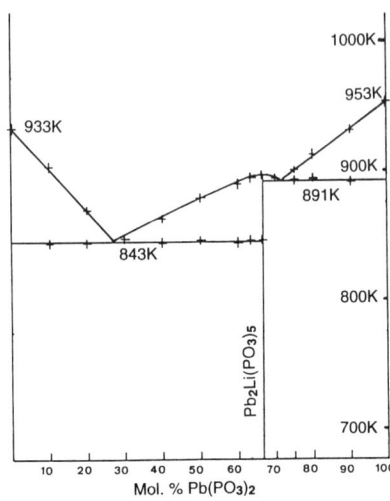

Figure 3.15. The LiPO₃–Pb(PO₃)₂ phase-equilibrium diagram.

laisoa,[616] and Averbuch-Pouchot and Durif[630] respectively. Crystal data for the zinc salt were reported by Averbuch-Pouchot et al.[631] The cadmium salt is a congruent-melting compound (mp = 1014 K) whereas the other three salts decompose at 1123 K(Mg), 1145 K(Co) and 993 K(Zn). From the various crystal data reported in the studies cited above these three compounds seem to be isomorphous, but no convincing structural data are available to confirm this assumption. A crystal structure for the magnesium salt was proposed by Shepelev et al.[632]

Table 3.7. Main Crystallographic data for Divalent Cation–Sodium Polyphosphates

Formula	a (Å) α (°)	b (Å) β (°)	c (Å) γ (°)	S.G.	Z	Reference(s)
CuNa₂(PO₃)₄	9.398(2) 95.62(5)	13.304(5)	7.717(2)	$C2/c$	4	638–640
MgNa₂(PO₃)₄	14.344(3)	14.258(3)	14.198(3)	$P2_12_12_1$	12	628
CoNa(PO₃)₃	14.266(3)	14.281(3)	14.255(3)	?	16	616
ZnNa(PO₃)₃	14.297(3)	14.297(3)	14.297(3)	?	16	631
MgNa(PO₃)₃	14.304(3)	14.183(5)	14.231(4)	$Pbca$	16	632
CdNa(PO₃)₃	14.62	14.62	14.78	?	16	630
CaNa(PO₃)₃	6.940(2) 98.57(1)	7.620(3) 97.08(1)	6.725(2) 83.64(1)	$P\bar{1}$	2	634, 635
SrNa(PO₃)₃	7.162(4) 99.68(5)	7.738(4) 97.29(5)	6.831(4)ʳ 83.84(5)	$P\bar{1}$	2	636
PbNa(PO₃)₃	7.159(2) 99.81(2)	7.769(3) 96.01(2)	6.791(2) 84.01(2)	$P\bar{1}$	2	637

Two of these salts (Zn, Co) exhibit a low-temperature modification corresponding to the $M_4Na_4(P_4O_{12})_3$ cyclotetraphosphate.

(ii) $M^{II}Na(PO_3)_3$ (M^{II} = Ca, Sr, Pb)

$CaNa(PO_3)_3$. The $NaPO_3-Ca(PO_3)_2$ phase-equilibrium diagram, first established by Morey,[633] was revised by Grenier *et al.*[634] These authors identified a new polyphosphate in this system: $CaNa(PO_3)_3$, an incongruent melting compound decomposing at 1013 K. This salt is isotypic with $PbLi(PO_3)_3$.[635] This revised diagram is presented in Fig. 3.18.

$SrNa(PO_3)_3$. $SrNa(PO_3)_3$ appears as an incongruent melting compound decomposing at 1023 K in the $NaPO_3-Sr(PO_3)_2$ phase-equilibrium diagram elaborated by Martin and Durif.[636] The crystal data reported by these authors showed this salt to be isotypic with $PbLi(PO_3)_3$.

$PbNa(PO_3)_3$. This compound does not appear in the $NaPO_3-Pb(PO_3)_2$ phase-equilibrium diagram (see Chapter 4, Fig. 4.2.5). It was first characterized during attempts to produce single crystals of $PbNa_4(P_3O_9)_2$. It is dimorphous; its first form is a cyclotriphosphate isotypic with $BaNaP_3O_9$. Its second form which was described by Prisset[637] is a long-chain polyphosphate isotypic with $PbLi(PO_3)_3$.

(iii) $M^{II}Na_2(PO_3)_4$ Polyphosphates

$MgNa_2(PO_3)_4$. This compound appears in the phase-equilibrium diagram $NaPO_3-Mg_2P_4O_{12}$ elaborated by Thonnerieux *et al.*[628] as a congruent-melting compound (mp = 1093 K). Crystal data reported by these authors are not convincing.

$CuNa_2(PO_3)_4$. In the $NaPO_3-Cu_2P_4O_{12}$ phase-equilibrium diagram established by Laügt *et al.*[638] (Fig. 3.16), $CuNa_2(PO_3)_4$, a congruent-melting compound (mp = 983 K) is the only intermediate compound. The atomic arrangement of this salt was determined by Laügt et al.[639,640]

3.2.3.3. Divalent Cation–Silver Polyphosphates

$M^{II}Ag(PO_3)_3$ (M^{II} = Co, Zn, Mg, Ni). These four isomorphous compounds have been characterized during the elaboration of the corresponding phase-equilibrium diagrams.

The $Co(PO_3)_2-AgPO_3$ phase-equilibrium diagram was first established by Rakotoma-hanina-Rolaisoa[616] and later reinvestigated by Bukhalova *et al.*[641] $CoAg(PO_3)_3$ is the only compound observed in this system. This incongruent melting polyphosphate decomposes at 863 K.

The $AgPO_3-Zn_2P_4O_{12}$ phase-equilibrium diagram was established by Savenkova *et al.*[642] The only intermediate compound found in the system, $ZnAg(PO_3)_3$, was erroneously described by the authors as a cyclotriphosphate. This incongruent-melting compound decomposes at 833 K.

The nickel salt, $NiAg(PO_3)_3$, characterized by de Pontcharra[643] during the investigation of the $AgPO_3-NiP_4O_{12}$ phase-equilibrium diagram decomposes at 975 K and is the only intermediate compound in this system.

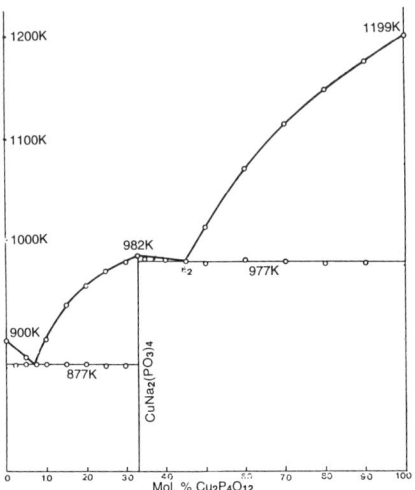

Figure 3.16. The NaPO$_3$–Ca(PO$_3$)$_2$ phase-equilibrium diagram.

MgAg(PO$_3$)$_3$ is the only compound observed in the AgPO$_3$–Mg$_2$P$_4$O$_{12}$ phase diagram established by Rakotomahanina-Rolaisoa.[616] It decomposes at 936 K. This diagram is reproduced in Fig. 3.17.

As in the case of the corresponding sodium salts, most of the crystal data reported for these four polyphosphates are wrong. Recently, Averbuch-Pouchot and Durif[644] reinves-

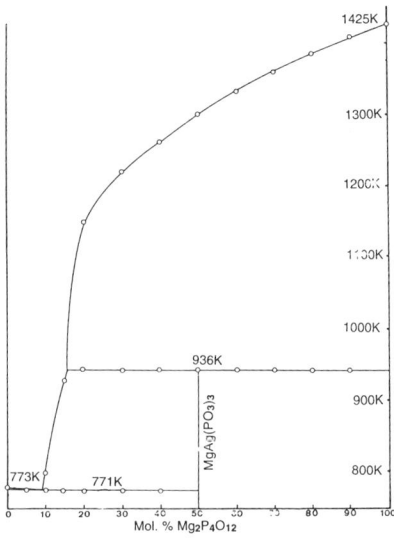

Figure 3.17. The AgPO$_3$–Mg$_2$P$_4$O$_{12}$ phase-equilibrium diagram.

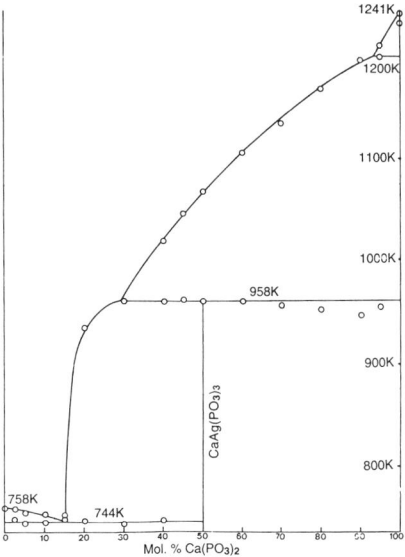

Figure 3.18. The $AgPO_3$-$Ca(PO_3)_2$ phase-equilibrium diagram. The $\alpha \rightarrow \beta$ transformation of $Ca(PO_3)_2$ is observed at 1200 K on the right-hand side of the diagram.

tigated this class of compounds, reported the crystal structure of the zinc salt, and, from relationships between the cubic unit cell of $Zn_4Na_4(P_4O_{12})$ and the cell dimensions of $ZnAg(PO_3)_3$, were able to explain the previous erroneous results.

$CaAg(PO_3)_3$. The $AgPO_3$–$Ca(PO_3)_2$ phase diagram elaborated by Henry and Durif[645] (Fig. 3.18) shows the existence of $CaAg(PO_3)_3$, an incongruent melting-compound, decomposing at 958 K and isotypic with $PbLi(PO_3)_3$.

$CuAg_2(PO_3)_4$. $CuAg_2(PO_3)_4$ the only $M^{II}Ag_2(PO_3)_4$ polyphosphate presently known. It was characterized during the elaboration of the $AgPO_3$–$Cu_2P_4O_{12}$ phase-equilibrium diagram (Fig. 3.19) by Laügt.[646] This incongruent melting compound decomposes at 835 K and is isotypic with $K_2Cu(PO_3)_4$.

Crystallographic Data. Table 3.8. contains the main crystallographic data for divalent cation–silver polyphosphates.

3.2.3.4. Divalent Cation–Potassium Polyphosphates

The divalent cation–potassium polyphosphates correspond to three types of formulas: $M^{II}K(PO_3)_3$, $M^{II}K_2(PO_3)_4$, and $M^{II}_2K(PO_3)_5$.

$BeK(PO_3)_3$ and $BeK_2(PO_3)_4$. The KPO_3–$Be(PO_3)_2$ phase diagram was investigated by Omezzine and Kbir-Ariguib.[647] $BeK(PO_3)_3$ and $BeK_2(PO_3)_4$ appear in this diagram as two incongruent-melting compounds decomposing at 1038 and 1044 K respectively. A crystal structure determination for $BeK(PO_3)_3$ was later performed by Durif *et al.*[648]

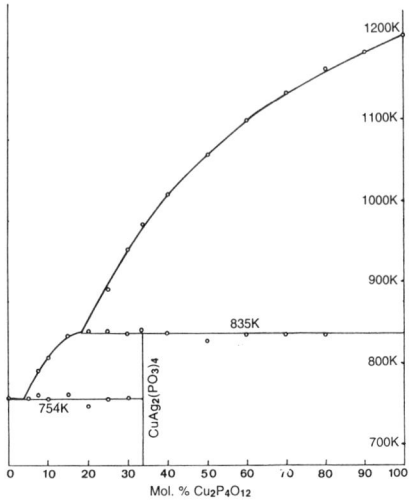

Figure 3.19. The $AgPO_3$–$Cu_2P_4O_{12}$ phase-equilibrium diagram.

$M^{II}K(PO_3)_3$ (M^{II} = Co, Mg, Ni, Zn). These four polyphosphates are isotypic with $NH_4Ni(PO_3)_3$.[649] They have all been characterized during the elaboration of the corresponding KPO_3–$M_2P_4O_{12}$ phase-equilibrium diagrams. Some of these diagrams are presented in Chapter 4.

The KPO_3–$Mg_2P_4O_{12}$-phase-equilibrium diagram was first determined by Andrieu and Diament[650] and later revised by Averbuch-Pouchot *et al.*[651]

The KPO_3–$Zn_2P_4O_{12}$ system was first investigated by Krivovyasov *et al.*[629] and the phase-equilibrium diagram was later revised by Averbuch-Pouchot *et al.*[631]

The KPO_3–$Co_2P_4O_{12}$ and KPO_3–$Ni_2P_4O_{12}$ phase-equilibrium diagrams were elaborated by Thonnerieux *et al.*[652] and De Pontcharra and Durif,[615] respectively.

With the exception of the nickel salt, these polyphosphates are the low-temperature forms of benitoite-like cyclotriphosphates MKP_3O_9. The transformation to the cyclo-

Table 3.8. Main Crystallographic Data for Divalent Cation–Silver Polyphosphates

Formula	a (Å) α (°)	b (Å) β (°)	c (Å) γ (°)	S.G.	Z	Reference
$ZnAg(PO_3)_3$	13.950	10.735	9.951	*Pcca*	8	644
$CoAg(PO_3)_3$	13.986	10.772	9.958	*Pcca*	8	644
$NiAg(PO_3)_3$	13.852	10.712	9.874	*Pcca*	8	644
$MgAg(PO_3)_3$	13.888	10.730	9.973	*Pcca*	8	644
$CuAg_2(PO_3)_4$	9.524(3)	13.180(4) 95.52(5)	7.771(3)	*C2/c*	4	646
$CaAg(PO_3)_3$	7.011 100.45	7.799 98.34	6.804 82.13	$P\bar{1}$	2	645

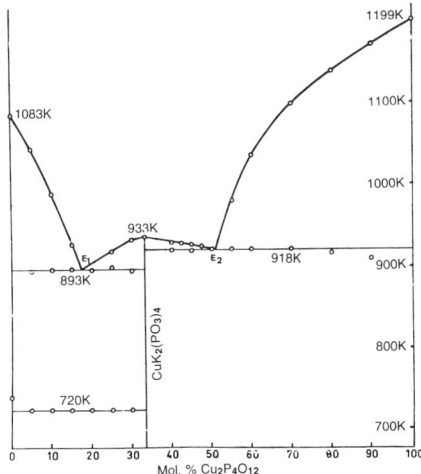

Figure 3.20. The KPO_3–$Cu_2P_4O_{12}$ phase-equilibrium diagram. The transformation of KPO_3 is observed at 720 K on the left-hand side of the diagram.

triphosphate occurs at 623 K for the zinc salt. Crystal data for these four polyphosphates were reported by Durif *et al.*[653]

The nickel salt decomposes at 1125 K, whereas the other three salts melt congruently at 1176 K (Mg), 1076 K (Co) and 908 K (Zn).

$M^{II}K_2(PO_3)_4$ (M^{II} = Co, Mg, Ni, Zn, Cu). These five polyphosphates are isotypic. The crystal structure of $CuK_2(PO_3)_4$ was determined by Tordjman *et al.*[654] and later refined by Laügt *et al.*[655]

In addition the atomic arrangement of $CoK_2(PO_3)_4$ has been refined by Laügt *et al.*[655] confirming the isotypy of this salt with $CuK_2(PO_3)_4$.

The first four salts of this series have been characterized during the elaboration of the corresponding phase diagrams discussed above, while $CuK_2(PO_3)_4$ is the only compound observed in the KPO_3–$Cu_2P_4O_{12}$ phase-equilibrium diagram elaborated by Laügt *et al.*[656] (Fig. 3.20). The zinc and copper salts melt at 898 and 933 K, respectively, while the three other salts decompose at 994 K (Co), 1036 K (Mg) and 1035 K (Ni).

$CdK_2(PO_3)_4$. $CdK_2Cd(PO_3)_4$ appears as an incongruent melting compound decomposing at 994 K in the KPO_3–$Cd(PO_3)_2$ phase equilibrium diagram elaborated by Mermet *et al.*[657] This salt is isotypic with $Mn(NH_4)_2(PO_3)_4$.

$Ba_2K(PO_3)_5$. $Ba_2K(PO_3)_5$ is the only compound appearing in the KPO_3–$Ba(PO_3)_2$ phase-equilibrium diagram established by Martin.[625] It decomposes at 1041 K. Its atomic arrangement, determined by Martin *et al.*[658,659] provided the first example of a long-chain anion with a period of ten tetrahedra.

$PbK_2(PO_3)_4$. The KPO_3–$Pb(PO_3)_2$ system has been investigated several times. The first phase-equilibrium diagram was reported by Mahama *et al.*[660] $PbK_2(PO_3)_4$, a congruent-

Table 3.9. Main Crystallographic Data for Divalent Cation–Potassium
Polyphosphates

Formula	a (Å) α (°)	b (Å) β (°)	c (Å) γ (°)	S.G.	Z	Reference(s)
BeK(PO$_3$)$_3$	7.844(1)	8.610(2)	11.399(4)	$P2_12_12_1$	4	648
CoK(PO$_3$)$_3$	10.18(1)	10.18(1)	6.983(5)	R3	3	653
MgK(PO$_3$)$_3$	10.17(1)	10.17(1)	7.011(5)	R3	3	653
NiK(PO$_3$)$_3$	10.09(1)	10.09(1)	6.975(5)	R3	3	653
ZnK(PO$_3$)$_3$	10.15(1)	10.15(1)	6.966(5)	R3	3	653
MgK$_2$(PO$_3$)$_4$	11.10(3)	12.56(4) 102.84(2)	7.612(3)	Cc	4	631
ZnK$_2$(PO$_3$)$_4$	11.12(3)	12.55(4) 102.37(2)	7.679(5)	Cc	4	631
NiK$_2$(PO$_3$)$_4$	11.09(1)	12.57(1) 103.10(5)	7.586(3)	Cc	4	615
CoK$_2$(PO$_3$)$_4$	11.127(1)	12.579(3) 102.89(1)	7.636(1)	Cc	4	652, 655
CuK$_2$(PO$_3$)$_4$	10.924(1)	12.227(1) 100.55(1)	7.900(1)	Cc	4	654–656
CdK$_2$(PO$_3$)$_4$	11.20(3)	13.06(3) 101.85(5)	7.782(5)	$P2_1/n$	4	657
Ba$_2$K(PO$_3$)$_5$	8.646(2)	7.329(1) 129.17(1)	13.884(2)	Pc	2	625, 658, 659
PbK$_2$(PO$_3$)$_4$	15.510(1)	15.550(1)	9.249(1)	Pbca	8	663

melting long-chain polyphosphate (mp = 842 K), is the only compound observed in this system. A revised version of this diagram by Grunze and Möwius[661,662] shows the existence of a second compound in this system: PbK$_4$(P$_3$O$_9$)$_2$, an incongruent-melting cyclotriphosphate.

The low-temperature form of PbK$_2$(PO$_3$)$_4$ is a cyclotetraphosphate, PbK$_2$P$_4$O$_{12}$ (see 4.3.2.3).

Crystal structure of PbK$_2$(PO$_3$)$_4$ was performed by Brunel-Laügt and Guitel.[663]

Crystallographic Data. Table 3.9 contains the main crystallographic data for divalent cation–potassium polyphosphates.

3.2.3.5. Divalent Cation–Ammonium Polyphosphates

The divalent cation–ammonium polyphosphates identified to date correspond to two types of chemical formulas: M^{II}NH$_4$(PO$_3$)$_3$ and M^{II}(NH$_4$)$_2$(PO$_3$)$_4$. All of them have been prepared by thermal or flux methods at relatively low temperatures.

M^{II}NH$_4$(PO$_3$)$_3$ (M^{II} = Ni, Mg, Co). These three polyphosphates are dimorphous. The rhombohedral form of these three salts was first described by Tordjman *et al.*[649] who reported their chemical preparation, crystal data, and crystal structure. The atomic arrangement in this class of compounds was determined by using the nickel salt.

Averbuch-Pouchot and Tranqui[664] reported the chemical preparation of and crystal data for the orthorhombic modification, and later Tranqui *et al.*[665] performed the determination of the atomic arrangement using the cobalt salt.

The cobalt and magnesium salts exist also as cyclotriphosphates isotypic with benitoite.

$ZnNH_4(PO_3)_3$. The only polyphosphate form of this salt is the orthorhombic one described by Averbuch-Pouchot and Tranqui[664] as an isotype of the cobalt salt. This compound exists also as a cyclotriphosphate isotypic with benitoite.

$CuNH_4(PO_3)_3$. The chemical preparation and crystal structure of $CuNH_4(PO_3)_3$ were reported by Tranqui et al.[666]

$Co(NH_4)_2(PO_3)_4$, $Mn(NH_4)_2(PO_3)_4$, and $Cd(NH_4)_2(PO_3)_4$. The chemical preparation of and crystal data for these three isotypic polyphosphates were reported by Averbuch-Pouchot and Durif.[667]

$Zn(NH_4)_2(PO_3)_4$. This polyphosphate described by Averbuch-Pouchot and Durif[667] is isotypic with $CuK_2(PO_3)_4$.

Crystallographic Data. Table 3.10 contains the main crystallographic data for divalent cation–ammonium polyphosphates.

3.2.3.6. Divalent Cation–Rubidium Polyphosphates

As in the case of divalent cation–ammonium polyphosphates, only two kinds of formulas have been observed for divalent cation–rubidium polyphosphates: $M^{II}Rb(PO_3)_3$ and $M^{II}Rb_2(PO_3)_4$.

$CuRb(PO_3)_3$. This polyphosphate, which is isotypic with the corresponding copper salt, was first observed by Laügt et al.[668] during the determination of the $RbPO_3$–$Cu_2P_4O_{12}$

Table 3.10. Main Crystallographic Data for Divalent Cation–Ammonium Polyphosphates

Formula	a (Å) α (°)	b (Å) β (°)	c (Å) γ (°)	S.G.	Z	Reference(s)
$NiNH_4(PO_3)_3$	10.130(5)	10.130(5)	7.098(2)	$R3$	3	649
$MgNH_4(PO_3)_3$	10.240(5)	10.240(5)	7.153(2)	$R3$	3	649
$CoNH_4(PO_3)_3$	10.264(5)	10.264(5)	7.119(2)	$R3$	3	649
$NiNH_4(PO_3)_3$	5.096(1)	11.89(2)	12.93(2)	$Pbcm$	4	664
$ZnNH_4(PO_3)_3$	5.125(2)	11.88(2)	12.92(2)	$Pbcm$	4	664
$MgNH_4(PO_3)_3$	5.159(2)	11.98(2)	12.95(2)	$Pbcm$	4	664
$CoNH_4(PO_3)_3$	5.142(2)	11.93(3)	12.95(3)	$Pbcm$	4	664, 665
$CuNH_4(PO_3)_3$	5.182(4)	11.544(8) 97.16(5)	13.06(1)	$P2_1/c$	4	666
$Mn(NH_4)_2(PO_3)_4$	11.297(4)	12.993(5) 101.50(5)	7.839(5)	$P2_1/n$	4 4	630, 667
$Co(NH_4)_2(PO_3)_4$	11.22(4)	12.80(5) 101.00(5)	7.720(5)	$P2_1/n$	4	667
$Cd(NH_4)_2(PO_3)_4$	11.34(4)	13.20(5) 101.00(5)	7.879(5)	$P2_1/n$	4	667
$Zn(NH_4)_2(PO_3)_4$	11.27(4)	12.79(5) 101.98(5)	7.758(5)	Cc	4	667

phase-equilibrium diagram. It is an incongruent-melting compound decomposing at 858 K. This diagram was later revised by Laügt.[595] The revised diagram showed the existence of an additional compound: $Cu_3Rb_2P_8O_{24}$, a cyclooctaphosphate. This diagram is presented in Chapter 4 (see Fig. 4.6.1).

$M^{II}Rb(PO_3)_3$ (M^{II} = Zn, Ni). These two polyphosphates are dimorphous. One of the crystalline forms is orthorhombic and isotypic with $CoNH_4(PO_3)_3$ whereas the second modification is rhombohedral, isotypic with $NiNH_4(PO_3)_3$. Both were characterized during the elaboration of the $RbPO_3$–$Zn_2P_4O_{12}$ and $RbPO_3$–$Ni_2P_4O_{12}$ phase-equilibrium diagrams, performed by Averbuch-Pouchot[669] and de Pontcharra[643] respectively, in which they appear as incongruent melting compounds decomposing at 872 K (Zn) and 1141 K (Ni). In the case of the nickel salt the transformation from the rhombohedral to the orthorhombic form occurs at about 623 K and is not reversible. The rhombohedral form of $ZnRb(PO_3)_3$ was characterized by Prisset.[637]

The $RbPO_3$–$Ni_2P_4O_{12}$ phase-equilibrium diagram is given in Fig. 3.21.

$MgRb(PO_3)_3$. This salt appears as an incongruent melting compound decomposing at 1139 K in the $RbPO_3$–$Mg_2P_4O_{12}$ system[670] given in Fig. 3.22. It is isotypic with $NH_4Ni(PO_3)_3$ and has a high-temperature form, a cyclotriphosphate, isotypic with the orthorhombic form of $MgNH_4P_3O_9$.

$CoRb(PO_3)_3$. This polyphosphate was identified as an isotype of the orthorhombic form of $CoNH_4(PO_3)_3$ by Rakotomahanina-Rolaisoa[616] during an investigation of the $RbPO_3$–$Co_2P_4O_{12}$ system. It decomposes at 1048 K.

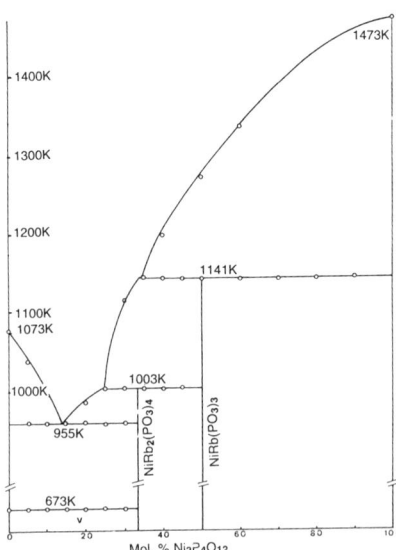

Figure 3.21. The $RbPO_3$–$Ni_2P_4O_{12}$ phase-equilibrium diagram. The transformation of $RbPO_3$ is observed at 673 K on the left-hand side of the diagram.

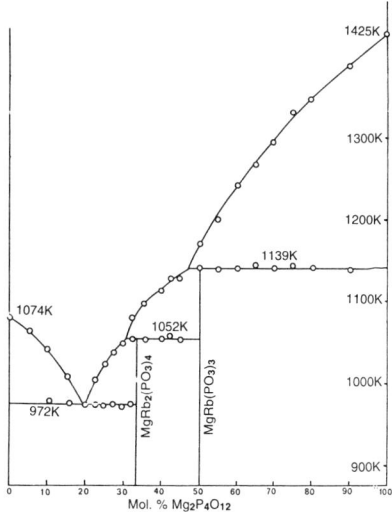

Figure 3.22. The $RbPO_3$–$Mg_2P_4O_{12}$ phase-equilibrium diagram.

$M^{II}Rb_2(PO_3)_4$ (M^{II} = Mg, Co, Zn, Ni, Cd, Ca). These six polyphosphates characterized during the elaboration of the corresponding phase-equilibrium diagrams are all isotypic with $Mn(NH_4)_2(PO_3)_4$. The $RbPO_3$–$Cd(PO_3)_2$ and $RbPO_3$–$Ca(PO_3)_2$ diagrams were elaborated by Mermet *et al.*[657] and Henry and Durif,[671] respectively; the corresponding diagrams for M^{II} = Mg, Co, Zn, and Ni have been referenced above.

$CdRb_2(PO_3)_4$ melts congruently at 1039 K whereas the other five compounds decompose at 1052 K (Mg), 862 K (Zn), 1003 K (Ni), 999 K (Co), 1015 K (Ca).

Crystallographic Data. Table 3.11 contains the main crystallographic data for divalent cation–rubidium polyphosphates.

3.2.3.7. Divalent Cation–Cesium Polyphosphates

$CdCs_2(PO_3)_4$ and $MnCs_2(PO_3)_4$. The cadmium salt was characterized by Averbuch-Pouchot[672] during the elaboration of the $CsPO_3$–$Cd(PO_3)_2$ phase diagram; it appears as a congruent-melting compound (mp = 985 K) and is isotypic with $(NH_4)_2Mn(PO_3)_4$.

Klinkert and Jansen[673] described the chemical preparation of $MnCs_2(PO_3)_4$ and performed a crystal structure determination for this salt, which is also isotypic with $Mn(NH_4)_2(PO_3)_4$.

$CuCs_2(PO_3)_4$ and $CuCs_4(PO_3)_6$. These two salts were characterized by Laügt and Martin[674] during the elaboration of the $CsPO_3$–$Cu_2P_4O_{12}$ phase equilibrium diagram. $Cs_4Cu(PO_3)_6$ is a congruent-melting compound (mp = 839 K) whereas $Cs_2Cu(PO_3)_4$ decomposes at 837 K. Crystal data were reported for $Cs_2Cu(PO_3)_4$. In addition, a cyclooctaphosphate, $Cu_3Cs_2P_8O_{24}$ appears in the system (see Section 4.6.2.3).

Table 3.11. Main Crystallographic Data for Divalent Cation–Rubidium Polyphosphates

Formula	a (Å) α (°)	b (Å) β (°)	c (Å) γ (°)	S.G.	Z	Reference(s)
MgRb(PO₃)₃	10.268(8)	10.268(8)	7.180(5)	$R3$	3	670
ZnRb(PO₃)₃[a]	10.243(6)	10.243(6)	7.128(5)	$R3$	3	637
NiRb(PO₃)₃[a]	10.161(1)	10.161(1)	7.138(1)	$R3$	3	643
CuRb(PO₃)₃	13.057(3)	11.543(3) 97.08(1)	5.178(2)	$P2_1/a$	4	603, 668
ZnRb(PO₃)₃[b]	5.126(3)	11.84(2)	12.91(2)	$Pbcm$	4	669
CoRb(PO₃)₃	5.170(2)	11.966(4)	13.031(5)	$Pbcm$	4	616
NiRb(PO₃)₃[b]	5.087(1)	11.820(3)	12.880(3)	$Pbcm$	4	643
MgRb₂(PO₃)₄	12.258(5)	12.831(5) 102.02(5)	7.789(3)	$P2_1/a$	4	670
CoRb₂(PO₃)₄	11.290(5)	12.905(5) 102.02(5)	7.804(2)	$P2_1/n$	4	616
ZnRb₂(PO₃)₄	11.22(1)	12.81(1) 102.08(5)	7.750(5)	$P2_1/n$	4	669
NiRb₂(PO₃)₄	11.227(3)	12.774(4) 102.12(4)	7.729(3)	$P2_1/n$	4	643
CdRb₂(PO₃)₄	11.38(3)	13.19(3) 101.94(5)	7.870(5)	$P2_1/n$	4	657
CaRb₂(PO₃)₄	11.436(4)	13.352(4) 101.89(5)	7.908(2)	$P2_1/n$	4	671

[a]Rhombohedral form.
[b]Orthorhombic form.

BaCs₄(PO₃)₆ and Ba₂Cs(PO₃)₅. The $CsPO_3$–$Ba(PO_3)_2$ phase-equilibrium diagram first elaborated by Tokman and Bukhalova,[675] was revised by Masse and Averbuch-Pouchot.[676] This revised diagram, given in Fig. 3.23, shows the existence of two polyphosphates: BaCs₄(PO₃)₆ and Ba₂Cs(PO₃)₅. BaCs₄(PO₃)₆ appears in the system as a congruent-melting compound (mp = 963 K). Its atomic arrangement was determined by Averbuch-Pouchot and Durif.[677] Ba₂Cs(PO₃)₅ is an incongruent-melting compound decomposing at 993 K. The crystal data reported by the authors showed this polyphosphate to be isotypic with Ba₂K(PO₃)₅.

PbCs₂(PO₃)₄. The $CsPO_3$–$Pb(PO_3)_2$ phase-equilibrium diagram was elaborated by Shpakova *et al.*[678] According to these authors PbCs(PO₃)₃, melting congruently at 814 K, is the only compound observed in this system. These results are inconsistent with those later published by Averbuch-Pouchot,[679] who observed two compounds in this system: Pb₂Cs₃(P₄O₁₂)(PO₃)₃, a mixed-anion phosphate, and PbCs₂(PO₃)₄, a polyphosphate.

Crystallographic Data. Table 3.12 contains the main crystallographic data for divalent cation–cesium polyphosphates.

3.2.3.8. Divalent Cation–Thallium Polyphosphates

CuTl(PO₃)₃. This polyphosphate isotypic with CuNH₄(PO₃)₃, was characterized during the elaboration of the $TlPO_3$–$Cu_2P_4O_{12}$ phase diagram by Laügt *et al.*[668] It is an incon-

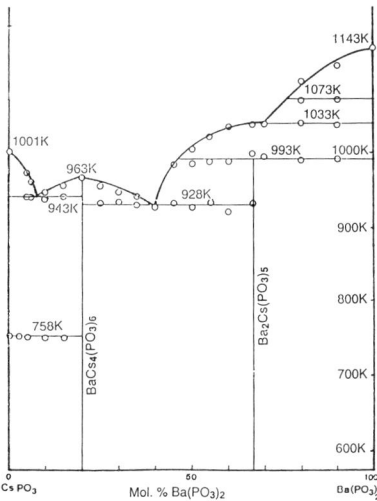

Figure 3.23. The CsPO₃–Ba(PO₃)₂ phase-equilibrium diagram. The transformation of CsPO₃ appears at 758 K on the left-hand side of the diagram.

gruent melting-compound decomposing at 871 K. This diagram was later revised by Laügt[680]; the revised diagram showed the existence of an additional compound: $Cu_3Tl_2P_8O_{24}$, a cyclooctaphosphate (see Section 4.6.2.3).

$NiTl(PO_3)_3$, and $CoTl(PO_3)_3$. These two polyphosphate were identified as isotypes of the orthorhombic form of $CoNH_4(PO_3)_3$ during investigations of the $TlPO_3–Ni_2P_4O_{12}$ and $TlPO_3–Co(PO_3)_2$ systems by De Pontcharra[643] and Rakotomahanina-Rolaisoa *et al.*[681] respectively. Both are incongruent-melting compounds, decomposing at 1031 K (Ni) and 970 K (Co).

$ZnTl(PO_3)_3$ and $ZnTl_2(PO_3)_4$. These two polyphosphates were characterized by Averbuch-Pouchot[669] during the elaboration of the $TlPO_3–Zn(PO_3)_2$ phase-equilibrium dia-

Table 3.12. Main Crystallographic Data for Divalent Cation–Cesium Polyphosphates

Formula	a (Å) α (°)	b (Å) β (°)	c (Å) γ (°)	S.G.	Z	Reference(s)
$MnCs_2(PO_3)_4$	11.546(1)	13.249(1) 101.97(1)	7.976(1)	$P2_1/n$	4	673
$CdCs_2(PO_3)_4$	11.65(5)	13.44(5) 102.02(5)	8.014(5)	$P2_1/n$	4	672
$CuCs_2(PO_3)_4$	8.998(6)	9.857(8) 91.56(5)	7.418(5)	$P2_1/a$		674
$Ba_2Cs(PO_3)_5$	8.444(3)	7.509(3) 125.31(10)	13.976(5)	Pc	2	676
$BaCs_4(PO_3)_6$	11.541(9)	11.541(9)	9.112(9)	$P3_1c$	2	676, 677

Table 3.13. Main Crystallographic Data for Divalent Cation–Thallium
Polyphosphates

Formula	a (Å) α (°)	b (Å) β (°)	c (Å) γ (°)	S.G.	Z	Reference(s)
CuTl(PO$_3$)$_3$	13.127(3)	11.617(3) 96.99(1)	5.157(2)	$P2_1/a$	4	668, 680
CoTl(PO$_3$)$_3$	5.115(2)	11.90(1)	12.99(1)	$Pbcm$	4	681
ZnTl(PO$_3$)$_3$	5.125(2)	11.90(1)	13.01(1)	$Pbcm$	4	669
NiTl(PO$_3$)$_3$	5.073	11.86	12.927	$Pbcm$	4	643
ZnTl$_2$(PO$_3$)$_4$	11.27(1)	12.82(1) 102.66(4)	7.737(3)	Cc	4	669
CdTl$_2$(PO$_3$)$_4$	11.44(2)	13.20(2) 102.68(5)	7.854(2)	Cc	4	682
MgTl$_2$(PO$_3$)$_4$	11.205(5)	12.83(1) 102.25(8)	7.755(5)	$P2_1/n$	4	681

gram. Both are incongruent melting compounds, decomposing at 842 and 755 K respectively. ZnTl(PO$_3$)$_3$ is isotypic with the orthorhombic form of CoNH$_4$(PO$_3$)$_3$ whereas ZnTl$_2$(PO$_3$)$_4$ is isotypic with CuK$_2$(PO$_3$)$_4$.

CdTl$_2$(PO$_3$)$_4$. This polyphosphate was first observed by Averbuch-Pouchot[682] during the elaboration of the TlPO$_3$–Cd(PO$_3$)$_2$ phase-equilibrium diagram. It appears as an incongruent-melting compound decomposing at 925 K and is isotypic with CuK$_2$(PO$_3$)$_4$.

MgTl$_2$(PO$_3$)$_4$. This polyphosphate was characterized by Rakotomahanina-Rolaisoa et al.[681] during the determination of the TlPO$_3$–Mg$_2$P$_4$O$_{12}$ phase-equilibrium diagram. This incongruent-melting compound decomposes at 763 K and the crystal data reported by the authors show it to be isotypic with Mn(NH$_4$)$_2$(PO$_3$)$_4$.

Crystallographic Data. The main crystallographic data for divalent cation–thallium polyphosphates are given in Table 3.13.

3.2.4. Trivalent Cation Long-Chain Polyphosphates

For the sake of clarity we must distinguish two kinds of trivalent cation polyphosphates. The first group comprises compounds of the general formula M(PO$_3$)$_3$ with M = M = Al, Cr, Ga, V, Mn, Fe, Rh, Ti, Mo, Sc, In, and Tl. The second group contains compounds of yttrium, the lanthanides and bismuth. In this second family of compounds two different formulas have been observed: MH(PO$_3$)$_3$ and M(PO$_3$)$_3$.

3.2.4.1. M(PO$_3$)$_3$ Polyphosphates for M = Al, Cr, Ga, V, Mn, Fe, Rh, Ti, Mo, Sc, In, and Tl

For a long time, trivalent cation polyphosphates M(PO$_3$)$_3$ containing the above-mentioned elements received relatively little attention. However, chemical preparation and proper characterization of some of them were already achieved last century. As early as 1847, Maddrell[6] described the chemical preparation of the Mn, Al, Fe, and Cr polyphosphates, and, in 1899, Johnson[683] confirmed the existence of Fe(PO$_3$)$_3$, Cr(PO$_3$)$_3$, and

$Al(PO_3)_3$. Nevertheless this class of compounds was not well understood until the early 1950s, when the development of chromatographic and crystallographic techniques helped to clarify rapidly the situation. Before beginning our survey of the current state of knowledge in this field, it is essential to mention the fundamental pioneering investigations performed by d'Yvoire[684] and Rémy and Boullé.[685] During their systematic studies of aluminum, iron, and chromium salts, they showed that these compounds are polymorphic and can adopt up to five crystalline forms, which they denoted A, B, C, D, and E. The A form corresponds to a cyclic tetrameric anion, P_4O_{12} and the B form to a cyclic hexameric anion, P_6O_{18}. It is well established that the C form corresponds to a long-chain polyphosphate. Very little is known concerning the D and E forms, whose atomic arrangements are still unknown. Some crystal data have been published for three salts belonging to the E form (Al, Cr, and Fe).[685,686] On the basis of a careful examination of these crystal data and of the thermal behavior of these three compounds, Bagieu-Beucher[686] concluded that these salts are probably cyclophosphates. In addition, a sixth form has been characterized by Bagieu-Beucher[687] in the case of the manganese salt. This last form is a polyphosphate. Most of the well-investigated compounds in this class of polyphosphates belong to the C form. The crystal structure of this form was first determined by Van Der Meer[688] with the aluminum salt and later reinvestigated by Middlemiss *et al.*[689] and Linde *et al.*[690] using the vanadium salt. Methods of chemical preparation and crystal data have been reported for most of the compounds belonging to the C form. $Ti(PO_3)_3$ was characterized by Liebau and Williams,[691] $Mo(PO_3)_3$ by Colani[692] and Douglass and Staritzky,[693] $Sc(PO_3)_3$ by Bagieu-Beucher[694] and Smolin *et al.*,[695] $In(PO_3)_3$ by Deichman *et al.*,[696] $Ga(PO_3)_3$ by Chudinova *et al.*[697] and Mel'nikov *et al.*[698] In the studies dealing with these compounds, various unit cells, sometimes corresponding to nonconventional space groups, were used. In a systematic investigation of this class of compounds Bagieu-Beucher[686] normalized the unit cells previously used and, in addition, characterized some additional compounds belonging to this form. Table 3.14 reports the main crystallographic features for the $M(PO_3)_3$ compounds belonging to the C form classified according to Bagieu-Beucher.[686]

Table 3.14. Unit-Cell Dimensions for the 12 Monoclinic Polyphosphates Belonging to the C Form[a]

Formula	a (Å)	b (Å)	c (Å)	β (°)	Reference(s)
$Al(PO_3)_3$	10.423(3)	18.687(2)	9.222(1)	98.37(1)	688
$Cr(PO_3)_3$	10.525	18.997	9.349	98.12	684, 685
$Ga(PO_3)_3$	10.510	18.910	9.319	98.25	697, 698
$V(PO_3)_3$	10.615	19.095	9.432	97.94	689, 690
$Fe(PO_3)_3$	10.633	19.108	9.437	98.19	684, 685
$Rh(PO_3)_3$	10.478	19.046	9.308	97.91	686
$Ti(PO_3)_3$	10.734	19.341	9.553	97.91	691
$Mo(PO_3)_3$	10.76(1)	19.48(3)	9.55(1)	97.6(1)	692, 693
$Sc(PO_3)_3$	10.844	19.582	9.694	97.92(5)	694, 695
$In(PO_3)_3$	10.907	19.621	9.674	97.85	696
$Tl(PO_3)_3$	10.987	19.867	9.721	97.40	686

[a] The common space group is $I2/a$ with $Z = 12$. The estimated standard deviations of the values for a, b, c, and β are 0.003, 0.005, 0.002, and 0.03, respectively, except in the case of the Ti and Sc salts, for which these values should be doubled.

Table 3.15. Main Crystallographic Features for the Three Representatives of the E Form and for $Mn(PO_3)_3$

Formula	a (Å)	b (Å)	c (Å)	S.G.	Z	Reference
$Al(PO_3)_3$	10.940(8)	10.940(8)	9.192(8)	Hexagonal	6	686
$Cr(PO_3)_3$	11.017(6)	11.017(6)	9.381(7)	Hexagonal	6	686
$Fe(PO_3)_3$	11.11(2)	11.11(2)	9.54(2)	Hexagonal	6	686
$Mn(PO_3)_3$	9.703(3)	10.667(3)	6.362(2)	*Pnaa*	4	687

During the reinvestigation of the crystal structure of $V(PO_3)_3$, Linde *et al.*[690] did not observe the pseudo thirding of the unit cell along the **b** direction as previously seen in all the polyphosphates belonging to the C form. From the authors' discussion, it seems that crystals prepared under equilibrium conditions do not exhibit the tripling of the unit cell along **b** direction attributed by these authors to a polytypism phenomenon.

The chemical preparation of $Mn(PO_3)_3$ was described by Bagieu-Beucher.[687] This author reported a complete description of the atomic arrangement in this polyphosphate, the only representative of the sixth form of the $M(PO_3)_3$ compounds. Table 3.15 contains the main crystallographic data for $Mn(PO_3)_3$ as well as for the three representatives of the E form.

The thermal behavior of some of these salts has been reported by several authors. A valuable survey of these various studies was provided by Bagieu-Beucher.[686] The same author also investigated the effect of high pressures on these compounds. From these observations it appears that the C form is by far the most stable, except in the case of aluminum and scandium salts, for which the A form seems to be the most stable:

$$Al(PO_3)_3(C) \overset{1283\ K}{\underset{(2\ hr)}{\longrightarrow}} Al_4(P_4O_{12})_3(A)$$

$$Sc(PO_3)_3(C) \overset{1318\ K}{\underset{(DTA)}{\longrightarrow}} Sc_4(P_4O_{12})_3(A)$$

Under prolonged heating at temperatures higher than 1273 K the polyphosphates belonging to the C form transform to the monophosphates according to the following scheme:

$$M(PO_3)_3 \rightarrow MPO_4 + P_2O_5$$

Some intermediate compounds can appear during this transformation—form A in the case of Al and Sc (see above) and diphosphates $M_4(P_2O_7)_3$ for M = Cr and In. Only in the case of Tl $(PO_3)_3$ is the C form stable up to its melting point (1063 K). The stabilities of forms A and B (cyclophosphates) are described in Chapter 4.

3.2.4.2. Ytrrium, Lanthanide and Bismuth Polyphosphates

Following the discovery of lanthanide ultraphosphates and of their interesting properties as efficient laser materials or phosphors, a large number of investigations were performed in the field of lanthanide phosphates. Most of these systematic studies were conducted in the Soviet Union.

(i) $M^{III}(PO_3)_3$ Compounds (M^{III} = Y and Ln)

A good number of $M^{III}(PO_3)_3$ compounds (M^{III} = Y and Ln from Sm to Lu) crystallize in the monoclinic form (C) already mentioned above whereas seven of them (Ln from La to Gd) are orthorhombic. Some others are polymorphic.

$Y(PO_3)_3$. The Y_2O_3–P_2O_5 system was investigated by Agrawal and Hummel.[489] $Y(PO_3)_3$ is a congruent-melting compound (mp = 1733 K).

$La(PO_3)_3$. Six intermediate compounds were found in the La_2O_3–P_2O_5 system studied by Park and Kreidler.[490] Among them, $La(PO_3)_3$ was shown to be an incongruent melting compound decomposing at 1508 K. The rate of vaporization losses of P_2O_5 leading to the monophosphate,

$$La(PO_3)_3 \rightarrow LaPO_4 + P_2O_5$$

were determined at 1173 and 1473 K.

A crystal structure determination for $La(PO_3)_3$ was performed by Matuszewski *et al.*[699] who observed a phase transition above 1366 K. This compound is isotypic with the corresponding neodymium salt. In order to obtain pure $La(PO_3)_3$, Chudinova *et al.*[700] investigated various routes for its chemical preparation and examined intermediate compounds formed during the optimized reaction between $NH_4H_2PO_4$ and La_2O_3. During an investigation of the La_2O_3–P_2O_5–H_2O system between 373 and 773 K Chudinova *et al.*[701] determined the crystallization region for $La(PO_3)_3$.

$Ce(PO_3)_3$. Bukhalova *et al.*[702] described the chemical preparation and properties of cerium polyphosphate. According to these authors, the melting point is observed at 1529 K and above 1203 K $Ce(PO_3)_3$ slowly loses mass, corresponding to the loss of P_2O_5:

$$Ce(PO_3)_3 \rightarrow CePO_4 + P_2O_5$$

The chemical preparation of various cerium phosphates, including $Ce(PO_3)_3$, was discussed by Tsuhako *et al.*[703] Vaivada and Konstant[704] investigated the reaction of CeO_2 and $(NH_4)H_2PO_4$ between 293 and 1173 K. For a P/Ce ratio <5 they observed the formation of $Ce(PO_3)_3$ and CeP_2O_7.

$Nd(PO_3)_3$. Crystal structure of $Nd(PO_3)_3$ was first described by Hong.[705] The formation of $Nd(PO_3)_3$ by reaction between Nd_2O_3 and $NH_4H_2PO_4$ was investigated by Chudinova *et al.*[452] The reaction begins at about 443 K and is completed at 773–823 K. Several intermediate compounds—$NdPO_4$, $NH_4NdP_4O_{12}$, and $NH_4NdHP_3O_{10}$—are observed during the reaction.

$Gd(PO_3)_3$. From the Gd_2O_3–P_2O_5 system, investigated by Agrawal and Hummel,[489] $Gd(PO_3)_3$ appears as a congruent–melting compound (mp = 1553 K). A transition is detected at about 1023 K.

$Eu(PO_3)_3$. This salt was characterized during an investigation of the Eu_2O_3–P_2O_5–H_2O system by Chudinova et al.[706] According to these authors, it undergoes a polymorphic transition at 1303 K.

$Er(PO_3)_3$. The crystal structure of $Er(PO_3)_3$ was determined by Dorokhova and Karpov.[707] Crystals were obtained by a hydrothermal process, which was not described by the authors. Tarasenkova et al.[708] identified two forms of $Er(PO_3)_3$ in their investigation of the K_2O–Er_2O_3–P_2O_5–H_2O system. They reported unit-cell dimensions and crystal morphologies.

$Yb(PO_3)_3$. The Yb_2O_3–P_2O_5–H_2O system was investigated between 373 and 673 K by Chudinova et al.[709] in order to determine the crystallization region of $Yb(PO_3)_3$.

The atomic arrangement of the monoclinic form was determined by Hong,[710] and a triclinic form was described by Rzaigui and Kbir-Ariguib.[711] Crystal structure of the latter form is not yet known.

(ii) BiH(PO3)4 and Bi(PO3)3

$Bi(PO_3)_3$ and $BiH(PO_3)_4$ were first characterized as polyphosphates by Chudinova et al.[712] during an investigation of the reaction of H_3PO_4 and Bi_2O_3 at various temperatures. Palkina and Jost determined the crystal structure of $Bi(PO_3)_3$,[713] as well as that of $BiH(PO_3)_4$.[714] The X-ray-photoelectron spectra of these two salts were reported by Chudinova.[715]

The thermal behavior of $BiH(PO_3)_4$ was investigated by Hilmer et al.[716] At 743 K the following decomposition is observed:

$$BiH(PO_3)_4 \rightarrow Bi(PO_3)_3 + HPO_3$$

This process is reversible when the temperature is decreased but on prolonged heating at 773–973 K bismuth polyphosphate reacts with the melt (HPO_3), with the formation of the ultraphosphate BiP_5O_{14}:

$$Bi(PO_3)_3 + 2HPO_3 \rightarrow BiP_5O_{14} + H_2O$$

(iii) LnH(PO3)4 Compounds

Acid polyphosphates of rare-earth metals were prepared by Palkina et al.[717] These authors reported the chemical preparation of these salts for Ln = Sm, Eu, Gd, Tb, Dy, Ho, and Er. These compounds are triclinic for Ln = Sm, Eu, and Gd and monoclinic for Ln = Gd, Tb, Dy, Ho, and Er. The gadolinium salt has been isolated in both forms. The atomic arrangement in the monoclinic erbium salt was described by the authors.

$EuH(PO_3)_4$. This salt was characterized by Chudinova et al.[706] during an investigation of the Eu_2O_3–P_2O_5–H_2O system between 373 and 673 K. According to these authors, this salt is isotypic with $BiH(PO_3)_4$ and decomposes into $Eu(PO_3)_3$ and EuP_5O_{14} at 843 K.

$$2\ EuH(PO_3)_4 \xrightarrow{843\ K} Eu(PO_3)_3 + EuP_5O_{14} + H_2O \xrightarrow{1243\ K} 2\ Eu(PO_3)_3 + P_2O_5$$

$ErH(PO_3)_4$. Crystal structure determination for this acid polyphosphate was performed by Palkina et al.[717]

Table 3.16.   Unit-Cell Dimensions of the Orthorhombic $Ln(PO_3)_3$ Polyphosphates[a]

Formula	a (Å)	b (Å)	c (Å)	Reference(s)
$La(PO_3)_3$	11.303(4)	8.648(5)	7.397(3)	686, 699, 718
$Ce(PO_3)_3$	11.236(4)	8.602(1)	7.349(2)	696
$Pr(PO_3)_3$	11.290(5)	8.641(4)	7.372(4)	686, 718
$Nd(PO_3)_3$	11.172(2)	8.533(2)	7.284(2)	686, 705, 718
$Sm(PO_3)_3$	11.089(4)	8.487(1)	7.237(2)	686
$Eu(PO_3)_3$	11.070(4)	8.463(2)	7.215(2)	686
$Gd(PO_3)_3$	11.050(4)	8.445(2)	7.191(2)	686

[a]The common space group is $C222_1$ with $Z = 4$.

(iv) General Investigations

Mel'nikov et al.[718] published crystal data for a number of $Ln(PO_3)_3$ polyphosphates. The phase transition temperatures and melting points of $Ln(PO_3)_3$ polyphosphates were determined by Balagina et al.[719] by Raman spectroscopy. Chudinova[720] discussed the crystalline modifications of rare-earth polyphosphates and cyclophosphates. Valtere[721] investigated the crystallization of La, Eu, Gd, and Y polyphosphates by the luminescent probe method. According to this author, the crystallization of some structural forms depends on the history of the specimen.

Gupta and Bhargava[722] investigated five polyphosphates—$Y(PO_3)_3$, $La(PO_3)_3$, $Ce(PO_3)_3$, $Pr(PO_3)_3$, and $Sm(PO_3)_3$—by differential thermal analysis and differential thermal gravimetry.

(v) Crystallographic Data

The main crystallographic data for Y, Ln, and Bi polyphosphates are gathered in Tables 3.16–3.18.

(vi) Hydrated Salts

As in some other sections of long-chain polyphosphate chemistry, some chemists claim to have characterized hydrates. Ezhova et al.[723] reported the preparation of hydrated

Table 3.17.   Main Crystallographic Data for the Y, and Ln Polyphosphates Belonging to the Monoclinic C Form[a]

Formula	a (Å)	b (Å)	c (Å)	β (°)	Reference(s)
$Y(PO_3)_3$	11.263(2)	19.93(1)	10.033(2)	97.62(4)	718
$Sm(PO_3)_3$	11.245(5)	19.86(1)	9.957(5)	97.57(4)	718
$Eu(PO_3)_3$	11.286(8)	19.74(2)	10.015(7)	97.23(4)	718
$Gd(PO_3)_3$	11.394(6)	20.31(1)	10.181(6)	96.81(4)	686, 718
$Dy(PO_3)_3$	11.358(7)	20.02(2)	10.169(7)	96.54(4)	718
$Ho(PO_3)_3$	11.288(3)	19.98(1)	10.147(3)	97.08(4)	686, 718
$Er(PO_3)_3$	11.306(4)	20.10(1)	10.076(3)	97.03(4)	708, 718
$Tm(PO_3)_3$	11.278(3)	20.01(1)	10.077(3)	97.02(4)	718
$Yb(PO_3)_3$	11.219(2)	19.983(3)	9.999(3)	97.30(2)	686, 710, 718
$Lu(PO_3)_3$	11.249(4)	19.86(1)	10.018(4)	97.42(4)	718

[a]The common space group is Ic with $Z = 12$.

Table 3.18. Main Crystallographic Data for Various Bi, and Ln Polyphosphates

Formula[a]	a (Å) α (°)	b (Å) β (°)	c (Å) γ (°)	S.G.	Z	Reference(s)
ErH(PO$_3$)$_4$	9.574(2) 100.95(1)	7.096(1)	13.637(3)	$P2_1/a$	4	717
Er(PO$_3$)$_3$	10.943(3) 91.82(2)	6.971(2)	9.670(2)	Pm	4	707
Yb(PO$_3$)$_3$	8.361(3) 86.02(3)	7.508(1) 103.35(6)	6.237(5) 90.92(1)	$P\bar{1}$ or $P1$	2	711
BiH(PO$_3$)$_4$	8.625(1) 112.17(1)	8.866(1) 108.54(1)	7.062(1) 98.49(1)	$P\bar{1}$	2	714, 717
Bi(PO$_3$)$_3$	13.732(2) 93.35(1)	6.933(1)	7.152(1)	$P2_1/a$	4	713

long-chain polyphosphates of lanthanum and neodymium. The formulas proposed by these authors for these two salts are La(PO$_3$)$_3$·6H$_2$O and Nd(PO$_3$)$_5$·7H$_2$O in a report on an investigation of the reaction of lithium polyphosphate with lanthanum and neodymium chlorides in aqueous solution at 273 K Ezhova *et al.*[724] described two hydrated polyphosphates, La(PO$_3$)$_3$·6H$_2$O and Nd(PO$_3$)$_3$·6H$_2$O. During an investigation at 273 K of the EuCl$_3$–NaPO$_3$–H$_2$O system Ezhova *et al.*[725] observed the formation of Eu(PO$_3$)$_3$·6H$_2$O. For all these hydrated salts, the degree of condensation remains to be confirmed by structural investigations.

3.2.5. Trivalent–Monovalent Cation Polyphosphates

Apart from the yttrium, bismuth, and lanthanide derivatives, trivalent–monovalent cation polyphosphates have not yet been thoroughly investigated. With the exception of some LnM$^I_2$(PO$_3$)$_5$ compounds the vast majority of (Y–Bi–Ln)–M^I polyphosphates corre-

Table 3.19. Main Geometrical Features of M^{III}M^I(PO$_3$)$_4$ Polyphosphates (M^I = Li, Na)

Formula	a (Å) α (°)	b (Å) β (°)	c (Å) γ (°)	S.G.	Z	Structure type	Reference(s)
GaLi(PO$_3$)$_4$	8.244(2)	9.044(2)	12.540(2)	$Pbna$	4		730
LaLi(PO$_3$)$_4$	16.53(3)	7.08(3) 126.42(5)	9.88(2)	$C2/c$	4	I	732
NdLi(PO$_3$)$_4$	16.408(3)	7.035(4) 126.38(5)	9.729(4)	$C2/c$	4	I	733, 734
ErNa(PO$_3$)$_4$	7.155(15)	12.987(7) 89.32(2)	9.662(32)	$P2_1/n$	4	II	740
CeNa(PO$_3$)$_4$	9.981	13.129 89.93	7.226	$P2_1/n$	4	II	738
NdNa(PO$_3$)$_4$	9.907(4)	13.10(1) 90.51(3)	7.201(3)	$\dot{P}2_1/n$	4	II	739
NdAg(PO$_3$)$_4$	9.947	13.17 90.48	7.271	$P2_1/n$	4	II	744

Table 3.20. Main Crystallographic Features of $M^{III}K(PO_3)_4$ and $M^{III}K_2(PO_3)_4$ Polyphosphates

Formula	a (Å) α (°)	b (Å) β (°)	c (Å) γ (°)	S.G.	Z	Structure type	Reference(s)
$ErK(PO_3)_4$	11.75(1)	10.325(2)	17.31(1)	$C222_1$	8	VII	759
$CeK(PO_3)_4$	7.236(1)	13.168(4) 90.46(2)	9.990(3)	$P2_1/n$	4	II	749
$CeK(PO_3)_4$	7.278(2)	8.478(4) 92.04(4)	8.037(4)	$P2_1$	2	III	748
$SmK(PO_3)_4$	7.97	8.24 92.0	7.27	$P2_1$	2	III	747
$NdK(PO_3)_4$	8.008(1)	8.438(1) 91.97(1)	7.280(1)	$P2_1$	2	III	751, 752
$YbK(PO_3)_4$	7.766(1)	8.853(1) 96.36(1)	14.831(2)	$P2_1/n$	4	V	760
$DyK(PO_3)_4$	7.97	8.47 91.0	7.31	$P2_1$	2	III	747
$HoK(PO_3)_4$	8.00	8.58 93.0	7.24	$P2_1$	2	III	747
$ErK(PO_3)_4$	7.285(1)	8.012(2) 91.96(2)	8.444(1)	$P2_1$	2	III	757
$ErK(PO_3)_4$	10.80(1)	8.859(7) 128.89(6)	12.70(1)	$P2_1/c$	4	IV	758
$BiK(PO_3)_4$	8.16	8.58 93.0	7.30	$P2_1$	2	III	747
$NdK_2(PO_3)_5$	8.430(1)	11.752(2) 90.69(1)	13.272(2)	Cc	4		754

spond to the general formula $M^{III}M^I(PO_3)_4$. The literature dealing with these compounds was rather confusing for a long time, but it is currently well established that the $LnM^I(PO_3)_4$ compounds can be classified into seven different structure types, nowadays usually denoted by roman numerals. This nomenclature, first proposed by Palkina et al.,[726] is today generally accepted. In addition, some of these compounds are polymorphic, and some of them can adopt a cyclophosphate arrangement.

Most of these compounds were discovered during investigations of numerous P_2O_5–M^I_2O–$M^{III}_2O_3$ systems, by Soviet and Tunisian researchers; flux methods were employed in the work done in the Soviet Union whereas the Tunisian investigations involved the systematic elaboration of the M^IPO_3–$M^{III}(PO_3)_3$ phase-equilibrium diagrams. We present a selection of these diagrams below. The main crystallographic data for trivalent–monovalent cation polyphosphates are gathered in Tables 3.19–3.21, and Table 3.22 shows the distribution of the $M^{III}M^I(PO_3)_4$ compounds among the seven structure types.

3.2.5.1. Trivalent Cation-Lithium Polyphosphates

(i) $M^{III}M^I(PO_3)_4$ polyphosphates (M^{III} = Al, Ga, and Cr)

The $LiPO_3$–$Al(PO_3)_3$ phase-equilibrium diagram was established by Plyshevskii et al.[727] No intermediate compound exists. The diagram is that of a simple eutectic system [7% $Al(PO_3)_3$ and 938 K].

Table 3.21. Main Crystallographic Features of $M^{III}M^I(PO_3)_4$ and $M^{III}M_2(PO_3)_5$ Polyphosphates (M^I = NH$_4$, Rb, Cs, and Tl)

Formula	a (Å) α (°)	b (Å) β (°)	c (Å) γ (°)	S.G.	Z	Structure type	Reference(s)
YNH$_4$(PO$_3$)$_4$	11.927(9)	10.357(8)	17.309(15)	$C222_1$	8	VII	762
BiNH$_4$(PO$_3$)$_4$	10.925(9)	9.034(8) 106.18(1)	10.438(9)	$P2_1/n$	4	IV	767
Ce(NH$_4$)$_2$(PO$_3$)$_5$	7.241(5) 90.35(5)	13.314(8) 107.50(5)	7.241(5) 90.28(5)	$P1$	2		765, 766
PrCs(PO$_3$)$_4$	7.159	9.190 99.66	8.809	$P2_1$	2	VI	760
NdTl(PO$_3$)$_4$	10.440(3)	8.050(2) 105.86(2)	11.007(2)	$P2_1/n$	4	IV	781
NdRb(PO$_3$)$_4$	10.461(3)	9.041(2) 106.16(2)	10.983(4)	$P2_1/n$	4	IV	764
NdCs(PO$_3$)$_4$	10.448(3)	9.039(3) 106.43(2)	11.233(2)	$P2_1/n$	4	IV	775
NdCs(PO$_3$)$_4$	7.123(2)	9.152(3) 99.72(3)	8.782(2)	$P2_1$	2	VI	773, 774
TbCs(PO$_3$)$_4$	7.032(1)	8.705(1) 100.0	9.051(1)	$P2_1$	2	VI	776, 777
PrCs(PO$_3$)$_4$	7.159	9.190 99.66	8.809	$P2_1$	2	VI	772
HoRb(PO$_3$)$_4$	10.266(4)	8.853(3) 106.28(3)	10.953(3)	$P2_1/n$	4	IV	769
TmRb(PO$_3$)$_4$	10.217(2)	8.803(2) 106.28(2)	10.928(2)	$P2_1/n$	4	IV	769
ErCs(PO$_3$)$_4$	10.215(2)	8.833(2) 106.32(2)	11.136(2)	$P2_1/n$	4	IV	769

Table 3.22. Distribution of the Seven Structure Types among the $M^{III}M^I(PO_3)_4$ Compounds[a]

MIII	Bi	Y	La	Ce	Pr	Nd	Sm	Eu	Gd	Tb	Dy	Ho	Er	Tm	Yb	Lu
Li			I	I	I	I	I	I	I	I	I	I	I	I	I	I
Na	II		II	II	II	II	II	II	II	II	II	II	II			
Ag					II											
K	III		III	III	III	III	III	III	III	III	III	III	III IV VII	V IV	V IV	V IV
NH$_4$	IV	VII														
Rb			VI	IV	VI IV	VI IV	VI IV	VI IV	VI IV	VI IV	VI IV	VI IV	VI IV	IV	IV	IV
Cs			IV VI	VI	IV VI	IV VI	IV VI	IV VI	IV VI	IV VI	IV VI	IV VI	IV VI			
Tl						IV										

[a]This table does not include all the assumptions found in scientific literature. For instance attribution of a compound to a given form based on the comparison of unindexed powder patterns have been rejected.

GaLi(PO$_3$)$_4$. The LiPO$_3$–Ga(PO$_3$)$_3$ phase-equilibrium diagram elaborated by Bukhalova *et al.*[728] is that of a simple eutectic system [10% Ga(PO$_3$)$_3$, 893 K]. Nevertheless, crystals of GaLi(PO$_3$)$_4$ were prepared by Chudinova *et al.*[729] and the crystal structure of this compound was determined by Palkina *et al.*[730]

CrLi(PO$_3$)$_4$. During an investigation of the P$_2$O$_5$–Li$_2$O–Cr$_2$O$_3$–H$_2$O system between 473 and 673 K Grunze and Chudinova[445] characterized this polyphosphate. These authors reported an indexed X-ray powder diagram but gave no information on the unit-cell dimensions.

(ii) Ln(Bi)Li(PO3)4 Polyphosphates

Chudinova and Vinogradova[731] investigated the P$_2$O$_5$–Li$_2$O–Ln(Bi)$_2$O$_3$ systems at various temperatures with a constant molar ratio (P$_2$O$_5$:Li$_2$O:Ln$_2$O$_3$ = 15:5:1) in the starting mixture. From their crystallographic measurements, the authors concluded that all the LnLi(PO$_3$)$_4$ compounds belong to the same structure type.

LaLi(PO$_3$)$_4$. The LiPO$_3$–La(PO$_3$)$_3$ phase-equilibrium diagram was established by Moktar *et al.*[732] The only compound observed in this system is LaLi(PO$_3$)$_4$, an incongruent-melting compound decomposing at 1233 K. From the crystal data reported by these authors, this polyphosphate is isotypic with NdLi(PO$_3$)$_4$.

NdLi(PO$_3$)$_4$. Nakano *et al.*[735] established the phase-equilibrium diagram for a portion of the P$_2$O$_5$–Li$_2$O–Nd$_2$O$_3$ system [LiPO$_3$–Nd(PO$_3$)$_3$] in order to determine the best conditions for the crystal growth of NdLi(PO$_3$)$_4$ single crystals. In this diagram, NdLi(PO$_3$)$_4$ appears as an incongruent-melting compound decomposing at 1243 K. Koizumi[733] and Hong[734] determined the atomic arrangement in this polyphosphate and Nakano *et al.*[735] described a method of crystal growth for the production of large single crystals.

Miscellaneous Investigations. Zorina *et al.*[736] investigated the photoluminescence and cathode luminescence of several series of solid solutions: Ln$_{(1-x)}$Eu$_x$Li(PO$_3$)$_4$ and Ln$_{(1-x)}$Tb$_x$Li(PO$_3$)$_4$, for Ln = La, Y, and Ce. The luminescence efficiency of the Tb-containing materials increases with the Ce content.

3.2.5.2. Trivalent Cation-Sodium Polyphosphates

GaNa(PO$_3$)$_4$. The NaPO$_3$–Ga(PO$_3$)$_3$ phase equilibrium diagram elaborated by Bukhalova *et al.*[728] is that of a simple eutectic system [7.5% Ga(PO$_3$)$_3$, 823 K] but the authors reported the existence of GaNa(PO$_3$)$_4$, also prepared by Chudinova *et al.*[729]

YNa(PO$_3$)$_4$ and LaNa(PO$_3$)$_4$. The NaPO$_3$–Y(PO$_3$)$_3$ and NaPO$_3$–La(PO$_3$)$_3$ phase-equilibrium diagrams were constructed by Fedorova *et al.*[737] YNa(PO$_3$)$_4$ and LaNa(PO$_3$)$_4$ are the only compounds appearing in these systems. The formation of glasses and optical properties were reported by the authors.

CeNa(PO$_3$)$_4$. The NaPO$_3$–Ce(PO$_3$)$_3$ phase-diagram reported in Fig. 3.24 was elaborated by Rzaigui *et al.*[738] CeNa(PO$_3$)$_4$, the only compound observed in this system, decomposes at 1138 K. Crystal data reported by the authors show that this polyphosphate belongs to form II.

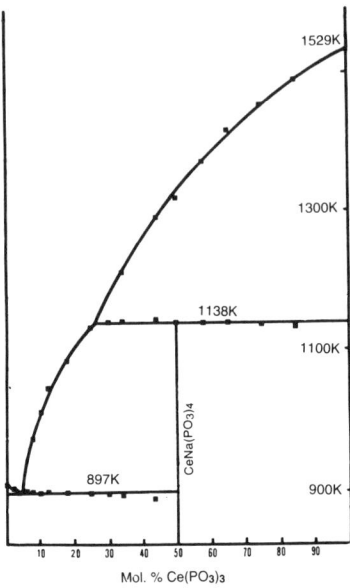

Figure 3.24. The NaPO₃–Ce(PO₃)₃ phase-equilibrium diagram.

NdNa(PO₃)₄ and ErNa(PO₃)₄. The atomic arrangement in NdNa(PO₃)₄ was determined by Koizumi,[739] and the chemical preparation and crystal structure of ErNa(PO₃)₄ were reported by Maksimova *et al.*[740] The two compounds are isotypic and belong to form II.

BiNa₃(PO₃)₆ or BiNa(PO₃)₄. Bukhalova *et al.*[741] observed only one intermediate compound, BiNa₃(PO₃)₆, in the NaPO₃–Bi(PO₃)₃ phase diagram that they elaborated. On the basis of the powder diffraction data reported by these authors and the examination of single crystals by Bagieu-Beucher,[742] this compound appears to be isotypic of NdNa(PO₃)₄, showing that the phase-equilibrium diagram needs to be revised.

Hydrated Lanthanide–Sodium Polyphosphates. In addition to the above well-characterized polyphosphates, some authors have described hydrated lanthanide–sodium polyphosphates. Ezhova *et al.*[743] reported the preparation of hydrated lanthanum–sodium and neodymium–sodium.polyphosphates. According to these authors, the formulas of these salts are LaNa(PO₃)₄·6H₂O, NdNa(PO₃)₄·7H₂O, LaNa₃(PO₃)₆·6H₂O and NdNa₃(PO₃)₆·9H₂O. The same authors also investigated the EuCl₃–NaPO₃–H₂O system at 273 K.[743] They claim to have characterized two new hydrated polyphosphates: EuNa₃(PO₃)₆·6H₂O and T159.

Eu₂Na(PO₃)₇·12H₂O. The degrees of condensation of these hydrated compounds remain uncertain.

3.2.5.3. Trivalent Cation–Silver Polyphosphates

NdAg(PO$_3$)$_4$. Trunov *et al.*[744] prepared single crystals of NdAg(PO$_3$)$_3$ during a study of the P$_2$O$_5$–Ag$_2$O–Nd$_2$O$_3$–H$_2$O system between 573 and 673 K. The best results were obtained for a starting mixture of components corresponding to a P$_2$O$_5$:Ag$_2$O:Nd$_2$O$_3$ molar ratio of 15:5:1. The authors performed a crystal structure determination, showing NdAg(PO$_3$)$_3$ to be isotypic with the corresponding sodium salt.

BiAg(PO$_3$)$_4$. An investigation of the AgPO$_3$–Bi(PO$_3$)$_3$ system by Bukhalova *et al.*[745] showed the existence of BiAg(PO$_3$)$_4$, identified as a long-chain polyphosphate by the authors. The compound melts congruently at 863 K, and crystals are columnar prisms.

3.2.5.4. Trivalent Cation–Potassium Polyphosphates

Chudinova *et al.*[746] investigated all the P$_2$O$_5$–K$_2$O–Ln(Bi)$_2$O$_3$ systems at various temperatures with a constant molar ratio in the starting mixture (P$_2$O$_5$:K$_2$O:Ln(Bi)$_2$O$_3$ = 15:5:1). Palkina *et al.*[747] reported crystal data for several LnK(PO$_3$)$_4$ polyphosphates of type III and identified the cyclotetraphosphate forms of EuK(PO$_3$)$_4$ and HoK(PO$_3$)$_4$.

CeK(PO$_3$)$_4$ and CeK$_2$(PO$_3$)$_5$. The KPO$_3$–Ce(PO$_3$)$_3$ phase-equilibrium diagram given in Fig. 3.25 was established by Rzaigui *et al.*[748] This diagram shows the existence of two incongruent-melting compounds: CeK(PO$_3$)$_4$, decomposing at 1153 K and CeK$_2$(PO$_3$)$_5$, decomposing at 1014 K.

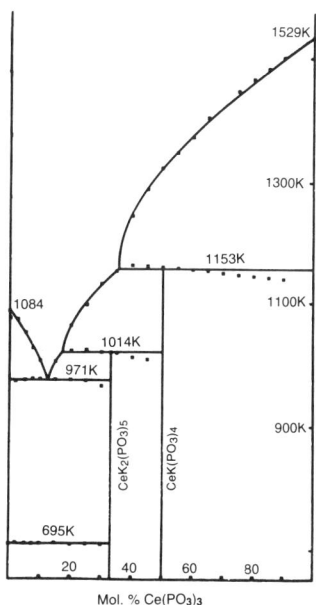

Figure 3.25. The KPO$_3$–Ce(PO$_3$)$_3$ phase-equilibrium diagram. The transformation of KPO$_3$ is observed at 695 K on the left-hand side of the diagram.

Crystals of $CeK(PO_3)_4$ were obtained by Linde et al.[749] by heating a mixture of H_3PO_4, K_2CO_3, and $CeCl_3$ having an initial molar ratio of 15:3:0.5. The authors performed the determination of the atomic arrangement, showing $CeK(PO_3)_4$ to be isotypic with $NdNa(PO_3)_4$.

$NdK(PO_3)_4$ and $NdK_2(PO_3)_5$. The synthesis and properties of single crystals of $NdK(PO_3)_4$ were described by Litvin et al.[750] Crystal structure of $NdK(PO_3)_4$ was determined simultaneously by Palkina et al.[751] and Hong.[752] The epitaxial growth of $NdK(PO_3)_4$ on $LaK(PO_3)_4$ substrates was reported by Myazawa et al.[753] This salt exists also as a cyclotetraphosphate, $NdKP_4O_{12}$.

Single crystals of $NdK_2(PO_3)_5$ were prepared by Palkina et al.[754] The authors reported a complete description of the atomic arrangement. $NdK_2(PO_3)_5$ is isotypic with $Ce(NH_4)_2(PO_3)_5$.

$EuK(PO_3)_4$. Synthesis of $EuK(PO_3)_4$ was described by Chudinova et al.[755] This salt exists also as a cyclotetraphosphate, $EuKP_4O_{12}$.

$ErK(PO_3)_4$. During a careful investigation of the $P_2O_5–K_2O–Er_2O_3–H_2O$ system, Tarasenkova et al.[756] identified four crystalline forms of $ErK(PO_3)_4$, three long-chain polyphosphates, and one cyclotetraphosphate. Unit-cell dimensions and morphologies were given by these authors. The crystal structure of form II was determined by Krutik et al.,[757] that of form IV by Dago et al.,[758] and that of form VII by Karpov and Dorokhova.[759]

$YbK(PO_3)_4$. The $P_2O_5–K_2O–Yb_2O_3–H_2O$ system was studied by Palkina et al.[760] These authors observed the crystallization of four compounds: $YbKHP_3O_{10}$, $Yb(PO_3)_3$, $YbKP_2O_7$, and $YbK(PO_3)_4$. In addition, they performed a crystal structure determination for $YbK(PO_3)_4$ (V).

$BiK(PO_3)_4$. The $KPO_3–Bi(PO_3)_3$ phase-equilibrium diagram was established by Bukhalova et al.[761] The incongruent-melting compound $BiK(PO_3)_4$, decomposing at 899 K, is the only intermediate observed in this system. This salt was later recognized as belonging to type III.[747]

3.2.5.5. Trivalent Cation–Ammonium Polyphosphates

$YNH_4(PO_3)_4$. This salt was characterized by Bagieu-Beucher and Guitel[762] who presented a detailed description of the atomic arrangement. This description was the first given for form VII.

$CeNH_4(PO_3)_4$ and $Ce(NH_4)_2(PO_3)_5$. Vaivada and Konstant[764] investigated the reaction between CeO_2 and $NH_4H_2PO_4$ between 293 and 1173 K. With a P/Ce ratio of 5 and between 653 and 873 K, mainly $Ce(NH_4)_2(PO_3)_5$, $CeNH_4P_4O_{12}$, and CeP_5O_{14} crystallize from the melts. CeP_2O_7 and $Ce(PO_3)_3$ appear at higher CeO_2 concentrations.

Rzaigui and Ariguib[763] characterized $CeNH_4(PO_3)_4$ and $Ce(NH_4)_2(PO_3)_5$ and described their chemical preparation. Crystal data were also reported by these authors. Vaivada and Konstant[764] also reported the chemical preparation of $Ce(NH_4)_2(PO_3)_5$, a long-chain polyphosphate. This compound transforms into the ultraphosphate CeP_5O_{14} at 823–943 K:

$$Ce(NH_4)_2(PO_3)_5 \rightarrow CeP_5O_{14} + 2NH_3 + H_2O$$

Crystal structure determinations for this salt have been performed twice: first by Palkina *et al.*[765] with the proper symmetry and unit cell but with an incorrect formula [Ce(NH$_4$)(PO$_3$)$_5$] and later by Rzaigui *et al.*[766] with the correct formula but an improper unit cell.

BiNH$_4$(PO$_3$)$_4$. Crystals of BiNH$_4$(PO$_3$)$_4$ were obtained by Averbuch-Pouchot and Bagieu-Beucher[767] during the course of an investigation of the P$_2$O$_5$–(NH$_4$)$_2$O–Bi$_2$O$_3$ system that was undertaken in order to optimize the conditions for crystal growth of BiNH$_4$HP$_3$O$_{10}$. The crystal structure was determined, and BiNH$_4$(PO$_3$)$_4$ was recognized as belonging to form IV.

3.2.5.6. Trivalent Cation–Rubidium Polyphosphates

Byrappa and Litvin[770] investigated the P$_2$O$_5$–Rb$_2$O–Ln$_2$O$_3$–H$_2$O systems within the temperature range 573–1073 K. They presented a composition diagram showing fields of crystallization for the different phases. The crystal chemistry of the various observed phases was discussed.

NdRb(PO$_3$)$_4$. The atomic arrangement in NdRb(PO$_3$)$_4$ was refined by Maksimova *et al.*[768] The luminescence properties of this salt were described by the authors.

HoRb(PO$_3$)$_4$ and TmRb(PO$_3$)$_4$. The atomic arrangements in these two polyphosphates were determined by Maksimova *et al.*[769]

3.2.5.7. Trivalent Cation–Cesium Polyphosphates

Vinogradova and Chudinova[779] performed a detailed investigation of the P$_2$O$_5$–Rb$_2$O–Ln$_2$O$_3$ and P$_2$O$_5$–Cs$_2$O–Ln$_2$O$_3$ systems. They discussed the stability of the compounds obtained in relation to the nature of the alkali metal and of the lanthanide cations. Crystal growth, crystal structures, and infrared spectra of LnCs(PO$_3$)$_4$ and LnCsP$_4$O$_{12}$ compounds were discussed by Byrappa *et al.*[780]

CeCs(PO$_3$)$_4$. The CsPO$_3$–Ce(PO$_3$)$_3$ phase-equilibrium diagram was established by Bukhalova *et al.*[771] CeCs(PO$_3$)$_4$, the only intermediate compound characterized in this system, decomposes at 1163 K. The authors did not report crystal data.

PrCs(PO$_3$)$_4$. The crystal structure of the monoclinic form of this salt was described by Palkina *et al.*[772]

NdCs(PO$_3$)$_4$. The chemical preparation and crystal structure of NdCs(PO$_3$)$_4$ were reported by Koizumi and Nakano.[773] This polyphosphate belongs to the form VI of the LnM(PO$_3$)$_4$ compounds. This crystal structure was reexamined by Maksimova *et al.*[774] These authors also investigated the luminescent properties of this salt at 77 K. Maksimova *et al.*[775] also determined the crystal structure of a second form of NdCs(PO$_3$)$_4$ corresponding to form IV of the LnM(PO$_3$)$_4$ compounds.

TbCs(PO$_3$)$_4$. This polyphosphate was first described as a triclinic compound by Palkina *et al.*[776] Later, its atomic arrangement was reexamined and described with a monoclinic symmetry by Palkina *et al.*[777]

ErCs(PO$_3$)$_4$. The crystal structure determination for ErCs(PO$_3$)$_4$ was performed by Maksimova *et al.*[775]

$BiCs(PO_3)_4$. This salt appears as a congruent-melting compound (mp = 853 K) in the $CsPO_3$–$Bi(PO_3)_2$ phase-equilibrium diagram elaborated by Bukhalova et al.[778]

3.2.5.8. Trivalent Cation–Thallium Polyphosphates

$NdTl(PO_3)_4$. Palkina et al.[781] described the chemical preparation of this compound and determined its crystal structure. This salt belongs to form IV of the $LnM(PO_3)_4$ compounds. The spectral luminescence properties of $NdTl(PO_3)_4$ have been investigated.[768]

3.2.5.9. Miscellaneous Investigations

The luminescence of Ce^{3+} in poly- and metaphosphates was investigated by Rzaigui et al.[782]

A good number of $LnM(PO_3)_4$ and $LnMP_4O_{12}$ compounds have several crystalline forms. Their thermal transformations were studied by Raman spectroscopy over a wide range of temperatures by Banishev et al.[783]

3.2.6. Trivalent-Divalent Cation Long-Chain Polyphosphates

The $Al(PO_3)_3$–$Ba(PO_3)_2$ system was studied by Kuz'menkov et al.[784] The phase-equilibrium diagram is of the eutectic type, with eutectic composition of 11.5 mol % $Al(PO_3)_3$ and 88.5 mole% $Ba(PO_3)_2$ at 1103 K.

3.2.7. Tetravalent Cation Long-Chain Polyphosphates

$Zr(PO_3)_4$ and $Hf(PO_3)_4$. The ZrO_2–P_2O_5 and HfO_2–P_2O_5 phase-equilibrium diagrams were elaborated by Mal'shikov and Bondar.[785,786] In both diagrams the polyphosphates are observed as incongruent-melting compounds decomposing at 873 (Zr) and 923 K (Hf). At 923 K, $Hf(PO_3)_4$ transforms into HfP_2O_7 releasing P_2O_5. The same type of decomposition is observed for the zirconium salt at 873 K. The atomic arrangement in the low-temperature form of $Zr(PO_3)_4$ was determined by Gorbunova et al.[787]

$Ce(PO_3)_4$. Masse and Grenier[788] reported the chemical preparation of the orthorhombic form of $Ce(PO_3)_4$. The crystal data given by the authors need some revision. The chemical preparation of various cerium condensed phosphates including $Ce(PO_3)_4$ was discussed by Tsuhako et al.[789]

$Th(PO_3)_4$. Masse and Grenier[788] reported the existence of three forms of $Th(PO_3)_4$ and their chemical preparation. Burdese and Borlera[790] also reported chemical preparation of and crystal data for $Th(PO_3)_4$. The same authors[791] investigated the UO_2–P_2O_5 and ThO_2–P_2O_5 systems. $U(PO_3)_4$ and $Th(PO_3)_4$ were shown to be isotypic, on the basis of their crystal data and their complete miscibility.

$U(PO_3)_4$. Masse and Grenier[788] described the chemical preparation of and crystal data for two forms of $U(PO_3)_4$. Linde et al.[792] described the chemical preparation and reported the atomic arrangement of the orthorhombic form of $U(PO_3)_4$. Burdese and Borlera[790] also described the chemical preparation of $U(PO_3)_4$ and reported crystal data. Baskin[793] described the chemical preparation and main properties of α- and β-$U(PO_3)_4$.

$Pu(PO_3)_3$. Douglass[794] reported crystallographic data for the orthorhombic forms of $Pu(PO_3)_4$ and $U(PO_3)_4$. The two compounds are isomorphous.

Table 3.23. Main Crystallographic Data for Tetravalent Cation Long-Chain Polyphosphates

Formula	a (Å) α (°)	b (Å) β (°)	c (Å) γ (°)	S.G.	Z	Reference
β-U(PO$_3$)$_4$	6.907(1)	14.947(2)	8.986(1)	*Pbcn*	4	320
α-U(PO$_3$)$_4$	23.42	13.02 90.0(2)	23.00	?	32	321
Zr(PO$_3$)$_4$	13.495(1)	28.799(2) 90.04(2)	8.658(1)	*Cc*	16	315
Zr(PO$_3$)$_4$	20.504	9.758	8.490	*Pca2$_1$*	8	313

Crystallographic Data. Table 3.23 contains crystal data for tetravalent cation long-chain polyphosphates.

3.2.8. Uranyl Polyphosphates

(UO$_2$)H(PO$_3$)$_3$ and (UO$_2$)(PO$_3$)$_2$. (UO$_2$)H(PO$_3$)$_3$ was prepared by Linde *et al.*[795] by heating a mixture of uranyl nitrate and H$_3$PO$_4$ (P/UO$_2$ = 4/6) at 613–653 K. The crystal structure was determined. According to these same authors, when this compound is heated at 1023 K, it is converted into a phosphate of tetravalent uranium corresponding to the formula U$_2$P$_6$O$_{19}$. (UO$_2$)H(PO$_3$)$_3$ was investigated by neutron diffraction by Sarin *et al.*[796]

The thermal stability of (UO$_2$)(PO$_3$)$_2$ was studied by Barten and Cordfunken.[335,336] According to Lavrov,[797] (UO$_2$)(PO$_3$)$_2$ is difficult to crystallize and is formed only over a small temperature range (553–573 K). This author reported a chemical preparation of pure (UO$_2$)(PO$_3$)$_2$.

Na(UO$_2$)(PO$_3$)$_3$. Na(UO$_2$)(PO$_3$)$_3$ was prepared by Linde *et al.*[798] by slowly heating a mixture of H$_3$PO$_4$, NaNO$_3$, and uranyl nitrate in the molar ratio 12:3:1 to 633 K and keeping it at this temperature for 5–6 days. The authors performed the crystal structure determination.

Crystal data for (UO$_2$)H(PO$_3$)$_3$ and Na(UO$_2$)(PO$_3$)$_3$ are given in Table 3.24.

3.3. Chemical Preparations of Long-Chain Polyphosphates

In most of the published characterizations of polyphosphates, a more or less detailed preparative procedure is reported, but unfortunately not always a reproducible one.

Table 3.24. Main Crystallographic Data for Uranyl Long-Chain Polyphosphates

Formula	a (Å) α (°)	b (Å) β (°)	c (Å) γ (°)	S.G.	Z	Reference(s)
(UO$_2$)H(PO$_3$)$_3$	9.8110(9)	20.814(3)	8.6947(4) 94.09(1)	*P2$_1$/b*	8	795, 796
Na(UO$_2$)(PO$_3$)$_3$	7.070(2)	13.481(5)	9.660(3) 105.48(2)	*P2$_1$/c*	4	798

As we have already mentioned a feature common to all long-chain polyphosphates is their insolubility in water. This property limits considerably the possibilities for synthesis of these compounds. In addition, all these compounds form glasses upon melting, inhibiting the use of some classical processes of crystal growth. Nevertheless, some more or less general procedures for their preparation can be described.

3.3.1. Thermal Dehydration of Dihydrogenomonophosphates

Most of the monovalent or divalent cation polyphosphates can be prepared as polycrystalline or glassy specimens by thermal dehydration of the corresponding dihydrogenomonophosphates:

$$MH_2PO_4 \rightarrow MPO_3 + H_2O$$

$$M(H_2PO_4)_2 \rightarrow M(PO_3)_2 + 2H_2O$$

In the case of some divalent cations (Cu, Mg, Ni, Co, etc.) the final product is the cyclotetraphosphate.

3.3.2. Thermal Methods

A good number of higher valency cation polyphosphates or mixed-cation polyphosphates have been prepared as polycrystalline samples by using various thermal processes, in most of cases very similar to that used, for instance, for the preparation of nickel–potassium polyphosphate:

$$NiCO_3 + 1/2K_2CO_3 + 3(NH_4)_2HPO_4 \rightarrow NiK(PO_3)_3 + 3/2CO_2 + 6NH_3 + 9/2H_2O$$

The reaction temperature must be optimized by successive attempts or determined from the phase-equilibrium diagram, when it exists.

3.3.3. Flux Methods

The preparation of single crystals is, as in the case of most of the non-water-soluble condensed phosphates, achieved through the use of flux methods. Generally, large excesses of H_3PO_4 or $(NH_4)_2HPO_4$ are used for these processes. Upon progressive heating, they transform into more or less viscous substances corresponding approximately to the formulas HPO_3 and $(NH_4)PO_3$.

At the end of the crystallization process, the excess flux is easily removed by hot water. One must nevertheless remember that the refinement of a flux process in order to achieve crystallization of a single phase involves the optimization of a number of parameters, such as initial ratios of the components, temperature, nature of the crucible, time of heating, pressure, nature of the atmosphere etc. and is always a long and tedious enterprise.

3.3.4. Hydrothermal Methods

A small number of long-chain polyphosphates, for instance, $FeLi(PO_3)_3$, have been prepared by hydrothermal processes.

3.4. Some Atomic Arrangements of Long-Chain Polyphosphates

3.4.1. Sodium Polyphosphate (A)

$$NaPO_3,^{511,512} \text{monoclinic}, P2_1/n, Z = 8$$

$$a = 12.12(4), \ b = 6.20(2), \ c = 6.99(3) \text{ Å}, \ \beta = 92.0(5)°$$

In this atomic arrangement, infinite $(PO_3)_n$ chains with a period of four tetrahedra spiral around the helical axes, parallel to the **b** direction. Figure 3.26, a projection along the **c** direction, shows how these chains are separated by corrugated layers of associated cations. Figure 3.27, a projection along the chain axis (**b**), probably illustrates more clearly the general organization of this arrangement. The two independent sodium atoms both have fivefold coordination, with Na–O distances ranging from 2.343 to 2.511 Å.

The corresponding silver polyphosphate, $AgPO_3$, is isotypic.[523]

3.4.2. Rubidium Polyphosphate

$$RbPO_3,^{31,529} \text{monoclinic}, P2_1/n, Z = 4$$

$$a = 12.123(2), b = 4.228(2), c = 6.479(2) \text{ Å}, \beta = 96.32(30)°$$

This very simple arrangement is represented in projection along the **c** axis in Fig. 3.28. Two phosphoric chains with a period of two tetrahedra cross the unit cell. These chains parallel to the **b** axis spiral around the screw axis. The coordination of the rubidium atom may be regarded as built by a distorted octahedron of oxygen atoms with a mean Rb–O distance of 2.96 Å and a seventh neighbor at a longer distance (3.20 Å).

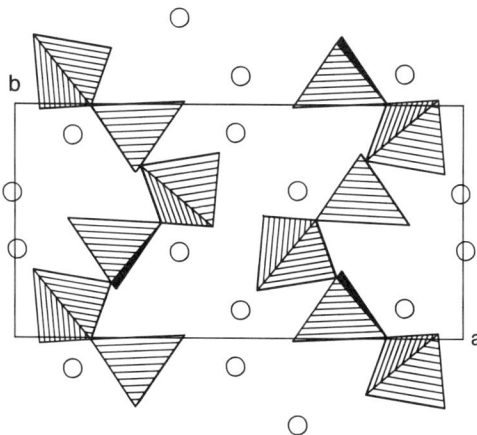

Figure 3.26. Projection, along the **c** axis, of the atomic arrangement of $NaPO_3$ (A). The open circles represent the sodium atoms.

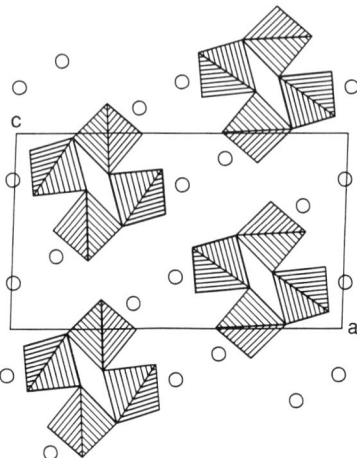

Figure 3.27. Projection, along the *b* axis, of the atomic arrangement in NaPO$_3$ (A). Empty circles figurate the sodium atoms.

Figure 3.28. Projection, along the *c* axis, of the atomic arrangement in RbPO$_3$. The open circles figurate the rubidium atoms.

3.4.3. Potassium–Silver Polyphosphate

KAg(PO$_3$)$_2$,[543] monoclinic, $P2_1/a$, $Z = 4$

$a = 7.490(6)$, $b = 13.175(10)$, $c = 6.037(5)$ Å, $\beta = 94.32(6)°$

The unit cell is crossed by two chains extending parallel to the **a** direction. The period of the chain corresponds to four tetrahedra but, owing to the presence of the *a* glide plane,

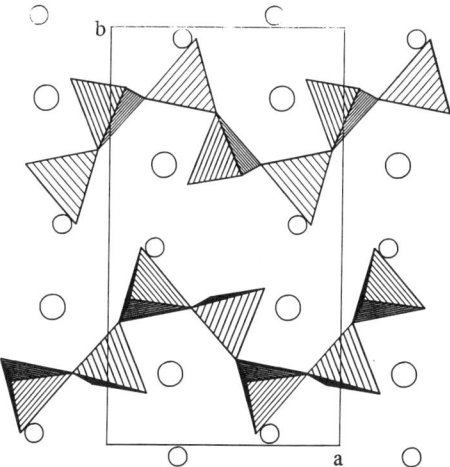

Figure 3.29. Projection of the atomic arrangement in $KAg(PO_3)_2$ along the c direction. The small open circles represent the silver atoms, and the larger ones the potassium atoms.

the chain is in fact built up by only two crystallographically independent tetrahedra. Figure 3.29, a projection along the c direction, shows the respective locations of phosphoric chains and associated cations. In this view, the zigzag configuration of the chain is clearly apparent in contrast with its rather straight appearance in Fig. 3.30, the projection of an isolated chain along the b direction.

The associated cations establish the three-dimensional cohesion between the phosphoric chains. Within a limit of 3.5 Å, the potassium atom has an eightfold oxygen coordination, with K–O distances ranging from 2.697 to 3.256 Å. The silver atom is surrounded by five oxygen atoms, with Ag–O distances ranging from 2.309 to 2.592 Å.

As is very often the case in the description of long-chain polyphosphates, the best way to comprehend the atomic arrangement is to examine its projection along the chain direction. For this structure, the projection along the **a** direction in Fig. 3.31 illustrates the validity of this statement, as it shows clearly how the phosphoric chains are intercalated between very corrugated layers built up by the associated cations.

The corresponding rubidium–silver, cesium–silver and thallium–silver polyphosphates are isotypic.[543]

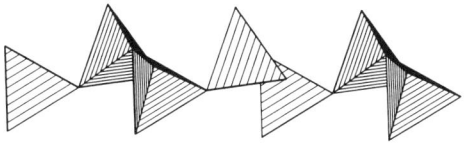

Figure 3.30. An isolated phosphoric chain as observed in $KAg(PO_3)_2$. This projection along the b direction shows the rather flat configuration of this anion.

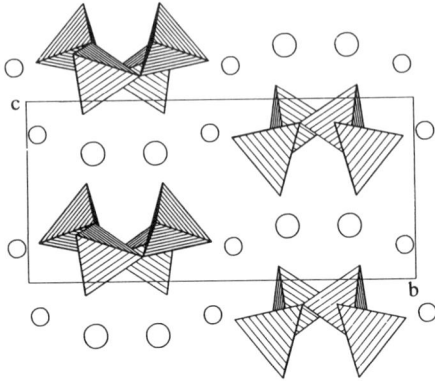

Figure 3.31. Projection of the atomic arrangement in $KAg(PO_3)_2$ along the *a* direction. The small open circles represent the silver atoms, and the larger ones the potassium atoms.

3.4.4. Beryllium Polyphosphate (II)

$$Be(PO_3)_2 \text{ (II)},^{560,561} \text{ monoclinic, } P2_1/n, Z = 4$$

$$a = 6.966(4), b = 12.875(8), c = 4.844(3) \text{ Å}, \beta = 106.73(2)°$$

As shown in Fig. 3.32, two very corrugated phosphoric chains cross the unit cell parallel to the **a** axis. They are interconnected in a three-dimensional array by the BeO_4 tetrahedra. Each BeO_4 tetrahedron shares its four corners with four adjacent PO_4 tetrahedra, with Be–P distances ranging from 2.901 to 2.974 Å. These rather short distances explain

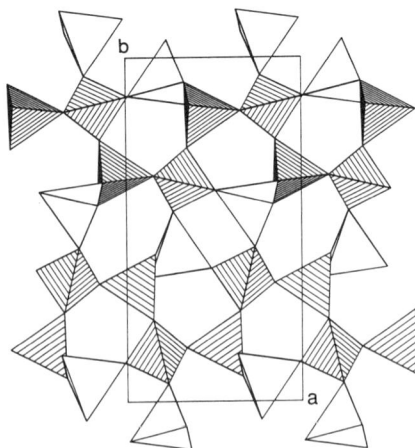

Figure 3.32. Projection of the structure of $Be(PO_3)_2$ along the *c* axis. The hatched tetrahedra represent the PO_4 groups, and the plain ones the BeO_4 tetrahedra.

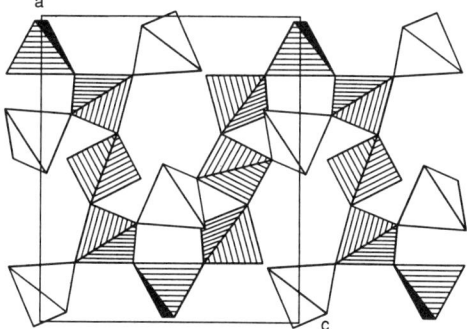

Figure 3.33. Projection of a part of the structure of $Be(PO_3)_2$ along the b axis. The hatched tetrahedra correspond to the PO_4 tetrahedra, and the plain ones represent the BeO_4 tetrahedra. This projection is limited to $0.250 < y < 0.875$.

the very compact nature of this atomic arrangement (17.27 Å^3 per oxygen). Within a BeO_4 tetrahedron, the Be–O distances range between 1.604 and 1.622 Å. All attempts to compare this three-dimensional network of tetrahedra with known forms of silica failed. The period of the phosphoric chain correspond to four tetrahedra.

3.4.5. Beryllium Polyphosphate (III)

$$Be(PO_3)_2 \text{ (III)},^{562,563} \text{orthorhombic, } C222_1, Z = 8$$

$$a = 9.968(4), b = 10.080(4), c = 8.692(2) \text{ Å}$$

Very corrugated phosphoric chains with a period of eight tetrahedra extend along the c axis. These chains are joined by BeO_4 tetrahedra to give a three-dimensional framework with a topology identical to that of the keatite form of silica. This arrangement provided the first example of a silica-like structure in which the silicon atoms are replaced in the ratio 1:2 instead of the 1:1 ratio commonly observed in, for instance, the $M^{III}PO_4$ derivatives. Figure 3.33 presents a projection of this atomic arrangement along the b axis.

Within the BeO_4 tetrahedron, the observed Be–O distances, ranging from 1.55 to 1.66 Å, cover a wider span than those reported for form II, described above. It should also be noted that the mean P–O–Be angle ($157.2°$) is considerably larger than the mean P–O–P angle ($138.6°$).

3.4.6. Zinc Polyphosphate

$$Zn(PO_3)_2,^{565,566} \text{monoclinic, } C2/c, Z = 2$$

$$a = 9.734(2), b = 8.889(2), c = 4.963(1) \text{ Å}, \beta = 108.49(5)°$$

The phosphoric chain extends parallel to the c axis with a period of two tetrahedra. The zinc atoms located on the twofold axes have an octahedral coordination. The ZnO_6 octahedra assemble so as to form chains parallel to the phosphoric chains, each ZnO_6 octahedron sharing two edges with its two adjacent neighbors. Figure 3.34, a projection

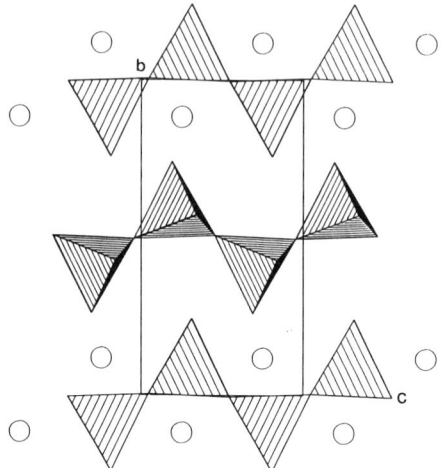

Figure 3.34. Projection of the structure of Zn(PO₃)₂ along the *a* axis. The open circles represent the zinc atoms.

along the **a** axis, shows the two phosphoric chains crossing the cell while Fig. 3.35. illustrates the very regular layer organization of this arrangement. In addition, Fig. 3.36. shows a projection along the a axis illustrating the organization of the ZnO_6 chains crossing the cell. This structure is probably the most compact ever observed in condensed phosphate chemistry, with a volume of 16.97 A^3 per oxygen atom. In this arrangement the Zn–Zn distances are rather short (3.252 Å) and are comparable to the Zn–P distances (3.138 and 3.339 Å). These short interatomic distances probably explain the compactness of this

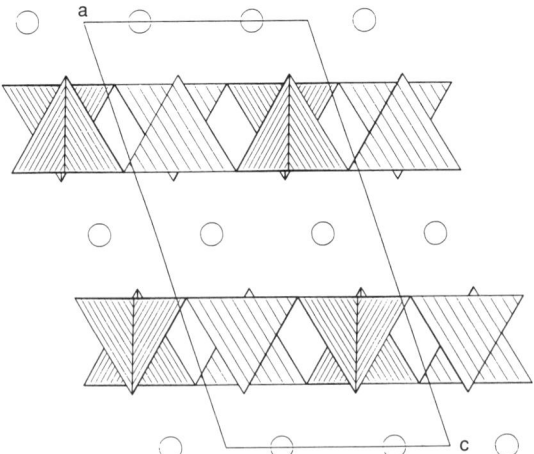

Figure 3.35. Projection of the structure of Zn(PO₃)₂ along the *b* axis, showing the very regular layer organization of this arrangement. The open circles represent the zinc atoms.

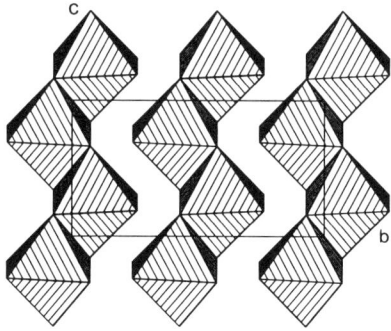

Figure 3.36. Projection along the *a* axis of the ZnO_6 chains crossing $Zn(PO_3)_2$ unit cell.

arrangement. Whithin the very regular ZnO_6 octahedron, the Zn–O distances vary from 2.004 to 2.224 Å.

Seven polyphosphates crystallize with this type of arrangement (see Table 3.3).

3.4.7. Calcium Polyphosphate

$$Ca(PO_3)_2,^{9,575} \text{ monoclinic, } P2_1/a, Z = 8$$

$$a = 16.960(9), b = 7.7144(2), c = 6.9963(2) \text{ Å}, \beta = 90.394(5)°$$

The chains extend along the **c** direction with a period of four tetrahedra. As shown in Fig. 3.37, a projection along the **b** axis, four of them cross the unit cell. The chains are connected by the calcium atoms. One of them is coordinated by eight oxygen atoms close

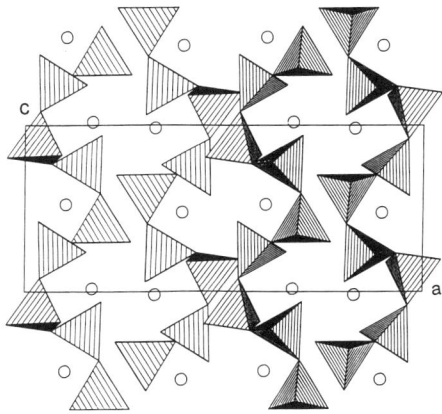

Figure 3.37. Projection, on the (a,c) plane, of the atomic arrangement in $Ca(PO_3)_2$. The open circles represent the calcium atoms.

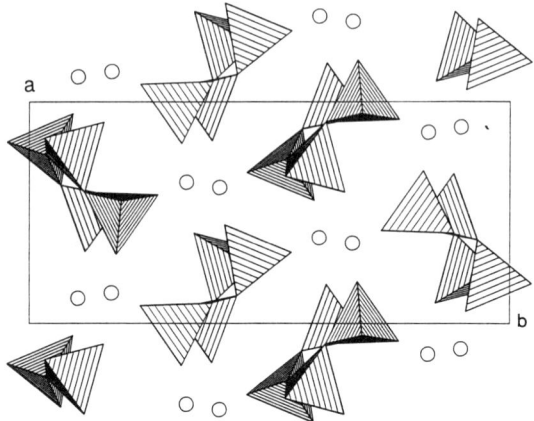

Figure 3.38. Projection, along the c axis, of the atomic arrangement in Ca(PO$_3$)$_2$, showing the layer organization between the phosphoric chains and the associated cations, represented by the open circles.

to a tetragonal antiprismatic arrangement, with Ca–O distances ranging from 2.381 to 2.696 Å; the second one is surrounded by seven oxygen atoms forming a capped trigonal prism, with Ca–O distances ranging from 2.338 to 2.637 Å. As is very often the case in the examination of long-chain polyphosphates, it is probably easier to understand the structure by considering its projection along the chain direction (Fig. 3.38).

Pb(PO$_3$)$_2$ and Sr(PO$_3$)$_2$ have the same type of structure.

3.4.8. Cadmium Polyphosphate (Low-Temperature Form)

Cd(PO$_3$)$_2$, low-temperature form,[580–582] orthorhombic, *Pbca*, $Z = 8$

$$a = 9.607(3), b = 13.70(1), c = 7.037(3) \text{ Å}$$

Chains with a period of four tetrahedra extend parallel to the **c** axis spiraling around the 2$_1$ axes. Figure 3.39, a projection along the **a** axis, gives a general view of this

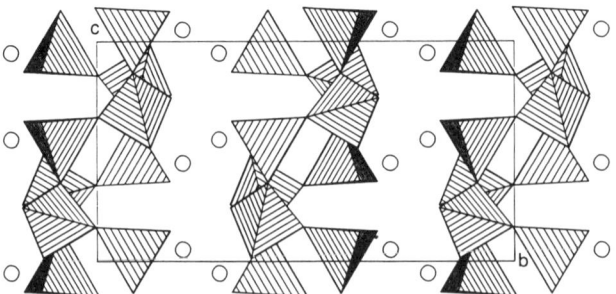

Figure 3.39. Projection, along the a axis, of the atomic arrangement in Cd(PO$_3$)$_2$. The open circles represent the cadmium atoms.

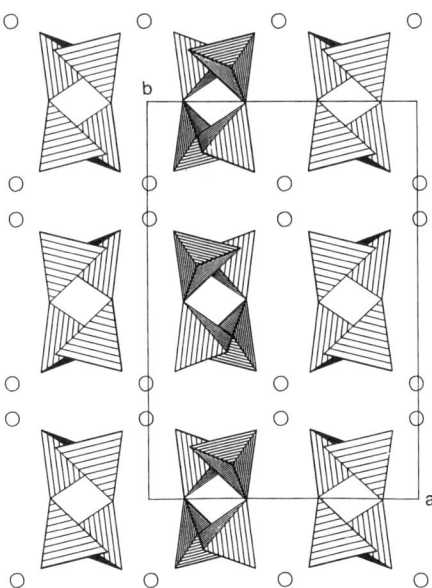

Figure 3.40. Projection of the structure of $Cd(PO_3)_2$ along the direction of the phosphoric chain (c), showing the layer organization of the arrangement.

arrangement, showing how thick layers containing the phosphoric chains alternate with layers of associated cations. Figure 3.40 shows the same arrangement projected along the direction of the chains, illustrating more clearly the layered organization of this structure. The connection between the chains is made by CdO_6 octahedra centered approximately in the planes $z = 1/4$ and $3/4$. Within these layers, the CdO_6 octahedra, by sharing edges form infinite chains parallel to the **c** axis. Figure 3.41, a projection along **b**, represents the organization of such chains in the layer centered by the plane $z = 1/4$. The CdO_6 octahedron is itself rather distorted, with Cd–O distances ranging from 2.171 to 2.445 Å and O–Cd–O angles ranging from 75.7 to 122.2°.

Mercury polyphosphate crystallizes with the same type of structure.

3.4.9. Cadmium Polyphosphate (High-Temperature Form)

$Cd(PO_3)_2$, high-temperature form,[583,584] orthorhombic, $P2_12_12_1$, $Z = 4$

$$a = 7.428(2), b = 7.360(2), c = 8.577(2) \text{ Å}$$

The unit cell is crossed by two phosphoric chains with a period of four tetrahedra, extending parallel to the **b** direction. The edge-sharing CdO_6 octahedra assemble so as to form chains parallel to the phosphoric chains but in planes $x = 0$ and $1/2$ (Fig. 3.41). Within the cadmium octahedra, the Cd–O distances range from 2.33 to 2.39 Å. Figure 3.42. shows a projection of this atomic arrangement along the **a** direction.

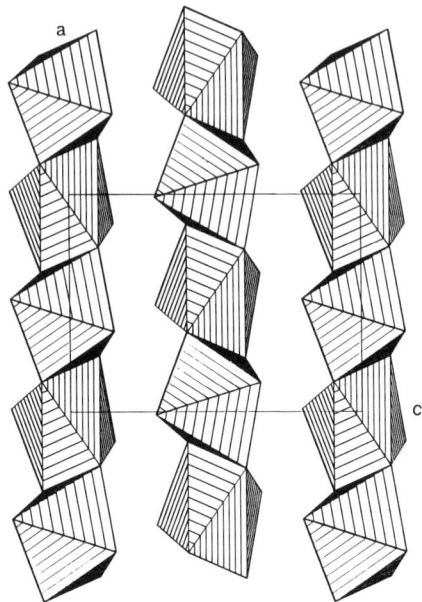

Figure 3.41. Infinite chains of CdO_6 edge-sharing octahedra in the structure of $Cd(PO_3)_2$ as observed in the layer centered by the plane $y = 1/4$.

The main feature of this atomic arrangement is the high thermal factors for the oxygen atoms of the P–O–P bonds ($B = 9$ and 11 Å^2), reflecting the metastability of this form of $Cd(PO_3)_2$. Some abnormal geometrical features showing the instability of this species may also be noted. The two P–O–P angles, 150.2 and 154.2°, are very significantly larger than those usually observed in condensed phosphate anions whereas the P–O distances involved

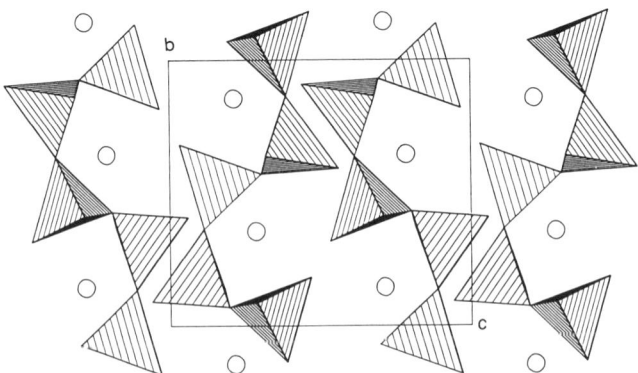

Figure 3.42. Projection, along the *a* direction, of the atomic arrangement in the high temperature form of $Cd(PO_3)_2$. The open circles represent the cadmium atoms.

in these bonds (1.524 < P–O < 1.555 Å) are much shorter than those normally expected (1.59–1.60 Å). The P–P distances are within the normal range of values.

3.4.10. β-Barium Polyphosphate

$$\beta\text{-Ba(PO}_3)_2,^{588}\text{orthorhombic, }P2_12_12_1, Z = 4$$

$$a = 4.510(2), b = 13.44(2), c = 8.360(5) \text{ Å}$$

Four chains with a period of two tetrahedra cross the unit cell parallel to the **a** axis. Figure 3.43 presents a projection of this arrangement along the chain direction. This drawing illustrates clearly the alternation of planes containing the chains and planes containing the barium atoms.

Figure 3.44 represents the same arrangement in projection along the **c** axis in order to show the four chains crossing the cell. The barium atom has an irregular eightfold coordination, with Ba–O distances ranging from 2.73 to 3.19 Å.

Figure 3.43. Projection of the structure of β-Ba(PO₃)₂ along the *a* axis. The open circles represent the barium atoms.

3.4.11. γ-Barium Polyphosphate

$$\gamma\text{-Ba(PO}_3)_2,^{587,589}\text{monoclinic, }P2_1/n, Z = 4$$

$$a = 9.692(8), b = 6.894(4), c = 7.512(5) \text{ Å}, \beta = 94.77(3)°$$

Two phosphoric chains with a period of four tetrahedra cross the unit cell parallel to the **b** direction. Figure 3.45, a projection along the direction of the chain, shows clearly the layer organization of this atomic arrangement. Layers containing the chains alternate with planes of barium atoms, both being almost perpendicular to the [110] direction. Figure

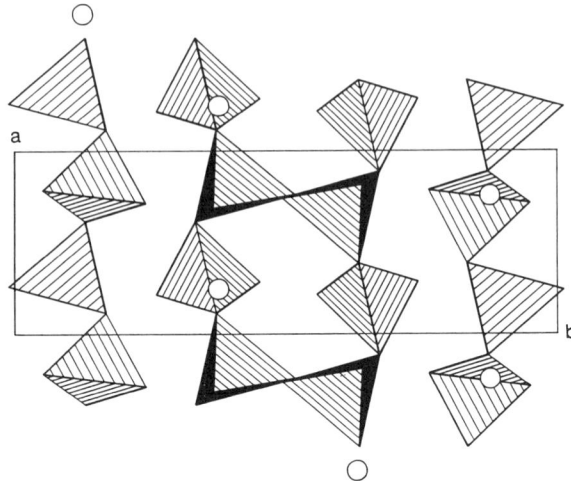

Figure 3.44. Projection of the structure of β-Ba(PO$_3$)$_2$ along the c axis. The open circles represent the barium atoms.

3.46, a projection along the **c** direction, shows the corrugated aspect of the phosphoric chain.

The barium atom is coordinated by eight oxygen atoms in a rather irregular arrangement, with Ba–O distances ranging from 2.740 to 3.064 Å and with an average Ba–O distance of 2.81 Å. These eight oxygen atoms build a distorted dodecahedron. Each such polyhedron shares two faces with its two adjacent neighbors, forming a chain parallel to the **b** direction. Within such a chain, the Ba–Ba distance is 4.118 Å. These chains of BaO$_8$ polyhedra are in fact not independent, since, in addition, each BaO$_8$ polyhedron shares two

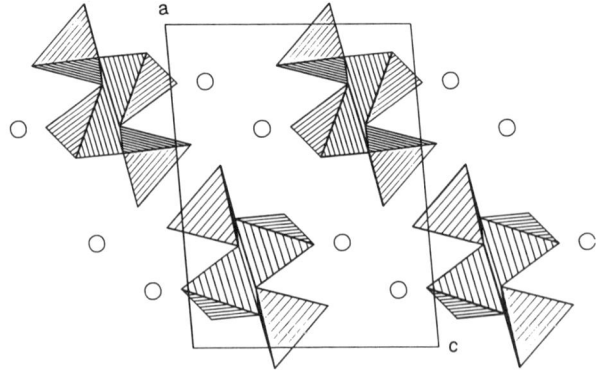

Figure 3.45. Projection of the structure of γ-Ba(PO$_3$)$_2$ along the b axis. The open circles represent the barium atoms.

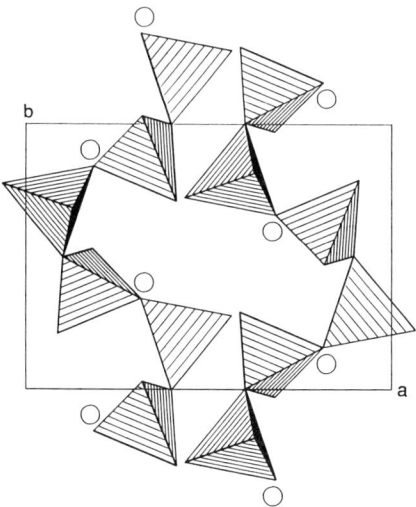

Figure 3.46. Projection of the structure of γ-Ba(PO$_3$)$_2$ along the c axis. The empty circles represent the barium atoms.

of its edges with two BaO$_8$ dodecahedra belonging to the two adjacent chains so that the layers containing the barium atoms in fact constitute a two-dimensional network of BaO$_8$ polyhedra.

3.4.12. Palladium Polyphosphate

Pd(PO$_3$)$_2$,[594] orthorhombic, *Pcnm*, $Z = 2$

$a = 4.233(3)$, $b = 4.571(1)$, $c = 12.472(3)$ Å

This very simple atomic arrangement is depicted by a projection along the **b** axis in Fig. 3.47, showing clearly its layered organization. Two phosphoric chains with a period of two tetrahedra cross the cell parallel to the **a** direction. Within the chain the phosphorus atom and the bonding oxygen atom are both located on twofold axes. These chains are interconnected in a three-dimensional array by the planar PdO$_4$ groups. These PdO$_4$ groups, located around the inversion center at 0,0,0, are centrosymmetric with Pd–O distances of 1.485 and 1.587 Å.

3.4.13. Beryllium–Potassium Polyphosphate

BeK(PO$_3$)$_3$,[648] orthorhombic, $P2_12_12_1$, $Z = 4$

$a = 7.844(1)$, $b = 8.610(2)$, $c = 11.399(4)$ Å

The phosphoric chains, with a period of six tetrahedra, zigzag along the **b** axis. Two such chains cross the unit cell. Figure 3.48, a projection along the **c** direction, shows the

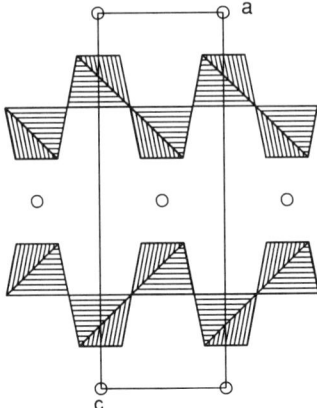

Figure 3.47. Projection, along the *b* axis, of the atomic arrangement in Pd(PO$_3$)$_2$. The open circles represent the palladium atoms.

general organization of this arrangement. The beryllium atom has tetrahedral oxygen coordination, and the BeO$_4$ tetrahedra link the phosphoric chains so as to build up a three-dimensional network of tetrahedra. This network can be schematically described as sheets rings made of six tetrahedra, either five PO$_4$ tetrahedra and one BeO$_4$ tetrahedron or four PO$_4$ and two BeO$_4$ tetrahedra. These sheets, perpendicular to the **c** axis, are interconnected by Be–O–P bonds. Figure 3.49 represents schematically the organization of these sheets.

Within this three-dimensional network of tetrahedra, the P–Be distances ranging from 2.813 to 3.070 Å, are quite similar to these P–P distances in the phosphoric chain (2.892–2.918 Å). Among the four P–O–Be angles observed in this network, three of them

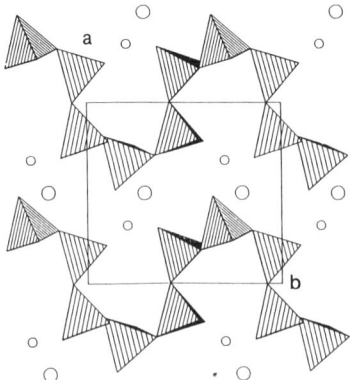

Figure 3.48. Projection, along the *c* axis, of the atomic arrangement in BeK(PO$_3$)$_3$. The large open circles represent the potassium atoms, and the smaller ones the beryllium atoms. This projection is restricted to $-0.05 < z < 0.55$.

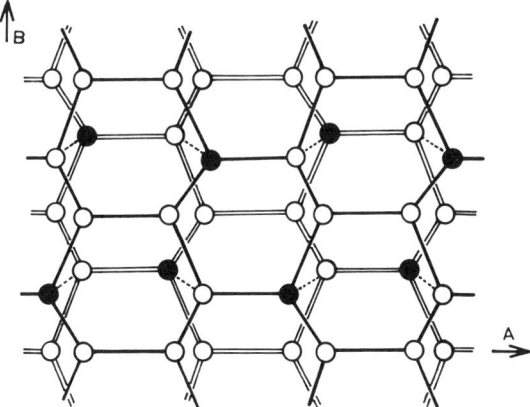

Figure 3.49. Schematic representation of the three-dimensional network of tetrahedra in BeK(PO₃)₃, projected along the *c* axis. Only the beryllium (filled circles) and the phosphorus atoms (open circles) are represented. Two layers are shown; the upper one is drawn with solid bonds, and the lower one with hollow bonds. The dotted lines represent the P–O–Be bonds connecting the two layers.

are quite comparable to the usual P–O–P angles measured in condensed phosphoric anions, but the fourth one is significantly larger (175.6°). In the rather regular BeO₄ tetrahedron, the Be–O distances range between 1.594 and 1.628 Å.

Within a range of 3.50 Å, the potassium atom has sevenfold coordination, with K–O distances ranging from 2.715 to 3.089 Å. Potassium atoms are located in channels parallel to the **c** direction.

3.4.14. Cobalt–Potassium Polyphosphate

CoK(PO₃)₃,[653] trigonal (rhombohedral), *R*3

$a = 10.209(5)$, $c = 6.984(3)$ Å, $Z = 3$ (hexagonal setting)

$a = 6.337(3)$ Å, $\alpha = 107.31(5)°$, $Z = 1$ (rhombohedral setting)

This atomic arrangement was first determined from powder diffraction data using the *R*32 space group; the correct space group was later proved from single-crystal data to be *R*3. This structure is shown in projection along the ternary axis in Fig. 3.50. The (PO₃)ₙ chains, with a period of three tetrahedra, spiral around one family of helical axes. In each chain period (three tetrahedra), there is so only one independent PO₄ tetrahedron. All the associated cations alternate along the threefold internal axes; both have octahedral coordination, with Co–O distances ranging from 2.071 to 2.110 Å and K–O distances between 2.765 and 2.878 Å. Figure 3.50 shows how each CoO₆ octahedron shares its six corners with six different PO₄ tetrahedra belonging to three different phosphoric chains. This arrangement creates relatively large empty channels located along the second set of helical axes.

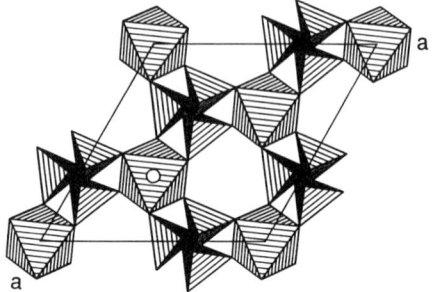

Figure 3.50. Projection, along the ternary axis, of the atomic arrangement in $CoK(PO_3)_3$. The hatched octahedra correspond to the CoO_6 groups, and the open circles represent the potassium atoms.

Crystal data for the ten compounds isotypic with $CoK(PO_3)_3$ are presented in Tables 3.9–3.11.

3.4.15. Copper–Ammonium Polyphosphate

$$CuNH_4(PO_3)_3, [666] \text{ monoclinic, } P2_1/c, Z = 4$$

$$a = 5.182(4), b = 11.544(8), c = 13.06(1) \text{ Å}, \beta = 97.16(5)°$$

The phosphoric chains, with a period of six tetrahedra extend along the **b** direction. Two of them cross the unit cell as shown by Fig. 3.51, a projection along the a direction. These chains are separated by layers of associated cations corrugated in the same manner.

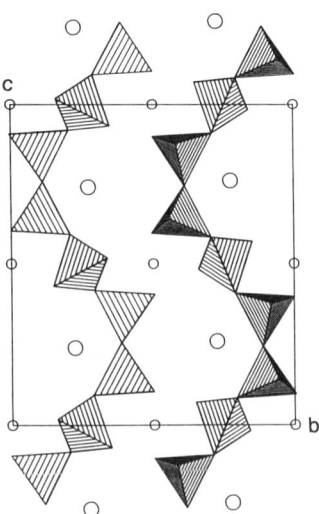

Figure 3.51. Projection, along the *a* direction, of the atomic arrangement in $CuNH_4(PO_3)_3$. The large open circles represent the ammonium groups, and the smaller ones the copper atoms.

In the latter last layers, the two independent copper atoms, both located on inversion centers are surrounded by six oxygen atoms, forming distorted octahedra with Cu–O distances ranging from 1.898 to 2.240 Å. The ammonium groups have a eightfold coordination, with N–O distances ranging from 2.87 to 3.56 Å.

Another way to describe this arrangement is to consider it as a three-dimensional network built up by the $(PO_3)_n$ chains and the CuO_6 octahedra, creating relatively large channels parallel to the a direction in which the ammonium groups are located.

This atomic arrangement is, in fact, a distortion of the orthorhombic structure described for some $M^{II}NH_4(PO_3)_3$ polyphosphates. This lowering of symmetry can probably be attributed to the asymmetric surroundings of the copper atom.

The corresponding thallium and rubidium salts are isotypic.

3.4.16. Barium–Copper Polyphosphate

$$Ba_2Cu(PO_3)_6,^{595,596} \text{monoclinic}, P2_1/a, Z = 4$$

$$a = 21.382(3), b = 7.286(1), c = 9.520(1) \text{ Å}, \beta = 97.96(2)°$$

Chains with a period of 12 tetrahedra cross the unit cell parallel to the **b** axis. These chains spiral around the 2_1 axis in such a manner as to form flattened helices. Figure 3.52, a projection of the atomic arrangement along the **b** axis, shows clearly the development of one chain period, its elongated aspect, and the layered organization of the arrangement. One chain has a 2_1 internal symmetry and is in fact built up by only six independent tetrahedra.

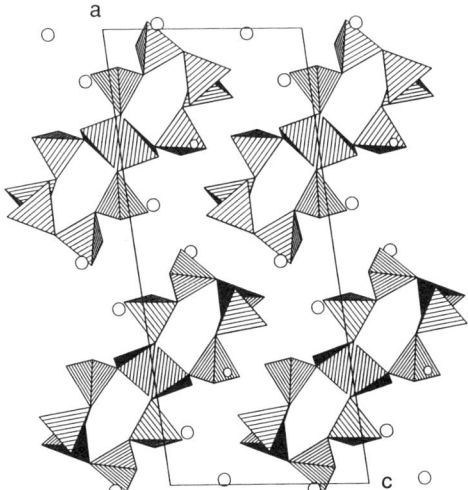

Figure 3.52. Projection, along the **b** axis, of the atomic arrangement in $Ba_2Cu(PO_3)_6$. The large open circles represent the barium atoms, and the smaller ones the copper atoms.

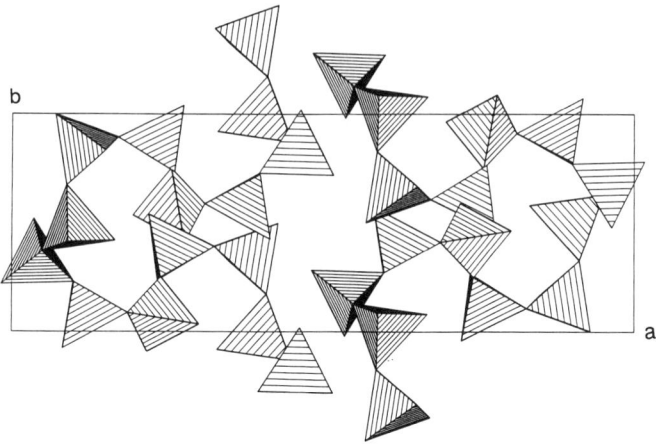

Figure 3.53. Projection, along the **c** axis, of the atomic arrangement in $Ba_2Cu(PO_3)_6$ showing the two phosphoric chains crossing the unit cell. The associated cations have been omitted.

Figure 3.53, a projection along the **c** axis, provides a representation of the two phosphoric chains crossing the unit cell. In spite of the layered aspect of the representation in Fig. 3.52, the associated cation polyhedra form a three-dimensional network. One of the barium atoms has a ninefold coordination, with Ba–O distances ranging from 2.60 to 3.12 Å, whereas the second one has eight neighbors with Ba–O distances ranging from 2.71 to 3.06 Å. The CuO_6 octahedron is very distorted, with four short Cu–O distances, ranging from 1.91 to 2.07 Å, and two long ones of 2.23 and 3.07 Å.

$Ba_2Zn(PO_3)_6$ is the only isotypic compound.[598]

3.4.17. Manganese(III) Polyphosphate

$Mn(PO_3)_3$,[687] orthorhombic, *Pnaa*, $Z = 4$

$a = 9.703(3)$, $b = 10.667(3)$, $c = 6.362(2)$ Å

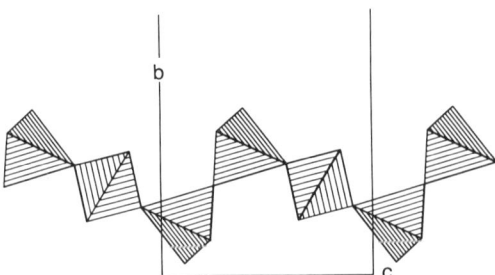

Figure 3.54. An isolated $(PO_3)_n$ chain, as observed in projection along the **a** direction, in the structure of $Mn(PO_3)_3$.

a

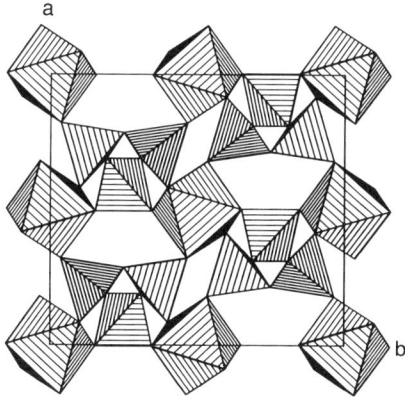

b

Figure 3.55. Projection of the structure of $Mn(PO_3)_3$ along the c direction, showing how the MnO_6 groups (hatched octahedra) provide the cohesion between the phosphoric chains.

Four infinite $(PO_3)_n$ chains cross the unit cell, extending along the c direction with a period of three tetrahedra. One of these three tetrahedra is located on a twofold axis, so the chain is in fact built up by only two crystallographically independent tetrahedra. An isolated chain is shown in Fig. 3.54, a projection along the **a** direction.

The manganese atoms, located on inversion centers, have a distorted octahedral coordination, with Mn–O distances ranging from 1.881 to 2.162 Å. These MnO_6 octahedra are isolated. As shown by Fig. 3.55, a projection of the arrangement along the direction of the chains, the cohesion between these chains is provided by the MnO_6 octahedra, each of them sharing its six corners with four different phosphoric chains.

3.4.18. Neodymium Polyphosphate

$$Nd(PO_3)_3,^{686,705,718} \text{orthorhombic}, C222_1, z = 4$$

$$a = 11.172(2), b = 8.533(2), c = 7.284(2) \text{ Å}$$

A projection of this noncentrosymmetric atomic arrangement along the **a** direction is presented in Fig. 3.56. Chains with a period of six tetrahedra zig-zag along the **c** direction. Whithin one period of the chain, two of the six phosphorus atoms lie on twofold axes so that this phosphoric anion is built up by only two crystallographically independent PO_4 tetrahedra.

The neodymium atoms, located on twofold axes, have their usual eightfold coordination; the eight oxygen neighbors build a rather regular dodecahedron, with Nd–O distances ranging from 2.33 to 2.67 Å. By sharing edges, these dodecahedra assemble themselves so as to build up infinite chains parallel to the **c** direction. Two such chains cross the unit cell, and the intrachain the Nd–Nd distance is 4.234 Å.

At least seven $Ln(PO_3)_3$ polyphosphates are known to crystallize with this type of structure. Their unit-cell dimensions are given in Table 3.16.

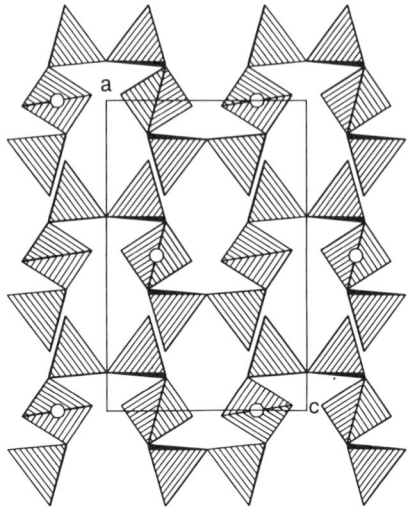

Figure 3.56. Projection of the structure of $Nd(PO_3)_3$ along the *b* direction. The open circles represent the neodymium atoms.

3.4.19. Gallium–Lithium Polyphosphate

$$GaLi(PO_3)_4,^{730} \text{ orthorhombic, } Pbna, Z = 4$$

$$a = 8.244(2), b = 9.044(2), c = 12.540(2) \text{ Å}$$

Two slightly corrugated infinite $(PO_3)_n$ chains cross the unit cell parallel to the **a** direction. The chains have a period of four PO_4 tetrahedra. As shown by Fig. 3.57 these

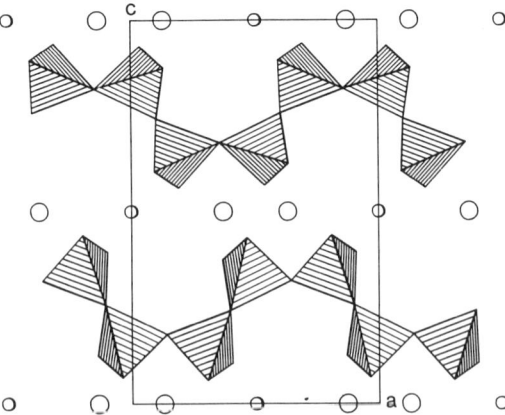

Figure 3.57. Projection, along the *b* direction, of the structure of $GaLi(PO_3)_4$. The small open circles represent the lithium atoms, and the larger ones the gallium atoms. Two lithium atoms are almost superimposed in this projection.

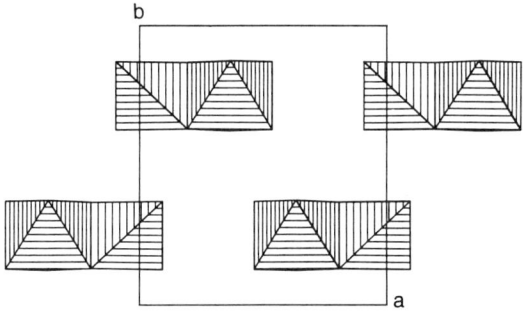

Figure 3.58. Distribution of the $GaLiO_8$ clusters inside a plane of associated cations in the structure of $GaLi(PO_3)_4$. This projection along the **c** direction is limited by $1/3 < z < 2/3$.

chains are separated by layers of associated cations in the planes $z = 0$ and $1/2$. Both the gallium and the lithium atoms are located on twofold axes. The gallium has sixfold coordination, with Ga–O distances ranging from 1.942 to 1.967 Å, while the lithium is located inside a distorted tetrahedron of oxygen atoms, with Li–O distances ranging from 1.892 to 2.048 Å. Within a plane of associated cations, these two polyhedra form isolated $GaLiO_8$ clusters by sharing a common edge. Inside such a cluster, the Ga–Li distance is rather short (3.030 Å). Figure 3.58, a projection along the **c** direction, shows the distribution

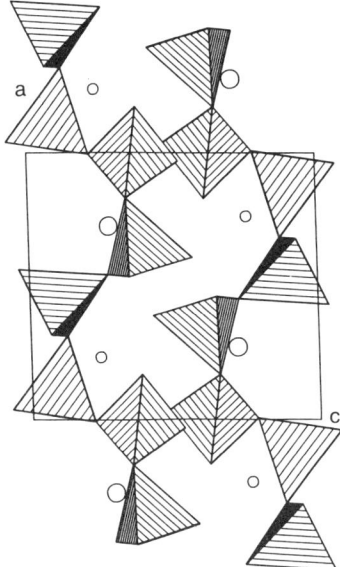

Figure 3.59. Projection, along the **b** direction, of the atomic arrangement in $NdK(PO_3)_4$. The large open circles represent the potassium atoms, and the small ones, the neodymium atoms.

of these $GaLiO_8$ clusters within a plane of associated cations. A $GaLiO_8$ dimer shares oxygen atoms with four different phosphoric chains.

3.4.20. Neodymium–Potassium Polyphosphate

$$NdK(PO_3)_4, {}^{751,752} \text{ monoclinic, } P2_1, Z = 2$$

$$a = 7.280(1), b = 8.438(1), c = 8.008(1) \text{ Å}, \beta = 91.97(1)°$$

Two infinite $(PO_3)_n$ chains with a period of four tetrahedra cross the unit cell parallel to the a direction. Figure 3.59, a projection along the b direction, shows the organization of these chains and of the associated cations.

The neodymium atoms are as usual, coordinated by eight oxygen atoms, building a slightly distorted dodecahedron. Within this NdO_8 group, the Nd–O distances range from 2.360 to 2.484 Å. The shortest Nd–Nd distance is 6.59 Å. Within a range of 3.20 Å, the potassium atoms also have eightfold, but less regular coordination, with K–O distances ranging from 2.774 to 3.196 Å.

Among the $LnM^I(PO_3)_4$ compounds that have been investigated, at least 11 crystallize with this type of structure (III), and all of these are potassium salts (Table 3.20). Some of them were characterized by simple comparison of unindexed powder diagrams so that unit-cell dimensions are not available.

3.5. Geometries of Long-Chain Polyphosphate Anions

At the beginning of this chapter, we gave a definition of one of the main geometrical features of a long-chain anion, its *period*. The longest chain period, that has been observed to date is 16 tetrahedra in form VII of the $LnM^I(PO_3)_4$ compounds.

A chain with a period of n tetrahedra may be built up by less than n crystallographically independent tetrahedra owing to the location of some of its components on symmetry elements, usually a twofold axis or mirror plane, or its arrangement around a 2_1 or 3_1 helical axis. The part of the chain corresponding to the linkage of the crystallographically independent tetrahedra is usually denoted as the *sub-period* or *pseudo-period*. In some cases the presence of internal symmetry in the chain induces deviations from the usual geometry.

From the examples of atomic arrangements of long-chain polyphosphates described in the last section, it may be noted that the chain anions are more or less corrugated. The same observation is made in the crystal chemistry of condensed silicates, where infinite $(SiO_3)_n$ anions are also commonly encountered. Liebau [503] proposed that the degree of stretching of a chain be represented numerically by a *stretching factor*, defined by the formula

$$Fs = l/L_t \cdot n$$

Where n is the period of the chain, l its length in angströms, and L_t is a constant taken as the value of l/n for the most highly stretched chain. For silicates, L_t was found to have a value of 2.70 Å, while in the case of phosphates we determined a value of 2.48 Å from the crystal structure of $Zn(PO_3)_2$.[565,566] It can be seen from this formulation that Fs will decrease from 1.00, for the most highly stretched chain to lower values for corrugated chains. In the case of silicates, Liebau[503] reported a minimum value of 0.234 from an examination of a great number of them.

In Table 3.25, we present the P–P distances and P–O–P and P–P–P angles as well as the periods and stretching coefficients for a limited number examples of phosphoric chains. A more exhaustive study involving the examination of the geometrical data for all the long-chain polyphosphates investigated to date is in preparation. Preliminary results

Table 3.25. Some Geometrical Features Observed in the Anions of Long-Chain Polyphosphates

Formula	P–P (Å)	P–O–P (°)	P–P–P (°)	Period	Fs	Reference(s)
$NaPO_3$ (A)	2.870	124.8	106.1	4	0.625	511, 512
	2.967	136.1	106.6			
$AgK(PO_3)_2$	2.919	129.6	92.6	4	0.755	543
	2.957	134.5	111.9			
$Be(PO_3)_2$ (II)	2.957	129.53	98.7	4	0.702	560, 561
	2.887	138.97	108.1			
$Be(PO_3)_2$ (III)	2.933	139.8	105.2	8	0.438	562, 564
	2.951	137.4	103.5			
$Zn(PO_3)_2$	2.946	135.5	114.7	2	1.000	565, 566
$Ca(PO_3)_2$	2.947	140.6	111.6	4	0.705	511, 575
	2.978	135.9	93.1			
	2.939	135.9	88.2			
	2.980	141.5	116.7			
$Cd(PO_3)_2$ (LT)[a]	2.934	133.3	102.0	4	0.709	580, 582
	2.875	128.0	105.8			
$Cd(PO_3)_2$ (HT)[b]	2.982	150.2	113.1	4	0.742	583, 584
	3.002	154.2	94.9			
β-$Ba(PO_3)_2$	2.898	124.6	102.2	2	0.909	588
		133.9				
γ-$Ba(PO_3)_2$	2.818	120.7	127.3	4	0.695	587, 589
	3.022	143.4	93.0			
$Pd(PO_3)_2$	2.872	129.6	94.9	2	0.853	594
$Mn(PO_3)_3$	3.008	146.2		3	0.855	687
	2.872	128.7				
$YNH_4(PO_3)_4$	2.895	130.8	?	16	0.436	762
	2.897	132.3	?			
	2.839	126.6	?			
	2.869	130.0	?			
$Ba_2Cu(PO_3)_6$	2.866	127.6	140.6	12	0.245	595, 596
	2.883	133.8	113.1			
	2.943	131.3	113.6			
	2.920	131.0	94.3			
	2.896	146.2	109.6			
	3.017	129.0	99.5			

[a]LT: Low-temperature form.
[b]HT: High-temperature form.

obtained during the course of this investigation show that, in spite of the great diversity of the long-chain anions, the observed values never depart significantly from those given for the selected examples in Table 3.25. In addition, these preliminary results confirm that there are no apparent meaningful correlations between the geometry of a chain and the nature of the associated cations.

Cyclophosphates

4.1. Introduction

As we have already mentioned, cyclophosphates are characterized by a cyclic P_nO_{3n} anion built by n corner-sharing PO_4 tetrahedra. Ring anions of this type are presently well characterized for n = 3, 4, 5, 6, 8, 10 and 12. For each family of cyclophosphates, corresponding to a particular value of n, the methods of chemical preparation are very specific, and, with a few exceptions, the development of these methods required the optimization of a procedure for the chemical preparation of an appropriate starting material. We will present for each family of cyclophosphate a short survey of its development and will describe as much detail as possible the specific methods of chemical preparation.

Systematic investigations of cyclotri- and cyclotetraphosphates have been performed during the past 30 years, so that the chemistry of these two families is now well established. On the other hands, the investigation of cyclohexaphosphates did not begin seriously before 1985, and that of cycloocta- and cyclodecaphosphates before 1990. The rare examples of larger rings, $P_{12}O_{36}$ for instance, are till now the unexpected fruit of various flux methods.

Similar ring anions built by corner-sharing XO_4 tetrahedra have been observed in several classes of compounds. As early as 1930, Zachariasen's[799] crystal structure of benitoite, $BaTiSi_3O_9$ established the existence of the Si_3O_9 ring in naturally occurring silicates. Since then, many Si_nO_{3n} ring anions have been characterized, including some having very large values of n. Similarly the germanates also provide also many examples of Ge_nO_{3n} rings. On the other hands, the occurrence of ring anions among arsenates is not common. Apart from cyclotetraarsenate, $Tl_4As_4O_{12}$, which is well known[800] as an isotype of the corresponding phosphate, the only example of an arsenate containing such a ring is $Cr_2H_2(As_2O_7)(As_4O_{12})$, a mixed-anion compound described by Jansen and Brachtel.[801] The rarity of arsenate rings can probably be explained by the great sensitivity of As–O–As bonds to hydrolysis.

Ring anions are also frequently observed in the crystal chemistry of borates but most of these are not composed entirely of BO_4 tetrahedra but instead contain both BO_4 and BO_3 groups. For instance, the $B_3O_3(OH)_5$ rings found in meyerhofferite, described by Christ and Clark[802] and in inyoite described by Clark,[803] are built from two BO_4 tetrahedra and one BO_3 triangle. Among vanadates, it is only very recently that structural evidence for the existence of a cyclic V_4O_{12} anion has been reported. SO_3 and CrO_3 also form Cr_3O_9 and S_3O_9 rings built by three corner-sharing tetrahedra. In this case, they are molecular compounds. A detailed survey of both molecular and anionic inorganic rings was published by Haiduc and Sowerby.[804] These authors also proposed a unified nomenclature for these entities.

In cyclophosphate chemistry as in almost all other areas of condensed phosphate chemistry, with some rare exceptions, the basic properties of most of the well-characterized compounds have still not been investigated.

As we will see later in this chapter the chemical preparations of cyclophosphates are very specific. Based on this fact, one might expect this class of compounds to exhibit specific properties. In fact, no general features are observed for cyclophosphates; as in many other chemical families their physicochemical behavior seems to be mainly dependent on the nature of the associated cations. Nevertheless, it may be noted that, as in the case of the long-chain polyphosphates, the acidic cyclophosphates are very rare, the only examples being $Na_2HP_3O_9$ and possibly $Na_2H_2P_4O_{12}$. However the latter compound has not yet been fully characterized.

The thermal behavior of cyclophosphates also seems to be very dependent on the nature of the associated cations. As far as we know, all monovalent cation cyclophosphates transform into long-chain polyphosphates when heated. Some salts of divalent cations, such as calcium, cadmium, strontium, barium, and lead cyclophosphates, also transform into long-chain polyphosphates, but the magnesium, cobalt, nickel, copper, and manganese salts are stable as cyclotetraphosphates up to their melting points. Among the mixed-cation cyclophosphates, the observed thermal behavior is highly variable. Some of them— $NaBaP_3O_9$, and $SrK_2P_4O_{12}$, for instance—are stable as cyclophosphates up to their melting points whereas others including evidently all the incongruent-melting salts, decompose, forming various species. For instance, when calcium–potassium cyclotetraphosphate, $CaK_2P_4O_{12}$, is heated to 973 K, it decomposes into a cyclotriphosphate and a long-chain polyphosphate:

$$CaK_2P_4O_{12} \rightarrow CaKP_3O_9 + KPO_3$$

The water-soluble cyclophosphates are generally stable in fairly neutral aqueous solutions at room temperature. The rate of hydrolysis is rather low under these conditions, and this stability seems to increase with ring size. Nevertheless, under less mild conditions, the cyclophosphates, like all the condensed phosphates, are hydrolyzed. For instance, potassium cyclotriphosphate is rapidly transformed into KH_2PO_4 when heated in a humid atmosphere:

$$K_3P_3O_9 + 3H_2O \rightarrow 3KH_2PO_4$$

The water solubility of alkali cyclophosphates does not seem to depend on the ring size. For instance, the recently characterized potassium cyclodecaphosphate, $K_{10}P_{10}O_{30}\cdot4H_2O$, has a very high water solubility by comparison with that of potassium cyclotri- and cyclotetraphosphates. For cyclophosphates with higher valency associated cations, quantitative measurements of water solubility are lacking.

4.2. Cyclotriphosphates

4.2.1. Introduction

As early as last century, several chemists suspected the existence of cyclic phosphoric anions and produced crystalline compounds containing this type of anion. For instance, sodium cyclotriphosphate seems to have been prepared in 1833 by Graham.[5] In the middle of last century Fleitmann and Henneberg,[7] using the Graham's principles, prepared several salts and double salts, today recognized as cyclotriphosphates. They also suggested the trimeric nature of the anion. During the second part of last century, some other chemists, including Lindbom,[12] Tammann,[16,17] and Von Knorre,[20] described a number of additional cyclotriphosphates, most of them were, clearly characterized as such more or less recently. During the first half of this century, cyclophosphate chemistry did not develop significantly because no suitable starting material was available in convenient amounts and sufficient purity for syntheses. The methods reported for the chemical preparation of sodium cyclotriphosphate, the parent compound of almost all the compounds prepared last century, were difficult, tedious, and, in many cases not reproducible. Nevertheless, as early as 1938, Boullé[805] succeeded in preparing silver cyclotriphosphate monohydrate, $Ag_3P_3O_9 \cdot H_2O$, and showed how this compound can be used to synthesize any of water-soluble cyclotriphosphates via a metathesis reaction. This reaction, known today as Boullé's process, has been extensively used and will be described in detail in Section 4.2.3, which is devoted to the general methods of preparation of cyclotriphosphates. It was not until the early 1950s that a careful study of the thermal reorganization of NaH_2PO_4 by Thilo and Grunze[806] led to a reproducible procedure for the production of pure $Na_3P_3O_9$ according to the following scheme:

$$3NaH_2PO_4 \rightarrow Na_3P_3O_9 + 3H_2O$$

The first structural evidence for the cyclic nature of the anion, already clearly suggested by Lindbom,[12] was reported by Eanes and Ondik[33] in 1962, in their publication on the crystal structure of $LiK_2P_3O_9 \cdot H_2O$. Since then more than 100 cyclotriphosphates have been characterized.

4.2.2. Present State of Cyclotriphosphate Chemistry

4.2.2.1. Alkali and Monovalent Cation Cyclotriphosphates

$Li_3P_3O_9 \cdot 3H_2O$. The only lithium cyclotriphosphate reported to date is the trihydrate, $Li_3P_3O_9 \cdot 3H_2O$. Its characterization was first reported by Eanes,[807] and its atomic arrangement was later determined by Masse *et al.*[808,809] Grenier and Durif[507] observed that the trihydrate is stable up to 403 K and that the loss of the water molecules occurs between 403 and 423 K. At temperatures higher than 423 K, it slowly transforms into the long-chain polyphosphate. This compound is normally prepared by Boullé's metathesis reaction.[805] The hydrolytic decyclization kinetics of this salt in aqueous solutions of LiOH, in the temperature range 313–353 K, has been studied by Sotnikova-Yuzhik *et al.*[810] The decyclization scheme is

$$Li_3P_3O_9 + 2LiOH \rightarrow Li_5P_3O_{10} + H_2O$$

and crystalline triphosphate, $Li_5P_3O_{10} \cdot 5H_2O$, is obtained in good yield.

No anhydrous lithium cyclotriphosphate has been observed during the process of thermal dehydration–condensation of LiH_2PO_4 (see Section 4.5).

$Na_2HP_3O_9$. Two crystalline forms of this salt were reported by Griffith.[811] The products obtained by this author were always polycrystalline. According to the author, form I is not water-soluble, whereas form II is. Crystals of $Na_2HP_3O_9$ (I) were obtained by Averbuch-Pouchot et al.[812] during a study of the P_2O_5–SrO–Na_2O system. The authors reported the chemical preparation of this salt and gave a detailed description of its atomic arrangement. This salt seems to be the only known example of an acidic cyclotriphosphate.

$Na_3P_3O_9 \cdot nH_2O$. Sodium cyclotriphosphate can be easily prepared by thermal condensation of NaH_2PO_4. The reaction scheme was given in Section 4.2.1, and details of this procedure are reported in Section 4.2.3.

A detailed crystal-chemical study of $Na_3P_3O_9$ and of its various hydrates was performed by Ondik and Gryder.[813] $Na_3P_3O_9$, $Na_3P_3O_9 \cdot H_2O$, $Na_3P_3O_9 \cdot 3/2H_2O$, $Na_3P_3O_9 \cdot 3H_2O$ and $Na_3P_3O_9 \cdot 6H_2O$ were clearly characterized, and, for the first time, unit-cell dimensions for these compounds were reported. The chemical preparation of the various hydrates and the conditions under which they stable were described. The authors observed neither the existence of the hemihydrate nor that of polymorphs for $Na_3P_3O_9$ and for $Na_3P_3O_9 \cdot H_2O$, which had been previously reported in the chemical literature. They also noticed a striking similarity between the unit-cell parameters of $Na_3P_3O_9$ and those of its monohydrate.

Crystal structure determinations for $Na_3P_3O_9$ and $Na_3P_3O_9 \cdot H_2O$ were performed some years later by Ondik.[814] The two atomic arrangements are closely related as was previously suggested by the similarity of their unit-cell dimensions.[813] In spite of this similarity, the water molecule of the monohydrate is not of a zeolitic nature but belongs to the coordination polyhedra of the sodium atoms.

The crystal structure of $Na_3P_3O_9 \cdot 6H_2O$, the hydrate crystallizing at room temperature, was determined by Tordjman and Guitel.[815] A detailed investigation of the dehydration of this salt was reported by Prodan and Pytlev.[816]

$Ag_3P_3O_9 \cdot nH_2O$. As early as 1848, the existence of $Ag_3P_3O_9 \cdot H_2O$ was reported by Fleitmann and Henneberg.[7] These authors also reported the existence, not yet confirmed, of the anhydrous salt. Later, Lindbom[12] and Knorre[20] also characterized the monohydrate, but Boullé[805] was the first to describe a reproducible procedure for the preparation of this salt and to demonstrate its possible use as a starting material for the preparation of water-soluble cyclotriphosphates. Since then, this salt has been extensively used in what is now known as the Boullé's process. This process is described in Section 4.2.3. The first crystallographic investigations of $Ag_3P_3O_9 \cdot H_2O$ were performed simultaneously by Eanes[817] and Grenier,[818] and its atomic arrangement was determined by Bagieu-Beucher et al.[819]

$K_3P_3O_9$ and $(NH_4)_3P_3O_9$. Both these salts were prepared by Boullé's process and investigated by Grenier[818] and Grenier and Durif.[820] The two salts are isotypic. Their atomic arrangement was determined by Bagieu-Beucher et al.[821] with the potassium salt.

$K_3P_3O_9$ transforms irreversibly into a long-chain polyphosphate on heating:

$$K_3P_3O_9 \rightarrow 3KPO_3$$

Table 4.2.1. Crystallographic Data for Monovalent Cation Cyclotriphosphates

Formula	a (Å) α (°)	b (Å) β (°)	c (Å) γ (°)	S.G.	Z	Reference(s)
$Li_3P_3O_9 \cdot 3H_2O$	12.511(4)	12.511(4)	5.594(2)	$R3$	3	807–809
$Na_2HP_3O_9$	7.788(5) 116.69(5)	7.809(5) 103.41(5)	7.129(5) 81.94(5)	$P\bar{1}$	2	812
$Na_3P_3O_9$	7.928(2)	13.214(3)	7.708(2)	$Pmcn$	4	814
$Na_3P_3O_9 \cdot H_2O$	8.500(1)	13.189(1)	7.558(1)	$Pmcn$	4	814
$Na_3P_3O_9 \cdot 6H_2O$	7.826(4) 112.79(3)	9.530(5) 106.10(3)	10.828(5) 88.55(3)	$P\bar{1}$	2	815
$Ag_3P_3O_9 \cdot H_2O$	7.78(1) 114.70(5)	9.24(1) 91.00(5)	7.78(1) 114.70(5)	$P\bar{1}$	2	817–819
$K_3P_3O_9$	11.074(8)	11.965(9) 102.18(3)	7.350(6)	$P2_1/n$	4	818, 820, 821
$(NH_4)_3P_3O_9$	11.515(6)	12.206(8) 101.63(5)	7.699(4)	$P2_1/n$	4	818–820
$Rb_3P_3O_9 \cdot H_2O^a$	10.240(12) 96.72(5)	7.778(8) 69.32(5)	7.813(8) 96.23(5)	$P\bar{1}$	2	818
$CS_3P_3O_9 \cdot H_2O$	10.610(6) 96.64(8)	7.966(4) 68.84(8)	8.172(5) 95.42(8)	$P\bar{1}$	2	818, 823
$Tl_3P_3O_9$	12.035(6) 90.63(4)	18.74(1)	8.940(3)	$P2_1/a$	8	818, 824

^{a}Triclinic setting was recalculated for comparison with $Cs_3P_3O_9 \cdot H_2O$.

This behavior is common to all monovalent cation cyclotriphosphates except the ammonium salt. In the case of the potassium salt, the effect of various factors on the kinetics of this thermal conversion were investigated by Dombrovskii and Koval.[822]

$Rb_3P_3O_9 \cdot H_2O$ and $Cs_3P_3O_9 \cdot H_2O$. These two salts are isotypic. Grenier[818] was the first to describe the chemical preparation of these two phosphates by Boullé's process and to report their main crystallographic features. The crystal structure for this type of compounds was later determined by Tordjman *et al.*[823] using the cesium salt.

$Tl_3P_3O_9$. This salt was first described by Grenier[818] and its crystal structure solved by Boudjada.[824]

Crystallographic Data. The main crystallographic data for monovalent cation cyclotriphosphates are given in Table 4.2.1.

Mixed Monovalent Cation Cyclotriphosphates

$LiK_2P_3O_9 \cdot H_2O$. Eanes and Ondik[33] prepared this compound by using Boullé's process and reported a complete determination of its atomic arrangement. This structural investigation was the first to provide the geometrical features of a P_3O_9 ring anion.

$Na_2KP_3O_9$, $Na_2RbP_3O_9$, $Na_2TlP_3O_9$, and $Na_2NH_4P_3O_9$. These four compounds are isotypic. The $NaPO_3$–KPO_3 phase-equilibrium diagram was first investigated by Tammann and Ruppelt.[825] They found this diagram to be of the eutectic type, but nevertheless they

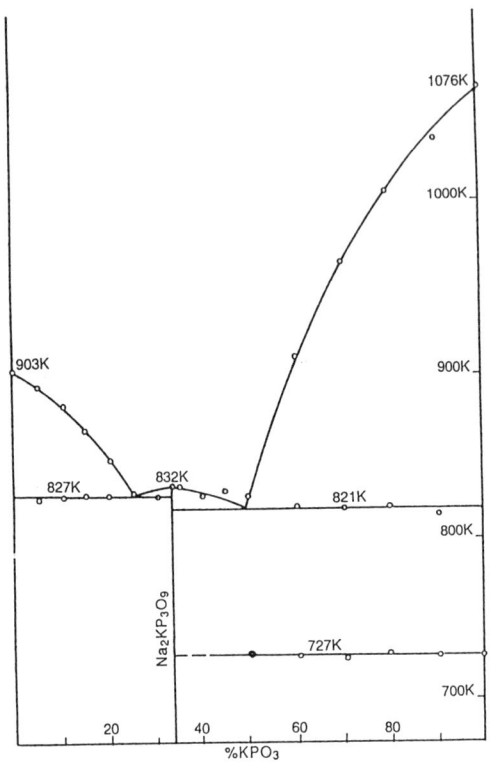

Figure 4.2.1. The $NaPO_3$–KPO_3 phase-equilibrium diagram as revised by Cavero-Ghersi and Durif. The transformation of KPO_3 is observed at 727 K on the right-hand side of the diagram.

reported the existence of $Na_2KP_3O_9$. Morey[826] redetermined this diagram and found $Na_3K(PO_3)_3$ to be the only compound in this system. Griffith and Van Wazer[827] characterized the latter compound as a cyclotriphosphate. Bukhalova and Mardirosova[828] did not confirm the Morey's result and found $Na_2KP_3O_9$ as the only compound in the system. Finally, this salt appears as a congruent-melting compound (mp = 832 K) in the $NaPO_3$–KPO_3 phase-equilibrium diagram (Fig. 4.2.1) elaborated by Cavéro-Ghersi and Durif.[829] Its atomic arrangement was determined by Tordjman *et al.*[830]

$Na_2RbP_3O_9$, characterized by Cavéro-Ghersi and Durif[831] during the elaboration of the $NaPO_3$–$RbPO_3$ phase-equilibrium diagram (Fig. 4.2.2), appears as an incongruent-melting compound decomposing at 785 K.

$Na_2TlP_3O_9$ and $Na_2NH_4P_3O_9$ were prepared by Cavéro-Ghersi[832] and identified as isotypes of $Na_2KP_3O_9$.

In spite of numerous similarities in the geometry of condensed silicate and phosphate anions, very few cases of isomorphism between silicates and phosphates have been reported. Thus we think that it is worth noting that the four compounds described above—$Na_2KP_3O_9$, $Na_2RbP_3O_9$, $Na_2TlP_3O_9$, and $Na_2NH_4P_3O_9$—are isotypic with some margarosanite-type silicates, such as $Ca_2BaSi_3O_9$ and $Ca_2PbSi_3O_9$.

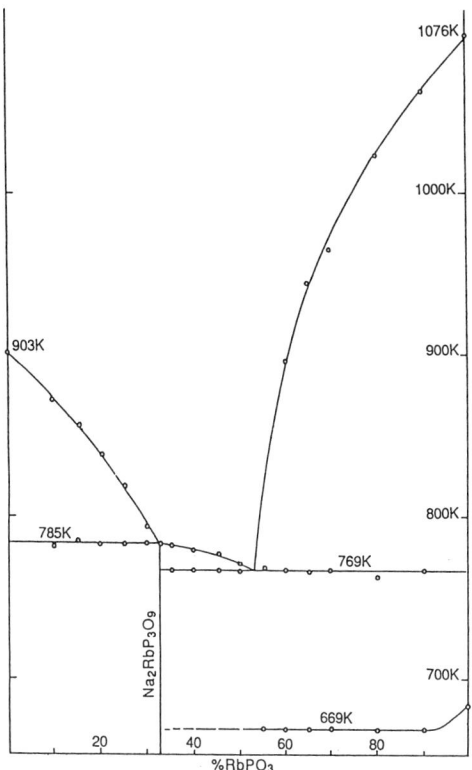

Figure 4.2.2. The $RbPO_3$–$NaPO_3$ phase-equilibrium diagram. The transformation of $RbPO_3$ is observed at 669 K on the right-hand side of the diagram.

$Na_2CsP_3O_9 \cdot 2H_2O$. This salt was first observed during attempts to prepare cesium cyclotriphosphate by Boullé's process using a silver cyclotriphosphate contaminated with sodium. Its crystal structure was determined by Boudjada.[824]

$Na_2LiP_3O_9 \cdot 4H_2O$. This mixed compound was accidentally prepared for the first time during a preparation of lithium cyclotriphosphate using an ion-exchange resin. A reproducible procedure for its preparation was later reported by Averbuch-Pouchot and Durif[833] who performed an accurate determination of its atomic arrangement.

Crystallographic Data. The main crystallographic data for mixed-monovalent cation cyclotriphosphates are given in Table 4.2.2.

4.2.2.2. Divalent Cation Cyclotriphosphates

Some anhydrous cyclotriphosphates of divalent cations have been reported in the chemical literature, but their existence has not yet been confirmed. To date, only cadmium, calcium, manganese, strontium, barium, and lead cyclotriphosphate hydrates have been clearly characterized.

Table 4.2.2. Crystallographic Data for Mixed Cation Cyclotriphosphates

Formula	a (Å) α (°)	b (Å) β (°)	c (Å) γ (°)	S.G.	Z	Reference(s)
$LiK_2P_3O_9 \cdot H_2O$	8.669(3)	14.497(4) 99.72(5)	7.634(6)	$P2_1/c$	4	33
$Na_2KP_3O_9$	6.886(2) 110.07(2)	9.494(3) 104.69(2)	6.797(2) 86.68(2)	$P\bar{1}$	2	829, 830
$Na_2RbP_3O_9$	7.010(2) 108.98(5)	9.542(3) 104.26(5)	6.783(3) 87.37(5)	$P\bar{1}$	2	831
$Na_2TlP_3O_9$	6.977 108.70	9.511 104.40	6.787 86.82	$P\bar{1}$	2	832
$Na_2NH_4P_3O_9$	6.918 106.87	9.412 106.87	7.006 88.09	$P\bar{1}$	2	832
$Na_2CsP_3O_9 \cdot 2H_2O$	11.393(6)	13.188(8) 102.31(4)	7.622(3)	$P2_1/a$	4	824
$Na_2LiP_3O_9 \cdot 4H_2O$	6.905(5) 95.00(5)	9.346(5) 104.36(5)	9.876(5) 107.75(5)	$P\bar{1}$	2	833

$Cd_3(P_3O_9)_2 \cdot 14H_2O$. The first evidence for the existence of a cadmium cyclotriphosphate was reported by Averbuch-Pouchot and Durif.[834] They described the chemical preparation of cadmium cyclotriphosphate tetradecahydrate, $Cd_3(P_3O_9)_2 \cdot 14H_2O$ by Boullé's process and the main crystallographic data for this salt. Later on, its atomic arrangement was determined by Averbuch-Pouchot et al.[835] and then reexamined at low temperature.[836] The physical properties of this salt were carefully investigated by Michot,[837] Simonot-Grange,[838] and Simonot-Grange and Michot.[839,840] These authors showed that, under dynamic vacuum, 8 of the 14 water molecules can be removed without any alteration of the atomic framework and that this phenomenon is reversible. They also investigated the decondensation of the tetradecahydrate in wet atmospheres at various temperatures.

$Cd_3(P_3O_9)_2 \cdot 10H_2O$. For reasons that are not yet clearly understood, Boullé's process, leading normally to the tetradecahydrate at room temperature, sometimes produces a decahydrate, $Cd_3(P_3O_9)_2 \cdot 10H_2O$. The crystals, thus obtained are invariably twinned. Good-quality crystals, suitable for a structural study, were obtained accidentally during attempts to prepare an yttrium–lithium–cadmium cyclotriphosphate. The crystal structure determined by Averbuch-Pouchot et al.[841] shows that two of the ten water molecules are of a zeolitic nature.

No relationship exists between the structures of the two hydrates, explaining why the decahydrate was never observed during dehydration experiments on the tetradecahydrate.

$Ca_3(P_3O_9)_2 \cdot 10H_2O$ and $Mn_3(P_3O_9)_2 \cdot 10H_2O$. $Ca_3(P_3O_9)_2 \cdot 10H_2O$ was apparently first prepared by Boullé[805] as an example of the use of his metathesis reaction. The author assigned this salt the formula of an henneahydrate. The existence of a manganese cyclotriphosphate crystallizing with 9 or 11 water molecules was reported at the end of last century by Tammann[17] and by Knorre.[20] El-Horr and Durif[842] prepared these two salts by the action of $H_3P_3O_9$ on the corresponding carbonates and showed that they are isotypic with the cadmium decahydrate salt. A heptahydrate of calcium, isotypic with the strontium salt described below, was prepared by Durif,[843] but its preparation was not reproducible.

$Sr_3(P_3O_9)_2 \cdot 7H_2O$. Ropp *et al.*[140] reported the preparation of an anhydrous strontium cyclotriphosphate, but there is no structural evidence for the existence of such a salt. The only reported cyclotriphosphate of strontium is a hydrate, $Sr_3(P_3O_9)_2 \cdot 7H_2O$, described by Durif *et al.*[844] These authors reported its chemical preparation by Boullé's process and its main crystallographic features. The atomic arrangement of this salt was later determined by Tordjman *et al.*[845]

$Ba_3(P_3O_9)_2 \cdot 6H_2O$ and $Ba_3(P_3O_9)_2 \cdot 4H_2O$. There is no evidence for the existence of an anhydrous barium cyclotriphosphate, although such a salt was reported by Fleitmann and Henneberg.[7] However—two hydrates $Ba_3(P_3O_9)_2 \cdot 6H_2O$ and $Ba_3(P_3O_9)_2 \cdot 4H_2O$—are well characterized.

$Ba_3(P_3O_9)_2 \cdot 6H_2O$ was originally prepared through the use of Boullé's process[805] by Grenier and Martin.[846] A detailed structural characterization was later reported by Masse *et al.*[847] Two of the six water molecules are of a zeolitic nature. Some time later, Averbuch-Pouchot and Durif[848] optimized a chemical preparation of this hydrate not involving Boullé's process. The thermal behavior of this salt was carefully investigated by Tacquenet[849] and by Thrierr-Sorel *et al.*[850]

$Ba_3(P_3O_9)_2 \cdot 4H_2O$ was prepared by Averbuch-Pouchot and Durif[848] during a detailed investigation of the reactions between $Na_3P_3O_9$ and $BaCl_2$ solutions at various concentrations. These authors reported a reproducible procedure for producing the tetrahydrate as polycrystalline samples or as single crystals. If a very concentrated aqueous solution of $BaCl_2$ (15 g of $BaCl_2 \cdot 2H_2O$ in 35 cm^3 of water) is slowly added to a concentrated aqueous solution of $Na_3P_3O_9$ (3 g of $Na_3P_3O_9$ in 25 cm^3 of water), without any mechanical stirring, one observes almost immediately the formation of a large amount of small tufts of lathlike needles. After filtration to remove the mother liquor, an equivalent volume of water is added to the solid phase, which is transformed within a few minutes into small crystals of $Ba_3(P_3O_9)_2 \cdot 4H_2O$. If the same experiment is performed with less concentrated solutions, the same quantities of starting materials being respectively dissolved in 50 cm^3 of water, the formation of large crystals of the tetrahydrate is observed after some hours. In this case, the preparation is sometimes contaminated by the hexahydrate in the form of large elongated prisms representing 10 to 20% of the product. All the experiments have been conducted at room temperature. Crystals of $Ba_3(P_3O_9)_2 \cdot 4H_2O$ appear as stout monoclinic prisms that are stable at room temperature. The authors[848] reported a description of the atomic arrangement.

For these two hydrates, firing at 873 K for some hours leads to the polyphosphate:

$$Ba_3(P_3O_9)_2 \cdot xH_2O \rightarrow 3Ba(PO_3)_2 + xH_2O \ (x = 4, 6)$$

$Pb_3(P_3O_9)_2 \cdot 3H_2O$. This salt was reported for the first time in 1848 by Fleitmann and Henneberg.[7] Later, Durif and Brunel-Laügt[851] improved its chemical preparation and reported its main crystallographic features. The crystal structure was determined by Brunel-Laügt *et al.*[852] The thermal dehydration of this salt was investigated by Kuz'menkov *et al.*[853]

Table 4.2.3. Crystallographic Data for Divalent Cation Cyclotriphosphates

Formula	a (Å) α (°)	b (Å) β (°)	c (Å) γ (°)	S.G.	Z	Reference(s)
$Cd_3(P_3O_9)_2 \cdot 14H_2O$	12.228(3)	12.228(3)	5.451(3)	$P\bar{3}$	1	834–836
$Cd_3(P_3O_9)_2 \cdot 10H_2O$	9.424(8)	17.87(1) 107.72(1)	7.762(7)	$P2_1/n$	2	841
$Ca_3(P_3O_9)_2 \cdot 10H_2O$	9.332(7)	18.13(1) 106.69(5)	7.841(5)	$P2_1/n$	2	842
$Mn_3(P_3O_9)_2 \cdot 10H_2O$	9.219(4)	17.733(8) 107.37(2)	7.644(3)	$P2_1/n$	2	842
$Sr_3(P_3O_9)_2 \cdot 7H_2O$	16.05(1)	12.33(1)	10.87(1)	$Pnma$	4	844, 845
$Ba_3(P_3O_9)_2 \cdot 6H_2O$	7.547(4) 108.58(8)	11.975(6) 100.35(8)	13.068(8) 95.54(8)	$P\bar{1}$	2	846, 847
$Ba_3(P_3O_9)_2 \cdot 4H_2O$	16.09(1)	8.368(5) 95.38(5)	7.717(3)	$C2/m$	2	848
$Pb_3(P_3O_9)_2 \cdot 3H_2O$	11.957(5)	11.957(5)	12.270(5)	$P4_12_12$	4	851, 852
$Ba_2Zn(P_3O_9)_2 \cdot 10H_2O$	26.52(3)	7.625(5) 100.93(5)	12.92(1)	$C2/c$	4	854

Other Divalent Cation Cyclotriphosphates. At the end of last century, $Zn_3(P_3O_9)_2 \cdot 9H_2O$, $Cu_3(P_3O_9)_2 \cdot 9H_2O$, and $Co_3(P_3O_9)_2 \cdot 9H_2O$ were reported by Tammann,[17] and $Fe_3(P_3O_9)_2 \cdot 12H_2O$ was reported by Lindbom.[12] However the existence of these salts has not yet been confirmed.

$Ba_2Zn(P_3O_9)_2 \cdot 10H_2O$. This salt was prepared by Durif *et al.*[854] using Boullé's process. These authors reported a complete description of its atomic arrangement. This salt is the only mixed-divalent cation cyclotriphosphate that has been reported in the chemical literature.

Crystallographic Data. The main crystallographic data for the divalent cation cyclotriphosphates are presented in Table 4.2.3.

4.2.2.3. Divalent–Monovalent Cation Cyclotriphosphates

The M^{II}–M^{I} cyclotriphosphates that have been investigated to date can be roughly classified into two groups according to their chemical formulas:

$$M^{II}M_4^{I}(P_3O_9)_2 \cdot nH_2O \quad \text{and} \quad M^{II}M^{I}P_3O_9 \cdot nH_2O$$

It should be noted that lithium has never been observed in this series of compounds.

Most of the anhydrous salts have been characterized during the elaboration of $M^{I}PO_3$–$M^{II}(PO_3)_2$ phase-equilibrium diagrams and prepared as single crystals by various flux methods, whereas the vast majority of the hydrates have been synthesized by Boullé's process[805] or by classical methods of aqueous chemistry.

(i) Anhydrous Divalent-Monovalent Cation Cyclotriphosphates

$CaNa_4(P_3O_9)_2$ and $CdNa_4(P_3O_9)_2$. Griffith[855] reported the chemical preparation of the calcium salt and, from various experiments, obtained proof of the cyclic nature of its anion.

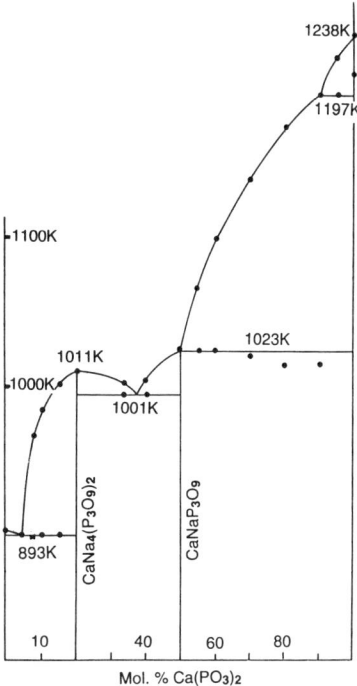

Figure 4.2.3. The $NaPO_3$–$Ca(PO_3)_2$ phase-equilibrium diagram. The $\alpha \rightarrow \beta$ transition of $Ca(PO_3)_2$ is observed on the right-hand side at 1213 K for pure $Ca(PO_3)_2$ and at 1197 K for the other concentrations.

Grenier *et al.*[856] presented a revised version (Fig. 4.2.3) of the $NaPO_3$–$Ca(PO_3)_2$ phase-equilibrium diagram, previously established by Morey,[857] and reported crystal data for $CaNa_4(P_3O_9)_2$, which is a congruent-melting compound (mp = 1006 K).

The cadmium salt, characterized by Averbuch-Pouchot and Durif[858] during the elaboration of the $NaPO_3$–$Cd(PO_3)_2$ phase-equilibrium diagram (Fig. 4.2.4), melts congruently at 968 K. From the crystal data reported by the authors, this salt seems to be isotypic with the corresponding calcium salt.

$PbNaP_3O_9$ and $PbNa_4(P_3O_9)_2$. The $NaPO_3$–$Pb(PO_3)_2$ phase-equilibrium diagram (Fig. 4.2.5), established by Mahama *et al.*[859] shows the existence of only one compound, $PbNa_4(P_3O_9)_2$, melting congruently at 913 K; its crystal structure was determined by Averbuch-Pouchot and Durif.[860] During attempts to prepare crystals of $PbNa_4(P_3O_9)_2$, the authors discovered another cyclotriphosphate, $PbNaP_3O_9$, not observed in the phase-equilibrium diagram and identified it as a cyclotriphosphate isotypic with the corresponding barium salt described below.

$BaNaP_3O_9$ and $BaNa_4(P_3O_9)_2$. These two compounds appear in the $NaPO_3$–$Ba(PO_3)_2$ phase-equilibrium diagram (Fig. 4.2.6) elaborated by Martin and Durif,[861] the first one as a congruent melting compound (mp = 939 K), and the second one as an incongruent-melt-

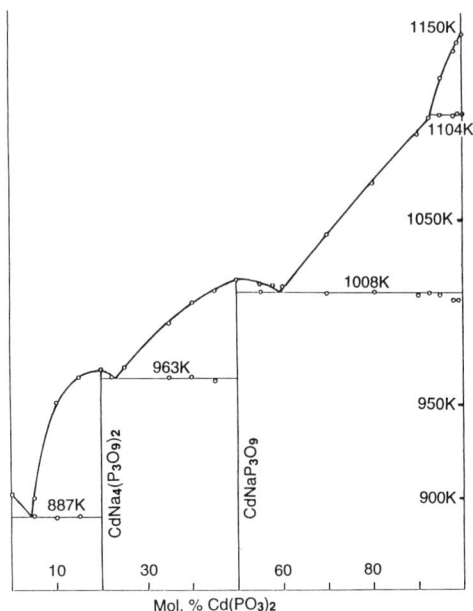

Figure 4.2.4. The NaPO₃–Cd(PO₃)₂ phase-equilibrium diagram. The reversible $\alpha \to \beta$ transition of Cd(PO₃)₂ is observed at 1104 K on the right-hand side of the diagram.

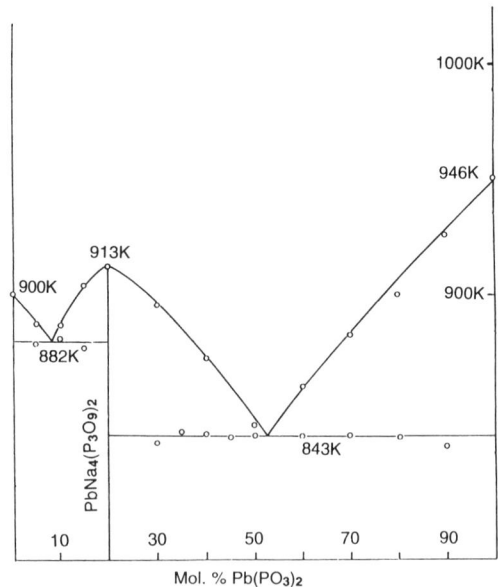

Figure 4.2.5. The NaPO₃–Pb(PO₃)₂ phase-equilibrium diagram.

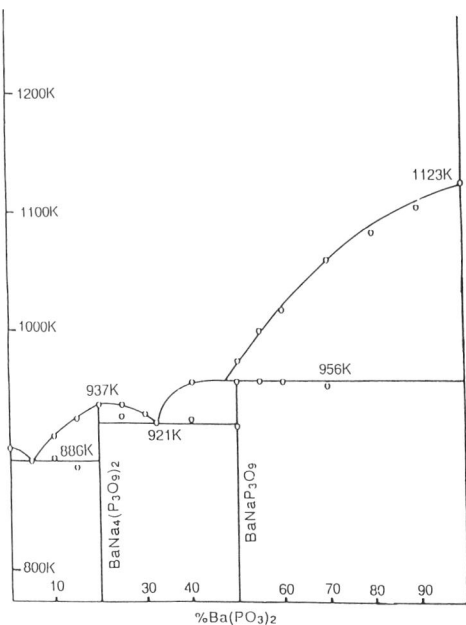

Figure 4.2.6. The $NaPO_3$–$Ba(PO_3)_2$ phase-equilibrium diagram.

ing compound decomposing at 956 K. The existence of the first one was mentioned as early as 1848 by Fleitmann and Henneberg,[7] and its atomic arrangement was determined by Martin and Mitschler.[862] The main crystallographic features of $BaNa_4(P_3O_9)_2$ and its chemical preparation were reported by Averbuch-Pouchot and Durif,[860] who identified this salt as an isotype of $PbNa_4(P_3O_9)_2$.

$CdAgP_3O_9$. The chemical preparation of this salt and crystal data have been reported by Pouchot *et al.*[863] who recognized it as an isotype of the mineral benitoite, $BaTiSi_3O_9$. $CdAgP_3O_9$ appears as an incongruent-melting compound decomposing at 923 K in the $AgPO_3$–$Cd(PO_3)_2$ phase-equilibrium diagram elaborated by Averbuch-Pouchot[864] (Fig. 4.2.7).

$BaAgP_3O_9$. Crystal data for this salt, prepared by heating the corresponding tetrahydrate, described below, at 623 K, were reported by Durif and Averbuch-Pouchot[865] who identified it as an isotype of $BaNaP_3O_9$.

$SrAgP_3O_9$. The $Sr(PO_3)_2$–$AgPO_3$ phase diagram was elaborated by Savenkova *et al.*[866] The incongruent-melting compound, $SrAgP_3O_9$, observed in this system is said to be a cyclotriphosphate, but no structural investigation has confirmed this assumption.

$CaKP_3O_9$ and $MgKP_3O_9$. These two salts were first characterized by Andrieu and Diament[867] during the elaboration of the KPO_3–$Mg_2P_4O_{12}$ and KPO_3–$Ca(PO_3)_2$ phase-equilibrium diagrams. They were soon recognized as isotypic with benitoite by Andrieu *et al.*[868]

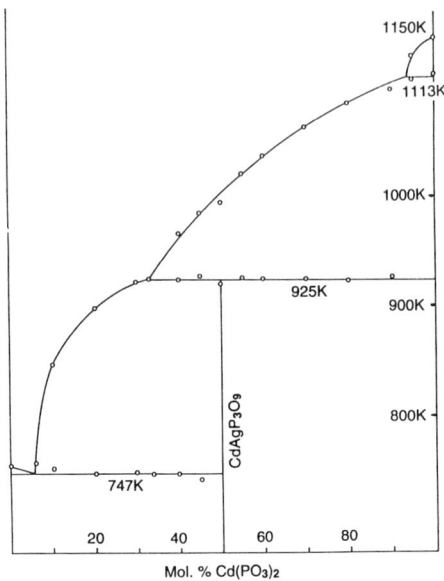

Figure 4.2.7. The AgPO$_3$–Cd(PO$_3$)$_2$-phase-equilibrium diagram. The reversible $\alpha \rightarrow \beta$ transformation of Cd(PO$_3$)$_2$ is observed at 1104 K on the right-hand side of the diagram. For pure Cd(PO$_3$)$_2$ the transition is measured at 1113 K.

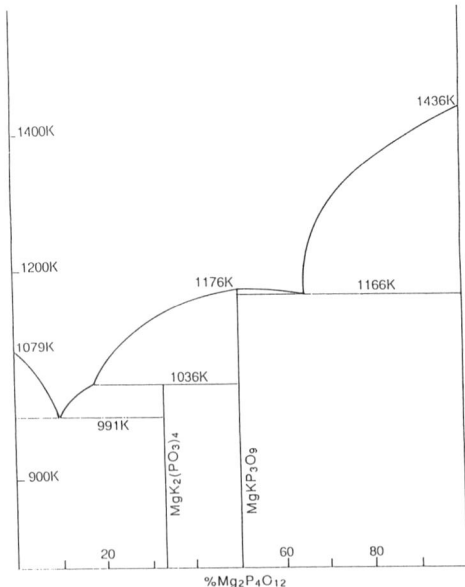

Figure 4.2.8. The KPO$_3$–Mg$_2$P$_4$O$_{12}$ phase-equilibrium diagram.

The KPO_3–$Mg_2P_4O_{12}$ phase-equilibrium diagram was subsequently revised by Averbuch-Pouchot et al.[869] This updated diagram, presented in Fig. 4.2.8, shows that $MgKP_3O_9$ is a congruent-melting compound (mp = 1176 K), as is $CaKP_3O_9$ (mp = 1123 K). Masse et al. [870] refined the crystal structure of the magnesium salt. In addition, an orthorhombic form of $CaKP_3O_9$ was prepared and described by Masse et al.[871] This second form is a distortion of the benitoite arrangement. The KPO_3–$Ca(PO_3)_2$ phase diagram is given in Fig. 4.2.9.

$ZnKP_3O_9$, $CoKP_3O_9$, $MnKP_3O_9$, and $CdKP_3O_9$. These four cyclotriphosphates were prepared by Andrieu et al.[868] and identified as isotypes of benitoite.

The zinc salt was said to be the only intermediate compound in the KPO_3–$Zn_2P_4O_{12}$ phase-equilibrium diagram elaborated by Krivoviazov et al.[206] This assumption was not confirmed during a reinvestigation of this system (Fig. 4.2.10) by Averbuch-Pouchot et al.,[869] who identified in the phase diagram a second compound, $ZnK_2(PO_3)_4$, a long-chain polyphosphate. $ZnKP_3O_9$ is a congruent-melting compound (mp = 1111 K).

The cobalt salt appears also as a congruent-melting compound (mp = 1069 K) in the KPO_3–$Co_2P_4O_{12}$ phase-equilibrium diagram elaborated by Rakotomahanina-Rolaisoa[872] (Fig. 4.2.11).

The cadmium salt is a congruent-melting compound (mp = 1021 K) in the KPO_3–$Cd(PO_3)_2$ phase-equilibrium diagram elaborated by Mermet et al.[873] (Fig. 4.2.12).

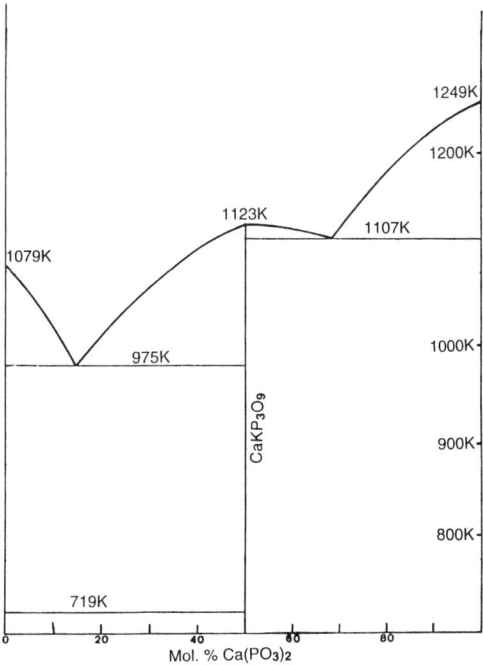

Figure 4.2.9. The KPO_3–$Ca(PO_3)_2$ phase-equilibrium diagram. The transformation of KPO_3 is observed at 719 K on the left-hand side of the diagram.

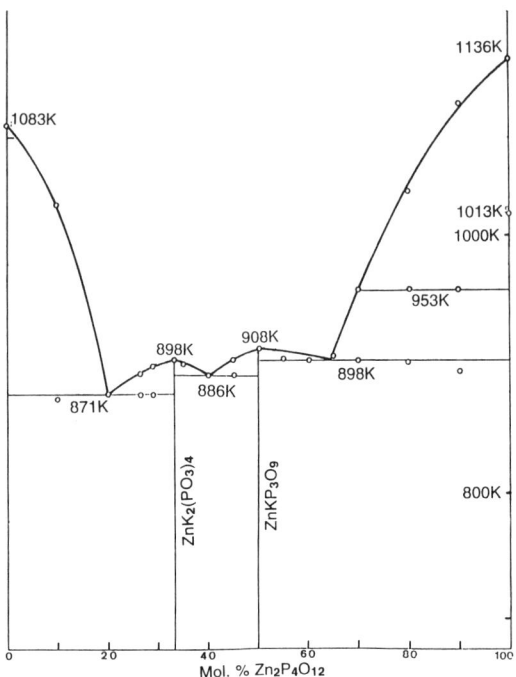

Figure 4.2.10. The KPO_3–$Zn_2P_4O_{12}$ phase-equilibrium diagram. On the right-hand side of the diagram, the transformation of $Zn_2P_4O_{12}$ into $Zn(PO_3)_2$ is observed at 1013 K for the pure salt and at 953 K for the other concentrations.

$HgKP_3O_9$. This salt was obtained during an investigation of the HgO–K_2O–P_2O_5 system by Averbuch-Pouchot and Durif.[874] These authors reported a complete determination of the atomic arrangement, showing this salt to be isotypic with $BaNaP_3O_9$.

$M^{II}NH_4P_3O_9$ (M^{II} = Zn, Co, Ca, Cd, Mg, Mn). Six cyclotriphosphates with the general formula reported above were synthesized by thermal methods and recognized as being isotypic with benitoite by Masse *et al.*[870] An accurate refinement of the crystal structure of the calcium–ammonium salt was performed later by Prisset.[637] Two of these salts, $MgNH_4P_3O_9$ and $CaNH_4P_3O_9$, are polymorphic. The magnesium salt can also crystallize as an orthorhombic compound. Its atomic arrangement, determined by Grenier and Masse,[875] is a distortion of the benitoite structure and is isotypic with the orthorhombic form of $CaKP_3O_9$. The second form of the calcium salt is monoclinic. Its structure, first determined by Grenier and Masse[875] and later reinvestigated by Masse *et al.*,[871] is also a distortion of the benitoite structure.

$MgRbP_3O_9$. This cyclotriphosphate is the high-temperature form of the long-chain poly-phosphate $MgRb(PO_3)_3$, characterized during the elaboration of the $RbPO_3$–$Mg_2P_4O_{12}$ phase-equilibrium diagram by Rakotomahanina-Rolaisoa.[876] It is isotypic with the orthorhombic form of $MgNH_4P_3O_9$.

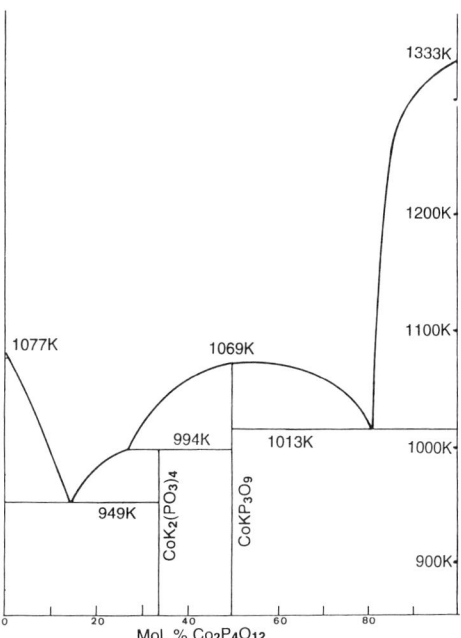

Figure 4.2.11. The KPO_3–$Co_2P_4O_{12}$ phase-equilibrium diagram.

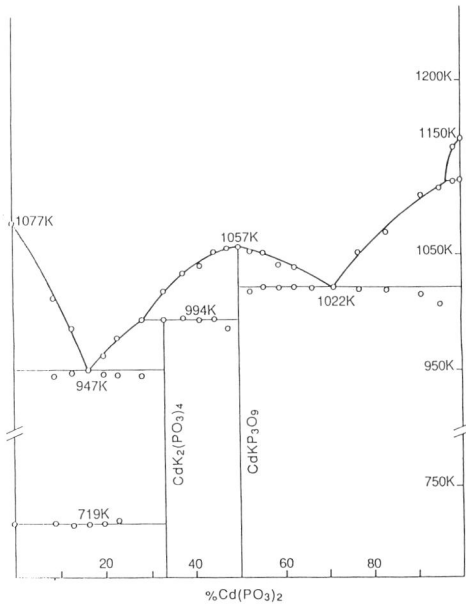

Figure 4.2.12. The KPO_3–$Cd(PO_3)_2$ phase-equilibrium diagram. The transformation of KPO_3 is observed at 719 K on the left-hand side of the diagram.

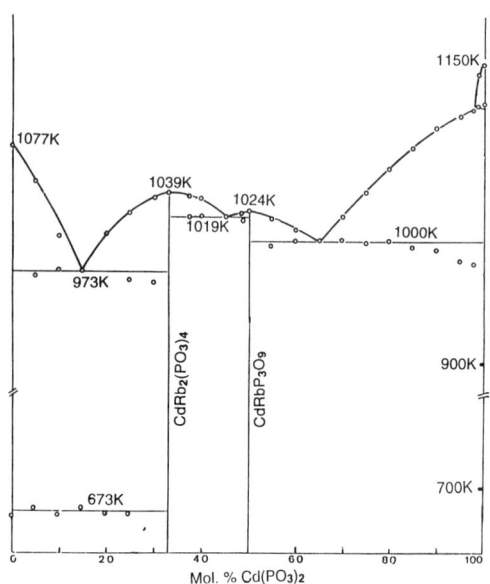

Figure 4.2.13. The RbPO₃–Cd(PO₃)₂ phase-equilibrium diagram. The transition of RbPO₃ is observed at 673 K on the left-hand side of the diagram, and that of Cd(PO₃)₂ at 1113 K on the right-hand side.

CdRbP$_3$O$_9$. The chemical preparation of CdRbP$_3$O$_9$ was described by Pouchot *et al.*[863] The crystal data reported by the authors show it to be isotypic with benitoite. This salt is a congruent-melting compound (mp = 1024 K) in the RbPO$_3$–Cd(PO$_3$)$_2$ phase-equilibrium diagram established by Mermet *et al.*[873] (Fig. 4.2.13).

CaRbP$_3$O$_9$. CaRbP$_3$O$_9$ appears as a congruent-melting salt (mp = 1180 K) in the RbPO$_3$–Ca(PO$_3$)$_2$ phase-equilibrium diagram elaborated by Henry and Durif[877] (Fig. 4.2.14). The crystal data reported by these authors indicate that this compound is isotypic with the monoclinic form of CaNH$_4$P$_3$O$_9$.

CdCsP$_3$O$_9$ and CaCsP$_3$O$_9$. The cadmium salt was observed by Averbuch-Pouchot[878] during the elaboration of the CsPO$_3$–Cd(PO$_3$)$_2$ phase-equilibrium diagram (Fig.4.2.15). It melts congruently at 996 K. The atomic arrangement was determined by Averbuch-Pouchot and Durif,[879] who confirmed that this salt is isotypic with the orthorhombic form of MgNH$_4$P$_3$O$_9$. The calcium salt prepared and described by Masse *et al.*[871] has the same type of structure.

SrCsP$_3$O$_9$. Tokman and Bukhalova[880] established the CsPO$_3$–Sr(PO$_3$)$_2$ phase-equilibrium diagram. The incongruent-melting compound, SrCsP$_3$O$_9$, observed in this system was said by these authors to be a cyclotriphosphate, but no structural work confirmed this assumption.

MgTlP$_3$O$_9$ and MgTl$_4$(P$_3$O$_9$)$_2$. These salts were characterized during the elaboration of the TlPO$_3$–Mg$_2$P$_4$O$_{12}$ phase-equilibrium diagram by Rakotomahanina-Rolaisoa *et al.*[881]

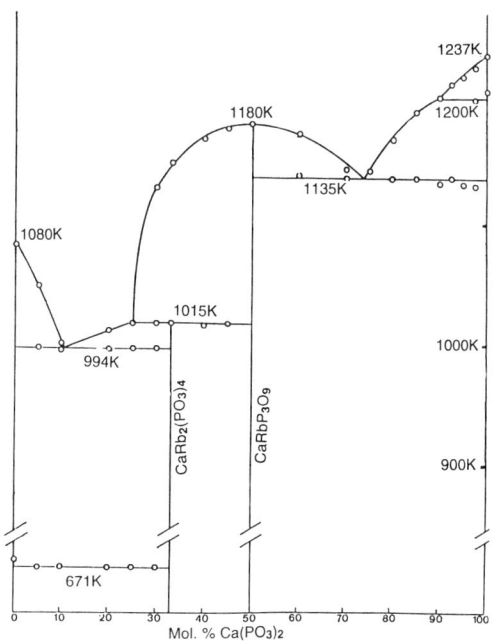

Figure 4.2.14. The RbPO₃–Ca(PO₃)₂ phase-equilibrium diagram. The $\alpha \rightarrow \beta$ transformation of Ca(PO₃)₂ is observed at 1200 K on the right-hand side of the diagram, and that of RbPO₃ is observed at 671 K on the left-hand side.

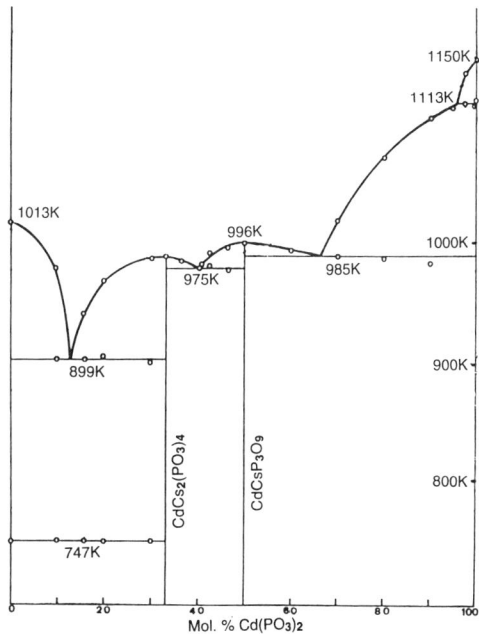

Figure 4.2.15. The CsPO₃–Cd(PO₃)₂ phase-equilibrium diagram. The transition of CsPO₃ is observed at 747 K on the left-hand side of the diagram, and that of Cd(PO₃)₂ is observed at 1113 K on the right-hand side.

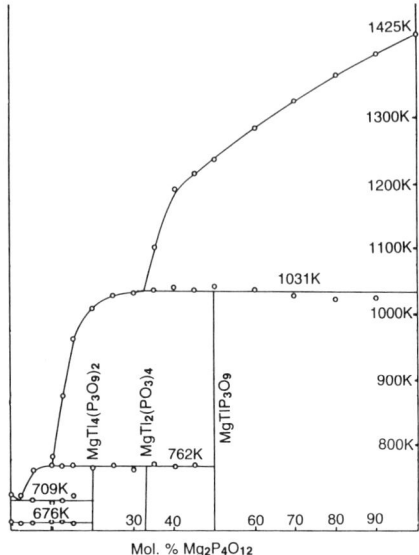

Figure 4.2.16. The TlPO₃–Mg₂P₄O₁₂ phase-equilibrium diagram. The transformation of Tl₄P₄O₁₂ into TlPO₃ is observed at 676 K on the left-hand side of the diagram.

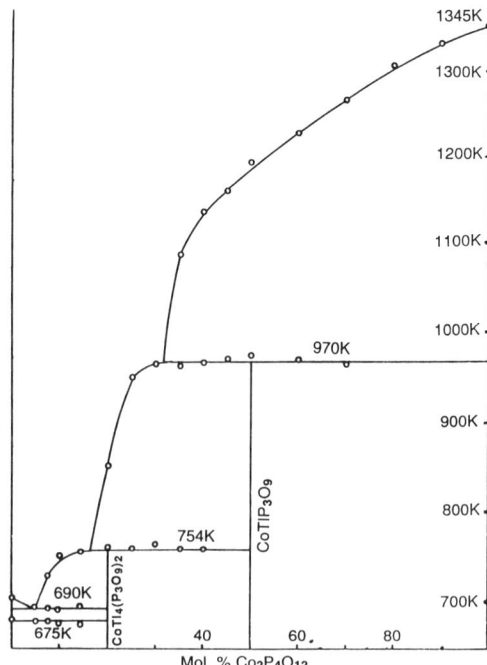

Figure 4.2.17. The TlPO₃–Co₂P₄O₁₂ phase-equilibrium diagram. The transformation of TlPO₃ is observed at 675 K on the left-hand side of the diagram.

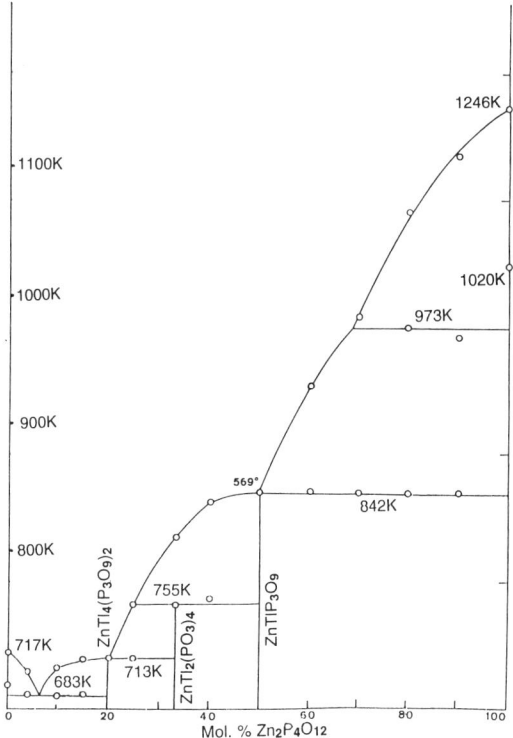

Figure 4.2.18. The TlPO₃–Zn(PO₃)₂ phase-equilibrium diagram. The transformation of $Zn(PO_3)_2$ is observed at 973 K on the right-hand side of the diagram and at 1020 K for pure $Zn(PO_3)_2$.

(Fig. 4.2.16). Both are incongruent-melting compounds; $MgTl_4(P_3O_9)_2$ decomposes at 763 K, and $MgTlP_3O_9$ at 1031 K. The latter compound is isotypic with the orthorhombic form of $MgNH_4P_3O_9$. A model for the atomic arrangement in $MgTl_4(P_3O_9)_2$ was proposed by Rakotomahanina-Rolaisoa.[872]

$CoTl_4(P_3O_9)_2$ and $ZnTl_4(P_3O_9)_2$. These salts were characterized during the elaboration of the $TlPO_3$–$Co_2P_4O_{12}$ and $TlPO_3$–$Zn(PO_3)_2$ phase-equilibrium diagrams by Rakotomahanina-Rolaisoa et al.[881] and Averbuch-Pouchot,[882] respectively. Both are incongruent-melting compounds; the cobalt salt decomposes at 754 K and the zinc salt at 713 K. They are isotypic with the corresponding magnesium salt. The two diagrams are presented in Fig. 4.2.17 and 4.2.18.

$CdTlP_3O_9$ and $CdTl_4(P_3O_9)_2$. The chemical preparation of $CdTlP_3O_9$ were reported by Pouchot et al.[863] The crystal data reported by these authors show this salt to be isotypic with benitoite. The second compound, $CdTl_4(P_3O_9)$, was not observed in the first version of the $TlPO_3$–$Cd(PO_3)_2$ phase-equilibrium diagram elaborated by Averbuch-Pouchot[864] but was clearly characterized in the revised version of this diagram (Fig. 4.2.19), produced by the same author,[883] and identified as an isotype of the corresponding magnesium salt.

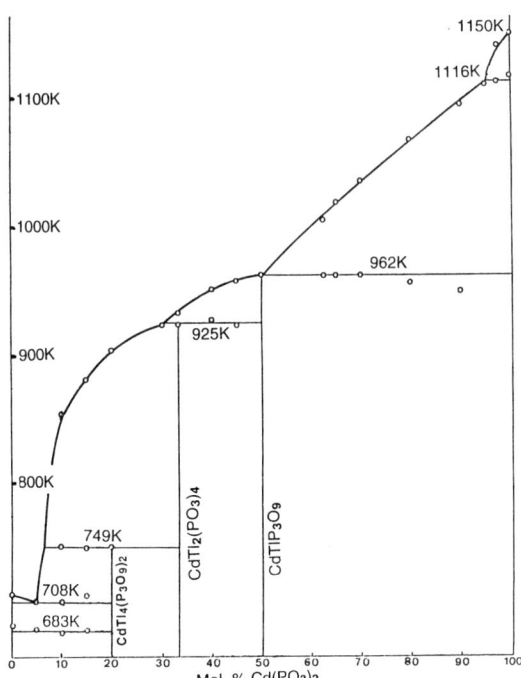

Figure 4.2.19. The TlPO₃–Cd(PO₃)₂ phase-equilibrium diagram. The transformation of TlPO₃ is observed at 683 K on the left-hand side of the diagram, and that of Cd(PO₃)₂ is observed at 1116 K on the right-hand side.

CdTlP₃O₉ and CdTl₄(P₃O₉)₂ appear as incongruent-melting compounds, decomposing at 962 K and 749 K, respectively.

CaTl₄(P₃O₉)₂ and CaTlP₃O₉. These salts were first observed by Rakotomahanina-Ro-laisoa *et al.*[881] during their elaboration of the TlPO₃–Ca(PO₃)₂ phase-equilibrium diagram (Fig. 4.2.20). CaTl₄(P₃O₉)₂ is an incongruent-melting compound, isotypic with MgTl₄(P₃O₉)₂ and decomposing at 793 K, whereas CaTlP₃O₉ is isotypic with the mono-clinic form of CaNH₄P₃O₉ and melts congruently at 1089 K.

HgNa₂(NH₄)₂(P₃O₉)₂. This salt was characterized during a study of the P₂O₅–HgO–Na₂O–(NH₄)₂O system by Averbuch-Pouchot and Durif.[884] The authors reported a complete determination of the atomic arrangement. The unit-cell dimensions of this compound are close to those measured for CaNa₄(P₃O₉)₂ and CdNa₄(P₃O₉)₂.

Crystallographic Data. Tables 4.2.4–4.2.6 contain report the main crystal data for all the anhydrous divalent–monovalent cation cyclotriphosphates presently known.

(ii) Hydrated Divalent–Monovalent Cation Cyclotriphosphates

CaNaP₃O₉·3H₂O and SrNaP₃O₉·3H₂O. The existence of these two salts was reported in 1875 by Lindbom.[12] Reproducible preparative procedures and crystal data were pub-

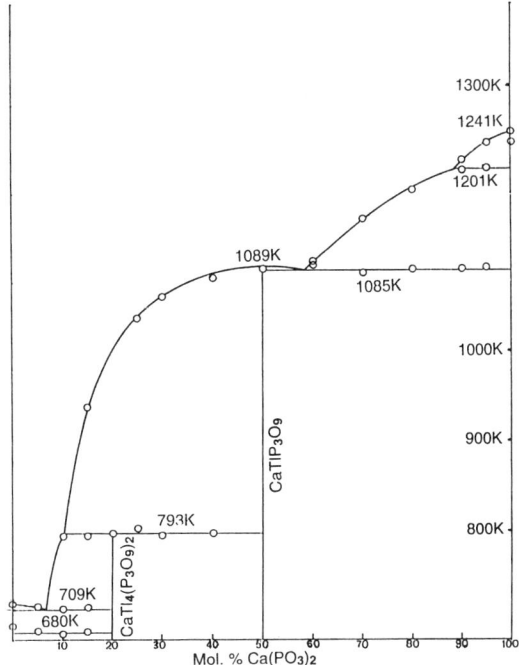

Figure 4.2.20. The $TlPO_3$–$Ca(PO_3)_2$ phase-equilibrium diagram. The transformation of $TlPO_3$ is observed at 680 K on the left-hand side of the diagram, and that of $Ca(PO_3)_2$ is observed at 1201 K on the right-hand side.

Table 4.2.4. Edge Lengths of the Hexagonal Unit Cell in the 15 Cyclotriphosphates Isotypic with Benitoite[a]

Formula	a (Å)	c (Å)	Reference(s)
$CdAgP_3O_9$	6.622(2)	9.921(5)	863
$CaKP_3O_9$	6.795(1)	10.336(1)	868
$MgKP_3O_9$	6.605(1)	9.772(1)	868
$ZnKP_3O_9$	6.606(1)	9.743(1)	868
$CoKP_3O_9$	6.637(1)	9.795(1)	868
$MnKP_3O_9$	6.686(1)	9.958(1)	868
$CdKP_3O_9$	6.780(1)	10.148(1)	868
$ZnNH_4P_3O_9$	6.718(3)	9.819(5)	870
$CoNH_4P_3O_9$	6.695(3)	9.819(5)	870
$CaNH_4P_3O_9$	6.887(3)	10.448(5)	637, 870
$MgNH_4P_3O_9$	6.698(3)	9.831(5)	870
$MnNH_4P_3O_9$	6.771(3)	10.026(5)	870
$CdNH_4P_3O_9$	6.870(3)	10.233(5)	870
$CdRbP_3O_9$	6.858(2)	10.211(5)	863
$CdTlP_3O_9$	6.845(2)	10.154(5)	863

[a]The common space groups is $P\bar{6}2c$; $Z = 2$.

Table 4.2.5. Main Crystallographic Data for $M^{II}M^{I}P_3O_9$ Cyclotriphosphates

Formula	a (Å) α (°)	b (Å) β (°)	c (Å) γ (°)	S.G.	Z	Reference(s)
$PbNaP_3O_9$	10.873(5)	12.11(5)	5.664(2)	$P2_12_12_1$	4	859
$BaNaP_3O_9$	11.134(8)	12.320(8)	5.802(3)	$P2_12_12_1$	4	861, 862
$BaAgP_3O_9$	11.05(1)	12.28(1)	5.903(5)	$P2_12_12_1$	4	865
$CaKP_3O_9$	7.316(4)	12.44(2)	9.955(6)	$Pmcn$	4	871
$HgKP_3O_9$	11.164(6)	12.46(1)	5.622(2)	$P2_12_12_1$	4	874
$MgNH_4P_3O_9$	7.22(1)	12.04(2)	9.33(1)	$Pmcn$	4	875
$CaNH_4P_3O_9$	7.446(5)	12.461(10)	10.050(10) 90.11(10)	$P2_1/n$	4	871, 875
$MgRbP_3O_9$	7.177(3)	11.981(4)	9.339(3)	$Pmcn$	4	876
$CaRbP_3O_9$	7.545(5)	12.51(2)	9.745(8) 90.00(5)	$P2_1/n$	4	877
$CdCsP_3O_9$	7.507(3)	12.68(2)	9.533(8)	$Pmcn$	4	878, 879
$CaCsP_3O_9$	7.523(4)	12.77(2)	10.056(6)	$Pmcn$	4	871
$MgTlP_3O_9$	7.185(5)	12.05(1)	9.331(5)	$Pmcn$	4	881
$CaTlP_3O_9$	7.471(5)	12.52(1)	9.913(5) 90.00(5)	$P2_1/n$	4	881

lished by Durif[885] for the calcium salt and by Martin and Durif[861] for the strontium one. The two salts are not isotypic. The atomic arrangement in $SrNaP_3O_9 \cdot 3H_2O$ was determined by Zilber *et al.*[886]; that in $CaNaP_3O_9 \cdot 3H_2O$ is still unknown. The influence of water vapor on the opening of the P_3O_9 ring during the thermal dehydration of $CaNaP_3O_9 \cdot 3H_2O$ was investigated by Simonot-Grange and Jamet.[887]

$BaNaP_3O_9 \cdot 4H_2O$ and $BaNaP_3O_9 \cdot 3H_2O$. The tetrahydrate was first reported by Fleitmann and Henneberg.[7] Its chemical preparation and crystallographic data for this salt were

Table 4.2.6. Main Crystallographic Data for $M^{II}M^{I}_4P_3O_9$ Cyclotriphosphates

Formula	a (Å) α (°)	b (Å) β (°)	c (Å) γ (°)	S.G.	Z	Reference(s)
$CaNa_4(P_3O_9)_2$	13.248(5)	8.120(3) 94.65(5)	14.384(6)	Cc or $C2/c$	4	856
$CdNa_4(P_3O_9)_2$	13.05(1)	7.952(3) 94.23(1)	14.19(1)	Cc or $C2/c$	4	858
$PbNa_4(P_3O_9)_2$	7.268(4) 121.52(5)	8.151(5) 102.06(5)	7.851(5) 73.00(5)	$P\bar{1}$	1	859, 860
$BaNa_4(P_3O_9)_2$	7.313(4) 121.38(5)	8.237(5) 102.66(5)	7.865(5) 72.23(5)	$P\bar{1}$	1	860, 861
$MgTl_4(P_3O_9)_2$	7.191(3)	7.191(3)	19.55(1)	$P3_1c$	2	881
$CoTl_4(P_3O_9)_2$	7.220(3)	7.220(3)	19.62(1)	$P3_1c$	2	881
$ZnTl_4(P_3O_9)_2$	7.213(2)	7.213(2)	19.625(15)	$P3_1c$	2	882
$CdTl_4(P_3O_9)_2$	7.320(5)	7.320(5)	19.81(1)	$P3_1c$	2	883
$CaTl_4(P_3O_9)_2$	7.389(3)	7.389(3)	19.99(1)	$P3_1c$	2	881
$HgNa_2(NH_4)_2(P_3O_9)_2$	13.524(8)	8.362(5) 92.58(5)	14.390(8)	$C2/c$	4	884

reported by Martin and Durif[861]; it is isotypic with the corresponding silver–barium salt described below. The trihydrate was obtained by crystallization of an aqueous solution of the tetrahydrate at 333 K. Averbuch-Pouchot and Durif[888] reported a complete description of its atomic arrangement.

$BaAgP_3O_9 \cdot 4H_2O$. This hydrate, prepared by Boullé's process, was described by Durif and Averbuch-Pouchot,[865] and its crystal structure was determined by Seethanen *et al.*[889]

$SrKP_3O_9 \cdot 3H_2O$. This salt was prepared by Martin[625] and characterized as an isotype of $SrNaP_3O_9 \cdot 3H_2O$.

$BaKP_3O_9 \cdot H_2O$. This salt was first characterized by Martin,[625] and its atomic arrangement was subsequently determined by Seethanen and Durif.[890]

$CaNH_4P_3O_9 \cdot 3H_2O$. This salt was prepared by Masse *et al.*[871] and recognized as an isotype of $CaNaP_3O_9 \cdot 3H_2O$. A hydrated calcium–ammonium cyclotriphosphate was reported by Feldman and Grunze.[891] Its state of hydration is not well defined, and, according to these authors, the Ca/NH_4 ratio is slightly different from unity.

$SrNH_4P_3O_9 \cdot 4H_2O$, $SrRbP_3O_9 \cdot 4H_2O$, and $SrTlP_3O_9 \cdot 4H_2O$. These three isotypic salts, prepared by Boullé's process, were described by Martin.[625] The atomic arrangement in this type of compounds is not yet known.

A strontium–ammonium trihydrate, $SrNH_4P_3O_9 \cdot 3H_2O$ was prepared by Takenaka *et al.*[892] These authors investigated its thermal decomposition both in a wet and in a dry atmosphere. The crystallization water is lost at 403 K, but up to 523 K the P_3O_9 ring anion is not cleaved. Above 523 K, ammonia is evolved.

$BaNH_4P_3O_9 \cdot H_2O$, $BaCsP_3O_9 \cdot H_2O$, and $BaTlP_3O_9 \cdot H_2O$. The three salts have been prepared by using Boullé's process. The preparation of $BaNH_4P_3O_9 \cdot H_2O$ was described by Durif *et al.*[893] who reported a complete description of the atomic arrangement. The cesium salt was prepared by Masse and Averbuch-Pouchot.[676] The crystal data reported by these authors indicates that this salt is isotypic with $BaNH_4P_3O_9 \cdot H_2O$. Durif *et al.*[893] reported the chemical preparation of $BaTlP_3O_9 \cdot H_2O$ and presented crystallographic data showing it to be isotypic with the ammonium and cesium salts. A second form of this salt, isotypic with the barium–potassium salt, was described by Martin.[625]

$NiNa_4(P_3O_9)_2 \cdot 6H_2O$ and $NiAg_4(P_3O_9)_2 \cdot 6H_2O$. The chemical preparation and crystal structure of these two isotypic salts were reported by Jouini and Dabbabi.[894–896]

$CuNa_4(P_3O_9)_2 \cdot 4H_2O$. The chemical preparation of this salt is easily performed by mixing aqueous solutions of $Na_3P_3O_9$ and $CuCl_2$ in a stoichiometric ratio and evaporating slowly at room temperature. The crystal structure was determined by Durif and Averbuch-Pouchot.[897]

$NiK_4(P_3O_9)_2 \cdot 7H_2O$ and $CoK_4(P_3O_9)_2 \cdot 7H_2O$. These two isotypic salts, prepared by Boullé's process, were described by Seethanen *et al.*[898] These authors reported a complete description of the atomic arrangement for the nickel compound.

CuK$_4$(P$_3$O$_9$)$_2$·4H$_2$O and Cu(NH$_4$)$_4$(P$_3$O$_9$)$_2$·4H$_2$O. Both these salts have been prepared by using Boullé's metathesis reaction. The atomic arrangement in the potassium salt was determined by Durif and Averbuch-Pouchot.[899] From the crystal data reported by these authors, the two salts are isotypic.

Ni(NH$_4$)$_4$(P$_3$O$_9$)$_2$·4H$_2$O and Co(NH$_4$)$_4$(P$_3$O$_9$)$_2$·4H$_2$O. The atomic arrangement in the nickel salt was determined by Jouini and Dabbabi.[900] In addition, the magnetic properties and IR spectrum of this salt were reported by these authors. The chemical preparation and crystal structure of Co(NH$_4$)$_4$(P$_3$O$_9$)$_2$·4H$_2$O were described by Belkhiria et al.[901] These two salts are isotypic with the corresponding copper–ammonium and copper–potassium compounds.[899]

CdK$_4$(P$_3$O$_9$)$_2$·2H$_2$O. The chemical preparation of this salt by Boullé's process was described by Averbuch-Pouchot.[902] This author performed a complete determination of the atomic arrangement.

ZnK$_4$(P$_3$O$_9$)$_2$·6H$_2$O and CoRb$_4$(P$_3$O$_9$)$_2$·6H$_2$O. From their crystal data and their atomic arrangements, these two salts appear to be isotypic. The zinc salt was first prepared and investigated by Seethanen et al.[903] These authors described this compound as a tetrahydrate. Subsequently, Belkhiria et al.[904] reported a detailed determination of the atomic arrangement of the cobalt salt, showing it to be a hexahydrate. In this salt, two of the six

Table 4.2.7. Main Crystallographic Data for M^{II}M^{I}P$_3$O$_9$·nH$_2$O Cyclotriphosphates

Formula	a (Å) α (°)	b (Å) β (°)	c (Å) γ (°)	S.G.	Z	Reference(s)
CaNaP$_3$O$_9$·3H$_2$O	14.923(8)	14.923(8)	10.148(10)	P6$_3$···	2	885
SrNaP$_3$O$_9$·3H$_2$O	16.167(8)	12.013(5)	10.645(5)	Pnma	8	861, 886
BaNaP$_3$O$_9$·4H$_2$O	21.33(8)	7.01(1) 122.18(10)	18.26(8)	C2/c	8	861
BaNaP$_3$O$_9$·3H$_2$O	7.067(3) 116.46(5)	9.071(3) 95.97(5)	9.906(4) 74.03(5)	P$\bar{1}$	2	888
BaAgP$_3$O$_9$·4H$_2$O	21.35(3)	7.163(3) 121.72(5)	18.35(2)	C2/c	8	865, 889
SrKP$_3$O$_9$·3H$_2$O	16.08(2)	12.24(1)	10.74(1)	Pnma	8	625
BaKP$_3$O$_9$·H$_2$O	7.34(1)	17.77(2) 95.24(5)	7.18(1)	P2$_1$/n	4	625, 890
CaNH$_4$P$_3$O$_9$·3H$_2$O	14.76(5)	14.76(5)	9.932(6)	P6$_3$···	8	871
SrNH$_4$P$_3$O$_9$·4H$_2$O	9.144(8)	15.31(1)	8.182(8)	Pna···	4	625
BaNH$_4$P$_3$O$_9$·H$_2$O	11.70(1)	12.12(1) 101.05(5)	7.559(5)	P2$_1$/n	4	893
SrRbP$_3$O$_9$·4H$_2$O	9.161(8)	15.32(1)	8.183(8)	Pna···	4	625
BaCsP$_3$O$_9$·H$_2$O	11.764(8)	12.292(8) 101.16(5)	7.681(5)	P2$_1$/n	4	676
SrTlP$_3$O$_9$·4H$_2$O	9.106(8)	15.296(10)	8.272(8)	Pna···	4	625
BaTlP$_3$O$_9$·H$_2$O (I)	11.76(1)	12.33(1) 100.92(5)	7.537(8)	P2$_1$/n	4	893
BaTlP$_3$O$_9$·H$_2$O (II)	7.39(1)	17.72(2) 95.45	7.21(1)	P2$_1$/n	4	625

Table 4.2.8. Main Crystallographic Data for $M^{II}M_4^I P_3O_9 \cdot nH_2O$ Cyclotriphosphates

Formula	a (Å) α (°)	b (Å) β (°)	c (Å) γ (°)	S.G.	Z	Reference(s)
$NiNa_4(P_3O_9)_2 \cdot 6H_2O$	9.186(2) 89.17(1)	8.020(2) 102.89(1)	6.838(1) 98.03(1)	$P\bar{1}$	1	894, 896
$NiAg_4(P_3O_9)_2 \cdot 6H_2O$	9.209(3) 89.15(2)	8.053(3) 102.94(1)	6.841(2) 97.24(1)	$P\bar{1}$	1	894, 895
$CuNa_4(P_3O_9)_2 \cdot 4H_2O$	7.907(5) 102.46(5)	8.364(5) 97.89(5)	7.122(5) 84.04(5)	$P\bar{1}$	1	897
$NiK_4(P_3O_9)_2 \cdot 7H_2O$	23.03(1)	11.882(4)	8.732(4)	$Fm2m$	4	898
$CoK_4(P_3O_9)_2 \cdot 7H_2O$	22.98(1)	11.869(2)	8.751(2)	$Fm2m$	4	898
$CuK_4(P_3O_9)_2 \cdot 4H_2O$	8.510(5)	14.303(8) 96.51(2)	8.487(5)	$P2_1/a$	2	899
$Cu(NH_4)_4(P_3O_9)_2 \cdot 4H_2O$	8.60(1)	14.65(2) 96.66(8)	8.79(1)	$P2_1/c$	2	899
$Ni(NH_4)_4(P_3O_9)_2 \cdot 4H_2O$	8.645(2)	14.698(3) 95.89(2)	8.774(2)	$P2_1/c$	2	894, 900
$Co(NH_4)_4(P_3O_9)_2 \cdot 4H_2O$	8.612(2)	14.698(3) 95.67(1)	8.809(2)	$P2_1/c$	2	901
$CdK_4(P_3O_9)_2 \cdot 2H_2O$	9.235(5) 96.38(1)	7.599(4) 103.90(1)	7.148(4) 102.06(1)	$P\bar{1}$	1	902
$ZnK_4(P_3O_9)_2 \cdot 4H_2O$	12.444(4)	10.978(2) 124.41(2)	9.624(3)	$C2/m$	2	903
$CoRb_4(P_3O_9)_2 \cdot 6H_2O$	13.216(3)	11.059(2) 126.71(1)	10.026(2)	$C2/m$	2	904
$NiCs_4(P_3O_9)_2 \cdot 6H_2O$	19.992(4)	6.500(2)	18.445(4)	$Pca2_1$ $Pcam$	4	894

water molecules are weakly bonded and have relatively high thermal factors. This fact explains why the zinc salt was incorrectly identified as a tetrahydrate in the earlier work.[903]

$NiCs_4(P_3O_9)_2 \cdot 6H_2O$. Jouini and Dabbabi[894] reported the chemical preparation of this salt and presented crystal data. Its atomic arrangement is still unknown.

Crystallographic Data. Tables 4.2.7 and 4.2.8 contain the main crystallographic data for hydrated divalent–monovalent cation cyclotriphosphates.

4.2.2.4. Trivalent and Trivalent–Monovalent Cation Cyclotriphosphates

Trivalent and trivalent–monovalent cation cyclotriphosphates have not yet been extensively investigated, and only a few of them have been well characterized.

$LnP_3O_9 \cdot nH_2O$. Serra and Giesbrecht[905] were the first to describe three rare-earth salts of general formula $TP_3O_9 \cdot 3H_2O$ with T = La, Ce, and Nd. Later, Birke and Kempe[908,910] described the chemical preparation and thermal behavior for $PrP_3O_9 \cdot 4H_2O$, $LaP_3O_9 \cdot 4H_2O$ and $ErP_3O_9 \cdot 4.4H_2O$. Bagieu-Beucher and Durif[906] prepared a series of three $TP_3O_9 \cdot 3H_2O$ compounds, with T = La, Ce, and Pr and reported their crystal data. The crystal structure of the cerium salt was determined by Bagieu-Beucher et al.[907] The main crystallographic features of these three salts are reported in Table 4.2.9. It is not possible to ascertain whether

Table 4.2.9. Edge Lengths of the Hexagonal Unit Cell
in $TP_3O_9 \cdot 3H_2O$ Cyclotriphosphates[a]

Formula	a (Å)	c (Å)
$LaP_3O_9 \cdot 3H_2O$	6.785(5)	6.112(5)
$CeP_3O_9 \cdot 3H_2O$	6.770(5)	6.079(5)
$PrP_3O_9 \cdot 3H_2O$	6.743(5)	6.048(5)

[a]The common space group is $P\bar{6}$; $Z = 1$.

these salts are similar to those reported in refs.[908,910] because of the lack of crystal data in the studies.

The thermal behavior of these three hexagonal $TP_3O_9 \cdot 3H_2O$ compounds was carefully investigated by Gobled.[911] They are not stable in air and decompose irreversibly according to the following scheme:

$$TP_3O_9 \cdot 3H_2O \rightarrow TPO_4 + P_2O_5 + 3H_2O$$

This investigation was conducted mainly with the lanthanum salt, the only one stable for more than a year.

Two other rare-earth compounds, $NdP_3O_9 \cdot 5H_2O$ and $SmP_3O_9 \cdot 5H_2O$, were described by Bagieu-Beucher.[912] All attempts to produce single crystals of a suitable size for a structural study failed. According to Bagieu-Beucher, $P2_1/a$ is the most probable space group. The monoclinic unit-cell dimensions of these compounds are reported below:

$NdP_3O_9 \cdot 5H_2O$ $a = 13.954$, $b = 11.500$, $c = 7.745$ Å, $\beta = 105.83°$

$SmP_3O_9 \cdot 5H_2O$ $a = 13.889$, $b = 11.496$, $c = 7.719$ Å, $\beta = 105.79°$

$M^{III}Na_3(P_3O_9)_2 \cdot 9H_2O$ (M^{III} = Sm, Eu, Gd, Dy, Ho, Er, Y, Bi). The bismuth salt was first prepared by Bagieu-Beucher and Durif[913] by the addition of solid bismuth nitrate or chloride to a saturated solution of $Na_3P_3O_9$. In such a process, crystal growth is very rapid. In some experiments, crystals up to 1 mm long were obtained within half an hour. These

Table 4.2.10. Unit-Cell Dimensions for
$M^{III}Na_3(P_3O_9)_2 \cdot 9H_2O$ Compounds[a]

Formula	a (Å)	c (Å)	Reference
$BiNa_3(P_3O_9)_2 \cdot 9H_2O$	30.845(15)	13.085(3)	913
$SmNa_3(P_3O_9)_2 \cdot 9H_2O$	31.07(1)	13.119(5)	914
$EuNa_3(P_3O_9)_2 \cdot 9H_2O$	31.05(6)	13.075(6)	914
$GdNa_3(P_3O_9)_2 \cdot 9H_2O$	30.99(1)	13.007(5)	914
$DyNa_3(P_3O_9)_2 \cdot 9H_2O$	30.99(1)	12.951(4)	914
$HoNa_3(P_3O_9)_2 \cdot 9H_2O$	30.95(1)	12.901(4)	914
$ErNa_3(P_3O_9)_2 \cdot 9H_2O$	30.99(1)	12.887(4)	914
$YNa_3(P_3O_9)_2 \cdot 9H_2O$	30.93(1)	12.904(5)	914

[a]$R,\bar{3}c$ is the common rhombohedral space group; $Z = 18$.

authors reported a complete description of the atomic arrangement. This bismuth salt is not stable for a long time. This instability can probably be explained by the fact that some of the water molecules are of a zeolitic nature. A series of seven compounds, isotypic with $BiNa_3(P_3O_9)_2 \cdot 9H_2O$, were prepared by Bagieu-Beucher.[914] Their chemical preparation is identical to that described for the bismuth salt. They are also not stable for a long time under normal conditions. The unit-cell dimensions for these eight compounds are presented in Table 4.2.10.

4.2.2.5. Organic and Alkali-Organic Cation Cyclotriphosphates

Recently, Boullé's metathesis reaction was extended to the syntheses of organic cation or mixed metal–organic cation cyclotriphosphates. Thus, glycinium, guanidinium, methyl-ammonium, ethylenediammonium, potassium–ethylenedi-ammonium and isopropylam-monium cyclotriphosphates have been clearly characterized.[915–920] All these salts are stable under normal conditions. In all cases, the atomic arrangements were accurately determined. The main crystallographic data for these salts are presented in Table 4.2.11.

4.2.2.6. Cyclotriphosphate Adducts with Telluric Acid

P_3O_9 ring anions appear also in adducts between monovalent cation cyclotriphos-phates and telluric acid. These compounds are easily prepared by slow evaporation, at room temperature, of aqueous solutions of the alkali cyclotriphosphates and telluric acid, prepared in a stoichiometric ratio. The atomic arrangements in these compounds have been described, and, in all cases, the phosphoric ring anions and the $Te(OH)_6$ groups coexist as independent units. As for all the other classes of phosphates forming adducts with telluric acid, no lithium salt was observed during these investigations.

Crystal data for two of these adducts, $K_3P_3O_9 \cdot Te(OH)_6 \cdot 2H_2O$ and $2Na_3P_3O_9 \cdot Te(OH)_6 \cdot 6H_2O$ were reported by Boudjada.[921] Their atomic arrangements were sub-sequently determined by Boudjada *et al.*[922,923]

Table 4.2.11. Main Crystallographic Data for Organic and Alkali-Organic Cation Cyclotriphosphates

Formula[a]	a (Å) α (°)	b (Å) β (°)	c (Å) γ (°)	S.G.	Z	Reference
$(EDA)_3(P_3O_9)_2$	15.558(8) 104.14(5)	10.450(6) 102.73(5)	7.639(4) 86.71 (5)	$P\bar{1}$	2	915
$(IPA)_3P_3O_9$	25.22(2)	12.22(2) 123.90(2)	15.45(2)	$C2$	8	916
$K(EDA)P_3O_9$	20.850(8)	9.044(4)	11.653(5)	$Ccca$	8	917
$(MA)_3P_3O_9$	12.144(7)	5.361(5) 97.32(8)	7.203(7)	$P2_1/n$	4	918
$(Gly)_3P_3O_9$	12.223(8)	14.52(1) 100.47(5)	10.229(7)	$P2_1/c$	4	919
$(Gua)_3P_3O_9 \cdot 2H_2O$	12.140(8)	15.183(8) 97.49(5)	10.706(5)	$P2_1/n$	4	920

[a]Abbreviations: EDA, ethylenediammonium, $[NH_3-(CH_2)_2-NH_3]^{2+}$; IPA, isopropylammonium, $[(CH_3)_2-CH-NH_3]^+$; MA, methylammonium, $(CH_3-NH_3)^+$; Gly, glycinium, $(NH_3-CH_2-COOH)^+$; Gua, guanidinium, $[C(NH2)3]^+$.

Table 4.2.12. Main Crystallographic Data for Cyclotriphosphate Adducts with Telluric Acid

Formula	a (Å) α (°)	b (Å) β (°)	c (Å) γ (°)	S.G.	Z	Reference
$K_3P_3O_9 \cdot Te(OH)_6 \cdot 2H_2O$	15.57(2)	7.438(6) 107.16(9)	14.85(1)	$P2_1/c$	4	922
$2Na_3P_3O_9 \cdot Te(OH)_6 \cdot 6H_2O$	11.67(1)	11.67(1)	12.12(1)	$P6_3/m$	2	923
$2(NH_4)_3P_3O_9 \cdot Te(OH)_6$	11.16(1)	11.16(1)	17.86(1)	$R\bar{3}$	3	924
$Rb_3P_3O_9 \cdot Te(OH)_6 \cdot H_2O$	15.564(6)	8.376(4) 113.33(2)	13.705(4)	$P2_1/a$	4	925
$Cs_3P_3O_9 \cdot Te(OH)_6 \cdot H_2O$	7.279(2)	13.984(8) 90.42(2)	17.071(4)	$P2_1/c$	4	926
$2Tl_3P_3O_9 \cdot Te(OH)_6$	11.168(1)	11.168(1)	11.733(3)	$P6_3/m$	2	824
$Cs_2NaP_3O_9 \cdot Te(OH)_6$	12.946(1)	9.174(1) 107.60(1)	13.406(1)	$C2/c$	4	824
$Na_3P_3O_9 \cdot K_3P_3O_9 \cdot Te(OH)_6$	18.42(1)	10.644(5) 119.76(5)	12.348(8)	$C2/c$	4	927

The following adducts between ammonium, rubidium, cesium and thallium cyclotriphosphates and telluric acid, corresponding to various formulas, have been prepared and investigated from a structural point of view:

- $2(NH_4)_3P_3O_9 \cdot Te(OH)_6$ (Boudjada *et al.*[924])
- $Rb_3P_3O_9 \cdot Te(OH)_6 \cdot H_2O$ (Boudjada and Durif [925])
- $Cs_3P_3O_9 \cdot Te(OH)_6 \cdot H_2O$ (Averbuch-Pouchot [926])
- $2Tl_3P_3O_9 \cdot Te(OH)_6$ (Boudjada [824])

In addition, two mixed-monovalent cation cyclotriphosphate–telluric acid adducts, $Cs_2NaP_3O_9 \cdot Te(OH)_6 \cdot$ and $Na_3P_3O_9 \cdot K_3P_3O_9 \cdot Te(OH)_6$ were studied by Boudjada[824] and Averbuch-Pouchot and Durif[927] respectively. The latter compound was first described with a monoclinic symmetry but was later proved to be trigonal.[928]

The main crystallographic data for cyclotriphosphate–telluric acid adducts are given in Table 4.2.12.

4.2.3. *Chemical Preparation of Cyclotriphosphates*

We report below some procedures commonly used for the preparation of cyclotriphosphates. Most of these preparative methods involve the use of sodium cyclotriphosphate as starting material. This material can now be prepared in large amounts, in a good state of purity and in high yield.

4.2.3.1. *Chemical Preparation of Sodium Cyclotriphosphate*

By heating NaH_2PO_4, in the temperature range 803–823 K, for at least five hours, one obtains $Na_3P_3O_9$ according to the following scheme:

$$3NaH_2PO_4 \rightarrow Na_3P_3O_9 + 3H_2O$$

The cyclotriphosphate obtained is sometimes contaminated with small amounts (usually less than 1%) of some insoluble sodium phosphates, mainly long-chain polyphosphates. Dissolution in water followed by filtration eliminates these polyphosphates. After the resulting solution has been kept at room temperature for some days, crystals of the hexahydrate, $Na_3P_3O_9 \cdot 6H_2O$, appear. Firing the hexahydrate at 623 K leads to very pure $Na_3P_3O_9$. A similar scheme of cyclization has been investigated for other alkali metals and for ammonium, but poor yields have been obtained.

4.2.3.2. The Boullé's Process

Boullé's metathesis reaction is widely used for the preparation of water-soluble cyclotriphosphates. The starting material is silver cyclotriphosphate monohydrate, $Ag_3P_3O_9 \cdot H_2O$. This sparingly water-soluble salt is easily prepared by adding an aqueous solution of silver nitrate ($\sim 0.1M$) to an aqueous solution of sodium cyclotriphosphate of approximately the same concentration. The reaction is

$$Na_3P_3O_9 + 3AgNO_3 \rightarrow Ag_3P_3O_9 \cdot H_2O + 3NaNO_3$$

It is recommended that an excess of silver nitrate be used to avoid the formation of a sodium-containing compound. The precipitation is achieved within one day. The monoclinic crystals obtained are not light-sensitive and are stable for years at room temperature.

As typical example of Boullé's process, we will describe here the preparation of potassium cyclotriphosphate, $K_3P_3O_9$. A slurry of $Ag_3P_3O_9 \cdot H_2O$ in water is slowly added to an aqueous solution of potassium chloride in a stoichiometric ratio. The reaction is

$$Ag_3P_3O_9 \cdot H_2O + 3KCl \rightarrow K_3P_3O_9 + 3AgCl + H_2O$$

After about one hour of mechanical stirring, the silver chloride is removed by filtration, and the resulting solution is evaporated, yielding crystals of $K_3P_3O_9$, or ethanol is added until precipitation of polycrystalline potassium cyclotriphosphate occurs. These operations must be conducted at room or lower temperature in order to avoid hydrolysis of the ring.

Most of the water-soluble cyclotriphosphates were prepared by this process.

Although this process as described seems rather simple, the chemical preparation of sodium-free silver cyclotriphosphate is not so easy, and some precautions are required. An excess of silver nitrate must be used, and the aqueous solution of sodium cyclotriphosphate must be added slowly to the aqueous solution of silver nitrate.

4.2.3.3. More Classical Methods

Classical methods of aqueous chemistry are in many cases valuable for the production of cyclotriphosphates. $CaNaP_3O_9 \cdot 3H_2O$ is, for instance, very easily prepared by adding, in the proper ratio, a $0.1M$ aqueous solution of $CaCl_2$ to an aqueous solution of $Na_3P_3O_9$ of the same concentration. $CaNaP_3O_9 \cdot 3H_2O$ is then precipitated by adding ethyl alcohol to the resulting solution.

4.2.3.4. Thermal Methods

Many cyclotriphosphates have been prepared as polycrystalline samples through the use of what are commonly called "thermal methods." A typical example is the preparation of most of the $M^{II}M^{I}P_3O_9$ benitoite-like compounds. They are very easily obtained in a good state of purity by heating, at relatively low temperatures (usually less than 773 K), a stoichiometric mixture of $(NH_4)_2HPO_4$ with the corresponding carbonates. The reaction scheme is, for instance,

$$3(NH_4)_2HPO_4 + CoCO_3 + 1/2K_2CO_3 \rightarrow CoKP_3O_9 + 3/2CO_2 + 6NH_3 + 9/2H_2O$$

The starting mixture must be slowly heated up to the optimum temperature with frequent homogenization.

4.2.3.5. Flux Methods

Flux methods are very frequently used in solid-state chemistry, mainly for growing single crystals. In the case of phosphates, the basic principle of this type of process is to use a starting mixture containing a large excess of H_3PO_4 or $(NH_4)_2HPO_4$. The composition of the starting mixture and the optimum temperature to be used are difficult parameters to determine. The optimization of a reproducible flux process for synthesizing a given material is in most cases a tedious and time-consuming enterprise. After some hours or days of heating at the appropriate temperature, crystals are formed in the melt. Excess phosphoric flux is then removed with hot water, and the crystalline part is separated by filtration. Because of this last step, this process is obviously not applicable to the preparation of water-soluble phosphates.

4.2.3.6. Ion-Exchange Resins

Cyclotriphosphoric acid, $H_3P_3O_9$, can be obtained by employing ion-exchange resins; Amberlite IR220 or IRN77 were extensively used. An aqueous solution of $Na_3P_3O_9$ is slowly passed through a column of resin, and the $H_3P_3O_9$ produced is immediately neutralized by reaction with a carbonate or hydroxide. $Na_2LiP_3O_9 \cdot 4H_2O$, $Mn_3(P_3O_9)_2 \cdot 10H_2O$, $Ca_3(P_3O_9)_2 \cdot 10H_2O$, and $K_3P_3O_9$, for instance, were prepared by this process.

4.2.3.7. Unconventional Methods

Very unconventional methods, specific to the preparation of a particular compound, have sometimes been reported, in various amounts of detail, in the chemical literature. Most of them have been elaborated by crystallographers in search of ways to produce single crystals. Descriptions of such methods were included in Section 4.2.2 and will not be repeated here.

4.2.3.8. Organic Cation Cyclotriphosphates

The recent investigation of the crystal chemistry of compounds of this type did not require the development of specific methods of preparation as most of these compounds can be synthesized by using the types of processes reported above. Boullé's process can be used when the corresponding organic chlorhydrate is available. Guanidinium cyclotriphosphate dihydrate, for instance, was prepared in this way:

$$3[C(NH_2)_3]Cl + Ag_3P_3O_9 \rightarrow [C(NH_2)_3]_3 \cdot P_3O_9 + 3AgCl$$

The direct action of cyclotriphosphoric acid, produced by the use of ion-exchange resins, on the required amine has also sometimes been used:

$$3R–NH_2 + H_3P_3O_9 \rightarrow (R–NH_3)_3P_3O_9$$

4.2.4. Some Atomic Arrangements of Cyclotriphosphates

4.2.4.1. Lithium Cyclotriphosphate Trihydrate

$$Li_3P_3O_9 \cdot 3H_2O,^{807-809} \text{trigonal (rhombohedral), } R3$$

$$a = 12.511(4), \quad c = 5.594(2) \text{ Å}, \quad Z = 3 \text{ (hexagonal setting)}$$

$$a = 7.460 \text{ Å}, \quad \alpha = 113.97°, \quad Z = 1 \text{ (rhombohedral setting)}$$

A projection along the **c** axis of the atomic arrangement is presented in Fig. 4.2.21. The P_3O_9 ring anion is located around the threefold axis and thus has threefold internal symmetry, which is relatively rare in the crystal chemistry of cyclotriphosphates. The lithium atoms are coordinated to three oxygen atoms and one water molecule in a tetrahedral configuration. By sharing corners these $LiO_3(H_2O)$ tetrahedra form infinite chains spiraling around one set of 3_1 helical axes. Within a $LiO_3(H_2O)$ tetrahedron, the three Li–O distances range from 1.916 to 1.984 Å, and the Li–H_2O distance is 2.036 Å.

4.2.4.2. Disodium Hydrogen Cyclotriphosphate

$$Na_2HP_3O_9,^{812} \text{triclinic, } P\bar{1}, Z = 2$$

$$a = 7.788(5), \quad b = 7.809(5), \quad c = 7.129(5) \text{ Å}$$

$$\alpha = 116.69(5), \quad \beta = 103.41(5), \quad \gamma = 81.94(5)°$$

As shown by Fig. 4.2.22, a projection of the atomic arrangement along the **a** axis P_3O_9 ring anions connected by hydrogen bonds form rows parallel to the **c** direction. This fact probably explains the relatively low solubility of this salt. Cohesion between these rows of P_3O_9 ring anions is provided by distorted NaO_6 octahedra, themselves interconnected

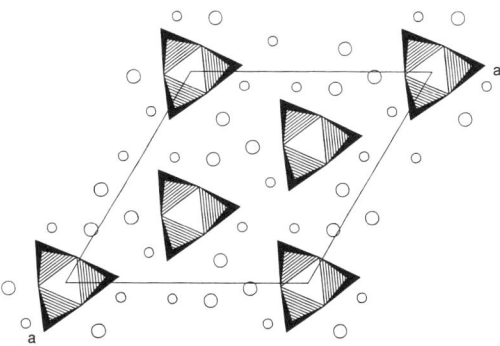

Figure 4.2.21. Projection, along the *c* axis, of the crystal structure of $Li_3P_3O_9 \cdot 3H_2O$. The small open circles represent the lithium atoms, and the larger ones the water molecules.

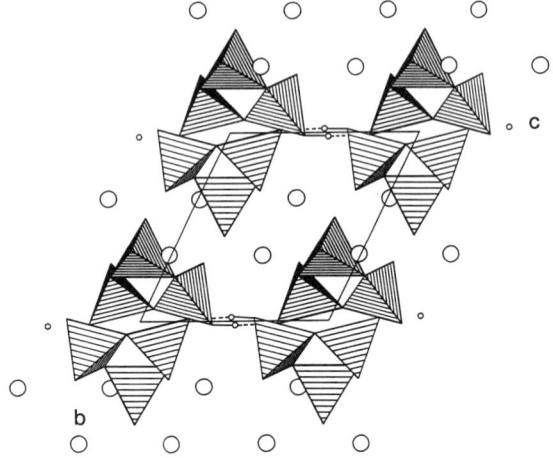

Figure 4.2.22. Projection, along the *a* axis, of the crystal structure of $Na_2HP_3O_9$. The open circles represent the hydrogen atoms, and the larger ones the sodium atoms.

by edge sharing so as to form monodimensional arrays also parallel to the **c** direction. Within the NaO_6 polyhedra, the Na–O distances range between 2.309 and 2.563 Å. The hydrogen bond connecting the ring anions has the following geometry:

$$O–H = 0.86 \text{ Å}, \quad H\cdots O = 1.590 \text{ Å}, \quad O–H\cdots O = 173°$$

This salt is the only known example of an acidic cyclotriphosphate.

4.2.4.3. Anhydrous Sodium Cyclotriphosphate

$$Na_3P_3O_9,[814] \text{ orthorhombic, } Pmcn, Z = 4$$

$$a = 7.928(2), \quad b = 13.214(3), \quad c = 7.708(2) \text{ Å}$$

Figure 4.2.23 presents the projection of this atomic arrangement along the **c** axis. The P_3O_9 anion has mirror symmetry and, according to Ondik,[814] a strong pseudo-3*m* symmetry, probably induced by the pseudo-hexagonal dimensions of the unit cell. Two crystallographically independent sodium atoms are present in this arrangement; one of them is located in the mirror plane. Both sodium atoms have fivefold coordination and, through edge and corner sharing, these NaO_5 polyhedra build a three-dimensional network. Within these NaO_5 polyhedra, the Na–O distances range between 2.337 and 2.466 Å.

4.2.4.4. Sodium Cyclotriphosphate Monohydrate

$$Na_3P_3O_9 \, H_2O,[814] \text{ orthorhombic, } Pmcn, Z = 4$$

$$a = 8.500(1), \quad b = 13.189(1), \quad c = 7.558(1) \text{ Å}$$

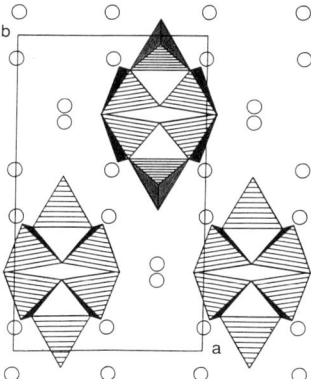

Figure 4.2.23. Projection, along the *c* axis, of the atomic arrangement in $Na_3P_3O_9$. The open circles represent the sodium atoms. Mirror planes lie along x = 1/4 and 3/4.

Figure 4.2.24 presents the projection of this atomic arrangement along the **c** axis. As previously mentioned this atomic arrangement is closely related to that of the anhydrous salt discussed above and can be described in almost the same words. The additional water molecule occupies a position on the mirror plane and provides sixfold coordination for both kinds of sodium atoms. Within the NaO_6 polyhedra, the Na–O distances range between 2.32 and 2.71 Å.

4.2.4.5. Cesium Cyclotriphosphate Monohydrate

$$Cs_3P_3O_9 \cdot H_2O,^{818,823} \quad \text{triclinic, } P\overline{1}, Z = 2$$

$$a = 10.610(6), \quad b = 7.966(4), \quad c = 8.172(5) \text{ Å}$$
$$\alpha = 96.64(8), \quad \beta = 68.84(8), \quad \gamma = 95.42(8)°$$

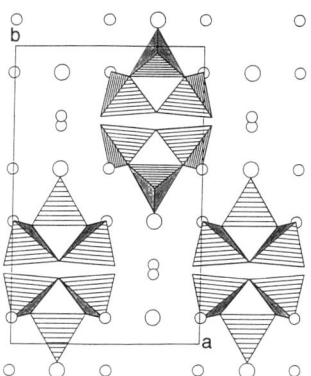

Figure 4.2.24. Projection, along the *c* axis, of the atomic arrangement in $Na_3P_3O_9 \cdot H_2O$. The smaller open circles represent the sodium atoms, and the larger ones the water molecules. Mirror planes lie along x = 1/4 and 3/4.

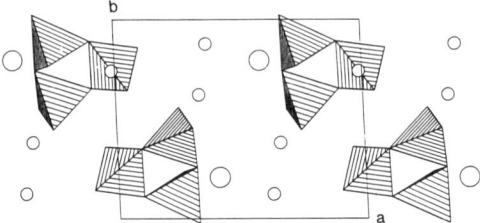

Figure 4.2.25. Projection, along the **c** direction, of the atomic arrangement in Cs₃P₃O₉·H₂O. The smaller open circles represent the cesium atoms, and the larger ones the water molecules.

A projection of this very simple arrangement along the **c** direction is presented in Fig. 4.2.25. The ring anion has no internal symmetry. The three independent cesium atoms have sixfold coordination, with Cs–O distances ranging from 3.005 to 3.364 Å. The water molecule is involved in two of the cesium coordination polyhedra.

The corresponding rubidium salt, $Rb_3P_3O_9 \cdot H_2O$[818] is isotypic.

4.2.4.6. Lithium–Potassium Cyclotriphosphate Dihydrate

$$LiK_2P_3O_9 \cdot 2H_2O,[33] \quad \text{monoclinic, } P2_1/c, Z = 4$$

$$a = 8.669(3), \quad b = 14.497(4), \quad c = 7.634(6) \text{ Å}, \beta = 99.72(5)°$$

The phosphoric ring anion has no internal symmetry. The projection of this structure along the **a** direction is shown in Fig. 4.2.26. The two potassium atoms have irregular coordination polyhedra. One of them is coordinated to one water molecule and six oxygen atoms from five different phosphoric groups. The second one is surrounded by eight oxygen atoms from four different P_3O_9 groups. The K–O distances range between 2.640 and 3.076 Å. Three oxygen atoms from three different phosphoric rings and one water molecule build

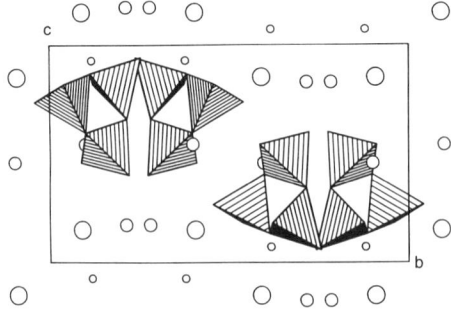

Figure 4.2.26. Projection, along the **a** axis, of the atomic arrangement in LiK₂P₃O₉·H₂O. The large open circles represent the potassium atoms, and the small ones represent the lithium atoms.

an almost regular tetrahedon around the lithium atom, with Li–O distances ranging from 1.930 to 1.968 Å. These LiO_4 tetrahedra have no common oxygen atoms.

4.2.4.7. Sodium–Potassium Cyclotriphosphate

$$Na_2KP_3O_9,^{829,830} \text{triclinic}, P\bar{1}, Z = 2$$

$$a = 6.886(2), \quad b = 9.494(3), \quad c = 6.797(2) \text{ Å}$$
$$\alpha = 110.07(2), \quad \beta = 104.69(2), \quad \gamma = 86.68(2) \text{ °}$$

A projection of this very simple arrangement along the **c** axis is depicted in Fig. 4.2.27. As is always the case in triclinic compounds the phosphoric ring anion has no internal symmetry. One of the two independent sodium atoms has a sixfold coordination, with Na–O distances ranging from 2.383 to 2.555 Å, whereas the second one has only five neighbors, with Na–O distances ranging from 2.370 to 2.574 Å. The potassium atom has ten neighbors, with K–O distances ranging from 2.690 to 3.356 Å. The NaO_6 and KO_{10} polyhedra build layers in the planes $y = \pm 0.83$ alternating with layers of NaO_5 polyhedra located in the plane $y = 0.50$. The centers of the phosphoric rings are situated halfway between these layers.

The corresponding sodium–rubidium,[831] sodium–ammonium,[843] and sodium–thallium[832] salts are isotypic. The most interesting feature of this series of compounds is that it represents one of the rare instances of structural analogy between cyclotrisilicates and cyclotriphosphates. All the members of this series are structurally similar to a group of silicates, $M_2NSi_3O_9$, with M = Ca and Mn and N = Pb and Ba. The substitution scheme is evidently:

$$Si_3O_9 \leftrightarrow P_3O_9$$

$$2Ca + Ba \leftrightarrow 2Na + K$$

4.2.4.8. Cadmium Cyclotriphosphate Tetradecahydrate

$$Cd_3(P_3O_9)_2 \cdot 14H_2O,^{834-836} \text{trigonal}, P\bar{3}, Z = 1$$

$$a = 12.228, c = 5.451 \text{ Å}$$

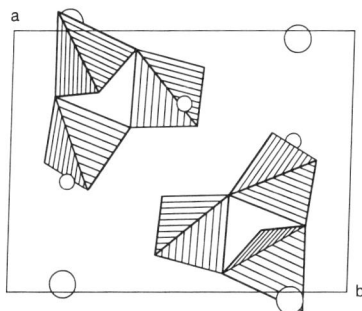

Figure 4.2.27. Projection, along the **c** direction, of the crystal structure of $Na_2KP_3O_9$. The smaller open circles represent the sodium atoms, and the larger ones represent the potassium atoms.

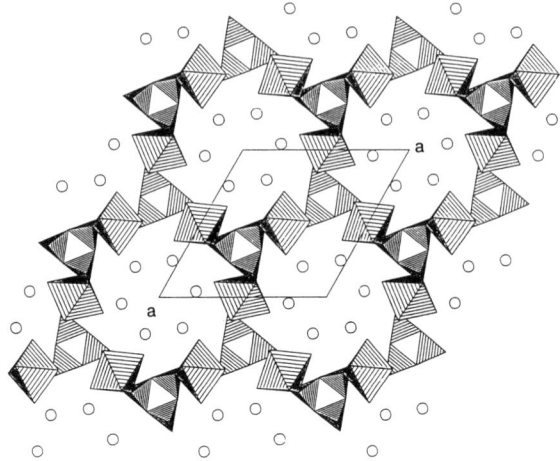

Figure 4.2.28. Projection of the crystal structure of $Cd_3(P_3O_9)_2 \cdot 14H_2O$ along the c axis. The open circles represent the zeolitic water molecules. The octahedra connecting the P_3O_9 groups are the $CdO_4(H_2O)_2$ polyhedra.

As shown in Fig. 4.2.28, a projection along the ternary axis, the P_3O_9 ring anions are located around the internal threefold axes at about $z = 1/2$ and thus have threefold internal symmetry. The cadmium atom is situated on the inversion center at $(1/2, 1/2, 0)$ and is coordinated to four oxygen atoms and two water molecules, forming an almost regular octahedron. Each cadmium octahedron shares two oxygen atoms with its two adjacent P_3O_9 groups.

Such a network generates wide hexagonal channels around the $\bar{3}$ axes. The eight water molecules not involved in the cadmium coordination were not localized during the first crystal structure investigation, conducted at room temperature.[846] Reinvestigation of the structure at 93 K[836] showed that six of the eight zeolitic water molecules line the channels, whereas the other two, which could not be localized even at this temperature, are probably on the $\bar{3}$ axis.

4.2.4.9. Barium Cyclotriphosphate Tetrahydrate

$$Ba_3(P_3O_9)_2 \cdot 4H_2O,^{848} \text{ monoclinic, } C2/m, Z = 2$$

$$a = 16.09(1), \quad b = 8.368(5), \quad c = 7.717(3) \text{ Å}, \quad \beta = 95.38(5)°$$

As shown in Fig. 4.2.29, a projection of the structure along the c axis, the ring anion has mirror symmetry. All the barium atoms are located in the mirror planes in $y = 0$ and $1/2$, but their crystallographic situations are different. Four of them occupy a fourfold position and have tenfold coordination, involving eight oxygen atoms and two water molecules. The two remaining ones are distributed statistically on a fourfold position close to an inversion center. They are located in a large oblong centrosymmetric cavity and have

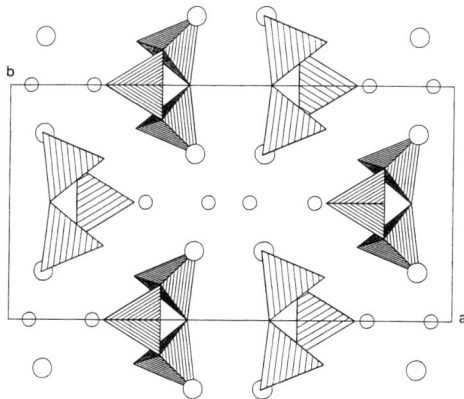

Figure 4.2.29. Projection, along the c axis, of the atomic arrangement in $Ba_3(P_3O_9)_2 \cdot 4H_2O$. The small open circles represent the barium atoms, and the large ones represent the water molecules. The pairs of barium atoms located close to the inversion centers correspond to the statistically distributed ones.

sevenfold coordination, involving five oxygen atoms and two water molecules. All the water molecules are involved in the associated-cation coordination polyhedra. Within these two kinds of BaO_n polyhedra, the Ba–O distances range between 2.718 and 3.073 Å.

4.2.4.10. Barium–Sodium Cyclotriphosphate

$$BaNaP_3O_9,^{861,862}\text{orthorhombic, }P2_12_12_1, Z = 4$$

$$a = 11.134(8), \quad b = 12.320(8), \quad c = 5.802(3) \text{ Å}$$

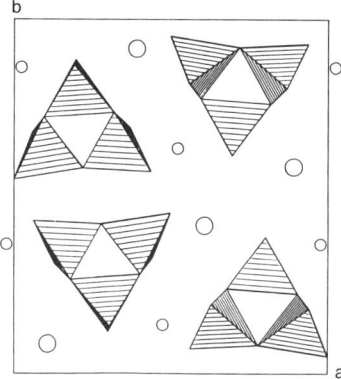

Figure 4.2.30. Projection of the atomic arrangement in $BaNaP_3O_9$ along the c direction. The smaller open circles represent the sodium atoms, and the larger ones the barium atoms.

Figure 4.2.30, a projection along the **c** direction of this vey simple atomic arrangement, shows the respective locations of the ring anions and the associated cations. The phosphoric ring anion has no internal symmetry.

The cohesion between rings is established by NaO_6 distorted octahedra and BaO_8 polyhedra. By sharing edges and corners, these two kinds of polyhedra form a three-dimensional network. Within the associated-cation polyhedra, Na–O distances range between 2.309 and 2.586 Å and Ba–O distances between 2.650 and 3.064 Å.

$PbNa(PO_3)_3$, $BaAg(PO_3)_3$ and $HgK(PO_3)_3$ are isotypic with $BaNa(PO_3)_3$. In the case of the mercury–potassium cyclotriphosphate the substitution scheme is

$$Ba \rightarrow K \quad \text{and} \quad Na \rightarrow Hg$$

4.2.4.11. Nickel–Potassium Cyclotriphosphate Heptahydrate

$$NiK_4(P_3O_9)_2 \cdot 7H_2O,^{898} \text{ orthorhombic, } Fm2m, Z = 4$$

$$a = 23.03(1), \quad b = 11.882(4), \quad c = 8.732(4) \text{ Å}$$

The ring anion has mirror symmetry, with one phosphorus atom, two external oxygen atoms and one bonding oxygen atom in the mirror plane. The nickel atoms located at 0,0,0 octahedrally coordinated to six water molecules. As shown in Fig. 4.2.31, these $Ni(H_2O)_6$ octahedra do not share any edges or corners. The Ni–H_2O distances range from 2.023 to 2.082 Å. Within a range of 3 Å, the potassium atoms have sixfold coordination, with K–O distances ranging between 2.694 and 2.826 Å. Through edge and corner sharing, these KO_6 polyhedra build layers parallel to the (a, c) planes. Such a layer is depicted in Fig. 4.2.32. This organization was incorrectly described in the original publication.[898]

4.2.4.12. Cadmium–Potassium Cyclotriphosphate Dihydrate

$$CdK_4(P_3O_9)_2 \cdot 2H_2O,^{902} \text{ triclinic, } P\bar{1}, Z = 1$$

$$a = 9.235(5), \quad b = 7.599(4), \quad c = 7.148(4) \text{ Å}$$
$$\alpha = 96.38(1), \quad \beta = 103.90(1), \quad \gamma = 102.06(1)°$$

Figure 4.2.31. Projection of the atomic arrangement in $NiK_4(P_3O_9)_2 \cdot 7H_2O$ along the **b** axis. The open circles represent the potassium atoms. The hatched octahedra represent the $Ni(H_2O)_6$ groups.

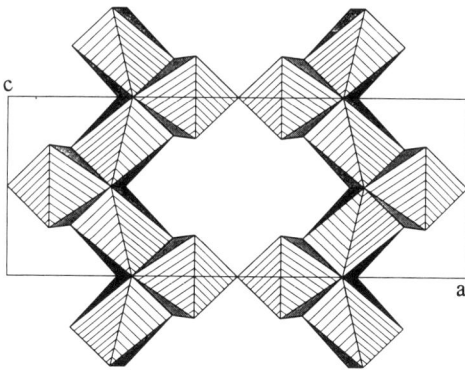

Figure 4.2.32. Projection, along the **b** axis, of a layer of edge- and corner-sharing KO_6 octahedra in $NiK_4(P_3O_9)_2 \cdot 7H_2O$.

The cadmium atom, located on the inversion center at the origin of the unit cell, has almost regular octahedral coordination, involving four oxygen atoms and two water molecules, with Cd–O distances ranging from 2.267 to 2.346 Å. Each CdO_6 octahedron shares its four oxygen atoms with the four adjacent P_3O_9 rings, giving rise to CdO_6–$(P_3O_9)_2$–CdO_6–$(P_3O_9)_2$ ribbons parallel to the **c** direction. Figure 4.2.33 shows the projection of this arrangement along the **b** axis. Within a range of 3.20 Å, one of the two

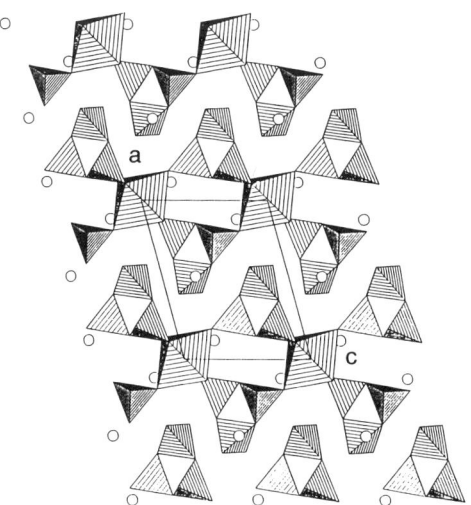

Figure 4.2.33. Projection of the atomic arrangement in $CdK_4(P_3O_9)_2 \cdot 2H_2O$ along the **b** direction. The open circles represent the potassium atoms.

independent potassium atoms has sevenfold coordination, the other one has ninefold coordination. Within the KO_7 polyhedron, the K–O distances range from 2.698 to 3.157 Å whereas in the KO_9 polyhedron they range between 2.736 and 3.182 Å. These potassium polyhedra establish the three-dimensional cohesion between the ring anions. The P_3O_9 phosphoric ring has no internal symmetry.

No isotypes of this compound have been identified.

4.2.4.13. Calcium–Ammonium Cyclotriphosphate

$$Ca(NH_4)P_3O_9,^{637,871} \text{ hexagonal, } P\bar{6}c2, Z = 2$$

$$a = 6.887(3), \quad c = 10.448(5) \text{ Å}$$

This cyclotriphosphate has the well-known structure of the mineral benitoite, Ba-$TiSi_3O_9$, as do the 14 other cyclotriphosphates included in Table 4.2.4. As depicted in Fig. 4.2.34, a projection along the **c** axis, the ring anions located around the $\bar{6}$ axis in $z = 1/4$ and 3/4 have 3/m internal symmetry. The divalent atoms, located on one of the threefold internal axes in $z = 0$ and 1/2, have regular octahedral coordination, whereas the monovalent cations, located on the other internal threefold axis at the same heights have a more distorted coordination geometry. Thus, the structure can be described as composed of layers of phosphoric groups alternating along the **c** axis with layers of associated cations. In the CaO_6 octahedron, the Ca–O distance is 2.329 Å, and the N–O distance is 2.921 Å.

4.2.4.14. Cerium Cyclotriphosphate Trihydrate

$$CeP_3O_9 \cdot 3H_2O,^{907} \text{ hexagonal, } P\bar{6}, Z = 1$$

$$a = 6.770(5), \quad c = 6.079(5) \text{ Å}$$

Figure 4.2.35 shows the projection, along the **c** axis, of the atomic arrangement of $CeP_3O_9 \cdot 3H_2O$. The P_3O_9 ring anion located around the $\bar{6}$ axis, has a planar configuration with 3/m internal symmetry, quite comparable to that already observed in the cyclotriphosphates isotypic with benitoite. The cerium atom, located on one of the $\bar{6}$ internal axes, has

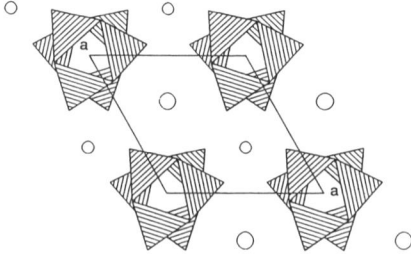

Figure 4.2.34. Projection, along the *c* axis, of the atomic arrangement in $CaNH_4P_3O_9$. The small open circles represent the calcium atoms, and the larger ones the ammonium groups.

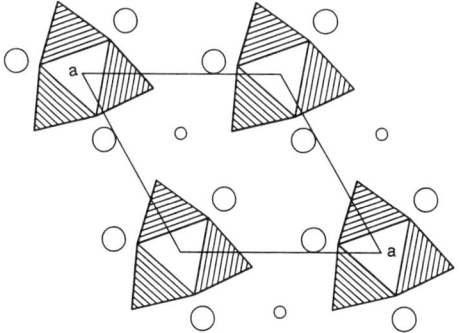

Figure 4.2.35. Projection of the atomic arrangement in $CeP_3O_9 \cdot 3H_2O$ along the *c* axis. The small open circles represent the cerium atoms, and the larger ones the water molecules.

very regular ninefold coordination, involving six oxygen atoms and three water molecules, with Ce–O distances ranging from 2.51 to 2.62 Å.

The corresponding lanthanum and praseodymium salts[906] are isotypic.

4.2.4.15. Sodium Cyclotriphosphate–Telluric Acid Adduct Hexahydrate

$$2Na_3P_3O_9 \cdot Te(OH)_6 \cdot 6H_2O, [923] \text{ hexagonal, } P6_3/m, Z = 2$$

$$a = 11.67(1), \quad c = 12.12(1) \text{ Å}$$

A projection of the atomic arrangement along the **c** axis is presented in Fig. 4.2.36. The structure can be described as a succession of two kinds of planes alternating perpendicular to the **c** axis. Planes of the first type, located in $z = 1/4$ and $3/4$, contain the sodium

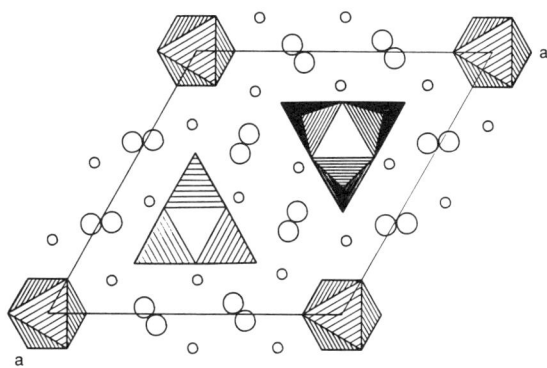

Figure 4.2.36. Projection of the atomic arrangement in $2Na_3P_3O_9 \cdot Te(OH)_6 \cdot 6H_2O$ along the *c* axis. Owing to the presence of mirror planes in $z = 1/4$ and $3/4$, two $Te(OH)_6$ groups and two P_3O_9 groups are superimposed in projection. The small open circles represent the sodium atoms, and the larger ones the water molecules.

atoms and the water molecules, while planes $z = 0$ and $1/2$ contain the $Te(OH)_6$ groups and the phosphoric ring anions. The latter groups, located around the $\bar{6}$ internal axes, have threefold internal symmetry while the $Te(OH)_6$ groups located at the origin have $\bar{6}$ internal symmetry with a Te–O distance of 1.923 Å. The two independent sodium atoms have very distorted octahedral coordination involving four oxygen atoms and two water molecules. Within these octahedra, the Na–O distances range from 2.367 to 2.504 Å.

4.2.4.16. Ammonium Cyclotriphosphate–Telluric Acid Adduct

$$Te(OH)_6 \cdot 2(NH_4)_3P_3O_9,^{924} \text{ trigonal (rhomboedral), } R\bar{3}$$

$$a = 11.16(1), \quad c = 17.86(1) \text{ Å}, \quad Z = 3 \text{ (hexagonal setting)}$$

$$a = 8.773(5) \text{ Å}, \quad \alpha = 100.37(5)°, \quad Z = 1 \text{ (rhomboedral setting)}$$

This arrangement can be easily described as a succession of two kinds of layers alternating perpendicular to the **c** axis. The first type of layer includes the P_3O_9 ring anions and the $Te(OH)_6$ groups, and the second type the ammonium groups. The layers of the first type are centered by planes $z = n \cdot c/3$ and the NH_4 layers by the planes $z = (2n+1)c/6$. Figure 4.2.37 shows a projection along the ternary axis of a part of this arrangement.

Both the $Te(OH)_6$ and the P_3O_9 groups are located around the threefold axis. As the Te atom is located at the origin of the unit cell, its six OH neighbors form a rather regular octahedron having $\bar{3}$ internal symmetry, with a Te–O distance of 1.912 Å. The P_3O_9 ring anion has threefold internal symmetry. The NH_4 group is a rather regular tetrahedron with N–H distances ranging between 0.84 and 1.03 Å and H–N–H angles ranging from 98 to 118°; within a range of 3.50 Å, it has sevenfold coordination, with N–O distances ranging from 2.812 to 3.152 Å.

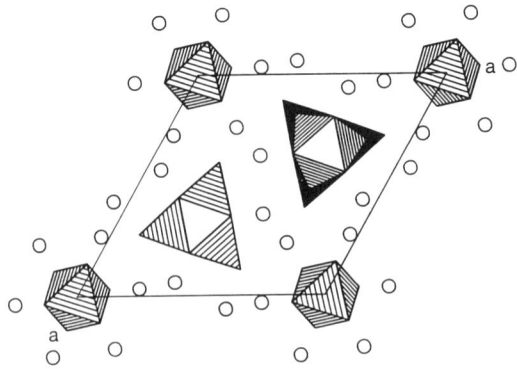

Figure 4.2.37. Projection of the atomic arrangement in $2(NH_4)_3P_3O_9 \cdot Te(OH)_6$. along the ternary axis. The projection is limited to $-0.25 < z < 0.25$. Thus, this projection includes one layer of the first type and its two adjacent NH_4 layers. The open circles represent the ammonium groups. The hydrogen atoms have been omitted.

4.2.4.17. Thallium Cyclotriphosphate–Telluric Acid Adduct

$$2Tl_3P_3O_9 \cdot Te(OH)_6,^{824} \text{ Hexagonal, } P6_3/m, Z = 2$$

$$a = 11.168(1), \quad c = 11.733(3) \text{ Å}$$

The unit-cell dimensions of this adduct are rather similar to those observed for $2Na_3P_3O_9 \cdot Te(OH)_6 \cdot 6H_2O$, and these two compounds have the same space group $P6_3/m$. Figure 4.2.38, a projection of this arrangement along the **c** direction, shows clearly that apart from some differences in the orientations of the anionic groups, the structure is strikingly similar to that of the sodium compound. As in the case of the sodium compound, one can describe the structure in terms of the stacking of two kinds of layers alternating perpendicular to the **c** direction. The first type of layer includes the phosphoric anion and the $Te(OH)_6$ groups while the second one contains the associated cations. The two layers of the first type crossing the unit cell are approximately centered by the planes $z = 0$ and $1/2$ while the layers of thallium are located in the mirror planes in $z = 1/4$ and $3/4$.

The tellurium atom located at $0,0,1/2$ is surrounded by six OH groups in a rather regular octahedron with $\bar{3}$ internal symmetry. Within this $Te(OH)_6$ group the Te–O distance is 1.907 Å. The P_3O_9 ring anion has a threefold internal symmetry. Within a range of 3.50 Å, the two independent thallium atoms, both located in mirror planes, have, respectively, sixfold and eightfold coordinations with Tl–O distances ranging from 2.738 to 3.248 Å.

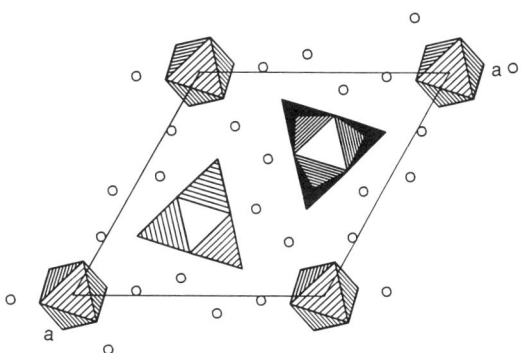

Figure 4.2.38. Projection, along the ternary axis, of the atomic arrangement in $2Tl_3P_3O_9 \cdot Te(OH)_6$. The open circles represent the thallium atoms.

4.2.4.18. Isopropylammonium Cyclotriphosphate

$$[NH_3(C_3H_7)]_2P_3O_9,^{916} \text{ monoclinic, } C2, Z = 8$$

$$a = 25.22(2), \quad b = 12.22(2), \quad c = 15.45(2) \text{ Å}, \quad \beta = 123.90(5)°$$

Two crystallographically independent P_3O_9 ring anions and six independent isopropylammonium groups are observed in this atomic arrangement. The phosphoric ring anions have no internal symmetry and are arranged in rows parallel to the **c** axis, these rows being organized in such a matter that they build large hexagonal channels. The internal face of

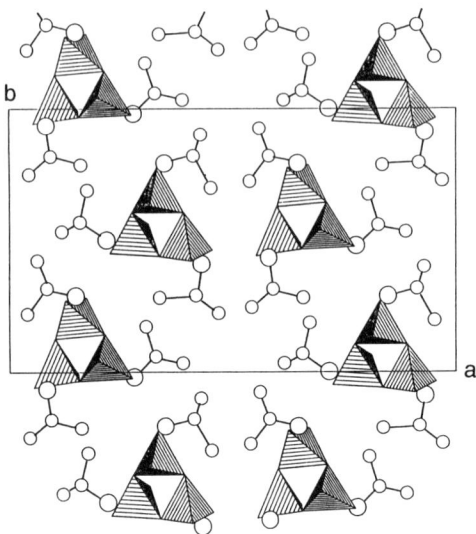

Figure 4.2.39. Projection, along the *c* axis, of the structure of [NH$_3$(C$_3$H$_7$)]$_2$P$_3$O$_9$. As anionic and organic groups are almost superimposed in projection, only one-half of the arrangement (0.50 < *z* < 1.00) is shown. The smaller open circles represent the carbon atoms, and the larger ones the nitrogen atoms.

the hexagonal channel is lined by the organic groups. Figure 4.2.39 gives a projection of this structure along the **c** axis.

The stability of such an arrangement results from a net of strong hydrogen bonds (N-H⋯O) connecting the NH$_3$ moities to the external oxygen atoms of the ring anions. In this set of H bonds, the N–O distances range from 2.726 to 3.038 Å, and the N–H⋯O angles from 145 to 175°.

4.2.4.19. Potassium–Ethylenediammonium Cyclotriphosphate

$$K[NH_3(CH_2)_2NH_3]P_3O_9,^{[917]} \text{ orthorhombic, } Ccca, Z = 8$$

$$a = 20.850(8), \quad b = 9.044(4), \quad c = 11.653(5) \text{ Å}$$

Figure 4.2.40 shows a projection of the atomic arrangement along the **b** axis. The structure can be described as a succession of layers perpendicular to the **c** axis. All the P$_3$O$_9$ ring anions are located in planes corresponding approximately to *z* = 0.25 and 0.75, while the ethylenediammonium groups and the potassium atoms alternate in planes corresponding approximately to *z* = 0 and 0.5. Figure 4.2.41, a projection along the **c** axis, shows the distribution inside these two kinds of layers.

The P$_3$O$_9$ ring anion has twofold symmetry through one of the phosphorus atoms and the opposite bonding oxygen atom. This type of internal symmetry is rare. The potassium atoms, located on a binary axis, have eightfold coordination, with K–O distances ranging from 2.726 to 3.029 Å. The ethylenediammonium group also has twofold internal symmetry. A set of strong hydrogen bonds connect the hydrogen atoms of the NH$_3$ groups to the

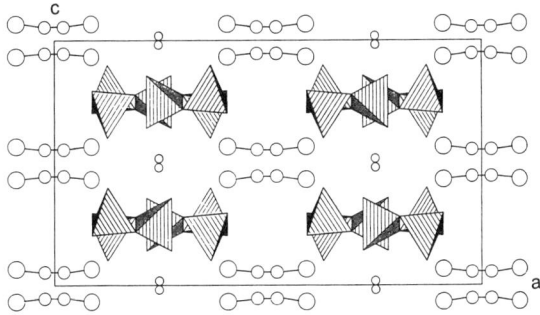

Figure 4.2.40. Projection of the atomic arrangement in K[NH₃(CH₂)₂NH₃]P₃O₉ along the *b* axis. The smallest circles represent the potassium atoms, the intermediate ones the carbon atoms, and the largest ones the nitrogen atoms.

external oxygen atoms of the phosphoric anion, with N–O distances ranging from 2.733 to 2.911 Å and N–H⋯O angles ranging from 157 to 163°.

4.2.5. Geometries of the Cyclotriphosphate Anion

P_3O_9 anions can adopt various internal symmetries—2, *m*, 3, 3/*m*—but an inspection of the 56 accurately investigated examples shows that the great majority of them, 41, have no internal symmetry. Among the remaining ones, two have twofold symmetry, six have mirror symmetry, five have threefold symmetry, and two have 3/*m* symmetry. There does not appear to be any correlation between the nature of the associated cations and the internal symmetry of the ring inside the atomic arrangement.

A survey of the P_3O_9 ring geometry is provided by Tables 4.2.13–4.2.16, listing numerical values of what we consider to be the main geometrical features of a phosphoric ring: the P–P distances and the P–P–P and P–O–P angles. In the following paragraphs, we report some average values for bond distances or bond angles measured in P_3O_9 ring anions. Some published data that have very poor accuracy have not been taken into account in these calculations.

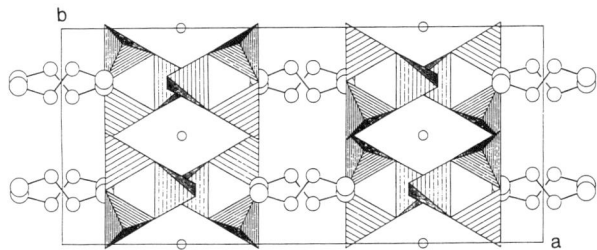

Figure 4.2.41. Projection of the atomic arrangement in K[NH₃(CH₂)₂NH₃]P₃O₉ along the *c* axis. The representation of the atoms is the same as in Fig. 4.2.40.

Table 4.2.13. Main Geometrical Features of P_3O_9 Ring Anions with $3/m$, 3, 2, or m Internal Symmetry

Formula	P–P (Å)	P–O–P (°)	P–P–P (°)	Reference
Rings with 3/m symmetry				
$CaNH_4P_3O_9$	2.963	137.1	60.0	637
$CeP_3O_9 \cdot 3H_2O$	2.856	130.7	60.0	907
Rings with 3 symmetry				
$Li_3P_3O_9.3H_2O$	2.915	130.9	60.0	808
$Cd_3(P_3O_9)_2 \cdot 14H_2O$	2.911	132.2	60.0	836
$2Na_3P_3O_9 \cdot Te(OH)_6 \cdot 6H_2O$	2.877	125.6	60.0	923
$2(NH_4)_3P_3O_9 \cdot Te(OH)_6$	2.896	128.3	60.0	924
$2Tl_3P_3O_9 \cdot Te(OH)_6$	2.886	127.4	60.0	824
Rings with m symmetry				
$Na_3P_3O_9$	2×2.887	2×126.3	59.9	814
	2.892	128.0	2×60.1	
$Na_3P_3O_9 \cdot H_2O$	2×2.873	2×125.4	59.2	814
	2.837	121.5	2×60.4	
$Ba_3(P_3O_9)_2 \cdot 4H_2O$	2×2.912	2×129.9	58.06	848
	2.826	123.4	2×60.97	
$CdCsP_3O_9$	2×2.864	2×126.1	62.6	879
	2.974	135.3	2×58.7	
$NiK_4(P_3O_9)_2 \cdot 7H_2O$	2×2.893	2×126.6	61.2	898
	2.944	131.6	2×59.4	
$CoRb_4(P_3O_9)_2 \cdot 6H_2O$	2×2.924	2×131.7	61.4	904
	2.988	136.0	2×59.3	
Rings with 2 symmetry				
$K(EDA)P_3O_9{}^a$	2×2.901	2×127.6	60.0	917
	2.900	129.1	2×60.0	
$Cs_2NaP_3O_9 \cdot Te(OH)_6$	2×2.864	2×124.4	60.7	824
	2.893	126.5	2×59.7	

^{a}EDA, ethylenediammonium, $[NH_3–(CH_2)_2–NH_3]^{2+}$.

The *P–P–P angles*, strictly equal to 60° in the rings having 3 or $3/m$ internal symmetry, never depart significantly from this value. We report below the ranges of values observed for the various categories of rings:

$58.1 < P–P–P < 62.6°$ in rings with mirror symmetry
$59.7 < P–P–P < 60.7°$ in rings with twofold symmetry
$58.6 < P–P–P < 61.4°$ in rings with no internal symmetry

The estimated standard deviation can be evaluated as 0.1.

The *P–P distances* fall within the range $2.818 < P–P < 2.974$ Å, with an overall average value of 2.890 Å. The ranges of values measured for the various categories of rings and the corresponding average values are as follows:

Table 4.2.14. Main Geometrical Data for P_3O_9 Rings with No Internal Symmetry[a]

Formula	P–P (Å)	P–O–P (°)	P–P–P (°)	Reference
$LiK_2P_3O_9 \cdot H_2O$	2.902	131.4	60.5	33
	2.901	128.3	59.7	
	2.921	127.1	59.8	
$Na_2HPO_3O_9$	2.854	124.8	60.8	812
	2.890	128.1	59.6	
	2.853	124.5	59.6	
$Na_3P_3O_9 \cdot 6H_2O$	2.880	128.3	60.1	815
	2.898	126.9	59.7	
	2.892	126.4	60.3	
$K_3P_3O_9$	2.890	125.6	61.1	821
	2.877	124.9	59.2	
	2.930	129.6	59.7	
$Tl_3P_3O_9$ [b]	2.85	122.0	61.0	824
	2.96	128.0	57.5	
	2.97	132.0	61.4	
	2.90	134.0	59.2	
	2.88	126.0	60.7	
	2.86	132.0	60.1	
$Cs_3P_3O_9 \cdot H_2O$	2.902	122.6	59.6	823
	2.935	130.6	60.8	
	2.900	129.8	59.6	
$Na_2KP_3O_9$	2.880	124.7	59.4	830
	2.911	127.4	60.8	
	2.871	124.5	59.7	
$CsNa_2P_3O_9 \cdot 2H_2O$	2.880	126.8	59.7	824
	2.866	126.1	60.2	
	2.880	126.9	60.2	
$Na_2LiP_3O_9 \cdot 4H_2O$	2.884	127.5	60.5	833
	2.896	128.0	60.1	
	2.864	126.1	59.4	
$Cd_3(P_3O_9)_2 \cdot 10H_2O$	2.909	131.3	58.7	841
	2.869	127.2	61.3	
	2.946	133.4	60.0	
$Sr_3(P_3O_9)_2 \cdot 7H_2O$	2.940	130.4	59.7	845
	2.868	124.7	58.9	
	2.893	126.7	61.4	
$Ba_3(P_3O_9)_2 \cdot 6H_2O$ [b]	2.892	126.7	59.4	847
	2.870	125.6	60.4	
	2.896	128.8	60.2	
	2.873	126.1	60.1	
	2.888	127.7	60.2	
	2.884	127.1	59.7	
$Pb_3(P_3O_9)_2 \cdot 3H_2O$	2.83	122.0	59.8	852
	2.83	122.0	60.4	
	2.85	128.0	59.8	
$Ba_2Zn(P_3O_9)_2 \cdot 10H_2O$	2.871	129.8	60.8	854
	2.899	128.8	60.1	
	2.918	123.5	59.1	

(continued)

Table 4.2.14. (continued)

Formula	P–P (Å)	P–O–P (°)	P–P–P (°)	Reference
PbNa$_4$(P$_3$O$_9$)$_2$	2.878	126.4	60.3	860
	2.894	127.5	60.1	
	2.898	126.7	59.6	
BaNaP$_3$O$_9$	2.856	128.5	60.4	862
	2.886	123.0	59.3	
	2.890	126.9	60.3	
HgKP$_3$O$_9$	2.847	128.3	61.0	874
	2.903	124.2	60.5	
	2.918	128.6	58.6	
Hg(NH$_4$)Na$_2$(P$_3$O$_9$)$_2$	2.840	123.5	62.0	884
	2.863	125.1	58.6	
	2.939	131.2	59.4	

[a]See also Tables 4.2.15 and 4.2.16.
[b]Two independent rings in the unit cell.

Table 4.2.15. Main Geometrical Features of P$_3$O$_9$ Ring Anions with No Internal Symmetry[a]

Formula	P–P (Å)	P–O–P (°)	P–P–P (°)	Reference
SrNaP$_3$O$_9$·3H$_2$O	2.879	126.8	59.9	886
	2.888	127.5	60.2	
	2.877	127.5	59.9	
BaNaP$_3$O$_9$·3H$_2$O	2.888	127.1	60.8	888
	2.882	128.8	59.5	
	2.921	124.9	59.7	
BaNH$_4$P$_3$O$_9$·H$_2$O	2.899	130.0	60.3	893
	2.872	126.6	59.4	
	2.900	129.7	60.3	
NiAg$_4$(P$_3$O$_9$)$_2$·6H$_2$O	2.877	128.8	59.7	895
	2.863	126.0	59.9	
	2.858	126.1	60.4	
NiNa$_4$(P$_3$O$_9$)$_2$·6H$_2$O	2.899	129.0	59.5	896
	2.866	124.8	59.7	
	2.860	125.8	60.8	
CuNa$_4$(P$_3$O$_9$)$_2$·4H$_2$O	2.892	127.5	60.3	897
	2.895	129.2	59.9	
	2.908	126.6	59.8	
CuK$_4$(P$_3$O$_9$)$_2$·4H$_2$O	2.878	126.3	60.7	899
	2.926	129.1	59.1	
	2.909	131.6	60.2	
Ni(NH$_4$)$_4$(P$_3$O$_9$)$_2$·4H$_2$O	2.910	128.3	60.4	900
	2.910	130.6	59.8	
	2.926	128.9	59.8	
Co(NH$_4$)$_4$(P$_3$O$_9$)$_2$·4H$_2$O	2.894	128.0	60.5	901
	2.907	129.2	60.0	
	2.921	130.7	59.5	
CdK$_4$(P$_3$O$_9$)$_2$·2H$_2$O	2.973	137.2	59.1	902
	2.930	129:2	59.7	
	2.913	132.4	61.2	
BiNa$_3$(P$_3$O$_9$)$_2$·9H$_2$O	2.848	124.5	59.5	913
	2.826	123.0	59.8	
	2.818	123.2	60.6	

[a]See also Tables 4.2.14 and 4.2.16.

Table 4.2.16. Main Geometrical Features of P_3O_9 Ring Anions with No Internal Symmetry[a]

Formula	P–P (Å)	P–O–P (°)	P–P–P (°)	Reference
(EDA)$_3$(P$_3$O$_9$)$_2$ [c]	2.907	129.7	59.9	915
	2.895	127.6	60.3	
	2.890	127.9	59.7	
	2.918	131.0	60.2	
	2.902	128.6	60.2	
	2.919	130.5	59.6	
(IPA)$_3$(P$_3$O$_9$)$_2$ [c]	2.874	128.3	59.6	916
	2.866	128.8	59.8	
	2.895	129.5	60.6	
	2.875	128.4	59.8	
	2.866	126.8	60.1	
	2.872	128.5	60.0	
(MA)$_3$P$_3$O$_9$	2.851	124.9	60.2	918
	2.874	127.0	59.5	
	2.876	126.9	60.3	
(Gly)$_3$P$_3$O$_9$	2.865	125.8	60.0	919
	2.883	128.3	60.3	
	2.872	128.7	59.7	
(Gua)$_3$P$_3$O$_9$·2H$_2$O	2.884	127.2	59.0	920
	2.902	124.5	60.8	
	2.851	128.6	60.2	
K$_3$P$_3$O$_9$·Te(OH)$_6$.2H$_2$O	2.894	129.1	60.2	922
	2.905	128.6	59.9	
	2.895	127.5	59.9	
Rb$_3$P$_3$O$_9$·Te(OH)$_6$·H$_2$O	2.889	127.0	60.6	925
	2.902	126.7	59.9	
	2.920	129.8	59.5	
Cs$_3$P$_3$O$_9$·Te(OH)$_6$·H$_2$O	2.837	127.2	60.6	926
	2.859	122.6	59.3	
	2.873	124.8	60.1	

[a]See also Tables 4.2.14 and 4.2.15.
[b]Abbreviations: EDA, ethylenediammonium, $[NH_3-(CH_2)_2-NH_3]^{2+}$; IPA, isopropylammonium, $[(CH_3)_2-CH-NH_3]^+$; MA, methylammonium, $(CH_3-NH_3)^+$; Gly, glycinium, $(NH_3-CH_2-COOH)^+$; Gua, guanidinium, $[C(NH_2)_3]^+$.
[c]Two independent rings in the unit cell.

Internal symmetry: 3/m	$2.856 < P–P < 2.963$ Å	Average 2.909 Å
3	$2.877 < P–P < 2.915$	2.897
2	$2.864 < P–P < 2.901$	2.887
m	$2.826 < P–P < 2.974$	2.898
None	$2.818 < P–P < 2.946$	2.887

For the numerical values given for the P–P distances, the estimated standard deviation can be evaluated as 0.005, but for most of the recent data this value is close to 0.001.

For the *P–O–P angles*, the overall average value is 127.9° with an estimated standard deviation of 0.1. We report below the average values calculated for the various categories rings:

Internal symmetry: 3/m   average   133.9°
                             3   128.9
                             2   126.6
                           m   128.2
                        None   127.5

Thus, for this very small ring, we do not observe the large deviations from the average values that we will see when examining the larger ones, mainly P_6O_{18} and P_8O_{24}, since the geometrical strain decreases with the ring size.

4.3. Cyclotetraphosphates

4.3.1. Introduction

The early development of cyclotetraphosphate chemistry was marked by much controversy, mainly regarding the degree of condensation of the phosphoric anion owing to the lack of techniques for proper structural characterization.

Maddrell[6] appears to were the first to prepare cyclotetraphosphates. He synthesized a series of compounds of general formula $MO \cdot P_2O_5$ with M = Mn, Cu, Co, Ni, Mg, etc., which he called "metaphosphates"; all these compounds were later recognized to be cyclotetraphosphates. Some years later, the investigation of these compounds was continued by several chemists, including Fleitmann,[8,9] Glätzel,[13] Tammann,[17] and Warschauer.[21] Using copper cyclotetraphosphate, $Cu_2P_4O_{12}$, these last authors synthezed the corresponding sodium salt by the following reaction:

$$Cu_2P_4O_{12} + 2Na_2S \rightarrow Na_4P_4O_{12} + 2CuS$$

Then, using the cyclotetraphosphate tetrahydrate obtained in this manner, they synthesized numerous other compounds, mainly alkali and double salts. Structural investigations carried out during the past 30 years confirmed that most of these compounds are really cyclotetraphosphates. In the absence of any means for a proper determination of the condensation state, Fleitmann, used the term "dimetaphosphates" to designate these salts. It seems that Glätzel[13] was the first to suggest the tetrameric and cyclic nature of the anion. was suspected. The term "tetrametaphosphates" coined by Glätzel[13] was used thereafter by chemists—Warschauer[21] for instance—but was unfortunately not always correctly applied.

The tedious and very time-consuming process described above for the preparation of the sodium salt was for a long time the only means of producing this starting material. It was not until the middle of this century that a more convenient process was elaborated, based upon the low-temperature (273 K) hydrolytic degradation of $P_4O_{10,}$ leading to cyclotetraphosphoric acid in very high yield, according to the following scheme:

$$P_4O_{10} + 2H_2O \rightarrow H_4P_4O_{12}$$

This reaction was mentioned as early as 1934 by Travers and Chu[929] and later by Topley,[42] Raistrick,[930] Bell *et al.*[931] Finally, it was carefully analyzed and optimized by Thilo and Wicker.[932]

The first structural proof of the cyclic nature of the P_4O_{12} anion was reported in 1937 by Pauling and Sherman,[29] when these authors performed the crystal structure determination of the aluminum salt $Al_4(P_4O_{12})_3$. Since then, more than 90 cyclotetraphosphates have been clearly characterized.

4.3.2. Present State of Cyclotetraphosphate Chemistry

4.3.2.1. Alkali and Monovalent Cation Cyclotetraphosphates

$Li_4P_4O_{12} \cdot nH_2O$. The first experimental evidence for the existence of a lithium cyclotetraphosphate was reported by Grunze and Thilo[933] and Grunze[934] during their studies of the thermal condensation of LiH_2PO_4. This salt recrystallizes as a tetrahydrate. In a recent study of the thermal condensation of LiH_2PO_4, Schülke and Kayser[505] characterized two forms of the anhydrous salt and, among the recrystallized salts, an octahydrate, a hexahydrate, two forms of tetrahydrate, and two forms of dihydrate.

A crystal-chemistry study, performed by Grenier and Durif,[507] concluded that lithium cyclotetraphosphate freshly prepared by Boullé's process is a hexahydrate. Based on thermogravimetric experiments conducted with various aged specimens, these authors suspected that some of the water molecules must be of a zeolitic nature, with an initial loss of two of them at 343 K, according to the following scheme:

$$Li_4P_4O_{12} \cdot 6H_2O \xrightarrow{343\ K} Li_4P_4O_{12} \cdot 4H_2O \xrightarrow{443\ K} Li_4P_4O_{12} \xrightarrow{623\ K} LiPO_3$$

In the first structural study conducted by Averbuch-Pouchot and Durif[935] with crystals prepared at room temperature according to the procedure described by Grenier and Durif,[507] this salt was identified as a pentahydrate; two of the five water molecules being of a zeolitic nature. The very high thermal factors observed for the two nonbonded water molecules correspond probably to partly occupied positions, thus leading to a chemical composition close to that of the tetrahydrate.

$Na_4P_4O_{12} \cdot nH_2O$. The anhydrous form of sodium cyclotetraphosphate, the monohydrate, and two forms of the tetrahydrate are now well characterized. A decahydrate is said to crystallize below 293 K,[27] but no structural data have been reported for this salt.

Single crystals of $Na_4P_4O_{12}$ and $Na_4P_4O_{12} \cdot H_2O$ were obtained by Wiench and Jansen[936] by crystallization from aqueous solutions to which were added higher alcohols. The monohydrate exists in the temperature range 373–393 K. The transition from the monohydrate to the anhydrous salt occurs topochemically and is reversible. A detailed description of the atomic arrangement of the monohydrate was reported by Wiench and Jansen. The anhydrous salt is isotypic with the monohydrate. This phenomenon is to be compared with the similar analogy observed between anhydrous sodium cyclotriphosphate and its monohydrate.

As we have already mentioned, $Na_4P_4O_{12} \cdot 4H_2O$ was prepared by several chemists last century and at the beginning of this century.[8,9,13,17,21] As early as 1937, Bonneman,[937] using Warschauer's process,[21] prepared sodium cyclotetraphosphate tetrahydrate and determined its molecular weight by a cryoscopic procedure. He also produced the first X-ray diffraction diagrams for both the anhydrous salt and one form of the tetrahydrate.

Two forms of the tetrahydrate were reported[931]; one is monoclinic and the other one is triclinic. The form normally obtained by crystallization from aqueous solutions at room temperature is the monoclinic one. The triclinic form is obtained when crystallization is carried out at 353–363 K.

Andress *et al.*[938] described the chemical preparation of the monoclinic form by Warschauer's process and were the first to report the unit-cell dimensions and space group for this salt. A detailed description of the preparation and purification of $Na_4P_4O_{12}\cdot4H_2O$ produced by the same procedure was published by Barney and Gryder.[939]

Bell *et al.*[931] investigated the low-temperature hydration of P_4O_{10} and were probably the first to describe the chemical preparation of $Na_4P_4O_{12}\cdot4H_2O$ based on the hydrolysis of P_4O_{10} to $H_4P_4O_{12}$ (see Section 4.3.1).

The crystal structure of the monoclinic form was determined by Ondik *et al.*,[940] and that of the triclinic form was solved by Ondik.[941] These two structures were reexamined by Averbuch-Pouchot and Durif[942] in order to localize the H atoms.

There was some controversy regarding the transformation of the monoclinic form of $Na_4P_4O_{12}\cdot4H_2O$ to the triclinic form. Earlier workers reported a transition temperature of 327 K,[931] but Ondik[941] could not corroborate this finding and concluded that a transition in the solid state seems unlikely from a comparison of the two arrangements. Attempts by Gros[943] to observe such a transition also failed. More recently, Griffith[944] repeated these experiments and observed that air-dried crystals of both forms do not transform but that for wet samples, containing 13% water, the transition between the monoclinic and the triclinic form occurs at about 327 K, accompanied by a small thermal effect. As observed earlier, this transformation is not reversible. Griffith concluded that the early observations were in fact correct since the decahydrate had been employed as starting material and this phosphate would probably retain a sufficient amount of residual water up to 327 K to initiate the transition.

The effect of water vapor on the transformation of $Na_4P_4O_{12}$ was investigated by Nariai *et al.*[945] The anhydrous salt was prepared by heating the tetrahydrate for two hours at 373 K and one hour at 423 K. The experiments were conducted both in a dry and in humid atmosphere. In both cases, at 723 K, the conversion into $Na_3P_3O_9$ is total, but two different forms of cyclotriphosphate are obtained.

$Na_2H_2P_4O_{12}$ or $Na_2P_4O_{11}$. The chemical preparation of this salt by the reaction of H_3PO_4 with NaH_2PO_4 at 673 K was described by Griffith[946]:

$$2NaH_2PO_4 + 2H_3PO_4 \rightarrow Na_2H_2P_4O_{12} + 4H_2O$$

According to this author, the melting or, more probably, the decomposition point of $Na_2H_2P_4O_{12}$ is close by 673 K. A more detailed procedure for the preparation of this salt was later reported by the same author.[947] This salt is almost insoluble in water and has a fibrous crystal habit. These two characteristics are rather surprising for an alkali cyclotetra-phosphate. In order to explain these properties, Gryder *et al.*[948] performed a crystal-lographic investigation of this material. They found the so-called $Na_2H_2P_4O_{12}$ to be an intimate intergrowth of two distinct crystalline forms, both monoclinic, and, on the basis of various considerations, concluded that the anionic framework must be built up by chains of P_4O_{12} rings sharing an oxygen atom, thus leading to the formula of an ultraphosphate,

$Na_2P_4O_{11}$. On the other hand, Jarchow[949] proposed a structural model based on the $Na_2H_2P_4O_{12}$ formula reported by Griffith. Owing to the poor quality of the experimental data, this work was inconclusive and needs to be redone.

$K_4P_4O_{12} \cdot 2H_2O$ and $K_4P_4O_{12} \cdot 4H_2O$. The existence of two forms of $K_4P_4O_{12} \cdot 4H_2O$ and two forms of the anhydrous salt, reported by Van Wazer,[27] has not been confirmed. To date, only one form of the dihydrate and one form of the tetrahydrate have been well characterized. Averbuch-Pouchot and Durif reported the chemical preparation and accurate determinations of the atomic arrangements of both the dihydrate[942] and the tetrahydrate.[950]

$(NH_4)_3HP_4O_{12} \cdot H_2O$ and $(NH_4)_4P_4O_{12}$. According to Waerstad et al.,[951] $(NH_4)_3HP_4O_{12} \cdot H_2O$ can be obtained by adding P_4O_{10} to a concentrated NH_4OH solution at a temperature lower than 278 K. Unit-cell dimensions and an indexed powder diagram are reported by the authors.

$(NH_4)_4P_4O_{12}$ was first investigated by Romers et al.[952] and discussed by Andress and Fischer.[953] The first valuable description of its atomic arrangement was reported by Romers et al.[954] This work was later on improved by Cruickshank,[955] and a reexamination of the crystal structure, including the localization of the H atoms, reported by Koster and Wagner.[956] The thermal decomposition of $(NH_4)_4P_4O_{12}$ was investigated by Takena et al.[957]

$Rb_4P_4O_{12} \cdot 4H_2O$ and $Cs_4P_4O_{12} \cdot 4H_2O$. These two isotypic compounds were prepared and characterized by Averbuch-Pouchot and Durif.[958] There is no evidence for the existence of other hydrates or of the anhydrous salts. It should be noted that by a careful hydrolysis of the rubidium salt, at room temperature, Averbuch-Pouchot and Durif[959] prepared a new form of rubidium dihydrogenomonophosphate, closely related to the monoclinic form of CsH_2PO_4 and identical to the intermediate form of RbD_2PO_4. The hydrolysis reaction is

$$Rb_4P_4O_{12} + 4H_2O \rightarrow 4RbH_2PO_4$$

$Tl_4P_4O_{12}$. The thermal dehydation of TlH_2PO_4 and of the corresponding arsenate was studied by Dostal et al.[960] and by Dostal and Kocman.[800,961] During these investigations, these authors characterized $Tl_4P_4O_{12}$ as a water-soluble cyclotetraphosphate identical to $Tl_4As_4O_{12}$ and studied the solid solutions of the two salts. $Tl_4P_4O_{12}$ converts into a long-chain polyphosphate at about 690 K. Its atomic arrangement was determined by Fawcett et al.[962,963]

Crystallographic Data. The main crystallographic data for the 13 monovalent cation cyclotetraphosphates that have been investigated to date are presented in Table 4.3.1.

Mixed-Monovalent Cation Cyclotetraphosphates

$Na_2K_2P_4O_{12} \cdot 2H_2O$. Two crystalline forms of $Na_2K_2P_4O_{12} \cdot 2H_2O$ were prepared by Averbuch-Pouchot and Durif[942]; one is triclinic and the other one tetragonal. The authors reported complete structural studies for the two forms.

Table 4.3.1. Main Crystallographic Data for Monovalent Cation Cyclotetraphosphates

Formula	a (Å) α (°)	b (Å) β (°)	c (Å) γ (°)	S.G.	Z	Reference(s)
$Li_4P_4O_{12} \cdot 5H_2O$	17.073(8)	17.029(8) 127.32(1)	13.554(6)	$C2/c$	8	935
$Na_4P_4O_{12}$	13.808(2)	13.633(2)	6.027(2)	$P2_12_12_1$	4	936
$Na_4P_4O_{12} \cdot H_2O$	13.654(2)	13.475(3)	6.291(3)	$P2_12_12_1$	4	936
$Na_4P_4O_{12} \cdot 4H_2O$ (I)	6.652 103.40	9.579 106.98	6.320 93.28	$P\bar{1}$	1	941, 942
$Na_4P_4O_{12} \cdot 4H_2O$ (II)	9.667	12.358 92.58	6.170	$P2_1/a$	2	940, 942
$Na_2H_2P_4O_{12}$	18.74	14.79	7.03	$Pbnm$	8	949
$K_4P_4O_{12} \cdot 2H_2O$	8.153(4) 97.33(5)	8.222(4) 95.46(5)	11.154(8) 88.92(5)	$P\bar{1}$	2	942
$K_4P_4O_{12} \cdot 4H_2O$	9.061(3)	9.061(3)	10.284(5)	$I\bar{4}$	2	950
$(NH_4)_3HP_4O_{12} \cdot H_2O$	11.144(4)	11.322(3) 90.67(3)	10.514(5)	$P2_1/n$	4	951
$(NH_4)_4P_4O_{12}$	10.433(2)	10.871(3)	12.588(3)	$Cmca$	4	952–955
$Rb_4P_4O_{12} \cdot 4H_2O$	9.163(3)	9.163(3)	21.356(8)	$P4_1$	4	958
$Cs_4P_4O_{12} \cdot 4H_2O$	9.466(5)	9.466(5)	21.933(9)	$P4_1$	4	958
$Tl_4P_4O_{12}$	7.635	7.635	11.087	$P\bar{4}2_1c$	2	963

$Na_2(NH_4)_2P_4O_{12} \cdot 2H_2O$ and $Na_2Rb_2P_4O_{12} \cdot 2H_2O$. These two compounds were synthesized by Averbuch-Pouchot and Durif[942] during the same investigation and are isotypic with the triclinic form of the sodium–potassium salt.

$Na_3CsP_4O_{12} \cdot 4H_2O$ and $Na_3CsP_4O_{12} \cdot 3H_2O$. When cesium carbonate is added to an aqueous solution of $Na_4P_4O_{12}$, acidified by the addition of a small amount of $H_4P_4O_{12}$, at room temperature, several successive precipitations occur. The first one corresponds to the monoclinic form of $Na_4P_4O_{12} \cdot 4H_2O$, the second one to $Na_3CsP_4O_{12} \cdot 4H_2O$, and the last one to $Na_3CsP_4O_{12} \cdot 3H_2O$. Averbuch-Pouchot and Durif[964] reported the chemical preparation and crystal structures of these last two hydrates.

Crystallographic Data. Table 4.3.2. contains the main crystallographic data for mixed-monovalent cation cyclotetraphosphates.

4.3.2.2. Divalent Cation Cyclotetraphosphates

(i) Anhydrous Divalent Cation Cyclotetraphosphates

$M_2P_4O_{12}$ (M = Zn, Ni, Co, Fe, Mg, Cu, Mn, Cd). These eight cyclotetraphosphates have the same type of atomic arrangement. Some of them were characterized during last century. $Cu_2P_4O_{12}$, for instance, was described as early as 1847 by Maddrell[6] and subsequently by Fleitmann[8,9] and Warschauer.[21] These authors suspected the tetrameric nature of the anion on the basis of hypotheses that would now not seem acceptable today. More recently, pioneering work in this field was done by Thilo and Grunze.[558] They studied the thermal condensation of the $M(H_2PO_4)_2$ monophosphates:

Table 4.3.2. Main Crystallographic Data for Mixed Monovalent Cation Cyclotetraphosphates

Formula	a (Å) α (°)	b (Å) β (°)	c (Å) γ (°)	S.G.	Z	Reference(s)
$Na_2K_2P_4O_{12} \cdot 2H_2O$ (I)	11.366(8) 90.07(5)	7.908(5) 106.85(5)	7.929(5) 95.66(5)	$P\bar{1}$	2	942
$Na_2K_2P_4O_{12} \cdot 2H_2O$ (II)	7.928(5)	7.928(5)	21.66(2)	$P4_1$	4	942
$Na_2(NH_4)_2P_4O_{12} \cdot 2H_2O$	11.547(8) 89.76(4)	8.012(6) 106.22(3)	8.044(5) 94.78(3)	$P\bar{1}$	2	942
$Na_2Rb_2P_4O_{12} \cdot 2H_2O$	11.577(9) 89.79(5)	8.006(6) 106.58(4)	8.032(7) 95.19(4)	$P\bar{1}$	2	942
$Na_3CsP_4O_{12} \cdot 3H_2O$	11.39(1)	10.92(1) 95.24(5)	11.81(1)	$P2_1/c$	4	964
$Na_3CsP_4O_{12} \cdot 4H_2O$	14.50(2)	7.804(3)	7.006(3)	$Imm2$	2	964

$$2M(H_2PO_4)_2 \rightarrow M_2P_4O_{12} + 4H_2O$$

The resulting compounds were shown to be isomorphous, from X-ray diffraction data, and very probably cyclotetraphosphates, from paper chromatography experiments. The spectrometric investigations of Steger[965] and Steger and Simon[966] confirmed these results.

Later, unit-cell parameters for this series of compounds were reported by Beucher,[593] Beucher and Grenier,[967] and Laügt *et al.*,[68] and these salts were confirmed to be isotypic. The atomic arrangement was finally determined by Laügt *et al.*[969] using the copper salt. With the exception of the cadmium salt, these compounds are prepared by using thermal or flux methods. In most cases, the crystals produced by flux methods are twinned.

This series of compounds was the subject of a great number of chemical, crystallographic, and physical studies, and we will briefly review these here.

The thermal transformation of $Mn(H_2PO_4)_2 \cdot 2H_2O$ and $Mg(H_2PO_4)_2 \cdot 2H_2O$ to $Mn_2P_4O_{12}$ and $Mg_2P_4O_{12}$ was studied in detail by Shchegrov *et al.*[970]

The mechanism of the thermal dehydration–cyclization of $Mg(H_2PO_4)_2 \cdot 2H_2O$, leading to magnesium cyclotetraphosphate, was investigated by Serazetdinov *et al.*[971] both in a vacuum and at atmospheric pressure. When dehydration occurred in a vacuum, they observed the formation of an amorphous phase, at about 773 K, prior to crystallization of the cyclotetraphosphate. This phase did not appear when the experiments were conducted at atmospheric pressure. A similar study was performed by Bekturov *et al.*[972] These authors characterized two crystalline forms of $Mg_2P_4O_{12}$ as final condensation products.

The chemical preparation of $Mn_2P_4O_{12}$ using manganese carbonate and ammonium dihydrogenomonophosphate as starting materials was analyzed by Bukhalova *et al.*[973]

$Fe_2P_4O_{12}$ was prepared for the first time by Bagieu-Beucher *et al.*[565] These authors reported crystal data showing this salt to be isotypic with the copper compound. This compound was also characterized by Genkina *et al.*[974] during a study of the LiF–Fe_2O_3(FeO)–P_2O_5–H_2O system under hydrothermal conditions at 703 K and 1000 atm.

The surface tension of molten $Zn(PO_3)_2$ was investigated by Krivovyazov *et al.*[567] According to these authors, the melting point is observed at 1135(2) K, and the transformation of $Zn_2P_4O_{12}$ to $Zn(PO_3)_2$ occurs at 913 K.

$Cd_2P_4O_{12}$ is a special case in this series. It cannot be prepared by a flux method similar to that used for the other $M_2P_4O_{12}$ salts. Its existence was reported for the first time by Thilo and Grunze[558] during a study of the thermal condensation of $Cd(H_2PO_4)_2 \cdot 2H_2O$. These authors presented the following scheme for this condensation:

$$Cd(H_2PO_4)_2 \cdot 2H_2O \xrightarrow{373 \ K} Cd(H_2PO_4)_2 \xrightarrow{453 \ K} CdH_2P_2O_7 \xrightarrow{473 \ K} Cd_2P_4O_{12}$$

According to these authors, the cyclotetraphosphate is transformed into a long-chain polyphosphate at about 873 K.

Some years later, the above scheme was studied by Ropp and Aia[585] by differential thermal analysis. For the first two steps, they found temperatures relatively close by those reported by Thilo and Grunze[558]: i.e., 403 K–433 K instead of 373 K and 483 K instead of 453 K. For the next step, corresponding to the condensation of the diphosphate into the cyclotetraphosphate, they gave a temperature of 683 K. According to the same authors, the cyclotetraphosphate thus obtained transforms at 823 K into a tetraphosphate, $Cd_3P_4O_{13}$.

In spite of these data, pure specimens of $Cd_2P_4O_{12}$ were not obtained until a procedure for their preparation was described by Laügt et al.[968] These authors used as starting material a not yet characterized hydrate of cadmium cyclotetraphosphate prepared at room temperature by the action of $H_4P_4O_{12}$ on cadmium carbonate. By slowly heating this hydrate up to 573 K and keeping it at this temperature for 30 min, they obtained pure specimens of $Cd_2P_4O_{12}$, not contaminated by the long-chain cadmium polyphosphate. In contradiction to the findings of the previous authors, they observed that upon prolonged heating at 573 K $Cd_2P_4O_{12}$ transforms irreversibly into the long-chain cadmium polyphosphate. It is worth reporting that by firing a phosphoarsenate, $Cd(H_2P_{1-x}As_xO_4)_2 \cdot 2H_2O$ ($x \sim 0.5$), at temperatures higher than 673 K the same authors obtained a well-crystallized mixture of $Cd_2P_4O_{12}$ and $Cd(AsO_3)_2$ without any contamination by the long-chain cadmium polyphosphate.

A number of additional crystallographic investigations were performed on this series of compounds:

- The crystal structure of $Mg_2P_4O_{12}$ was refined from single-crystal X-ray data by Nord and Lindberg.[975]
- The crystal structure of $Co_2P_4O_{12}$ was refined by Nord[976] using neutron diffraction and a powdered specimen.
- $Ni_2P_4O_{12}$ was prepared as a powder sample and its crystal structure was refined by neutron diffraction by Nord.[977] The same author[978] investigated $NiCoP_4O_{12}$ and $NiZnP_4O_{12}$ samples by neutron diffraction, showing that in both solid solutions, nickel preferentially enters the centrosymmetric crystallographic site.
- Crystal structure determination for $Fe_2P_4O_{12}$ was performed by Genkina et al.[979] The authors confirmed that this salt is isotypic with the copper salt. The atomic arrangement was later determined again by Nord et al.[980]

Except in the case of the cadmium salt, a high-pressure form was observed for the $M_2P_4O_{12}$ compounds by Bagieu-Beucher et al.[565] For a time, this high-pressure form was thought to also be a cyclotetraphosphate, but was later recognized as a long-chain polyphosphate by Averbuch-Pouchot *et al.*[981]

Magnetic properties of $M_2P_4O_{12}$ for M = Mn, Co, Ni, and Cu were investigated by Gunsser *et al.*[982]

$Mg_{(2-x)}Zn_xP_4O_{12}$, $Mg_{(2-x)}Ca_xP_4O_{12}$ and $Ni_{(2-x)}Mg_xP_4O_{12}$ solid solutions were investigated by Trojan and coworkers[983,984] in order to examine their potential use as pigments.

$Ca_2P_4O_{12}$. This salt was obtained by Schneider and Jost[985] during a study of the thermal behavior of the tetrahydrate. The authors reported only its unit-cell dimensions.

Crystallographic Data. The main crystallographic data for anhydrous divalent cation cyclotetraphosphates are given in Table 4.3.3.

(ii) Hydrated Divalent Cation Cyclotetraphosphates

$M_2P_4O_{12}\cdot 8H_2O$ (M = Zn, Mg, Fe, Co, Cu). $Zn_2P_4O_{12}\cdot 8H_2O$ was first characterized by Averbuch-Pouchot,[986] who reported its chemical preparation and a complete description of its atomic arrangement. Later, the corresponding magnesium, iron, cobalt, and copper salts were prepared and recognized as being isotypic with $Zn_2P_4O_{12}\cdot 8H_2O$ by Foumakoye.[987]

The thermal behavior of a magnesium heptahydrate, $Mg_2P_4O_{12}\cdot 7H_2O$, both in a dry and in a humid atmosphere, was studied by Nariai *et al.*[988] The hydration state observed by the authors can be easily explained by the fact that in the atomic arrangement of the octahydrate, as described by Averbuch-Pouchot,[986] two of the eight water molecules are

Table 4.3.3. Main Crystallographic Data for Anhydrous Divalent Cation Cyclotetraphosphates

Formula	a (Å) α (°)	b (Å) β (°)	c (Å) γ (°)	S.G.	Z	Reference(s)
$Cu_2P_4O_{12}$	12.56(1)	8.088(6) 118.58(2)	9.574(8)	$C2/c$	4	967, 969
$Mg_2P_4O_{12}$	11.77(1)	8.287(4) 118.87(2)	9.949(10)	$C2/c$	4	967
$Ni_2P_4O_{12}$	11.65(1)	8.241(4) 118.46(2)	9.857(6)	$C2/c$	4	967
$Co_2P_4O_{12}$	11.815(5)	8.310(8) 118.68(2)	9.339(10)	$C2/c$	4	967
$Zn_2P_4O_{12}$	11.78(1)	8.302(6) 118.81 (2)	9.927(8)	$C2/c$	4	967
$Mn_2P_4O_{12}$	12.08(1)	8.471(6) 119.29(2)	10.171(8)	$C2/c$	4	967
$Fe_2P_4O_{12}$	11.952(2)	8.359(2) 118.76(5)	9.932(2)	$C2/c$	4	565, 968, 979
$Cd_2P_4O_{12}$	12.319(5)	8.631(3) 119.33(5)	10.382(6)	$C2/c$	4	968
$Ca_2P_4O_{12}$	8.02(1) 97.4(1)	10.42(1) 109.8(1)	7.20(1) 90.4(1)	$P1$ or $P\bar{1}$	2	985

of a zeolitic nature. The loss of the water molecules is observed, in two steps, at 368 and 428 K and the final product is said to be the long-chain zinc polyphosphate, $Zn(PO_3)_2$.

$Mn_2P_4O_{12}\cdot 10H_2O$, $Ni_2P_4O_{12}\cdot 10H_2O$, and $Cd_2P_4O_{12}\cdot 8H_2O$. These three hydrates were characterized by Lavrov *et al.*[989] They are prepared by the action of $H_4P_4O_{12}$ on an aqueous solution of the corresponding divalent cation perchlorates.

$Mn_3H_2(P_4O_{12})_2\cdot 19/2H_2O$. Nariai *et al.*[990] described the chemical preparation of this acidic manganese salt and investigated the effect of water vapor on its thermal behavior. The crystallization water is lost at 418 K, and the transformation into long-chain polyphosphate is complete at 723 K. The formation of various oligophosphates is observed between 438 and 538 K.

$Ca_2P_4O_{12}\cdot 4H_2O$ and $Ca_2P_4O_{12}\cdot H_2O$. A crystal structure determination for the tetrahydrate was performed by Schneider *et al.*[991] In spite of a close similarity to the corresponding lead salt, the authors could not conclude that these salts are isotypic after a comparison of the two atomic arrangements. The existence of a second form of $Ca_2P_4O_{12}\cdot 4H_2O$ was reported by Schülke,[992] but, to date no crystal data have been reported.

The thermal behavior of the monoclinic ($P2_1/n$) $Ca_2P_4O_{12}\cdot 4H_2O$ was investigated by Schneider and Jost.[985,993] The following scheme was reported by the authors:

$$Ca_2P_4O_{12}\cdot 4H_2O \xrightarrow{393\ K} Ca_2P_4O_{12}\cdot H_2O \xrightarrow{493\ K} Ca_2P_4O_{12} \xrightarrow{733\ K} \text{amorphous phase}$$

$$\text{amorphous phase} \xrightarrow{793\ K} xCa(PO_3)_2 \xrightarrow{813-893\ K} \beta\text{-}Ca(PO_3)_2 \xrightarrow{1123\ K} \beta\text{-}Ca_2P_2O_7$$

Each intermediate phase is crystallographically related to β-$Ca(PO_3)_2$ long-chain polyphosphate. It is of interest to note that the intermediate amorphous phase does not interrupt the oriented course of the reaction. The relationship between the crystallographic orientation of the educts and products was reported for all steps of the reaction, as well as the unit cells for each cyclotetraphosphate. The last step corresponds to a loss of P_2O_5.

The thermal behavior of $Ca_2P_4O_{12}\cdot 7/2H_2O$ was investigated by Nariai *et al.*[988] The release of the water molecules occurs at about 423 K, and the transformation into the anhydrous long-chain polyphosphate is observed at 862 K.

$Sr_2P_4O_{12}\cdot 6H_2O$. This hydrate was obtained by Durif and Averbuch-Pouchot[994] by the action of $H_4P_4O_{12}$ on a solution of strontium nitrate. These authors performed a complete determination of its atomic arrangement.

The investigation, both in a dry and in a humid atmosphere, of the thermal behavior of $Sr_2P_4O_{12}\cdot 11/2H_2O$, probably identical to the hexahydrate, was performed by Nariai *et al.*[988] The loss of the water molecules occurs in two steps, at 393 and 433 K, and the final calcination product, $Sr(PO_3)_2$, crystallizes at 663 K.

$Pb_2P_4O_{12}\cdot 4H_2O$, $Pb_2P_4O_{12}\cdot 3H_2O$, and $Pb_2P_4O_{12}\cdot 2H_2O$. The tetrahydrate was carefully investigated by Worzala,[995] who reported a detailed determination of its atomic arrangement. This salt is of fundamental interest for the preparation, by a process of dehydration and recondensation, of lead cyclooctaphosphate, an important starting material for the production of other cyclooctaphosphates (see Section 4.6). The dihydrate, which is formed in the first step in this process, was obtained as single crystals by Worzala.[996] They are formed topotactically by heating the tetrahydrate at 373 K. The author reported a complete description of the atomic arrangement. The trihydrate was prepared by Klinkert and Jansen[997] by using a gel technique. These authors reported a detailed determination of the atomic arrangement. This hydrate does not appear in the process of recondensation of the lead tetrahydrate.

Nariai *et al.*[998] investigated the influence of water vapor on the thermal decomposition of the dihydrate. The main loss of crystallization water occurs at 373 K, and the conversion into cyclooctaphosphate occurs above 573 K. An intermediate step between 423 and 443 K shows the formation of various oligophosphates.

$Ba_2P_4O_{12} \cdot 4H_2O$. This hydrate was characterized by Nariai *et al.*[988] The investigation, both in a dry and in a humid atmosphere, of its thermal behavior shows that crystallization water is released at about 423 K. At 783 K, the long-chain polyphosphate is observed after an intermediate step corresponding to an amorphous phase containing mainly oligophosphates. No crystal data exist for this compound.

Crystallographic Data. Table 4.3.4 contains the main crystallographic data for hydrated divalent cation cyclotetraphosphates.

Table 4.3.4. Main Crystallographic Data for Hydrated Divalent Cation Cyclotetraphosphates

Formula	a (Å) α (°)	b (Å) β (°)	c (Å) γ (°)	S.G.	Z	Reference(s)
$Zn_2P_4O_{12} \cdot 8H_2O$	8.610(5) 96.09(5)	7.137(5) 105.99(5)	7.108(5) 100.49(5)	$P\bar{1}$	1	986
$Mg_2P_4O_{12} \cdot 8H_2O$	8.740(5) 96.37(5)	7.139(5) 106.38(5)	7.103(5) 100.60(5)	$P\bar{1}$	1	987
$Fe_2P_4O_{12} \cdot 8H_2O$	8.643(5) 96.31(5)	7.175(5) 105.44(5)	7.114(5) 100.72(5)	$P\bar{1}$	1	987
$Co_2P_4O_{12} \cdot 8H_2O$	8.768(5) 92.43(5)	6.855(5) 105.83(5)	7.274(5) 104.44(5)	$P\bar{1}$	1	987
$Cu_2P_4O_{12} \cdot 8H_2O$	8.526(5) 93.20(5)	7.058(5) 105.58(5)	7.155(5) 100.59(5)	$P\bar{1}$	1	987
$Ca_2P_4O_{12} \cdot 4H_2O$	7.668(1)	12.895(1) 107.00(1)	7.144(1)	$P2_1/n$	1	985, 991
$Ca_2P_4O_{12} \cdot H_2O$	7.72(1) 95.9(1)	10.52(1) 105.1(1)	7.15(1) 83.9(1)	$P1$ or $P\bar{1}$	2	985
$Sr_2P_4O_{12} \cdot 6H_2O$	6.644(3) 101.62(5)	7.365(4) 109.98(5)	8.618(4) 95.65(5)	$P\bar{1}$	2	994
$Pb_2P_4O_{12} \cdot 4H_2O$	8.07(2) 108.2(3)	11.76(3)	7.50(2)	$P2_1/n$	2	995
$Pb_2P_4O_{12} \cdot 3H_2O$	7.864(3) 97.42(2)	9.144(3) 100.63(2)	10.216(3) 114.92(2)	$P\bar{1}$	2	997
$Pb_2P_4O_{12} \cdot 2H_2O$	8.02(2) 98.8(2)	10.58(2) 108.7(2)	7.53(2) 82.6(3)	$P\bar{1}$	2	996

4.3.2.3. Divalent–Monovalent Cation Cyclotetraphosphates

(i) Anhydrous Divalent–Monovalent Cation Cyclotetraphosphates

$Zn_4Na_4(P_4O_{12})_3$, $Co_4Na_4(P_4O_{12})_3$ and $Ni_4Na_4(P_4O_{12})_3$. The chemical preparation of these three isotypic compounds was reported by Averbuch-Pouchot and Durif.[644] The zinc salt was used by these authors to determine the atomic arrangement. They discussed the close relationship between this structure and that of $Al_4(P_4O_{12})_3$.

$SrNa_2P_4O_{12}$. This compound was not observed in the $NaPO_3$–$Sr(PO_3)_2$ phase-equilibrium diagram elaborated by Martin.[625] It was characterized for the first time by Averbuch-Pouchot and Durif[999] during attempts to prepare $SrNa_4(P_3O_9)_2$ single crystals by a flux method. These authors reported its chemical preparation and a complete description of its atomic arrangement.

$SrK_2P_4O_{12}$. In the KPO_3–$Sr(PO_3)_2$ phase-equilibrium diagram elaborated by Bukhalova and Tokman,[1000] $SrK_2P_4O_{12}$, the only compound observed in this system, is shown as an incongruent-melting salt. However, in the revised diagram reported by Martin[625] (Figure 4.3.1), $SrK_2P_4O_{12}$ appears as a congruent-melting compound (mp = 977 K). The chemical preparation and crystal data of this salt were reported by Durif et al.,[1001] and a crystal structure determination was performed by Tordjman et al.[1002]

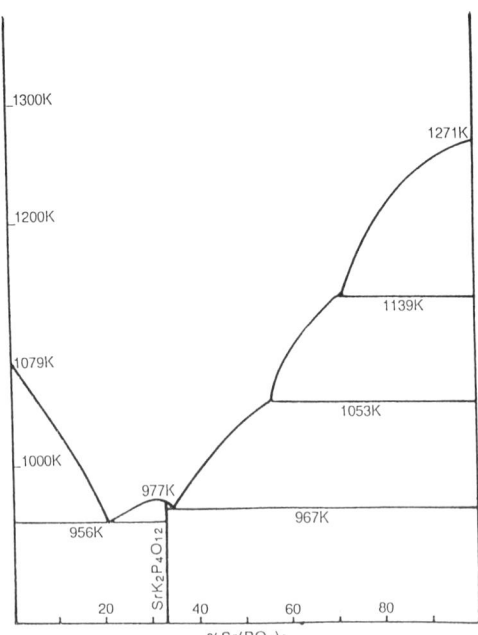

Figure 4.3.1. The KPO_3–$Sr(PO_3)_2$ phase-equilibrium diagram. Two transformations of $Sr(PO_3)_2$ are observed at 1053 and 1139 K on the right-hand side of the diagram.

Eight tetragonal $M^{II}M_2^IP_4O_{12}$ cyclotetraphosphates were clearly characterized as isotypic with this salt.

$CaK_2P_4O_{12}$. Cavéro-Ghersi and Durif[1003] reported the chemical preparation of this salt. On the basis of their crystal data for this salt, they recognized it as isotypic with $SrK_2P_4O_{12}$. At 973 K, it decomposes as follows:

$$CaK_2P_4O_{12} \rightarrow CaKP_3O_9 + KPO_3$$

This explains why this salt does not appear in the KPO_3–$Ca(PO_3)_2$ phase-equilibrium diagram, established almost simultaneously by Gill and Taylor[1004] and Andrieu and Diament.[867]

$PbK_2P_4O_{12}$. The existence of $PbK_2P_4O_{12}$ and crystallographic investigations of this salt were reported by Cavéro-Ghersi and Durif.[1003] This compound, isotypic with $SrK_2P_4O_{12}$, appears as the low-temperature form of the $PbK_2(PO_3)_4$ long-chain polyphosphate observed by Mahama *et al*.[660] during the elaboration of the KPO_3–$Pb(PO_3)_2$ phase-equilibrium diagram. At 810 K, it transforms irreversibly into the long-chain polyphosphate:

$$PbK_2P_4O_{12} \rightarrow PbK_2(PO_3)_4$$

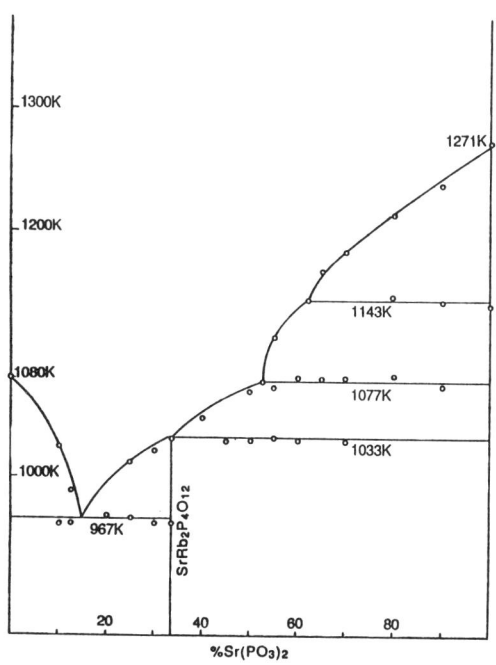

Figure 4.3.2. The $RbPO_3$–$Sr(PO_3)_2$ phase-equilibrium diagram. The two phase transformations of $Sr(PO_3)_2$ appear at 1077 and 1143 K on the right-hand side of the diagram.

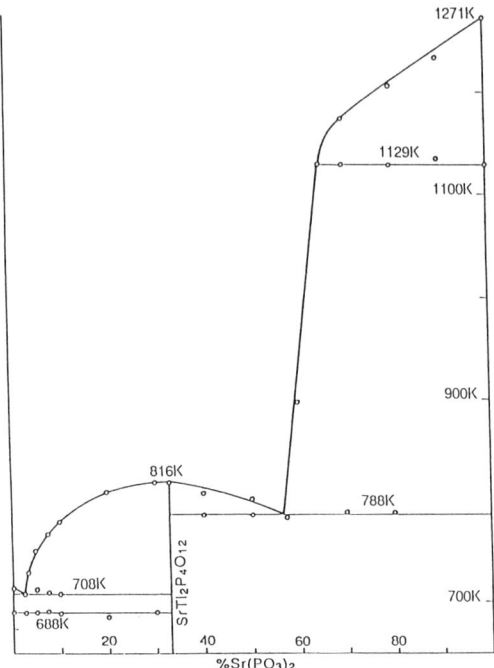

Figure 4.3.3. The $TlPO_3$–$Sr(PO_3)_2$ phase-equilibrium diagram. The transformation of $RbPO_3$ appears at 688 K on the left-hand side of the diagram, and that of $Sr(PO_3)_2$ at 1129 K on the right-hand side.

This transformation was reinvestigated by Schneider[1005] in a dry and in a humid atmosphere. In dry atmosphere, he confirmed the previous result but in moist atmosphere observed the following scheme:

$$PbK_2P_4O_{12} \rightarrow Pb_2P_2O_7 + \text{amorphous phase}$$

$Sr(NH_4)_2P_4O_{12}$, $SrRb_2P_4O_{12}$, and $SrTl_2P_4O_{12}$. These three salts were described by Durif *et al.*[1001] as being isotypic with $SrK_2P_4O_{12}$. The ammonium salt was prepared by a thermal method, and the other two appear as the only compounds in the $RbPO_3$–$Sr(PO_3)_2$ and $TlPO_3$–$Sr(PO_3)_2$ phase-equilibrium diagrams elaborated by Martin,[625] which are presented in Fig. 4.3.2 and 4.3.3. $SrTl_2P_4O_{12}$ is a congruent-melting compound (mp = 816 K), whereas $SrRb_2P_4O_{12}$ decomposes at 1033 K.

The thermal behavior of $Sr(NH_4)_2P_4O_{12}$ was investigated in both in a dry and in a humid atmosphere by Takenaka *et al.*[1006] The final crystalline product is the long-chain strontium polyphosphate, β-$Sr(PO_3)_2$.

$Pb(NH_4)_2P_4O_{12}$, $PbRb_2P_4O_{12}$, and $PbTl_2P_4O_{12}$. Cavéro-Ghersi and Durif[1003] reported the chemical preparation and crystal data for these three compounds, which are isotypic with $SrK_2P_4O_{12}$.

$SrCs_2P_4O_{12}$ and $Sr_3Cs_4H_2(P_4O_{12})_3$. The first compound appears as an incongruent-melting salt in the $CsPO_3$–$Sr(PO_3)_2$ phase diagram elaborated by Tokman and Bukhalova.[880] To date, there is no structural evidence to confirm the ring nature of the anion assumed by these authors. The second compound was obtained by Averbuch-Pouchot and Durif[1007] during an attempt to produce single crystals of $SrCs_2P_4O_{12}$. It was reported to be a sparingly water-soluble salt. A detailed determination of its atomic arrangement was reported by the authors, who discussed the close similarity between this structure and that of $Al_4(P_4O_{12})_3$.

Crystallographic Data. The main crystallographic data for anhydrous divalent–monovalent cation cyclotetraphosphates are presented in Table 4.3.5.

(ii) Hydrated Divalent–Monovalent Cation Cyclotetraphosphates

$CaNa_2P_4O_{12}\cdot11/2H_2O$ and $SrNa_2P_4O_{12}\cdot6H_2O$. The chemical preparation and crystal structures of these two salts were reported by Averbuch-Pouchot and Durif[1008] and Durif et al.[1009] respectively. No relationship exists between these two atomic arrangements. The crystal structure determination for the calcium salt provided the first example of a cyclotetraphosphate containing two crystallographically independent P_4O_{12} groups.

$NiK_2P_4O_{12}\cdot7H_2O$ and $Ni(NH_4)_2P_4O_{12}\cdot7H_2O$. These two isotypic salts were prepared by Jouini et al.[1010] The atomic arrangement, determined by these authors with the ammonium salt, shows that three of the water molecules are of a zeolitic nature.

$NiK_2P_4O_{12}\cdot5H_2O$ and $CoK_2P_4O_{12}\cdot5H_2O$. These two isotypic salts were obtained by dehydration of the corresponding heptahydrates. A crystal structure determination was performed by Jouini et al.[1011] using the cobalt salt. In addition, a detailed IR study of both

Table 4.3.5. Main Crystallographic Data for Anhydrous Divalent–Monovalent Cation Cyclotetraphosphates

Formula	a (Å) α (°)	b (Å) β (°)	c (Å) γ (°)	S.G.	Z	Reference(s)
$Zn_4Na_4(P_4O_{12})_3$	14.580	14.580	14.580	$I\bar{4}3d$	4	644
$Co_4Na_4(P_4O_{12})_3$	14.593	14.593	14.593	$I\bar{4}3d$	4	644
$Ni_4Na_4(P_4O_{12})_3$	14.446	14.446	14.446	$I\bar{4}3d$	4	644
$SrNa_2P_4O_{12}$	9.838(5)	9.838(5)	5.003(3)	$P4/nbm$	2	999
$SrK_2P_4O_{12}$	7.445(5)	7.445(5)	10.17(2)	$I\bar{4}$	2	1001, 1002
$CaK_2P_4O_{12}$	7.364(2)	7.364(2)	9.899(5)	$I\bar{4}$	2	1003
$PbK_2P_4O_{12}$	7.434(2)	7.434(2)	10.208(4)	$I\bar{4}$	2	1003
$Sr(NH_4)_2P_4O_{12}$	7.575(5)	7.575(5)	10.26(2)	$I\bar{4}$	2	1001
$SrTl_2P_4O_{12}$	7.608(5)	7.608(5)	10.25(2)	$I\bar{4}$	2	1001
$SrRb_2P_4O_{12}$	7.585(5)	7.585(5)	10.28(2)	$I\bar{4}$	2	1001
$Pb(NH_4)_2P_4O_{12}$	7.550(2)	7.550(2)	10.350(8)	$I\bar{4}$	2	1003
$PbRb_2P_4O_{12}$	7.567(2)	7.567(2)	10.357(8)	$I\bar{4}$	2	1003
$PbTl_2P_4O_{12}$	7.591(3)	7.591(3)	10.356(8)	$I\bar{4}$	2	1003
$Sr_3Cs_4H_2(P_4O_{12})_3$	15.455(5)	15.455(5)	15.455(5)	$I\bar{4}3d$	4	1007

Table 4.3.6. Main Crystallographic Data for Hydrated Divalent–Monovalent Cation Cyclotetraphosphates

Formula	a (Å) α (°)	b (Å) β (°)	c (Å) γ (°)	S.G.	Z	Reference(s)
$CaNa_2P_4O_{12}\cdot1\frac{1}{2}H_2O$	27.88(10)	7.536(5)	7.378(5)	$Pma2$	4	1008
$SrNa_2P_4O_{12}\cdot6H_2O$	7.332(5)	7.663(5)	14.408(8)	$I2mm$	2	1009
$NiK_2P_4O_{12}\cdot7H_2O$	13.820(3) 98.13(4)	9.640(5) 97.31(4)	7.450(2) 102.85(4)	$P\bar{1}$	2	1010
$Ni(NH_4)_2P_4O_{12}\cdot7H_2O$	13.841(3) 98.05(4)	9.621(5) 97.25(4)	7.482(2) 103.01(4)	$P\bar{1}$	2	1010
$CoK_2P_4O_{12}\cdot5H_2O$	12.955(2)	16.294(2) 92.17(1)	7.432(2)	$P2_1/c$	4	1011
$Ca_4K_4(P_4O_{12})_3\cdot8H_2O$	20.38(1)	12.683(5) 89.31(5)	7.830(2)	$P2_1/a$	2	1008
$Ca(NH_4)_2P_4O_{12}\cdot2H_2O$	16.783(10)	10.888(6) 90.82(8)	7.913(2)	$P2_1/n$	4	535, 1012

compounds was reported by these authors, and a tentative explanation for the reorganization subsequent to the dehydration of the heptahydrate was proposed.

$Ca_4K_4(P_4O_{12})_3\cdot8H_2O$. The chemical preparation and crystal structure of this salt were reported by Averbuch-Pouchot and Durif.[1008]

$Ca(NH_4)_2P_4O_{12}\cdot2H_2O$. This salt was first prepared by Cavéro-Ghersi,[535] who reported its main crystallographic features. Later, its crystal structure was determined by Tordjman *et al.*[1012]

Crystallographic Data. Table 4.3.6 contains the main crystallographic data for hydrated divalent–monovalent cation cyclotetraphosphates.

4.3.2.4. Trivalent Cation Cyclotetraphosphates

All of the six trivalent cation cyclotetraphosphates that have been well characterized to date exhibit the same type of structure (Table 4.3.7). This structure was first investigated by Hendricks and Wyckoff[1013] and then described in 1937 by Pauling and Sherman[29] for

Table 4.3.7. Edge Lengths of the Cubic Unit Cell in Trivalent Cation Cyclotetraphosphates[a]

Formula	a (Å)	Reference(s)
$Al_4(P_4O_{12})_3$	13.730	29, 1013
$Cr_4(P_4O_{12})_3$	13.912	227
$Ti_4(P_4O_{12})_3$	13.82	691
$Fe_4(P_4O_{12})_3$	14.013(7)	684
$Sc_4(P_4O_{12})_3$	14.363(5)	1014–1017
$Yb_4(P_4O_{12})_3$	14.66(1)	720

[a]The common space group is $I\bar{4}3d$; $Z = 4$.

the aluminum salt, $Al_4(P_4O_{12})_3$. The latter study provided the first structural proof for the existence of cyclic phosphoric anions.

$Fe_4(P_4O_{12})_3$ and $Cr_4(P_4O_{12})_3$. During an investigation of the Al_2O_3–P_2O_5 and Fe_2O_3–P_2O_5 systems, d'Yvoire[684] identified $Fe_4(P_4O_{12})_3$ as being isotypic with the aluminium salt. The chromium salt was characterized by Rémy and Boullé[227] during their investigation of the Cr_2O_3–P_2O_5 system.

$Ti_4(P_4O_{12})_3$. The chemical preparation and unit-cell dimensions of the titanium salt were reported by Liebau and Williams.[691]

$Sc_4(P_4O_{12})_3$. Bagieu-Beucher[1014] reported the chemical preparation of $Sc_4(P_4O_{12})_3$ together with crystal data for this salt. The crystal structure was refined by Bagieu-Beucher and Guitel,[1015] after a first unsuccessful attempt by Wappler.[1016] $Sc_4(P_4O_{12})_3$ was reinvestigated by Mezentseva *et al.*[1017] who reported density, crystallographic, optical, and IR data.

$Yb_4(P_4O_{12})_3$. Chudinova[720] described the chemical preparation of this salt and gave an indexed powder X-ray diagram for this salt.

4.3.2.5. Trivalent–Monovalent Cation Cyclotetraphosphates

Since the characterization of stable rare-earth ultraphosphates, LnP_5O_{14}, by Beucher[1018] and the discovery of their application as efficient laser materials, a large number of investigations were performed on Ln_2O_3–P_2O_5 and Ln_2O_3–M_2O–P_2O_5 systems. Most of the new materials characterized during these studies were long-chain polyphosphates with the exception of some $LnMP_4O_{12}$ compounds that we describe in this section.

Crystal data for the cubic and monoclinic compounds described below are presented in Tables 4.3.8 and 4.3.9 respectively.

$PrNH_4P_4O_{12}$ and $NdNH_4P_4O_{12}$. The chemical preparation of monoclinic $PrNH_4P_4O_{12}$ and $NdNH_4P_4O_{12}$ was reported by Masse *et al.*[1019] These authors performed a detailed determination of the atomic arrangement in the praseodymium salt and showed that the two compounds are isotypic.

$CeNH_4P_4O_{12}$. This cyclotetraphosphate is dimorphous. One of its forms is cubic, while the other one is monoclinic and isotypic with $PrNH_4P_4O_{12}$.

Vaivada and Konstant[764] reported the chemical preparation of $CeNH_4P_4O_{12}$ and studied its thermal behavior by DTA and X-ray diffraction. At 873 K, this compound decomposes according to the following scheme:

Table 4.3.8. Unit-Cell Dimensions of Cubic Trivalent–Monovalent Cyclotetraphosphates[a]

Formula	a (Å)	Reference(s)
$CeNH_4P_4O_{12}$	15.23(1)	763, 1020
$NdCsP_4O_{12}$	15.223(3)	1024
$NdRbP_4O_{12}$	15.241(7)	1022

[a]The common space group is $I\overline{4}3d$; $Z = 12$.

Table 4.3.9. Unit-Cell Dimensions of Monoclinic
Trivalent–Monovalent Cyclotetraphosphates[a]

Formula	a (Å)	b (Å) β (°)	c (Å)	Reference(s)
$PrNH_4P_4O_{12}$	7.916(5)	12.647(10) 110.34(8)	10.672(9)	1019
$NdNH_4P_4O_{12}$	7.881(8)	12.55(1) 110.80(10)	10.65(1)	1019
$CeNH_4P_4O_{12}$	7.930(3)	12.634(5) 110.05(3)	10.699(5)	763
$NdKP_4O_{12}$	7.888	12.447 112.70	10.770	708, 750
$HoKP_4O_{12}$	7.798(1)	12.310(1) 112.63(1)	10.511(1)	747
$EuKP_4O_{12}$	7.76	12.03 112.5	10.96	1020
$NdRbP_4O_{12}$	7.845(2)	12.691(3) 112.34(1)	10.688(3)	1021
$SmRbP_4O_{12}$	7.868(2)	12.735(3) 111.25(2)	10.589	1023

[a]The common group is $C2/c$; $Z = 4$.

$$2CeNH_4P_4O_{12} \rightarrow Ce(PO_3)_3 + CeP_5O_{14} + 2NH_3 + H_2O$$

The same authors[704] investigated the reaction between CeO_2 and $NH_4H_2PO_4$ between 293 and 1173 K. For initial P/Ce ratios of > 5, mainly $CeNH_4P_4O_{12}$, $Ce(NH_4)_2(PO_3)_5$ and CeP_5O_{14} are obtained. At higher CeO_2 concentrations the most stable phases are CeP_2O_7 and $Ce(PO_3)_3$. Rzaigui and Ariguib[763] described the chemical preparation of the two crystalline forms of $CeNH_4P_4O_{12}$ and reported their crystal data. Rzaigui et al.[1020] reported an accurate structure determination for the atomic arrangement of the cubic form, showing its structural analogy with $Al_4(P_4O_{12})_3$.

$LaNH_4P_4O_{12}$. This cyclotetraphosphate was characterized by Chudinova et al.[700] as an intermediate compound during a study of the preparation of lanthanum polyphosphate using $NH_4H_2PO_4$ and lanthanum oxide or lanthanum oxalate as starting materials. The authors did not report any crystal data for this salt.

$NdKP_4O_{12}$, $HoKP_4O_{12}$ and $EuKP_4O_{12}$. Tarasenkova et al.[708] characterized $NdKP_4O_{12}$ during a study of the Nd_2O_3–K_2O–P_2O_5–H_2O system between 573 and 773 K. They reported unit-cell dimensions and morphology, showing this salt to be isotypic of this salt with $PrNH_4P_4O_{12}$. Litvin et al.[750] described the synthesis, crystal structure and properties of single crystals of $NdKP_4O_{12}$. The crystal structure of $HoKP_4O_{12}$ was established by Palkina et al.,[747] and the synthesis of $EuKP_4O_{12}$ was described by Chudinova et al.[755] Both compounds are isotypic with $PrNH_4P_4O_{12}$.

$NdRbP_4O_{12}$. Like $CeNH_4P_4O_{12}$, this salt is dimorphous. The crystal structure of the monoclinic form was refined by Koizumi and Nakano.[1021] These authors described the procedure used for crystal growth. The cubic form was prepared by Dorokhova *et al.*[1022] who reported a detailed determination of the atomic arrangement.

$SmRbP_4O_{12}$. Palkina *et al.*[1023] described this salt and analyzed the structural differences in some isotypic compounds.

Other Investigations of $LnRbP_4O_{12}$ Compounds

Byrappa and Litvin[770] investigated the $Rb_2O-Ln_2O_3-P_2O_5-H_2O$ systems within the temperature range 573–1073 K. A composition diagram showing fields of crystallization for the different phases was given. The crystal chemistry of both $LnRbP_4O_{12}$ and $LnRb(PO_3)_4$ compounds was discussed.

$NdCsP_4O_{12}$. The cubic atomic arrangement of this salt was determined by Palkina *et al.*[1024]

Other Investigations of $LnCsP_4O_{12}$ Compounds

Crystal growth, crystal structures, and IR spectra of $LnCs(PO_3)_4$ and $LnCsP_4O_{12}$ compounds were discussed by Byrappa *et al.*[780]

$InNaP_4O_{12}$. During an investigation of the $P_2O_5-Na_2O-In_2O_3-H_2O$ system, in the temperature range 423–773 K, Avaliani *et al.*[251] characterized a new diphosphate, $InNa(H_2P_2O_7)_2$, and identified a new cyclotetraphosphate, $InNaP_4O_{12}$, during an investigation of its thermal behavior:

$$InNa(H_2P_2O_7)_2 \xrightarrow{588-678 \text{ K}} InNaP_4O_{12} \xrightarrow{973-1023 \text{ K}} In(PO_3)_3 + \text{melt}$$

$ErKP_4O_{12}·6H_2O$. Palkina *et al.*[1025] prepared several $LnMP_4O_{12}·nH_2O$ compounds by reaction of the monovalent cation cyclotetraphosphate with the rare-earth nitrate. The erbium–potassium cyclotetraphosphate hexahydrate was investigated in more details by these authors who reported a complete description of the atomic arrangement of this monoclinic $C2/c$ compound. The dimensions of its tetramolecular unit cell are:

$$a = 8.643(2), \quad b = 12.015(8), \quad c = 14.909(5) \text{ Å}, \quad \beta = 90.65(5)°$$

4.3.2.6. Tetravalent Cation Cyclotetraphosphates

Only two tetravalent cation cyclotetraphosphates are known: $SnP_4O_{12}·4H_2O$ and UP_4O_{12}.

$SnP_4O_{12}·4H_2O$. The chemical preparation of $SnP_4O_{12}·4H_2O$ was described by Nariai *et al.*[1026] These authors investigated the effect of water vapor on the thermal decomposition of this salt.

UP_4O_{12}. Linde *et al.*[1027] proposed a crystal structure for this compound, whose unit-cell dimensions are reported below:

$$a = 8.145(3), \quad b = 8.272(3), \quad c = 8.653(3) \text{ Å}$$

$$\alpha = 117.83(3), \quad \beta = 117.24(3), \quad \gamma = 91.02(3)°$$

The space group is $P\bar{1}$ with $Z = 2$.

4.3.2.7. Cyclotetraphosphate Adducts with Telluric Acid

Results in this field are relatively sparse by comparison with those obtained during the systematic investigation of the cyclotriphosphate–telluric acid adducts that we described in Section 4.2.2.6. Only three adducts have been clearly characterized. Their atomic arrangements were determined accurately and show that, as in all the other classes of adducts between telluric acid and phosphates, the $Te(OH)_6$ octahedra and the phosphoric groups coexist as independent units in the atomic arrangements.

The main crystallographic features for these three adducts are given in Table 4.3.10.

$(NH_4)_4P_4O_{12}\cdot 2Te(OH)_6\cdot 2H_2O$. This adduct was described by Durif et al.[1028]

$K_4P_4O_{12}\cdot Te(OH)_6\cdot 2H_2O$ and $Rb_4P_4O_{12}\cdot Te(OH)_6\cdot 2H_2O$. These two adducts, reported by Averbuch-Pouchot and Durif[1029,1030] are isotypic. The crystal structure was determined with the potassium salt.

Table 4.3.10. Main Crystallographic Data for Adducts between Alkali Cyclotetraphosphates and Telluric Acid

Formula	a (Å) α (°)	b (Å) β (°)	c (Å) γ (°)	S.G.	Z	Reference(s)
$(NH_4)_4P_4O_{12}\cdot 2Te(OH)_6\cdot 2H_2O$	11.845(6) 66.28(5)	8.554(5) 95.91(5)	7.433(5) 76.00(5)	$P\bar{1}$	1	1028
$K_4P_4O_{12}\cdot Te(OH)_6\cdot 2H_2O$	9.731(5)	11.43(1) 99.45(5)	17.16(1)	$C2/c$	4	1029, 1030
$Rb_4P_4O_{12}\cdot Te(OH)_6\cdot 2H_2O$	10.049(3)	11.73(1) 99.16(5)	17.34(1)	$C2/c$	4	1030

4.3.2.8. Organic and Metal-Organic Cyclotetraphosphates

Recently, a number of organic and metal–organic cation cyclotetraphosphates were characterized. Two rather similar methods were used for their preparation:

1. Action of $H_4P_4O_{12}$ acid, obtained from $Na_4P_4O_{12}\cdot 4H_2O$ by the use of an ion-exchange resin, on the appropriate aminocompound
2. Addition of the proper quantity of P_4O_{10} to an iced aqueous solution of the amine

In both cases, the reaction is of the same type:

$$4R-NH_2 + H_4P_4O_{12} \rightarrow (R-NH_3)_4P_4O_{12}$$

The first process has been mainly used by a group in Tunisia, and the second one by the French investigators. Most of the compounds that we report below are stable for years under normal conditions. In all cases, the atomic arrangements were determined very accurately.

Glycinium cyclotetraphosphate, $(NH_3CH_2COOH)_4P_4O_{12}$, and ethanolammonium cyclotetraphosphate, $[NH_3(CH_2)_2OH]_4P_4O_{12}$, were described by Averbuch-Pouchot *et al.*[1031,1032]

Investigations of $H_4P_4O_{12}$–$NH_2(CH_2)_2NH_2$–MO–H_2O systems have recently been performed. In a survey of these studies, Averbuch-Pouchot *et al.*[1033] showed that the following compounds characterized during these investigations belong to four new structure types:

- $(C_2N_2H_{10})_2P_4O_{12}\cdot H_2O$ described by Averbuch-Pouchot *et al.*[1034]
- A series of seven isomorphous compounds of general formula $M(C_2N_2H_{10})_3$-$(P_4O_{12})_2\cdot 14H_2O$, with M = Cu, Zn, Mg, Cd, Co, Ni, and Mn, described by Averbuch-Pouchot and Durif[1035]
- $Ca(C_2N_2H_{10})P_4O_{12}\cdot 15/2H_2O$, investigated by Averbuch-Pouchot et al.[1036]
- $Sr(C_2N_2H_{10})P_4O_{12}\cdot 5H_2O$, prepared and investigated by Bagieu-Beucher *et al.*[1037]

Some ethylenediammonium–alkali cyclotetraphosphates have also been described:

- Two isomorphous compounds, $Na_2(C_2N_2H_{10})P_4O_{12}\cdot 2H_2O$ and $K_2(C_2N_2H_{10})$-$P_4O_{12}\cdot 2H_2O$ prepared by Bdiri and Jouini.[1038] This structure type was investigated by Jouini[1039] using the sodium salt.
- $NaLi(C_2N_2H_{10})P_4O_{12}\cdot 3H_2O$, synthesized and investigated by Bdiri and Jouini.[1040]

Lastly, 1–3 diammoniumpropane cyclotetraphosphate, $(C_3N_2H_{12})_2P_4O_{12}\cdot 2H_2O$,[1041] and diethylenetriammonium cyclotetraphosphate, $(C_4N_3H_{15})_2P_4O_{12}\cdot 4H_2O$,[1042] were described by Bdiri and Jouini.

The main crystallographic data for the organic cation and metal–organic cation cyclotetraphosphates are presented in Tables 4.3.11 and 4.3.12 respectively.

Table 4.3.11. Main Crystallographic Data for Organic Cation Cyclotetraphosphates

Formula	a (Å) α (°)	b (Å) β (°)	c (Å) γ (°)	S.G.	Z	Reference(s)
$(Gly)_4P_4O_{12}$	7.988(5) 111.64(5)	8.449(5) 105.27(5)	9.739(5) 99.40(5)	$P\bar{1}$	1	1031
$(Gua)_4P_4O_{12}\cdot 4H_2O$	9.634(1)	18.112(8) 103.86(5)	7.292(3)	$P2_1/a$	2	920
$(EA)_4P_4O_{12}$	10.746(8)	10.746(8)	10.071(8)	$I\bar{4}$	2	1032
$(EDA)_2P_4O_{12}\cdot H_2O$	13.168(9)	8.599(6)	15.152(10)	$Pcca$	4	1034
$(DAP)_2P_4O_{12}\cdot 2H_2O$	14.627(1)	13.391(1)	10.068(1)	$Pbca$	4	1041
$(DETA)_2P_4O_{12}\cdot 4H_2O$	7.966(1) 116.1(1)	22.830(2)	7.708(1)	$P2_1/n$	2	1041, 1042

[a]Abbreviations: Gly, glycinium, $(NH_3–CH_2–COOH)^+$; Gua, guanidinium, $[C(NH_2)_3]^+$; EA, ethanolammonium, $[NH_3–(CH_2)_2–OH]^+$; EDA, ethylenediammonium, $[NH_3–(CH_2)_2–NH_3]^{2+}$; DAP, 1,3-diammoniumpropane, $[NH_3–(CH_2)_3–NH_3]^{2+}$; DETA, diethylenetriammonium, $[NH_3–(CH_2)_2–NH–(CH_2)_2–NH_3]^{2+}$.

Table 4.3.12. Main Crystallographic Data for Metal–Organic Cation
Cyclotetraphosphates

Formula[a]	a (Å) α (°)	b (Å) β (°)	c (Å) γ (°)	S.G.	Z	Reference
$Cu(EDA)_3(P_4O_{12})_2 \cdot 14H_2O$	13.162(8)	13.301(8) 106.69(1)	12.308(8)	$P2_1/n$	2	1035
$Zn(EDA)_3(P_4O_{12})_2 \cdot 14H_2O$	12.902(8)	13.187(8) 106.04(1)	12.303(8)	$P2_1/n$	2	1035
$Mg(EDA)_3(P_4O_{12})_2 \cdot 14H_2O$	13.101(8)	13.292(8) 108.20(1)	12.465(8)	$P2_1/n$	2	1035
$Cd(EDA)_3(P_4O_{12})_2 \cdot 14H_2O$	13.193(8)	13.309(8) 107.13(1)	12.466(8)	$P2_1/n$	2	1035
$Co(EDA)_3(P_4O_{12})_2 \cdot 14H_2O$	13.146(8)	13.296(8) 106.77(1)	12.258(8)	$P2_1/n$	2	1035
$Ni(EDA)_3(P_4O_{12})_2 \cdot 14H_2O$	12.916(8)	13.189(8) 106.94(1)	12.363(8)	$P2_1/n$	2	1035
$Mn(EDA)_3(P_4O_{12})_2 \cdot 14H_2O$	13.098(8)	13.278(8) 107.15(1)	12.424(8)	$P2_1/n$	2	1035
$Ca(EDA)P_4O_{12} \cdot \frac{15}{2}H_2O$	14.611(3)	18.709(3)	7.81(2)	$Pnma$	4	1036
$Sr(EDA)P_4O_{12} \cdot 5H_2O$	17.863(15)	15.317(13)	13.109(10)	$Pbca$	8	1037
$Na_2(EDA)P_4O_{12} \cdot 2H_2O$	7.797(1)	14.657(1) 91.39(1)	12.916(1)	$C2/c$	4	1038, 1039
$K_2(EDA)P_4O_{12} \cdot 2H_2O$	8.172(1)	14.690(1) 91.19(1)	13.421(1)	$C2/c$	4	1038
$NaLi(EDA)P_4O_{12} \cdot 3H_2O$	13.135(3)	7.737(2)	15.478(4)	$Pcca$	4	1040

[a]EDA, ethylenediammonium, $[NH_3–(CH_2)_2–NH_3]^{2+}$.

4.3.3. Chemical Preparation of Cyclotetraphosphates

We report here the most common processes used for the preparation of cyclotetra-phosphates.

In relation to what we described for the chemical preparation of cyclotriphosphates two differences should be noted:

1. As in the case of cyclotriphosphates and other cyclophosphates, the most important starting material for syntheses is the sodium salt. However in the case of cyclotetra-phosphates, this compound cannot be prepared by thermal condensation of NaH_2PO_4 through the hypothetical reaction scheme

$$4NaH_2PO_4 \rightarrow Na_4P_4O_{12} + 4H_2O$$

2. Boullé's metathesis reaction using the silver salt as starting material has almost never been used for the production of cyclotetraphosphates.

As for the cyclotriphosphates, the most used starting material for the preparation of cyclotetraphosphates is the sodium salt, here $Na_4P_4O_{12} \cdot 4H_2O$. For a long time, the only cyclotetraphosphates that could be prepared in a reproducible way were the members of the $M_2P_4O_{12}$ series (M = Cu, Mg, Co, etc.) They are simply prepared by a thermal method, leading to well-crystallized samples. A mixture of a salt of the required divalent cation and a large excess of phosphoric acid is heated for approximately one day at a temperature of 627–733 K. After the appearance of a crystalline mass inside this bath, the excess of phosphoric flux is removed by hot water, and the crystalline $M_2P_4O_{12}$ thus obtained is separated by filtration. One of these compounds, $Cu_2P_4O_{12}$, was extensively used up to the end of the 1950s to prepare the sodium salt through the process we describe below.

4.3.3.1. Preparation of $Na_4P_4O_{12} \cdot 4H_2O$ by an Exchange Reaction

Finely divided $Cu_2P_4O_{12}$ is added to an aqueous solution of Na_2S. The mixture is mechanically stirred until the following reaction is complete:

$$Cu_2P_4O_{12} + 2Na_2S \rightarrow Na_4P_4O_{12} + 2CuS$$

The temperature of the mixture must be kept close to 273 K during the reaction to prevent the hydrolysis of the P_4O_{12} groups. Then, the insoluble copper sulfide is removed by filtration, and the resulting solution is evaporated at room temperature or alcohol or sodium chloride is added to precipitate $Na_4P_4O_{12} \cdot 4H_2O$.

4.3.3.2. Preparation of $H_4P_4O_{12}$ by Hydrolysis of P_4O_{10}

When added to ice water, P_4O_{10} decomposes to cyclotetraphosphoric acid according to the following equation:

$$P_4O_{10} + 2H_2O \rightarrow H_4P_4O_{12}$$

Very good yields of cyclotetraphosphoric acid. (>75%) can be obtained.

If a stoichiometric amount of P_4O_{10} is added to an iced aqueous solution of Na_2CO_3 or $NaHCO_3$, one directly obtains a solution of sodium cyclotetraphosphate. Yields are also very good.

Most of the recently characterized alkali or mixed-alkali cyclotetraphosphates and a number of organic cation derivatives were prepared by using this process.

4.3.3.3. Thermal Processes

Many anhydrous cyclotetraphosphates were prepared by heating, at relatively low temperatures (623–723 K), a stoichiometric mixture of the appropriate starting materials. $SrK_2P_4O_{12}$, for instance, can be obtained easily according to the following reaction:

$$SrCO_3 + 4(NH_4)_2HPO_4 + K_2CO_3 \rightarrow SrK_2P_4O_{12} + 2CO_2 + 8NH_3 + 6H_2O$$

Most of the $M^{II}K_2P_4O_{12}$ compounds have been simply prepared as polycrystalline samples by using this process.

4.3.3.4. Flux Methods

Most of the cyclotetraphosphates characterized during the elaboration of the M^IPO_3–$M^{II}(PO_3)_2$ phase-equilibrium diagrams and, in general, all insoluble cyclotetraphosphates were synthesized as single crystals by using flux methods similar to those described for the preparation of cyclotriphosphates. As mentioned before, the main feature of this type of process is the use of a large excess of a phosphoric flux. A typical run of such a process for the synthesis of $SrNa_2P_4O_{12}$ is described here. In 8.6 cm^3 of H_3PO_4 (85%) are added

0.5 g of $SrCO_3$ and 3.5 g of Na_2CO_3. The resulting mixture is then heated at 623 K for one day. After removal of the excess of phosphoric flux by hot water, a crystalline product containing more than 80% of $SrNa_2P_4O_{12}$ is gathered.

4.3.3.5. Classical Methods

As in the case of cyclotriphosphates, classical methods of aqueous chemistry were frequently used in the preparation of cyclotetraphosphates. $Sr(NH_4)_2P_4O_{12}$, for instance, is easily prepared by mixing 0.1N aqueous solutions of $(NH_4)_4P_4O_{12}$ and $SrCl_2$. The same process has been used for the preparation of most of the cyclotetraphosphates listed in Table 4.3.6.

4.3.3.6. Ion-Exchange Resins

Cyclotetraphosphoric acid produced by passing an aqueous solution of $Na_4P_4O_{12}$ through an ion-exchange column is commonly used in the preparation of cyclotetraphosphates. Calcium and strontium cyclotetraphosphates have, for instance, been synthesized by adding cyclotetraphosphoric acid prepared in this manner to aqueous solutions of the corresponding nitrates. Usually, the acid produced is added immediately to an appropriate mixture of carbonates or hydroxides.

4.3.3.7. Hydrothermal Methods

The use of hydrothermal methods is not common in the field of condensed phosphates, but they have sometimes been employed. $Fe_2P_4O_{12}$, for instance, has been prepared under hydrothermal conditions at 703 K and 1000 atm.

4.3.3.8. Unconventional Methods

Some have reported specific methods for the preparation of particular compounds. These methods have been described in Section 4.3.2.

4.3.3.9. Organic Cation Cyclotetraphosphates

The procedures employed for the chemical preparation of organic cation cyclotetraphosphates are not fundamentally different from those described above. Most organic cation cyclotetraphosphates have been prepared by the action of $H_4P_4O_{12}$ on the corresponding amine or its carbonate. Schematically, the reaction is

$$4 \, R–NH_2 + H_4P_4O_{12} \rightarrow (R–NH_3)_4P_4O_{12}$$

In addition, some success has been achieved in attempts to use the poorly characterized precipitated silver cyclotetraphosphate in order to generalize Boullé's process to the preparation of organic cation cyclotetraphosphates. In one of the procedures described for the synthesis of guanidinium cyclotetraphosphate tetrahydrate, this silver salt was used assuming its chemical formula to be that of a tetrahydrate.

4.3.4. Atomic Arrangements of Some Cyclotetraphosphates

4.3.4.1. Lithium Cyclotetraphosphate Pentahydrate

$$Li_4P_4O_{12}\cdot 5H_2O,^{935} \text{ monoclinic, } C2/c, Z = 8$$

$$a = 17.073(6), \quad b = 17.029(6), \quad c = 13.554(6) \text{ Å}, \quad \beta = 127.32(1) \,°$$

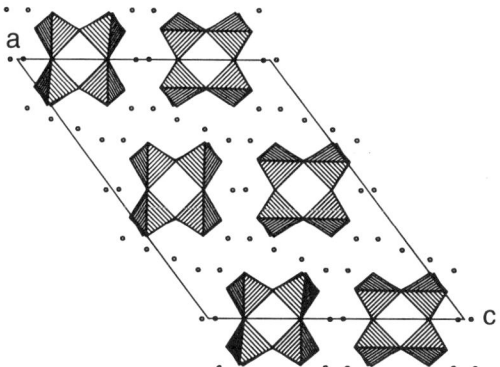

Figure 4.3.4. Projection, along the **b** axis, of the atomic arrangement in $Li_4P_4O_{12} \cdot 5H_2O$. The two independent P_4O_{12} anions are superimposed in this projection. The open circles represent the lithium atoms. The water molecules have been omitted.

Two crystallographically independent P_4O_{12} ring anions coexist in this large unit cell; both have twofold internal symmetry. These two rings have, in fact, a strong pseudo-fourfold symmetry. Figure 4.3.4 shows a projection along the **b** axis of part of this arrangement, showing the respective locations of the P_4O_{12} rings and Li atoms.

Three of the four independent Li atoms have tetrahedral coordination, and the fourth one has five neighbors, forming a square pyramid. These LiO_4 and LiO_5 polyhedra link themselves so as to form large rings located around the twofold axes.The global formula of such a ring corresponds to $Li_8O_{16}(H_2O)_6$. LiO_5 polyhedra share common edges with each of the two adjacent LiO_4 tetrahedra while two adjacent LiO_4 tetrahedra share a corner. Within such a ring, the Li–O distances range from 1.857 to 2.233 Å, and the Li–H_2O distances from 1.964 to 2.025 Å.

Two of the five water molecules are not bonded to the associated cations. They have large thermal factors, and, as in many similar cases, it is difficult to decide whether the magnitude of these factors is due to the mobility of this type of water or to partly occupied crystallographic sites.

Thus, the atomic arrangement of this cyclotetraphosphate may be schematically represented as a stacking of alternate P_4O_{12} and $Li_8O_{16}(H_2O)_6$ rings along the twofold axes. Another way to consider this arrangement is given by an additional projection, along the **a** direction (Fig. 4.3.5), showing that it can also be described as alternating layers of ring anions and of lithium atoms perpendicular to the **b** direction.

In addition, it should be noted that the five water molecules build pentagonal rings around the **c** axis. In this type of ring, the average O(W)–O(W)–O(W) angle is 107.6°, and thus these rings are flat within the experimental error. The O(W)–O(W) distances in the ring range from 2.693 to 2.989 Å. Figure 4.3.6 shows the distribution of these rings of water molecules inside the unit cell.

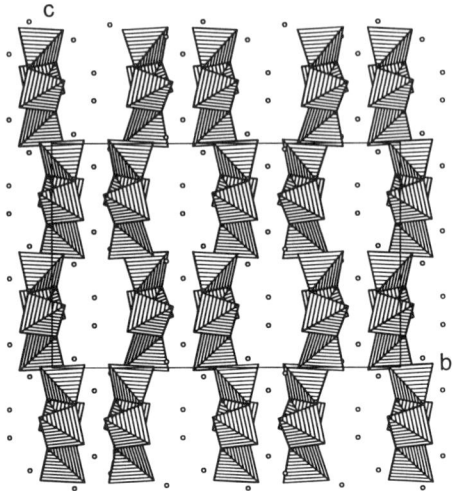

Figure 4.3.5. Projection, along the *a* direction, of the atomic arrangement in $Li_4P_4O_{12}·5H_2O$. As in Fig.4.3.4., the water molecules have been omitted.

4.3.4.2. *Anhydrous Sodium Cyclotetraphosphate and Monohydrate*

$Na_4P_4O_{12}·H_2O$,[936] orthorhombic, $P2_12_12_1$, $Z = 4$

$$a = 13.654(2), \quad b = 13.475(3), \quad c = 6.291(3) \text{ Å}$$

Figure 4.3.7 is a projection, along the **c** axis, of the atomic arrangement. Within a range of 3 Å, the four independent sodium atoms have six oxygen neighbors in a distorted octahedral arrangement, with Na–O distances ranging from 2.295 to 2.774 Å. The NaO_6 polyhedra establish the cohesion between the P_4O_{12} groups. The latter groups, despite having no internal symmetry, are very regular with, according to Wiench and Jansen, a very strong pseudo-D_{2d} conformation. Their mean planes are almost perpendicular to the **c** axis. The crystal structure of the anhydrous salt is similar.

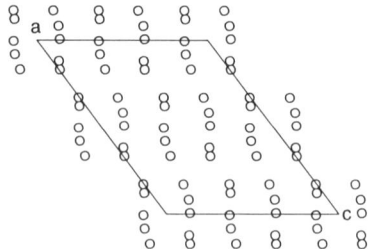

Figure 4.3.6. A projection of the $Li_4P_4O_{12}·5H_2O$ structure, along the *b* direction, showing only the water molecules and providing a profile view of the $(H_2O)_5$ rings.

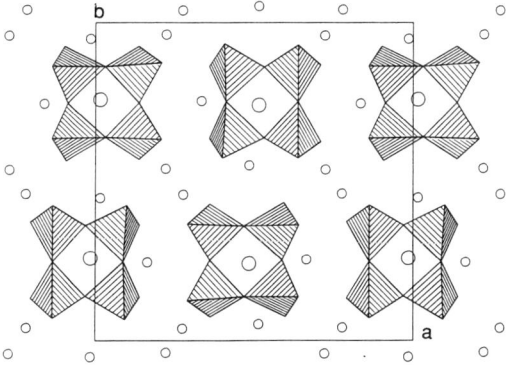

Figure 4.3.7. Projection of the atomic arrangement of $Na_4P_4O_{12} \cdot H_2O$ along the c axis. The small open circles represent the sodium atoms, and the large ones represent the water molecules.

4.3.4.3. Potassium Cyclotetraphosphate Tetrahydrate

$$K_4P_4O_{12} \cdot 4H_2O,^{950} \text{ tetragonal, } I\bar{4}, Z = 2$$

$$a = 9.061(3), \quad c = 10.284(5) \text{ Å}$$

This atomic arrangement can be described as the succession of two types of layers alternating perpendicular to the fourfold axis. The P_4O_{12} groups are located around the $\bar{4}$ axis and form layers centered by the planes $z = 0$ and $1/2$ as shown in Fig. 4.3.8, a projection along the b axis. The other components of this arrangement, the potassium atoms and the water molecules, form a second type of layers centered by the planes $z = (2n+1)/8$. Each phosphoric layer is sandwiched between two layers of the second type. Figure 4.3.9, a projection along the c axis, shows the respective locations of P_4O_{12} ring anions, potassium atoms, and water molecules in a portion of the arrangement including a phosphoric layer

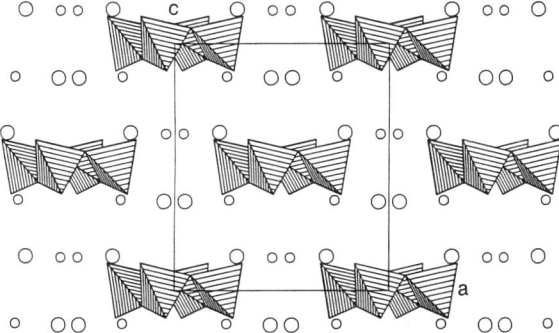

Figure 4.3.8. A projection, along the b direction, of the atomic arrangement in $K_4P_4O_{12} \cdot 4H_2O$ showing the succession of the two kinds of layers perpendicular to the c axis. The smaller circles represent the potassium atoms, and larger ones the water molecules. The hydrogen atoms have been omitted.

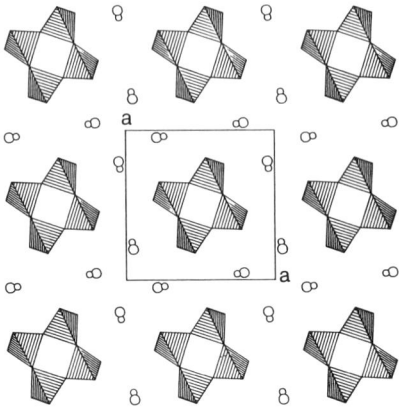

Figure 4.3.9. A projection, along the *c* direction, of a portion of the K₄P₄O₁₂·4H₂O atomic arrangement, (0.33 < z < 0.66), showing a layer of phosphoric ring anions and of its two adjacent layers of potassium atoms and water molecules. The representation of the atoms is the same as in Fig. 4.3.8.

and its two adjacent layers of potassium atoms and water molecules. The potassium atom in general position has, within a range of 3.50 Å, sevenfold coordination, involving four oxygen atoms and three water molecules with K–O distances ranging from 2.698 to 3.146 Å. A strong net of hydrogen bonds interconnects the water molecules with some external oxygen atoms of the P_4O_{12} ring, with $H_2O\cdots O$ distances of 2.764 and 2.835 Å and O–H$\cdots$O angles of 162 and 169°.

4.3.4.4. Ammonium Cyclotetraphosphate

$$(NH_4)_4P_4O_{12},^{952-956} \text{ orthorhombic, } Cmca, Z = 4$$

$$a = 10.433(2), \quad b = 10.871(3), \quad c = 12.588(3) \text{ Å}$$

Figure 4.3.10. Projection of the atomic arrangement in $(NH_4)_4P_4O_{12}$ along the *b* axis. The open circles represent the ammonium groups.

Figure 4.3.10 shows the projection of the atomic arrangement along the **b** axis. The ring anion has $2/m$ symmetry; two of its phosphorus atoms are located in the mirror plane, and the other two lie on a twofold axis. These ring anions have their centers in the planes $y = 0$ and $y = 1/2$ and form, in these planes, a pseudo-square-centered arrangement. The ammonium groups are located approximately in the planes $y = 1/4$ and $3/4$; eight of them are located in the mirror planes, and the other eight on the twofold axes. Thus, the atomic arrangement may be considered as formed by NH_4 and P_4O_{12} layers alternating perpendicular to the **b** direction.

One of the NH_4 groups is coordinated to eight oxygen atoms, and the other one to six oxygen atoms. Within these two polyhedra, the N–O distances range from 2.79 to 3.38 Å. A network of hydrogen bonds connects the ammonium groups to external oxygen atoms of the P_4O_{12} ring, with N–O distances ranging from 2.79 to 3.09 Å and N–H···O angles ranging from 147 to 172°.

4.3.4.5. Cesium Cyclotetraphosphate Tetrahydrate

$$Cs_4P_4O_{12}\cdot4H_2O,^{958} \text{ tetragonal, } P4_1, Z= 4$$

$$a = 9.466(5), \quad c = 21.933(9) \text{ Å}$$

The atomic arrangement can be easily described as a succession of layers perpendicular to the **c** axis and separated by a distance of $c/12$. The P_4O_{12} ring anions are in layers centered by the planes $z = (3n+2)/12$. Mixed layers containing both cesium atoms and water molecules correspond to $z = 3n/12$ and $(3n+1)/12$. Figure 4.3.11 shows a projection of this

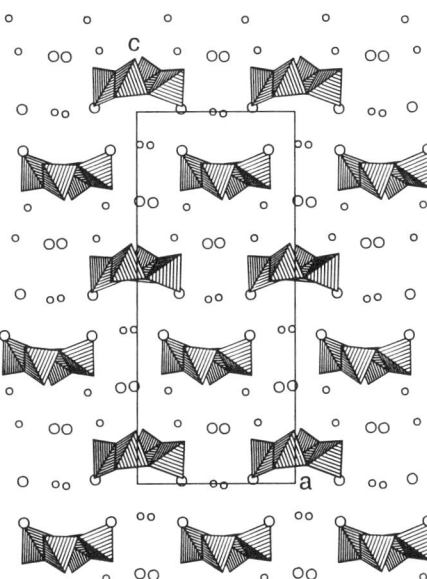

Figure 4.3.11. Projection, along the **b** direction, of the atomic arrangement in $Cs_4P_4O_{12}\cdot4H_2O$. The smaller circles represent the cesium atoms, and the larger ones the water molecules.

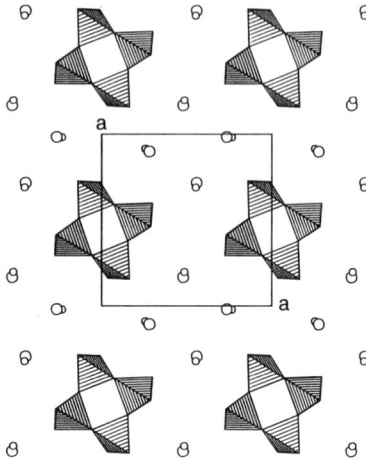

Figure 4.3.12. A projection, along the c direction, of a portion of the $Cs_4P_4O_{12}\cdot4H_2O$ atomic arrangement, $(0.05 < z < 0.31)$, showing a layer of phosphoric ring anions and the two adjacent layers of cesium atoms and water molecules. The representation of the atoms is the same as in Fig. 4.3.11.

arrangement along the **b** direction. As in the case of $K_4P_4O_{12}\cdot4H_2O$, each phosphoric layer is sandwiched between two layers of cesium atoms and water molecules. Figure 4.3.12, a projection along the **c** axis, provides a view of the respective locations of P_4O_{12} groups, cesium atoms, and water molecules in a portion of the arrangement including a layer of phosphoric groups and its two adjacent (Cs + H_2O) layers.

Cesium coordination polyhedra located in layers separated by $c/12$ share oxygen atoms whereas only water molecules are shared between two layers separated by $c/6$. Within a range of 3.50 Å, three of the four independent cesium atoms have a eightfold coordination, and the fourth one has nine neighbors. Within these polyhedra, the Cs–O distances range between 3.110 and 3.480 Å. The phosphoric ring anion has no internal symmetry.

The corresponding rubidium salt, $Rb_4P_4O_{12}\cdot4H_2O$,[958] is isotypic.

4.3.4.6. Thallium Cyclotetraphosphate

$$Tl_4P_4O_{12},^{[962,963]} \text{ tetragonal, } P\bar{4}2_1c, Z = 2$$

$$a = 7.635(5), \quad c = 11.087(7) \text{ Å}$$

The ring anions, being centered around the $\bar{4}$ axes, in 1/2,1/2,0 and 0,0,1/2, have $\bar{4}$ internal symmetry and are built by only one crystallographically independent PO_4 tetrahedron. The P_4O_{12} anions form layers extending through the crystal structure, in the planes $z = 0$ and $z = 1/2$. The two independent thallium atoms are both located on binary axes and, within a range of 3.5 Å, have eight and tenfold coordination, respectively, with Tl–O distances ranging from 2.70 to 3.44 Å. Figure 4.3.13 shows a projection of this atomic arrangement along the **c** axis.

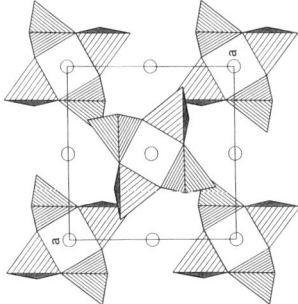

Figure 4.3.13. Projection of the atomic arrangement in $Tl_4P_4O_{12}$ along the *c* axis. The open circles represent the thallium atoms.

4.3.4.7. Sodium–Potassium Cyclotetraphosphate Dihydrate (Tetragonal)

$$Na_2K_2P_4O_{12} \cdot 2H_2O,^{942} \text{ tetragonal, } P4_1, Z = 4$$

$$a = 7.928(5), \quad c = 21.66(2) \text{ Å}$$

In this arrangement, the P_4O_{12} ring anion has no internal symmetry and is perpendicular to the fourfold axis. Layers of KO_8 and NaO_6 polyhedra and water molecules alternate with layers of the phosphoric groups perpendicular to the **c** axis. In fact, here three types of layers be distinguished. Layers of the first type contain the P_4O_{12} groups and are sandwiched as in some previous examples by two layers containing the associated cations

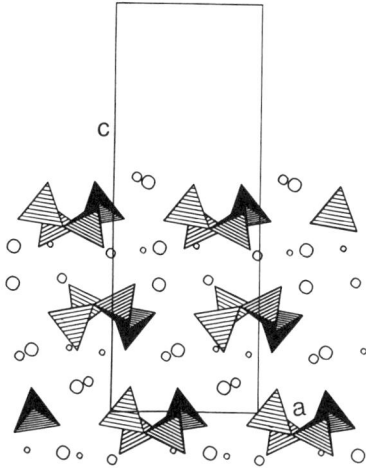

Figure 4.3.14. Projection, along the *a* direction, of the $Na_2K_2P_4O_{12} \cdot 4H_2O$ atomic arrangement. The largest circles represent the water molecules, the intermediate ones the potassium atoms, and the smallest ones the sodium atoms.

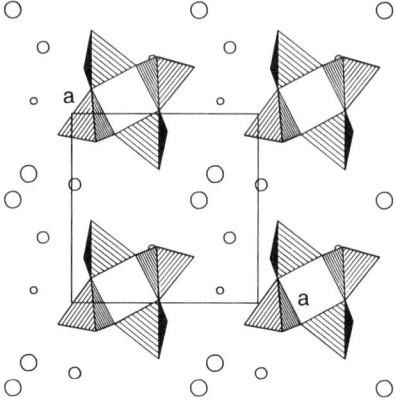

Figure 4.3.15. A projection, along the *c* direction, of a portion of the $Na_2K_2P_4O_{12} \cdot 2H_2O$ atomic arrangement, ($-0.12 < z < 0.12$), showing a layer of phosphoric ring anions and of the two adjacent layers of potassium and sodium atoms and water molecules. The representation of the atoms is the same as in Fig. 4.3.14.

and the water molecules. Here, one of these layers contain potassium atoms and water molecules, whereas the second one contains sodium atoms, potassium atoms, and water molecules. Figure 4.3.14, a projection along the **a** direction, shows this succession of layers. Within the sodium coordination polyhedra, the Na–O distances range from 2.384 to 2.573 Å whereas in the KO_8 polyhedra the K–O distances range from 2.734 to 3.147 Å. Figure 4.3.15, a projection along the **c** direction, shows the respective locations of the components for a layer of P_4O_{12} groups and its two adjacent layers of water molecules and associated cations.

4.3.4.8. Zinc Cyclotetraphosphate Octahydrate

$$Zn_2P_4O_{12} \cdot 8H_2O,^{986} \text{ triclinic, } P\bar{1}, Z = 1$$

$$a = 8.610(5), \quad b = 7.137(5), \quad c = 7.108(5) \text{ Å}$$

$$\alpha = 96.09(5), \quad \beta = 105.99(5), \quad \gamma = 100.49(5)°$$

A projection of the atomic arrangement of this salt, along the **a** axis, is given in Fig. 4.3.16. As shown by this figure, the structure can be described as layers of independent ZnO_6 octahedra spreading in the (a, b) planes, forming a two-dimensional face-centered network. These layers are themselves linked by the P_4O_{12} groups. The P_4O_{12} ring anions and the two kinds of ZnO_6 octahedra are centrosymmetric. One of the zinc polyhedra is built up by four oxygen atoms and two water molecules, while the second one is built up by four water molecules and two oxygen atoms. Within these octahedra, the Zn–O distances range from 2.011 to 2.212 Å. A $ZnO_4(H_2O)_2$ octahedron shares four oxygen atoms with two adjacent P_4O_{12} groups, whereas a $ZnO_2(H_2O)_4$ octahedron shares two. In addition, a water molecule not involved in the zinc coordination is located between the ZnO_6 layers. As shown by Fig. 4.3.17, a projection of this arrangement along the **c** direction, this

Figure 4.3.16. Projection, along the *a* direction, of the atomic arrangement in $Zn_2P_4O_{12}\cdot8H_2O$. The hatched octahedra represent the ZnO_6 groups, and the open circles represent the non bonded water molecules. The hydrogen atoms have been omitted.

organization creates large channels parallel to the **c** direction containing the nonbonded water molecules. A complete description of the hydrogen-bond network was given by Averbuch-Pouchot.[986]

Four other cyclotetraphosphates are isotypic with $Zn_2P_4O_{12}\cdot8H_2O$. Their unit-cell dimensions are reported in Table 4.3.4.

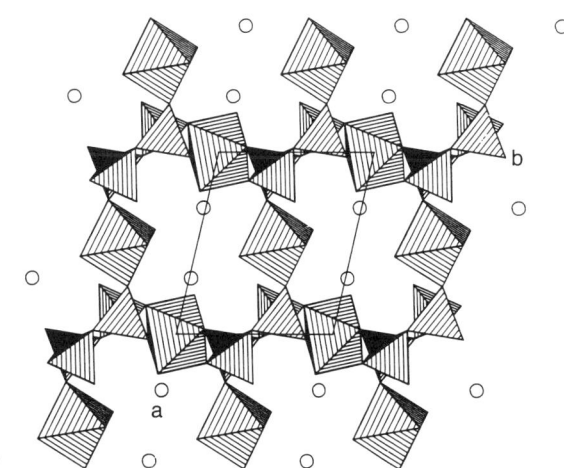

Figure 4.3.17. Projection, along the *c* direction, of the atomic arrangement in $Zn_2P_4O_{12}\cdot8H_2O$. The hatched octahedra represent the ZnO_6 groups, and the open circles the non bonded water molecules. The hydrogen atoms have been omitted.

4.3.4.9. Strontium Cyclotetraphosphate Hexahydrate

$$Sr_2P_4O_{12} \cdot 6H_2O,^{994} \text{ triclinic, } P\bar{1}, Z = 2$$

$$a = 6.664(3), \quad b = 7.365(4), \quad c = 8.618(4) \text{ Å}$$

$$\alpha = 101.62(5), \quad \beta = 109.98(5), \quad \gamma = 95.65(5)°$$

The centrosymmetric phosphoric ring anion is located around the inversion center at the origin of the unit cell. The strontium atoms have eightfold coordination, involving five oxygen atoms and three water molecules, with Sr–O distances ranging from 2.541 to 2.683 Å. These $SrO_5(H_2O)_3$ polyhedra are linked pairwise with a common edge so as to form finite centrosymmetric $Sr_2O_8(H_2O)_6$ groups around the inversion center located at 1/2,0,1/2. Figure 4.3.18, a projection along the **a** axis, shows the whole atomic arrangement while a projection of the same arrangement along the **c** axis is presented in Fig. 4.3.19 in order to show the flat aspect of the ring anions and the layer organization of the P_4O_{12} groups in the (110) planes.

4.3.4.10. Strontium–Sodium Cyclotetraphosphate

$$SrNa_2P_4O_{12},^{999} \text{ tetragonal, } P4nbm, Z = 2$$

$$a = 9.838(5), \quad c = 5.003(3) \text{ Å}$$

This salt is the only known example of an $MNa_2P_4O_{12}$ compound. Figure.4.3.20 shows the projection of its atomic arrangement along the **c** axis. All the associated cations are located in the plane $z = 1/2$. The sodium atoms, located on sites of $2/m$ symmetry, are surrounded by distorted oxygen octahedra having a pseudo-threefold axis parallel to the **c**

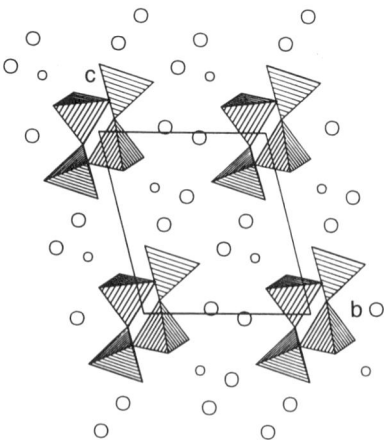

Figure 4.3.18. Projection, along the *a* direction, of the atomic arrangement in $Sr_2P_4O_{12} \cdot 6H_2O$. The larger circles represent the water molecules, and the smaller ones the strontium atoms.

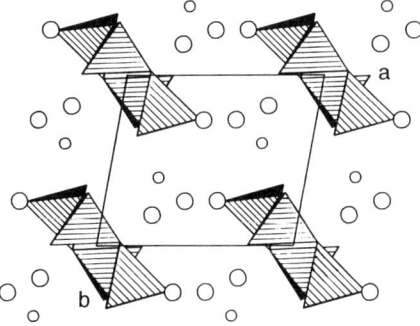

Figure 4.3.19. Projection, along the c direction, of the atomic arrangement in $Sr_2P_4O_{12} \cdot 6H_2O$. The larger circles represent the water molecules, and the smaller ones the strontium atoms.

direction, whereas the strontium atoms, located on sites of 42 symmetry, are on the centers of square antiprisms, sharing four of their corners with the four neighboring NaO_6 octahedra. These polyhedra build infinite layers perpendicular to the c axis. The P_4O_{12} ring anions, located around the $\bar{4}$ axes in $z = 0$, are thus halfway between the associated-cation polyhedra layers; they have $\bar{4}2m$ internal symmetry.

The Sr–O distance in the SrO_8 square antiprism is 2.572 Å. Within the NaO_6 octahedron, the Na–O distances are 2.685 and 2.334 Å.

4.3.4.11. Strontium–Potassium Cyclotetraphosphate

$$SrK_2P_4O_{12},^{1002} \text{tetragonal, } I\bar{4}, Z = 2$$

$$a = 7.445(5), \quad c = 10.17(2) \text{ Å}$$

Figure 4.3.21 shows the projection of the atomic arrangement along the c axis. Each unit cell contains two P_4O_{12} groups located around the $\bar{4}$ axes and centered at 0, 0, 0 and

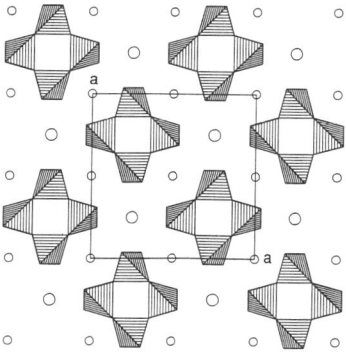

Figure 4.3.20. Projection of the atomic arrangement in $SrNa_2P_4O_{12}$ along the c axis. The larger open circles represent the strontium atoms, and the smaller ones the sodium atoms.

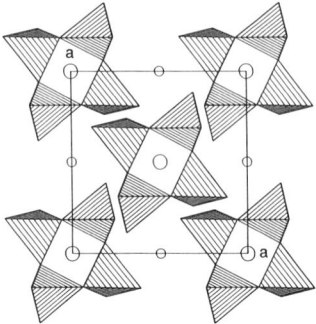

Figure 4.3.21. Projection, along the *c* axis, of the atomic arrangement in $SrK_2P_4O_{12}$. The small open circles represent the strontium atoms, and the large ones the potassium atoms.

1/2, 1/2, 1/2. These rings, having $\overline{4}$ internal symmetry, are built by only one independent PO_4 tetrahedron; they alternate with potassium atoms along the $\overline{4}$ axes. The strontium atoms, at 0, 1/2, 1/4 and 0, 1/2, 3/4, are located on another series of $\overline{4}$ axes. The strontium atom has eight oxygen neighbors, and the potassium atom ten. Within these various coordination polyhedra, the K–O distances range between 2.74 and 3.30 Å, and the Sr–O distances between 2.56 and 2.65 Å. A strong structural similarity exists between this arrangement and that previously described for $Tl_4P_4O_{12}$. Nine cyclotetraphosphates that crystallize with this type of structure are presently known; their unit-cell dimensions are given in Table 4.3.5.

4.3.4.12. Calcium–Sodium Cyclotetraphosphate Hydrate

$$CaNa_2P_4O_{12}\cdot 11/2H_2O,^{1008} \text{ orthorhombic, } Pma2, Z = 4$$

$$a = 27.88(10), \quad b = 7.536(5), \quad c = 7.378(5) \text{ Å}$$

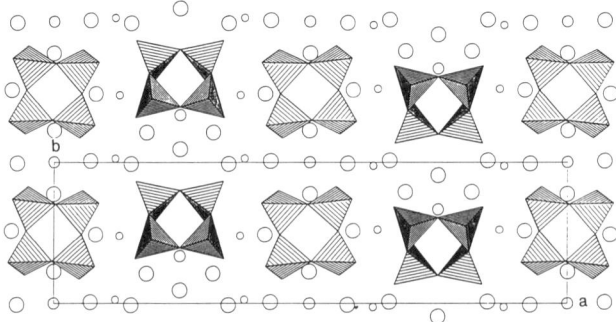

Figure 4.3.22. Projection of the atomic arrangement in $CaNa_2P_4O_{12}\cdot 11/2H_2O$ along the *c* axis. The smallest open circles represent the sodium atoms, the intermediate ones represent the calcium atoms, and the largest ones represent the water molecules.

Figure 4.3.22 shows a projection of the atomic arrangement along the **c** axis. The main feature of this structure is the coexistence, in the atomic arrangement, of two crystallographically independent P_4O_{12} anions with different internal symmetries. One of them has twofold internal symmetry, and the other one has mirror symmetry. This salt provided the first example of a P_4O_{12} group with m symmetry. As can be seen in Fig. 4.3.22, the mirror plane contains two opposite bonding oxygen atoms. The two independent calcium atoms have eightfold coordination, involving four oxygen atoms and four water molecules, while the two independent sodium atoms are located within distorted octahedra built by four oxygen atoms and two water molecules. Inside these coordination polyhedra, the Ca–O distances range from 2.345 to 2.507 Å, and the Na–O distances from 2.284 to 2.705 Å. This compound is one of the rare examples of a cyclotetraphosphate containing two P_4O_{12} ring anions with different internal symmetries.

4.3.4.13. Calcium–Potassium Cyclotetraphosphate Octahydrate

$$Ca_4K_4(P_4O_{12})_3 \cdot 8H_2O,^{[1008]} \text{monoclinic, } P2_1/a, Z = 2$$

$$a = 20.38(1), \quad b = 12.683(5), \quad c = 7.830(2) \text{ Å}, \quad \beta = 89.31(5)°$$

The main feature of this atomic arrangement is the coexistence of two crystallographically independent P_4O_{12} rings. The first one has no internal symmetry, whereas the second one is centrosymmetric. A projection of a part of this arrangement is shown in Fig. 4.3.23. The two independent calcium atoms have a sevenfold coordination, involving six oxygen atoms and one water molecule, with Ca–O distances ranging from 2.339 to 2.471 Å. Within a range of 3.5 Å, one of the potassium atoms has sevenfold coordination, including two water molecules and the second sixfold coordination, also including two water molecules. Within these KO_n coordination polyhedra the K–O distances range from 2.749 to 3.027 Å.

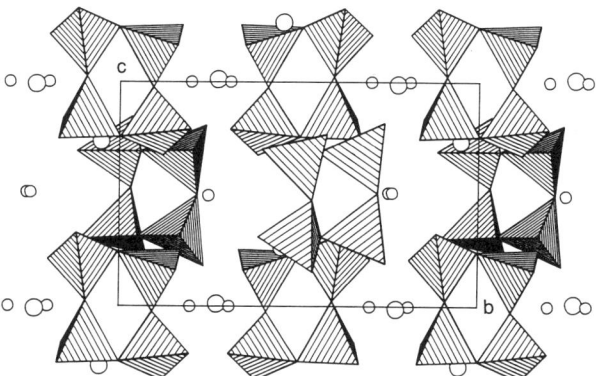

Figure 4.3.23. Projection of a portion of the atomic arrangement in $Ca_4K_4(P_4O_{12})_3 \cdot 8H_2O$ along the *a* axis. The small open circles represent the calcium atoms, and the larger ones the potassium atoms. The water molecules have been omitted.

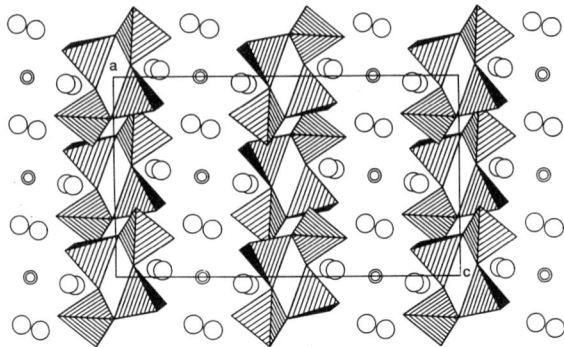

Figure 4.3.24. Projection of the atomic arrangement in $ErKP_4O_{12}\cdot6H_2O$ along the **b** axis. The largest open circles represent the water molecules, the smallest ones the erbium atoms, and the intermediate ones the potassium atoms.

4.3.4.14. Erbium–Potassium Cyclotetraphosphate Hexahydrate

$$ErKP_4O_{12}\cdot6H_2O,^{1025}\text{ monoclinic, } C2/c, Z = 4$$

$$a = 8.643(2), \quad b = 12.015(8), \quad c = 14.909(5) \text{ Å}, \quad \beta = 90.65(5)°$$

Figure 4.3.24, a projection of this structure along the **b** axis, shows the layer organization of this arrangement. Planes $z = 0$ and $1/2$, containing the centrosymmetric P_4O_{12} ring anions, alternate with planes containing the erbium and potassium atoms in $z = 1/4$ and $3/4$. The erbium and potassium atoms, which are superimposed in the drawing, alternate along rows parallel to the **b** axis. The ErO_8 polyhedron is a bicapped distorted trigonal prism, formed by four oxygen atoms and four water molecules. The irregular KO_8 polyhedron is also built from four oxygen atoms and four water molecules. These two types of polyhedra share edges so as to form chains parallel to the **b** axis.

4.3.4.15. Potassium Cyclotetraphosphate–Telluric Acid Dihydrate

$$K_4P_4O_{12}\cdot Te(OH)_6\cdot 2H_2O,^{1029}\text{ monoclinic, } C2/c, Z = 4$$

$$a = 9.731(5), \quad b = 11.43(1), \quad c = 17.16(1) \text{ Å}, \quad \beta = 99.45(5)°$$

Figure 4.3.25, a projection of this structure along the **b** axis, shows the typical layer organization of this arrangement. Layers containing P_4O_{12} rings and one kind of potassium atoms alternate with layers containing by the $Te(OH)_6$ groups, the second kind of potassium atoms, and water molecules.

The tellurium atom, located on the inversion center at 0,0,0 is, as usual, surrounded by its six OH groups in a rather regular octahedral arrangement. The Te–O distances range from 1.908 to 1.920 Å, the O–Te–O angles from 88.5 to 89.8°, and the Te–O–H angles from 108 to 113°. One of the three independent potassium atoms occupies a general position while the other two are located on twofold axes. Within a range of 3.50 Å, the potassium atom located in general position has 7-fold coordination, involving six oxygen

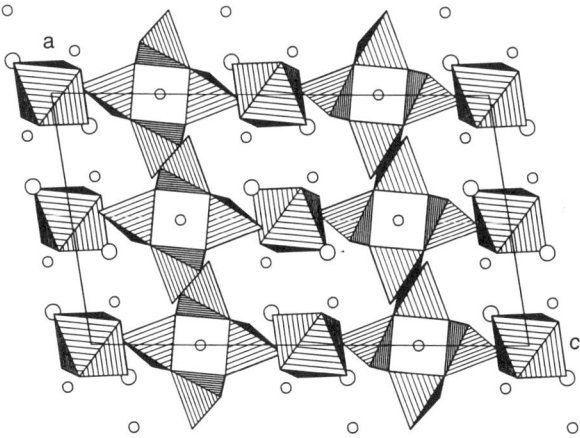

Figure 4.3.25. Projection, along the *b* direction, of the structure of $K_4P_4O_{12} \cdot Te(OH)_6 \cdot 2H_2O$ structure. The hatched octahedra represent the $Te(OH)_6$ groups; the smaller circles represent the potassium atoms, and the larger ones the water molecules. The two independent potassium atoms located on binary axes are superimposed in projection.

atoms and one water molecule, while the other two have 8- and 12-fold coordination. Within these various KO_n polyhedra, the K–O distances range between 2.734 and 3.373 Å.

Another, and, perhaps more illustrative way to describe this structure is to consider its projection along the **a** direction, as given in Fig. 4.3.26. From this projection, one can see that this arrangement is probably better described as a succession of corrugated layers perpendicular to the binary axis.

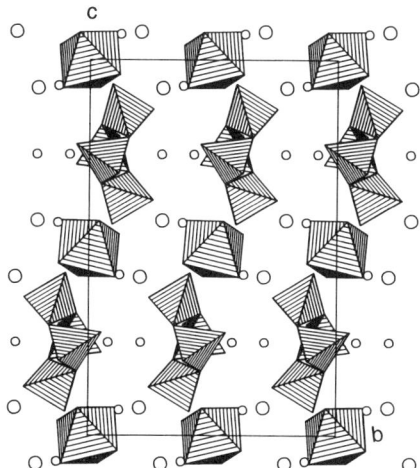

Figure 4.3.26. Projection, along the *a* direction, of the structure of $K_4P_4O_{12} \cdot Te(OH)_6 \cdot 2H_2O$. The representation of the atoms is the same as in Fig. 4.3.25

The P_4O_{12} ring anion, located around a twofold axis, is built up by only two independent PO_4 tetrahedra.

The corresponding rubidium adduct is isotypic.

4.3.5. The P₄O₁₂ Ring Anions

One can imagine many geometrical conformations for the P_4O_{12} ring anion, but we will limit our discussion to the geometries that have actually been observed. To date, 49 P_4O_{12} groups were investigated with high or acceptable accuracy. As was done in Section 4.2.5 for the cyclotriphosphates, we present here, in several tables (Tables 4.3.13–4.3.16.)

Table 4.3.13. Main Geometrical Features of Centrosymmetric P_4O_{12} Rings

Formula	P–P (Å)	P–O–P (°)	P–P–P (°)	Reference
$Na_4P_4O_{12} \cdot 4H_2O^a$	2.915	134.5	97.8	942
	2.964	130.6	82.2	
$Na_4P_4O_{12} \cdot 4H_2O^b$	2.981	137.3	96.1	942
	2.905	129.4	83.9	
$CsNa_3P_4O_{12} \cdot 3H_2O^c$	2.972	136.0	93.8	964
	2.988	136.5	86.2	
	2.978	137.7	86.1	
	2.991	138.4	93.9	
$Cu_2P_4O_{12}$	2.958	136.3	83.9	969
	2.960	137.7	96.1	
$Zn_2P_4O_{12} \cdot 8H_2O$	2.910	131.1	82.9	986
	2.981	137.4	97.1	
$Sr_2P_4O_{12} \cdot 6H_2O$	2.951	131.6	81.5	994
	2.916	129.3	98.5	
$Ca_2P_4O_{12} \cdot 4H_2O$	2.876	135.3	98.3	991
	2.955	126.8	81.7	
$Ca_4K_4(P_4O_{12})_3 \cdot 8H_2O$	2.956	136.9	95.0	1008
	2.966	133.4	85.0	
$PrNH_4P_4O_{12}$	2.953	138.2	86.6	1019
	2.985	134.4	93.4	
$ErKP_4O_{12} \cdot 6H_2O$	2.912	130.5	96.7	1025
	2.930	133.0	83.3	
$(NH_4)_4P_4O_{12} \cdot 2Te(OH)_6 \cdot 2H_2O$	2.961	136.5	86.6	1028
	2.767	124.1	93.4	
$(Gly)_4P_4O_{12}{}^d$	2.986	137.5	84.2	1031
	2.941	132.7	95.8	
$(DETA)_2P_4O_{12} \cdot 4H_2O$	2.922	131.6	81.6	1042
	2.941	133.9	98.4	
$Pb_2Cs_3(P_4O_{12})(PO_3)_3{}^c$	2.955	134.2	83.8	679
	2.952	135.0	96.2	
	2.937	132.4	99.2	
	2.905	130.4	80.8	
$KAl_2H_2(P_3O_{10})(P_4O_{12})$	2.953	136.2	84.6	1190
	2.966	137.8	95.4	

[a]Monoclinic.
[b]Triclinic.
[c]Two independent rings in the unit cell.
[d]Gly, glycinium, $(NH_3-CH_2-COOH)^+$.
[e]DETA, diethylenetriammonium, $[NH_3-(CH_2)_2-NH-(CH_2)_2-NH_3]^{2+}$.

Table 4.3.14. Main Geometrical Features of P_4O_{12} Rings with No Internal Symmetry

Formula	P–P (Å)	P–O–P (°)	P–P–P (°)	Reference
$Na_4P_4O_{12} \cdot H_2O$	2.964	135.9	90.5	936
	3.015	137.4	90.2	
	2.993	136.2	90.2	
	2.999	132.5	89.1	
$K_4P_4O_{12} \cdot 2H_2O$	2.973	133.4	84.5	942
	2.925	130.0	84.8	
	2.923	130.0	85.0	
	2.948	131.0	83.5	
$Cs_4P_4O_{12} \cdot 4H_2O$	2.953	133.0	89.2	958
	2.968	133.4	88.8	
	2.943	131.6	89.6	
	2.951	131.2	89.0	
$Na_2K_2P_4O_{12} \cdot 2H_2O^a$	2.928	128.9	85.1	942
	2.912	129.9	84.7	
	2.937	128.7	84.9	
	2.916	128.2	85.4	
$Na_2K_2P_4O_{12} \cdot 2H_2O^b$	2.910	128.3	85.3	942
	2.916	128.5	84.7	
	2.933	129.9	85.2	
	2.925	129.2	85.0	
$Ni(NH_4)_2P_4O_{12} \cdot 7H_2O$	2.943	134.1	86.7	1010
	2.944	132.7	91.4	
	2.937	133.7	86.8	
	2.946	133.5	91.2	
$Cu(EDA)P_4O_{12} \cdot 14H_2O^c$	2.945	131.8	90.4	1035
	2.929	130.4	88.1	
	2.975	135.4	89.5	
	2.945	132.3	89.0	
$Sr(EDA)P_4O_{12} \cdot 5H_2O$	2.968	134.2	89.3	1037
	2.917	129.4	87.9	
	2.902	127.7	90.8	
	2.925	130.6	89.0	
$Pb_2P_4O_{12} \cdot 3H_2O$	2.912	131.5	90.0	997
	2.898	130.5	90.0	
	2.907	131.3	89.6	
	2.921	130.0	90.0	
$CoK_2(P_4O_{12}) \cdot 5H_2O$	2.922	129.3	90.2	1011
	2.953	131.9	89.4	
	2.925	131.0	90.5	
	2.935	131.1	89.8	
$Ca_4K_4(P_4O_{12})_3 \cdot 8H_2O$	2.906	128.4	89.8	1008
	2.892	131.2	89.2	
	2.915	128.0	89.4	
	2.901	128.2	89.8	
$Ca(NH_4)_2(P_4O_{12}) \cdot 2H_2O$	2.904	134.4	88.0	1012
	2.947	129.5	90.2	
	2.891	127.7	89.3	
	2.893	127.3	91.0	

[a] Tetragonal.
[b] Triclinic.
[c] EDA, ethylenediammonium, $[NH_3–(CH_2)_2–NH_3]^{2+}$.

Table 4.3.15. Geometrical Data for P_4O_{12} Anions with Higher Symmetries[a]

Formula	P–P (Å)	P–O–P (°)	P–P–P (°)	Reference
Rings with 2 symmetry				
$Li_4P_4O_{12} \cdot 5H_2O$[b]	2.941	133.6	90.1	935
	2.914	130.3	89.7	
	2.895	128.7	89.4	
	2.916	130.2	90.6	
$K_4P_4O_{12} \cdot Te(OH)_6 \cdot 2H_2O$	2.908	129.7	84.6	1029
	2.909	129.8	84.4	
$(EDA)P_4O_{12} \cdot H_2O$	2.963	133.9	87.7	1034
	2.958	133.0	90.0	
$Na_2(EDA)P_4O_{12} \cdot 2H_2O$	2.950	131.4	88.7	1039
	2.949	132.8	88.5	
$NaLi(EDA)P_4O_{12} \cdot 3H_2O$	2.918	132.1	89.9	1040
	2.919	132.1	89.9	
$CaNa_2P_4O_{12} \cdot 1\frac{1}{2}H_2O$	2.989	138.3	90.3	1008
	2.939	132.3	89.6	
Rings with $\overline{4}$ symmetry				
$K_4P_4O_{12} \cdot 4H_2O$	2.944	132.6	90.0	950
$Zn_4Na_4(P_4O_{12})_3$	2.887	128.8	88.4	644
$Cs_4Sr_3H_2(P_4O_{12})_3$	2.905	129.3	89.8	1007
$Ce(NH_4)P_4O_{12}$	2.867	125.1	89.3	1020
$Sc_4(P_4O_{12})_3$	2.961	137.0	80.4	1015
$Tl_4P_4O_{12}$	2.979	133.2	86.5	963
$(EA)_4P_4O_{12}$[d]	2.935	131.9	89.8	1032

[a]See also Table 4.3.16.
[b]Two independent rings in the unit cell.
[c]EDA, ethylenediammonium, $[NH_3–(CH_2)_2–NH_3]^{2+}$.
[d]EA, ethanolammonium, $[NH_3–(CH_2)_2–OH]^+$.

their main geometrical features (P–P distances and P–O–P and P–P–P angles) of the P_4O_{12} ring in these cyclotetraphosphates.

The P_4O_{12} rings observed to date exhibit eight types of internal symmetry: with $\overline{1}$, 2, $2/m$, m, mm, $\overline{4}$, $\overline{4}2m$, or no internal symmetry. The number of representatives in each group is very variable; there is only one example of a ring with $2/m$ symmetry whereas 17 have $\overline{1}$ symmetry.

Apart from some rare exceptions, the *P–P–P angles*, strictly equal to 90° in rings with mm or $\overline{4}2m$ symmetries, never depart significantly from this value. Their overall average value is 88.3°. For the calculation of this average, rings with $\overline{1}$, m, and $2/m$ internal symmetries have not been taken into account because in these cases the two different P–P–P angles are complementary. Rings with mm and $\overline{4}$ $2m$ have also been excluded from this calculation, since as we have already mentioned their angles are strictly equal to 90° for symmetry reasons. Thus, the average P–P–P value reported above is strongly weighted by values observed in rings of low internal symmetry. The largest P–P–P angle (97.8°) was observed in the monoclinic form of $Na_4P_4O_{12} \cdot 4H_2O$. As in the case of the cyclotriphosphates, the estimated standard deviations for the reported P–P–P values are less than 0.1, but for recent structural determinations this value is very often less than 0.03.

Table 4.3.16. Geometrical Data for P_4O_{12} Anions with Higher Symmetries[a]

Formula	P–P (Å)	P–O–P (°)	P–P–P (°)	Reference
Ring with 2/m symmetry				
$(NH_4)_4P_4O_{12}$	2.926	131.2	97.7	956
			82.3	
Rings with m symmetry[b]				
$Ca(EDA)P_4O_{12}\cdot 1\frac{1}{2}H_2O$	2.935	130.9	90.2	1036
	2.917	132.6	89.8	
	2.951	134.3		
$CaNa_2P_4O_{12}\cdot 1\frac{1}{2}H_2O$	2.904	135.3	90.5	1008
	2.930	131.4	89.5	
	2.953	131.3		
$CsNa_3P_4O_{12}\cdot 4H_2O$	3.012	129.8	90.0	964
	2.929	139.8		
$SrNa_2P_4O_{12}\cdot 6H_2O$	2.971	135.1	90.0	1009
	2.994	138.3		
Ring with $\overline{4}2m$ symmetry				
$SrNa_2P_4O_{12}$	2.956	132.3	90.0	999

[a]See also Table 4.3.15.
[b]EDA, ethylenediammonium, $[NH_3–(CH_2)_2–NH_3]^{2+}$.

The *P–P distances* in cyclotetraphosphates are quite comparable to those observed in all the other kinds of cyclophosphates, with an overall average of 2.938 Å. The observed values for cyclotetraphosphates fall within the following range:

$$2.867 < P–P < 3.015 \text{ Å}$$

The estimated standard deviations for the reported P–P distances are generally 0.005, but are often close to 0.001 in the most recent data.

The *P–O–P angles* also do not depart significantly from the overall average value of 131.8°. The observed values range from 124.1 to 139.8°. The estimated standard deviation for these values is generally less than 0.1.

As in the case of cyclotriphosphates, there is no apparent correlation between the ring symmetry and the nature of the associated cations.

P_4O_{12} anions are also found in some mixed-anion phosphates, as $Pb_2Cs_3(P_4O_{12})$-$(PO_3)_3$ and $KAl_2H_2(P_3O_{10})(P_4O_{12})$. This type of phosphates will be examined in Chapter 5, but the numerical data for the corresponding rings were included in Table 4.3.13.

4.4. Cyclopentaphosphates

Cyclopentaphosphates are very rare compounds, and only four of them have been well characterized.

Thilo and Schülke[1043] showed that sodium phosphate glasses with a P/Na ratio close to one contain various cyclophosphates $Na_nP_nO_{3n}$, with n value varying from 3 to 8. After

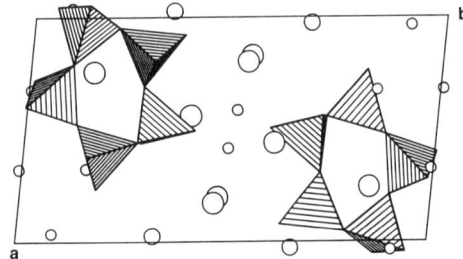

Figure 4.4.1. Projection, along the **c** direction, of the atomic arrangement in $Na_4NH_4P_5O_{15}\cdot4H_2O$. In order of dreasing size, the open circles represent the water molecules, the ammonium groups, and the sodium atoms.

fractionating solutions of these salts using acetone and hexamminecobalt(III), they were able to isolate the sodium salt $Na_5P_5O_{15}\cdot4H_2O$. They also prepared the corresponding barium and silver salts, $Ba_5(P_5O_{15})_2\cdot10H_2O$ and $Ag_5P_5O_{15}\cdot2.6H_2O$.

The only structural determination of a cyclopentaphosphate was performed by Jost[1044] with $Na_4NH_4P_5O_{15}\cdot4H_2O$ obtained during the above experiments. This salt is triclinic $(P\bar{1})$ with the following bimolecular unitcell:

$$a = 8.73(2), \quad b = 15.66(3), \quad c = 6.81(2) \text{ Å}$$

$$\alpha = 93.9(4), \quad \beta = 106.1(4), \quad \gamma = 95.1(3)°$$

Its very simple arrangement is shown in Figure. 4.4.1, which represents a projection along the **c** axis, and the main features of this unique ring anion are presented in Table 4.4.1. In this table, it may be noted that whereas the P–P distances (average value, 2.914 Å) and P–O–P angles (130.5°) are similar to those observed in cyclotri- and cyclotetraphosphates, the P–P–P angles deviate very significantly from their average value in this ring (104.0°) or from the ideal value (108°) for a planar regular ring. As will be seen below when we examine larger rings, the amplitudes of these deviations increase with an increase in the size of the ring. According to the author, the accuracies obtained in the determination of the ring geometry are 0.01 Å for distances and 0.3° for angles.

Two of the sodium atoms have distorted octahedral coordination, one fivefold coordination in a tetragonal pyramidal arrangement, and the fourth one has an irregular sixfold coordination. The author had some difficulty in distinguishing between water molecules and ammonium groups and thus in properly assigning their respective locations.

Table 4.4.1. Main Geometrical
Features of the P_5O_{15} Ring Anion

P–P–P (°)	P–O–P (°)	P–P (Å)
95.8	134.4	2.989
113.7	132.7	2.810
94.0	130.1	2.958
114.0	132.0	2.882
102.6	123.4	2.930

4.5. Cyclohexaphosphates

4.5.1. Introduction

The chemistry of cyclohexaphosphates was slow to develop and even today is relatively poor by comparison with that of cyclotri- and cyclotetraphosphates. The lack, for a long time, of a reliable process for producing large amounts of a starting material was probably the main reason for this slow development. The first evidence for the existence of a P_6O_{18} ring anion was reported by Thilo and Schülke in 1963,[1045,1046] and the first structural evidence was obtained in 1965 when Jost described the crystal structure of the sodium salt $Na_6P_6O_{18} \cdot 6H_2O$.[34,1047]

In the same year, Griffith and Buxton[1048] described for the first time a reproducible procedure for the preparation of useful amounts of sodium or lithium cyclohexaphosphates. It is worth noting that, despite this possibility for producing a suitable starting material, all the cyclohexaphosphates that where clearly characterized during the 20 years following the publication of this procedure were discovered by indirect methods. Thus, $Cu_2Li_2P_6O_{18}$ was identified during attempts to prepare single crystals of the long-chain polyphosphate, $CuLi(PO_3)_3$, characterized during the determination of the $LiPO_3$–$Cu_2P_4O_{12}$ phase-equilibrium diagram,[608,1049] and $Cr_2P_6O_{18}$ and $Cs_2(UO_2)_2P_6O_{18}$ were identified during systematic investigations of the Cr_2O_3–P_2O_5[1050] and Cs_2O–UO_2–P_2O_5[1051] systems by the use of various flux methods.

Recently, Schülke and Kayser[1052] reported a significant improvement in the Griffith–Buxton process, so that large amounts of $Li_6P_6O_{18}$ can now be prepared in a reproducible way and in almost quantitative yields. This process will be described in detail in Section 4.5.3.

Still more recently, Averbuch-Pouchot[1053] described the chemical preparation and accurate structural characterization of $Ag_6P_6O_{18} \cdot H_2O$, thereby opening the way for the use of metathesis reactions, derived from Boullé's process,[805] for the preparation of water-soluble cyclohexaphosphates.

4.5.2. Present State of Cyclohexaphosphate Chemistry

A good number of cyclohexaphosphates have been well characterized, mainly from a structural point of view, but the relatively small amount of information that is available concerning their chemical behavior does not allow for any meaningful comparison of their basic chemical properties with those of other kinds of cyclophosphates.

One can nevertheless note that P_6O_{18} rings seem relatively more resistant to hydrolysis and temperature than smaller phosphoric ring anions, but this statement is more an impression drawn from the author's experience rather than a well-established scientific fact based on quantitative experiments.

In this domain, as in many others in condensed phosphate chemistry, there is a need for basic experiments providing information on the fundamental chemical properties of these compounds.

4.5.2.1. Alkali and Monovalent Cation Cyclohexaphosphates

$Li_6P_6O_{18} \cdot nH_2O$. As mentioned above, Griffith and Buxton[1048] were the first to describe a process for the chemical preparation of the anhydrous salt. Later, Schülke and Kayser[1064] improved this process and reported the existence of one monohydrate, two forms of the tetrahydrate, and two forms of the anhydrous salt. The crystal structure of a pentahydrate was described by Trunov *et al.*[1054] and that of a more hydrated salt, $Li_6P_6O_{18}\cdot(5+n)H_2O$, by Bagieu-Beucher and Rzaigui.[1055] The solubility of $Li_6P_6O_{18}$ in water and in various aqueous–methanolic mixtures was investigated by Borodina and Chudinova.[1056] The effect of water vapor on the formation of lithium cyclohexaphosphate by dehydration-condensation of $Li_3HP_2O_7\cdot H_2O$ was described by Nariai *et al.*[1057]

The existence of the hexahydrate, reported by several authors as the normal hydrate crystallizing at room temperature, has not yet been confirmed.

$Na_6P_6O_{18}\cdot 6H_2O$. The hexahydrate is the only well-characterized sodium salt. Its crystal structure was described by Jost.[34] This compound is now easily prepared from the corresponding lithium salt by using ion-exchange resins or by taking advantage of the insolubility of lithium fluoride, according to the following reaction:

$$Li_6P_6O_{18} + 6NaF \rightarrow Na_6P_6O_{18} + 6LiF$$

$Ag_6P_6O_{18}\cdot H_2O$ and $Ag_6P_6O_{18}\cdot 2.2H_2O$. The monohydrate is presently the only well-characterized silver cyclohexaphosphate. It was first described by Averbuch-Pouchot,[1053] who reported a detail procedure for its chemical preparation and a complete determination of its atomic arrangement. The 2.2 hydrate was reported by Thilo and Schülke[1058] in 1965. The silver salts are of fundamental interest for the preparation of water-soluble cyclohexaphosphates by a metathesis reaction derived from Boullé's process[805] (see Section 4.5.3).

$K_6P_6O_{18}$ and $K_6P_6O_{18}\cdot 3H_2O$. Two crystalline potassium cyclohexaphosphates are now well known, the anhydrous salt and the trihydrate. The chemical preparation and indexed X-ray powder diffraction diagrams of these two salts were reported by Chudinova *et al.*[1059]

Crystal structure determinations for the anhydrous salt were performed almost simultaneously by Averbuch-Pouchot[1060] and by Kholodkovskaya *et al.*[1061] The latter authors also determined the crystal structure of the trihydrate.[1062]

The thermal behavior of the trihydrate was investigated by Chudinova *et al.*[1063] These authors reported the following scheme:

$$K_6P_6O_{18}\cdot 3H_2O \xrightarrow{363\ K} K_6P_6O_{18} \xrightarrow{553\ K} KPO_3$$

$(NH_4)_6P_6O_{18}\cdot H_2O$. The chemical preparation and crystal structure of this salt were reported by Averbuch-Pouchot.[1064] The atomic arrangement is similar to that reported for $Ag_6P_6O_{18}\cdot H_2O$.

The thermal decomposition of $(NH_4)_6P_6O_{18}\cdot 3/2H_2O$ was investigated by Takenaka *et al.*[1065] From the data given by these authors, it is not possible to decide whether this hydrate is similar to the monohydrate or belongs to a new structure type. Long-chain ammonium polyphosphate, $(NH_4)PO_3$, is observed in the final stage of the thermal decomposition.

$Rb_6P_6O_{18}\cdot 6H_2O$, $Cs_6P_6O_{18}\cdot 6H_2O$, $Rb_6P_6O_{18}$, and $Cs_6P_6O_{18}$. The two hydrates were prepared by Averbuch-Pouchot and Durif,[1066] using Boullé's reaction; they are triclinic and isotypic. These authors reported a detailed determination of the atomic arrangement

Table 4.5.1. Crystallographic Data for Monovalent Cation Cyclohexaphosphates

Formula	a (Å) α (°)	b (Å) β (°)	c (Å) γ (°)	S.G.	Z	Reference(s)
$Li_6P_6O_{18} \cdot 5H_2O$	9.490(1) 107.16(1)	8.069(1) 113.84(1)	7.810(1) 65.19(1)	$P\bar{1}$	1	1054
$Li_6P_6O_{18} \cdot (5+n)H_2O$	15.429(8)	11.794(5) 115.95(5)	14.369(8)	$C2/m$	4	1055
$Na_6P_6O_{18} \cdot 6H_2O$	11.58(2)	18.54(4)	10.48(2)	$Ccmb$	4	34
$Ag_6P_6O_{18} \cdot H_2O$	14.807(10)	14.807(10)	6.597(7)	$R\bar{3}$	3	1053
$K_6P_6O_{18}$	15.753(6)	15.753(6)	15.753(6)	$Pa3$	8	1060, 1061
$K_6P_6O_{18} \cdot 3H_2O$	6.803(5)	17.45(2) 107.18 (6)	9.195(8)	$P2_1/m$	2	1062
$(NH_4)_6P_6O_{18} \cdot H_2O$	15.445(10)	15.445(10)	7.553(7)	$R\bar{3}$	3	1064
$Rb_6P_6O_{18} \cdot 6H_2O$	9.662(7) 111.02(5)	9.640(6) 107.83(5)	8.704(3) 60.14(5)	$P\bar{1}$	1	1066, 1067
$Cs_6P_6O_{18} \cdot 6H_2O$	9.904(9) 111.53(2)	9.898(9) 106.53(2)	9.013(8) 60.08(3)	$P\bar{1}$	1	1066, 1067
$Rb_6P_6O_{18}$	16.23	16.23	16.23	$Pa3$	8	1059, 1063
$Cs_6P_6O_{18}$	25.16	25.16	25.16	?	?	1059, 1063

using the cesium salt. Crystal structures of these two salts were also determined by Kholodkovskaya *et al.*[1067] who also reported their chemical preparation and indexed X-ray powder diffraction diagrams. The thermal behavior of these two salts was investigated by Chudinova *et al.*[1063] The following transformation were observed:

$$Rb_6P_6O_{18} \cdot 6H_2O \xrightarrow{323\ K} Rb_6P_6O_{18} \cdot nH_2O \xrightarrow{373\ K} Rb_6P_6O_{18} \xrightarrow{573\ K} RbPO_3$$

$$Cs_6P_6O_{18} \cdot 6H_2O \xrightarrow{363\ K} Cs_6P_6O_{18} \cdot nH_2O \xrightarrow{413\ K} Cs_6P_6O_{18} \xrightarrow{553\ K} CsPO_3$$

According to Chudinova *et al.*,[1059,1063] both anhydrous salts are cubic, the rubidium salt being isotypic with $K_6P_6O_{18}$.

Crystallographic Data. The main crystallographic data for the monovalent cation cyclohexaphosphates described above are given in Table 4.5.1.

Mixed-Monovalent Cation Cyclohexaphosphates

Some mixed-monovalent cation cyclohexaphosphates have also been recently characterized. They correspond to two different types of stoichiometries for the associated cations:

- a 1:1 stoichiometry, leading to the formula $A_3B_3P_6O_{18} \cdot nH_2O$
- a 1:2 stoichiometry corresponding to $A_4B_2P_6O_{18} \cdot nH_2O$ compounds.

$Li_3Na_3P_6O_{18} \cdot 12H_2O$. The existence of this salt was first reported by Schülke and Kayser.[1052] Later, Averbuch-Pouchot[1068] gave a detailed description of its atomic arrangement.

$Li_3K_3P_6O_{18} \cdot H_2O$ and $Ag_3(NH_4)_3P_6O_{18} \cdot H_2O$. These two trigonal (rhombohedral) salts are isotypic. Both were prepared and investigated by Averbuch-Pouchot,[1068,1069] who performed their crystal structure determinations. The first one is simply synthesized by mixing aqueous solutions of the two cyclohexaphosphates in stoichiometric proportions, and the second one is prepared by adding an aqueous solution of silver nitrate to an aqueous solution of ammonium cyclohexaphosphate in the proportions required by the reaction:

$$(NH_4)_6P_6O_{18} + 3AgNO_3 \rightarrow Ag_3(NH_4)_3P_6O_{18} \cdot H_2O + 3NH_4NO_3$$

The silver–ammonium salt, which is the less water-soluble product, crystallizes first by slow evaporation of the solution at room temperature.

$Ag_4Li_2P_6O_{18} \cdot 2H_2O$. This triclinic salt was prepared by Averbuch-Pouchot and Durif[1070] according to the process described for the synthesis of the silver–ammonium salt. The authors performed a detailed determination of the atomic arrangement.

$Li_2(NH_4)_4P_6O_{18} \cdot 4H_2O$ and $Li_2K_4P_6O_{18} \cdot 4H_2O$. The chemical preparation and crystal structure of the lithium–ammonium salt were reported by Elmokhtar et al.[1071] Later, Averbuch-Pouchot[1072] prepared the lithium–potassium salt and identified it as an isotype.

$Na_2(NH_4)_4P_6O_{18} \cdot 2H_2O$ and $Na_2Tl_4P_6O_{18} \cdot 2H_2O$. These two isotypic salts were prepared and investigated by Averbuch-Pouchot and Durif,[1073] who reported the crystal structure of the sodium–ammonium compound. The thallium–sodium cyclohexaphosphate was prepared by a procedure similar to that described for the synthesis of $Ag_3(NH_4)_3P_6O_{18} \cdot H_2O$ except that thallium nitrate was employed, whereas the sodium–ammonium salt is simply prepared by mixing solutions of the two alkali cyclohexaphosphates.

$Na_4Rb_2P_6O_{18} \cdot 6H_2O$ and $Na_4Cs_2P_6O_{18} \cdot 6H_2O$. These two triclinic isotypic cyclohexaphosphates were studied by Averbuch-Pouchot and Durif,[1073] who determined the atomic arrangement in this type of structure with the sodium–rubidium salt.

Crystallographic Data. Table 4.5.2 contains the main crystallographic data for the mixed-monovalent cation cyclohexaphosphates.

4.5.2.2. Divalent Cation Cyclohexaphosphates

Nine divalent cation cyclohexaphosphates were characterized, but only four of them were investigated from a structural point of view.

Five cyclohexaphosphates, $Mn_3P_6O_{18} \cdot 9H_2O$, $Co_3P_6O_{18} \cdot 14H_2O$, $Ni_3P_6O_{18} \cdot 17H_2O$, $Cu_3P_6O_{18} \cdot 14H_2O$, and $Cd_3P_6O_{18} \cdot 16H_2O$, were prepared and characterized by Lazarevski et al.[1074] They were synthesized by the action of cyclohexaphosphoric acid on aqueous solutions of the corresponding perchlorates. The acid was produced from solutions of Li or Na phosphates through the use of ion-exchange resins. According to the X-ray powder diagrams given by the authors, it seems that the cobalt and copper salts, crystallizing with the same degree of hydration, are not isotypic. The thermal behavior of the cadmium, copper, cobalt, nickel, manganese, and barium salts was also investigated by the same authors.[1074–1076]

Table 4.5.2. Crystallographic Data for Mixed-Monovalent Cation Cyclohexaphosphates

Formula	a (Å) α (°)	b (Å) β (°)	c (Å) γ (°)	S.G.	Z	Reference(s)
$Li_3Na_3P_6O_{18}\cdot12H_2O$	10.474(8)	10.474(8)	41.68(5)	$R\bar{3}c$	6	1068
$Li_3K_3P_6O_{18}\cdot H_2O$	15.047(8)	15.047(8)	12.779(8)	$R\bar{3}$	6	1068
$Li_2K_4P_6O_{18}\cdot2H_2O$	9.088(3)	15.820(5) 106.53(8)	7.763(6)	$P2_1/c$	2	1072
$Ag_3(NH_4)_3P_6O_{18}\cdot H_2O$	15.172(5)	15.172(5)	13.994(5)	$R\bar{3}$	6	1069
$Ag_4Li_2P_6O_{18}2H_2O$	8.408(2) 107.47(3)	7.602(2) 106.09(3)	7.566(2) 72.64(3)	$P\bar{1}$	1	1070
$Li_2(NH_4)_4P_6O_{18}\cdot4H_2O$	9.429(2)	15.824(3) 106.26(1)	7.931(1)	$P2_1/c$	2	1071
$Na_2(NH_4)_4P_6O_{18}\cdot2H_2O$	13.363(7)	11.580(12) 101.87(5)	6.809(5)	$P2_1/n$	2	1073
$Na_2Tl_4P_6O_{18}\cdot2H_2O$	13.215(8)	11.583(8) 101.43(5)	6.874(8)	$P2_1/n$	2	1073
$Na_4Rb_2P_6O_{18}\cdot6H_2O$	7.532(3) 113.92(4)	9.752(3) 102.29(4)	8.730(3) 85.00(4)	$P\bar{1}$	1	1073
$Na_4Cs_2P_6O_{18}\cdot6H_2O$	7.653(4) 114.64(4)	9.59(4) 102.47(4)	8.740(4) 84.94(4)	$P\bar{1}$	1	1073

In the case of $Cu_3P_6O_{18}\cdot14H_2O$, at high temperature, the P_6O_{18} ring is destroyed with formation of the cyclotetraphosphate.

$$2Cu_3P_6O_{18}\cdot14H_2O \rightarrow 3Cu_2P_4O_{12} + 14H_2O$$

Single crystals of this salt were later prepared by Averbuch-Pouchot[1077] by adding aqueous solutions of $CuCl_2$ [or $Cu(NO_3)_2$] to an aqueous solution of $Li_6P_6O_{18}\cdot6H_2O$ or by adding an aqueous solution of copper nitrate to a solution of guanidinium cyclohexaphosphate in a stoichiometric ratio. After some days of evaporation of these solutions at room temperature, crystals of $Cu_3P_6O_{18}\cdot14H_2O$ appear as large turquoise calcite-like pseudo-rhombohedra. The author reported a complete determination of the atomic arrangement, confirming the hydration state previously observed.[1074–1076]

$Mn_3P_6O_{18}\cdot6H_2O$ and $Cd_3P_6O_{18}\cdot6H_2O$. The cadmium salt was prepared by Averbuch-Pouchot[1078] by using Boullé's process.[805] A complete description of its atomic arrangement was reported by the author. The manganese salt, prepared by the same procedure was characterized by Averbuch-Pouchot and Durif[1079] as an isotype of the cadmium salt.

The effect of water vapor on the thermal behavior of $Cd_3P_6O_{18}\cdot6H_2O$ was investigated by Nariai et al.[1080] The main weight loss was observed at 460 K and then decomposition occurred. At 588 K, the compound transformed into a mixture of mono- and diphosphates and at higher temperatures into the long-chain polyphosphate, $Cd(PO_3)_2$. The cyclotetraphosphate, $Cd_2P_4O_{12}$, was observed during the last step of reorganization.

$Ba_3P_6O_{18}\cdot9H_2O$ and $Ba_3P_6O_{18}\cdot8H_2O$. Thilo and Schülke[1058] and Lazarevski et al.[1081] obtained $Ba_3P_6O_{18}\cdot9H_2O$ as a white crystalline substance from the reaction between

Table 4.5.3. Crystallographic Data for Divalent Cation

Formula	a (Å) α (°)	b (Å) β (°)	c (Å) γ (°)	S.G.	Z	Reference
$Mn_3P_6O_{18} \cdot 6H_2O$	14.836(8)	14.836(8)	15.781(8)	$R\overline{3}$	6	1079
$Cd_3P_6O_{18} \cdot 6H_2O$	15.056(8)	15.056(8)	16.080(8)	$R\overline{3}$	6	1078
$CU_3P_6O_{18} \cdot 14H_2O$	10.944(8) 110.49(5)	7.539(4) 110.14(5)	8.974(4) 77.82(5)	$P\overline{1}$	1	1077
$Ba_3P_6O18 \cdot 8H_2O$	20.98(2) 119.56(3)	7.227(3)	17.44(1)	C2/c	4	1082

aqueous solutions of $BaCl_2$ and $Na_6P_6O_{18}$. Recently, Rzaigui et al.[1082] described the chemical preparation of the octahydrate and determined its crystal structure.

Crystallographic Data. Table 4.5.3 gives the main crystal features of the three well investigated divalent cation cyclohexaphosphates.

4.5.2.3. Divalent–Monovalent Cation Cyclohexaphosphates

With the exception of $Cu_2Li_2P_6O_{18}$ and $Cs_2(UO_2)_2P_6O_{18}$, all the well-characterized bivalent–monovalent cation cyclohexaphosphates are hydrates corresponding to the general formula: $M_2^{II}M_2^{I}P_6O_{18} \cdot nH_2O$ and belonging to seven different structure types.

$Cu_2Li_2P_6O_{18}$. This salt was characterized during the elaboration of the $LiPO_3$–$Cu_2P_4O_{12}$ phase-equilibrium diagram, when attempts were made to prepare single crystals of the $CuLi(PO_3)_3$ polyphosphate. Laügt et al.[1083] observed a low-temperature form of this polyphosphate and characterized it by chromatographic analysis as a cyclohexaphosphate. According to these authors, this salt transforms into the polyphosphate at 773 K. The atomic arrangement of the cyclohexaphosphate was later determined by Laügt and Durif.[608]

Most of the hydrates that we describe here have been prepared as polycrystalline samples by very classical methods of aqueous chemistry corresponding to the general reaction scheme:

$$M_6^I P_6O_{18} + 2M^{II}(NO_3)_2 \rightarrow M_2^I M_w^{II} P_6O_{18} \cdot nH_2O + 4M^I NO_3$$

All of them are sparingly water-soluble, and in some cases the production of single crystals suitable for structural investigations was achieved by unorthodox processes.

$Zn_2Li_2P_6O_{18} \cdot 10H_2O$ and $Mn_2Li_2P_6O_{18} \cdot 10H_2O$. These two isotypic triclinic compounds were prepared by Averbuch-Pouchot,[1084] who determined the atomic arrangement in this type of structure with the manganese salt.

$Ca_2Li_2P_6O_{18} \cdot 8H_2O$ and $Ca_2Na_2P_6O_{18} \cdot 8H_2O$. These two isotypic salts are also triclinic. They have been prepared by adding solid gypsum to an aqueous solution of the alkali cyclohexaphosphate. The crystal structure was performed by Averbuch-Pouchot and Durif[1085] with the calcium–lithium compound.

$Cd_2Na_2P_6O_{18} \cdot 14H_2O$. The chemical preparation and crystal structure of this triclinic compound were reported by Averbuch-Pouchot.[1086]

The thermal behavior of an undecahydrate, $Cd_2Na_2P_6O_{18} \cdot 11H_2O$, was investigated by Nariai *et al.*[1080] This compound is probably similar to the tetradecahydrate since nonbonded water molecules were observed in its atomic arrangement by Averbuch-Pouchot.[1086] The water is eliminated in two steps, between 371 and 406 K and then the compound reorganizes, at about 603 K, into the long-chain polyphosphate, $CdNa(PO_3)_3$. As in many similar cases, an intermediate amorphous phase is observed.

$Cu_2(NH_4)_2P_6O_{18} \cdot 8H_2O$. This salt can be prepared in a polycrystalline state by classical methods, but the production of single crystals suitable for structural investigations was difficult to optimize. Averbuch and Durif[1087] described the following process. To an almost saturated aqueous solution of $(NH_4)_6P_6O_{18} \cdot H_2O$, the required amount of copper hydroxy-carbonate $[CuCO_3.Cu(OH)_2]$ is added. Drops of concentrated hydrochloric acid are added to this slurry until the hydroxycarbonate has completely disappeared. Crystals of $Cu_2(NH_4)_2P_6O_{18} \cdot 8H_2O$ appear immediately as diamondlike plates. This compound is very sparingly water-soluble.

The thermal decomposition of a hydrate said to be $Cu_2(NH_4)_2P_6O_{18} \cdot 8.5H_2O$ was investigated by Takenaka *et al.*[1088] The crystallization water is lost below 423 K. Loss of ammonia does not occur below 553 K, and up to 653 K the only crystalline phase observed is $Cu_2P_2O_7$. At higher temperatures, 703–783 K, copper(II) cyclotetraphosphate, $Cu_2P_4O_{12}$, is observed.

$Ca_2(NH_4)_2P_6O_{18} \cdot 6H_2O$. This salt is easily obtained as large orthorhombic prisms after some days of evaporation, at room temperature, of an aqueous solution of ammonium cyclohexaphosphate to which solid gypsum has been added. Averbuch-Pouchot[1089] performed the determination of its crystal structure.

A hydrate identified as $Ca_2(NH_4)_2P_6O_{18} \cdot 7H_2O$ was prepared and studied by Takenaka *et al.*[1088] According to these authors all the water molecules are lost between room temperature and 450 K, leading to an anhydrous cyclohexaphosphate, $Ca_2(NH_4)_2P_6O_{18}$. At temperatures higher than 553 K ammonia evolved. Comparative experiments were conducted in dry and humid atmospheres.

$Zn_2(NH_4)_2P_6O_{18} \cdot 8H_2O$ and $Cd_2(NH_4)_2P_6O_{18} \cdot 9H_2O$. The chemical preparation and crystal structure of the zinc salt was described by Averbuch-Pouchot and Durif.[1090] The nonisotypic cadmium salt was investigated by Averbuch-Pouchot, Tordjman and Durif.[1091]

$Sr_2(NH_4)_2P_6O_{18} \cdot 7H_2O$. This salt was characterized by Takenaka *et al.*[1092] These authors reported an investigation of its thermal behavior both in a dry and a humid atmosphere. The loss of the water molecules is observed in several steps between 403 and 473 K and several of them seem to be of a zeolitic nature. The final calcination product is β-$Sr(PO_3)_2$. No crystal data exist for this compound.

Crystallographic Data. Table 4.5.4 contains the main crystallographic data for the bivalent–monovalent cation cyclohexaphosphates.

Table 4.5.4. Crystallographic Data for Divalent–Monovalent Cation Cyclohexaphosphates

Formula	a (Å) α (°)	b (Å) β (°)	c (Å) γ (°)	S.G.	Z	Reference(s)
$Cs_2(UO_2)_2P_6O_{18}$	6.988(2) 104.25(2)	10.838(4)	13.309(3)	$P2_1/n$	4	1051
$Cu_2Li_2P_6O_{18}$	9.485(2) 111.73(2)	9.419(2) 106.25(2)	9.379(2) 106.80(2)	$P\bar{1}$	2	1049
$Zn_2Li_2P_6O_{18}\cdot10H_2O$	7.160(5) 118.49(5)	9.741(7) 110.57(5)	9.928(6) 86.96(5)	$P\bar{1}$	1	1084
$Mn_2Li_2P_6O_{18}\cdot10H_2O$	7.286(5) 118.31(5)	9.761(7) 110.62(5)	10.026(6) 86.27(5)	$P\bar{1}$	1	1084
$Ca_2Li_2P_6O_{18}\cdot8H_2O$	7.767(2) 105.17(4)	10.144(3) 102.76(4)	7.225(2) 84.95(4)	$P\bar{1}$	1	1085
$Ca_2Na_2P_6O_{18}\cdot8H_2O$	8.031(5) 105.69(5)	10.296(9) 103.27(5)	7.279(5) 85.30(5)	$P\bar{1}$	1	1085
$Cd_2Na_2P_6O_{18}\cdot14H_2O$	7.709(1) 108.25(5)	11.028(6) 110.06(5)	9.231(2) 79.77(5)	$P\bar{1}$	1	1086
$Cu_2(NH_4)_2P_6O_{18}\cdot8H_2O$	7.413(3) 116.23(5)	9.334(4) 107.98(5)	9.634 83.10(5)	$P\bar{1}$	1	1087
$Ca_2(NH_4)_2P_6O_{18}\cdot6H_2O$	12.821(6)	12.537(6)	7.029(2)	$P2_12_12$	2	1089
$Zn_2(NH_4)_2P_6O_{18}\cdot8H_2O$	8.717(7) 104.82(2)	10.297(8) 111.03(2)	7.409(7) 70.96(2)	$P\bar{1}$	1	1090
$Cd_2(NH_4)_2P_6O_{18}\cdot9H_2O$	9.126(6) 107.95(6)	9.581(8) 112.93(6)	8.993(4) 61.90(5)	$P\bar{1}$	1	1091

4.5.2.4. Trivalent Cation Cyclohexaphosphates

$Cr_2P_6O_{18}$. This salt was first obtained by Rémy and Boullé [227,1093] during a systematic investigation of chromium phosphates. Later, Bagieu-Beucher and Guitel[1050] reported the preparation of single crystals and performed the determination of its atomic arrangement. According to these authors, this cyclophosphate is stable at 1273 K but transforms rapidly at 1373 K into the long-chain polyphosphate (C form). Under high pressures, a similar transformation is observed at lower temperature:

$$Cr_2P_6O_{18} \xrightarrow[\text{40 Kbar}]{\text{1073 K}} Cr(PO_3)_3(C)$$

$Ga_2P_6O_{18}$. The existence of $Ga_2P_6O_{18}$ was reported by Chudinova et al.[258] during a study of the NH_4–P_2O_5–Ga_2O_3–H_2O system. The most stable compound in this system is $GaNH_4HP_3O_{10}$, a triphosphate which, when heated at about 723 K, transforms into $Ga_2P_6O_{18}$:

$$2GaNH_4HP_3O_{10} \xrightarrow{723 \text{ K}} Ga_2P_6O_{18} + 2H_2O + 2NH_3$$

This salt is isotypic with the corresponding chromium salt.

$Al_2P_6O_{18}$. Crystalline samples of aluminum cyclohexaphosphate were obtained by Kanene *et al.*[1094] by heating Al_2O_3 and $NH_4H_2PO_4$ at 823 K for 4 h. When heated at 1223 K–1273 K, this compound decomposes into aluminum cyclotetraphosphate:

$$2Al_2P_6O_{18} \rightarrow Al_4(P_4O_{12})_3$$

Based on crystal data, the authors postulated that this salt is isotypic with the corresponding chromium compound.

$[Ga(OH)_2]_6P_6O_{18}$. During an investigation of the $GaCl_3$–$Na_6P_6O_{18}$ system, Lazarevski *et al.*[1081] chemically characterized this salt. Its thermal behavior was later investigated by the same authors.[1075]

$Cr_2P_6O_{18} \cdot 21H_2O$. Rzaigui[1095] prepared this hydrate and described its main chemical and physical properties. The crystal structure, determined by Bagieu-Beucher *et al.*[1096] shows that 9 of the 21 water molecules are of a zeolitic nature. This finding is consistent with the thermal behavior of this compound reported by Rzaigui.[1095]

$Y_2P_6O_{18} \cdot 18H_2O$. This hydrate was prepared by Lazarevski *et al.*[1074] by the action of YCl_3 on $Na_6P_6O_{18}$ in aqueous solution, and its thermal behavior was investigated by the same authors.[1087]

$Ce_2P_6O_{18} \cdot 10H_2O$. Rzaigui[1097] prepared single crystals of this salt and described its main chemical properties. Bagieu-Beucher and Rzaigui[1098] performed the crystal structure determination. Four of the ten water molecules are not involved in the cerium coordination polyhedra.

$Nd_2P_6O_{18} \cdot 12H_2O$. The crystal structure of this salt, prepared by reaction of $NdCl_3$ with an aqueous solution of $Li_6P_6O_{18}$, was determined by Trunov *et al.*[1099]

$Yb_2P_6O_{18} \cdot 16H_2O$. A crystal structure determination for ytterbium cyclohexaphosphate hexadecahydrate was performed by Bagieu-Beucher and Rzaigui.[1100] Nine of the water molecules are not involved in the associated-cation coordination polyhedra.

$Mn_3Cs_3(P_6O_{18})_2$. A trivalent–monovalent cation cyclohexaphosphate, $Mn_3Cs_3(P_6O_{18})_2$, was characterized by Guzeeva and Tananaev[277] during an investigation of the MnO_2–Cs_2O–P_2O_5–H_2O system between 423 and 637 K. No crystal data were reported by these authors.

Crystallographic Data. The main crystallographic data for trivalent cation cyclohexaphosphates of are given in Table 4.5.5.

$Cs_2(UO_2)_2P_6O_{18}$. This salt was characterized by Linde *et al.*[1051] during an investigation of the compounds obtained by heating mixtures of H_3PO_4, $CsNO_3$, and $UO_2(NO_3)_2 \cdot 6H_2O$ under various conditions. The authors reported the crystal structure of this salt.

4.5.2.5. Organic Cation and Organometallic Cyclohexaphosphates

Several organic cation cyclohexaphosphates have been well characterized. All of them were synthesized by Boullé's process,[805] and in all cases the atomic arrangements were accurately determined.

Table 4.5.5. Crystallographic Data for Trivalent Cation Cyclohexaphosphates

Formula	a (Å) α (°)	b (Å) β (°)	c (Å) γ (°)	S.G.	Z	Reference(s)
$Cr_2P_6O_{18}$	8.311(4)	15.221(8) 105.85(5)	6.220(3)	$P2_1/a$	2	1050
$Ga_2P_6O_{18}$	8.293	15.196 105.89	6.188	$P2_1/a$	2	258
$Al_2P_6O_{18}$	8.201(3)	15.062(3) 105.13(3)	6.101(2)	$P2_1/a$	2	1094
$Cr_2P_6O_{18} \cdot 21H_2O$	19.05(1)	19.05(1)	19.05(1)	$P\bar{4}3n$	8	1096
$Ce_2P_6O_{18} \cdot 10H_2O$	13.522(5)	13.105(9)	6.938(3)	$P2_12_12$	2	1098
$Nd_2P_6O_{18} \cdot 12H_2O$	9.149(2)	11.693(3) 96.92(1)	11.959(3)	$P2_1/c$	2	1099
$Yb_2P_6O_{18} \cdot 16H_2O$	16.019(8)	19.99(1)	9.699(5)	$P2_12_12_1$	4	1100

$(NH_3OH)_6P_6O_{18} \cdot 4H_2O$. Hydroxylammonium cyclohexaphosphate tetrahydrate was prepared and described by Durif and Averbuch-Pouchot.[1101] This salt is very stable under normal conditions.

$(C_2N_2H_{10})_3P_6O_{18} \cdot 2H_2O$. The ethylenediammonium cyclohexaphosphate dihydrate prepared and investigated by Durif and Averbuch-Pouchot[1102] appears to be as an interesting starting material for the syntheses of some species of complex cyclohexaphosphates.

$(C_2H_5NH_3)_6P_6O_{18} \cdot 4H_2O$. Ethylammonium cyclohexaphosphate tetrahydrate was prepared and described by Averbuch-Pouchot and Durif.[1103] This salt is stable for months at room temperature.

$(N_2H_5)_2(N_2H_6)_2P_6O_{18}$. During the chemical preparation of this compound, the expected product was a hydrazinium cyclohexaphosphate, $(N_2H_6)_3P_6O_{18}$. In fact, the crystal structure performed by Averbuch-Pouchot and Durif[1103] shows clearly that the final product is a monohydrazinium–dihydrazinium salt.

$M(C_2N_2H_{10})_2P_6O_{18} \cdot 6H_2O$ (M = Cu, Co, Ni, Mg, Zn, Fe). These series of six isotypic organometallic cation compounds was characterized by Averbuch-Pouchot and Durif.[1104] These compounds are prepared by adding stoichiometrically an aqueous solution of the divalent cation nitrate to an aqueous solution of ethylenediammonium (EDA) cyclohexaphosphate. Sulfate was used in the case of the iron salt. The reaction is

$$(EDA)_3P_6O_{18} + M(NO_3)_2 \rightarrow M(EDA)_2P_6O_{18} + (EDA)(NO_3)_2$$

If concentrated solutions are used, large, stout monoclinic prisms of $M(EDA)_2P_6O_{18} \cdot 6H_2O$ appear in the solution within some hours. Their structure was determined with the copper salt by Durif and Averbuch-Pouchot.[1102]

Crystallographic Data. Table 4.5.6 gives the main crystallographic data for organic and organometallic cation cyclohexaphosphates.

Table 4.5.6. Crystallographic Data for Organic and Organometallic Cation Cyclohexaphosphates

Formula[a]	a (Å) α (°)	b (Å) β (°)	c (Å) γ (°)	S.G.	Z	Reference(s)
$(EDA)_3P_6O_{18}\cdot 2H_2O$	11.064(5)	12.317(5) 90.53(5)	9.342(5)	$P2_1/n$	2	1102
$(NH_3OH)_6P_6O_{18}\cdot 4H_2O$	10.365(5) 108.39(5)	9.278(4) 100.30(5)	7.280(3) 96.02(5)	$P\bar{1}$	1	1101
$Cu(EDA)_2P_6O_{18}\cdot 6H_2O$	13.378(8)	11.574(6) 103.15(3)	8.687(3)	$P2_1/a$	2	1102
$Co(EDA)_2P_6O_{18}\cdot 6H_2O$	13.193(8)	11.604(6) 102.22(5)	8.616(3)	$P2_1/a$	2	1104
$Ni(EDA)_2P_6O_{18}\cdot 6H_2O$	13.165(8)	11.579(8) 102.61(5)	8.612(4)	$P2_1/a$	2	1104
$Mg(EDA)_2P_6O_{18}\cdot 6H_2O$	13.234(5)	11.630(5) 102.45(5)	8.674(2)	$P2_1/a$	2	1104
$Zn(EDA)_2P_6O_{18}\cdot 6H_2O$	13.217(5)	11.617(5) 102.44(5)	8.670(2)	$P2_1/a$	2	1104
$Fe(EDA)_2P_6O_{18}\cdot 6H_2O$	13.233(5)	11.674(5) 102.06(5)	8.646(2)	$P2_1/a$	2	1104
$(N_2H_5)_2(N_2H_6)_2P_6O_{18}$	8.175(8) 105.05(2)	7.926(8) 102.08(2)	8.457(7) 86.42(2)	$P\bar{1}$	1	1103
$(C_2H_5NH_3)_6P_6O_{18}\cdot 4H_2O$	16.80(1) 109.66(2)	23.88(1)	10.623(8)	$P2_1/a$	4	1103

[a]EDA, ethylenediammonium, $[NH_3-(CH_2)_2-NH_3]^{2+}$.

4.5.2.6. Cyclohexaphosphate–Telluric Acid Adducts

Formation of adducts with telluric acid was observed for all kinds of alkali phosphates, condensed or not, and was extensively investigated by the author during the past 15 years. Adducts with telluric acid are simply prepared by adding an aqueous solution of telluric acid to a solution of the alkali phosphate in the proper stoichiometry. All the compounds listed in Table 4.5.7 were prepared by this simple procedure, and for all of them the atomic arrangements were determined.

$(NH_4)_6P_6O_{18}\cdot Te(OH)_6\cdot 2H_2O$. This salt was recently prepared by Averbuch-Pouchot and Durif.[1105] The crystal structure determination performed by these authors shows the existence of two statistically distributed water molecules.

$K_6P_6O_{18}\cdot 2Te(OH)_6\cdot 3H_2O$. The layered arrangement of this trigonal (rhombohedral) adduct was established by Averbuch-Pouchot and Durif.[1106]

$Rb_6P_6O_{18}\cdot 3Te(OH)_6\cdot 4H_2O$ and $Cs_6P_6O_{18}\cdot 3Te(OH)_6\cdot 4H_2O$. These two isotypic adducts were recently prepared by Averbuch-Pouchot and Durif.[1106] These authors determined the atomic arrangement in this type of structure with the rubidium salt.

$(EDA)_3P_6O_{18}\cdot 2Te(OH)_6\cdot 2H_2O$. This salt, prepared and described by Averbuch-Pouchot and Durif,[1107] is one of the rare examples of an adduct between an organic cation condensed phosphate and telluric acid.

Table 4.5.7. Crystallographic Data for Cyclohexaphosphate
Adducts with Telluric Acid

Formula	a (Å) α (°)	b (Å) β (°)	c (Å) γ (°)	S.G.	Z	Reference(s)
$(NH_4)_6P_6O_{18} \cdot Te(OH)_6 \cdot 2H_2O$	9.899(4) 109.53(6)	11.042(7) 106.74(6)	7.632(9) 100.91(4)	$P\bar{1}$	1	1105
$K_6P_6O_{18} \cdot 2Te(OH)_6 \cdot 3H_2O$	13.084(5)	13.084(5)	34.80(2)	$R\bar{3}c$	6	1106
$Rb_6P_6O_{18} \cdot 3Te(OH)_6 \cdot 4H_2O$	11.222(8) 111.11(2)	8.077(6) 104.66(2)	11.731(9) 83.25(2)	$P\bar{1}$	1	1106
$Cs_6P_6O_{18} \cdot 3Te(OH)_6 \cdot 4H_2O$	11.549(8) 111.08(5)	8.228(4) 103.13(5)	11.946(6) 82.26(5)	$P\bar{1}$	1	1106
$(EDA)_3P_6O_{18} \cdot 2Te(OH)_6 \cdot 2H_2O^a$	10.945(3) 90.90(5)	11.252(3) 92.97(5)	8.042(7) 116.82(5)	$P\bar{1}$	1	1107

^{a}EDA, ethylenediammonium, $[NH_3–(CH_2)_2–NH_3]^{2+}$.

4.5.2.7. Cyclohexaphosphate Anions in Mixed-Anion Compounds

P_6O_{18} ring anions have up to now been observed in three series of mixed-anion salts. The main crystal data measured for these mixed-anion compounds are presented in Table 4.5.8.

$Mg_2(NH_4)_4P_6O_{18}(C_2O_4) \cdot 6H_2O$ and $Mn_2(NH_4)_4P_6O_{18}(C_2O_4) \cdot 6H_2O$. These two cyclohexaphosphate oxalates were prepared by Averbuch-Pouchot and Durif.[1108] They are isotypic. Their preparation is difficult and in some cases requires some months. A very dilute aqueous solution of the divalent cation oxalate is added dropwise (a few drops a day) to a solution of the alkali phosphate kept at room temperature. After some months,

Table 4.5.8. Main Crystallographic Data for Mixed Anion Compounds
Containing a P_6O_{18} Group

Formula	a (Å) α (°)	b (Å) β (°)	c (Å) γ (°)	S.G.	Z	Reference(s)
$Mg_2(NH_4)_4P_6O_{18}(C_2O_4) \cdot 6H_2O$	9.656(4) 98.88(5)	9.642(4) 106.21(5)	7.618(3) 100.37(5)	$P\bar{1}$	1	1108
$Mn_2(NH_4)_4P_6O_{18}(C_2O_4) \cdot 6H_2O$	9.747(3) 99.25(5)	9.751(3) 105.88(5)	7.689(3) 100.08(5)	$P\bar{1}$	1	1108
$(NH_4)_6P_6O_{18} \cdot NH_4Cl \cdot H_2O$	6.783(3) 101.48(2)	10.101(8) 90.84(3)	19.33(1) 107.31(2)	$P\bar{1}$	2	1110
$(NH_4)_6P_6O_{18} \cdot NH_4Br \cdot H_2O$	6.678(5) 100.58(5)	10.11(1) 91.34(5)	19.30(2) 107.23(5)	$P\bar{1}$	2	1110
$(NH_4)_6P_6O_{18} \cdot NH_4I \cdot H_2O$	14.96(1)	24.82(1) 91.86(5)	6.710(6)	$P2_1/n$	2	1110
$K_6P_6O_{18} \cdot KNO_3 \cdot H_2O$	9.700(7) 107.23(6)	9.749(8) 106.37(6)	6.543(5) 75.88(6)	$P1$	1	1109
$(NH_4)_6P_6O_{18} \cdot NH_4NO_3 \cdot H_2O$	10.090	10.146	6.710	$P1$	1	1109
$Rb_6P_6O_{18} \cdot RbNO_3 \cdot H_2O$	9.983 106.88	10.012 106.63	6.759 75.61	$P1$	1	1109

very sparingly water-soluble crystals of the phosphate oxalates appear in the solution. The crystal structure of this type of compounds was solved by using the manganese salt.

$M_6P_6O_{18}\cdot MNO_3\cdot H_2O$ (M = K, NH_4, Rb). Recently, Averbuch-Pouchot and Durif[1109] reported the existence of a series of three isotypic cyclohexaphosphate nitrates of general formula $M_6P_6O_{18}\cdot MNO_3\cdot H_2O$, where M = K, NH_4, and Rb. These salts are easily prepared by slow evaporation at room temperature of an aqueous solution of the two components in a stoichiometric ratio. The crystal structure of this new type of compounds was determined with the potassium salt.

$(NH_4)_6P_6O_{18}\cdot NH_4X\cdot H_2O$ (X = F, Cl, Br, I). The first evidence for the existence of a cyclohexaphosphate halide was obtained several years ago during the final stage of crystallization of aqueous solutions of ammonium cyclohexaphosphate monohydrate, $(NH_4)_6P_6O_{18}\cdot H_2O$, prepared by a metathesis reaction using the corresponding silver salt as starting material. The formation of the cyclohexaphosphate halide was attributed to the use of an impure silver cyclohexaphosphate or to a small excess of ammonium chloride during the metathesis reaction. The crystals obtained were few and of poor quality. Nevertheless, a rapid structural investigation was performed at that time to confirm the chemical formula. Subsequently, a reproducible procedure for the preparation of the chloride derivative was elaborated, and, later on, the corresponding bromide derivative was proved to be isostructural. The crystal structure of the chloride derivative was solved by Averbuch-Pouchot and Durif.[1109] The iodide derivative, whose crystal structure is very similar to that of the chloride derivative was investigated by the same authors.[1110] A fluoride derivative has also been characterized but is still under investigation.

4.5.3. Chemical Preparations of Cyclohexaphosphates

Because the investigation of cyclophosphates is a very recently developed branch of cyclophosphate chemistry, it seems useful to report in detail the main chemical processes used today for the preparation of cyclohexaphosphates.

4.5.3.1. The Griffith–Buxton Process

Griffith and Buxton[1048] were the first to optimize the cyclization-condensation reaction in the Li_2O–P_2O_5 system leading to the formation of lithium cyclohexaphosphate. The exact procedure as described by these authors is given here. To 115.3 g of H_3PO_4 (85%) is added 51.7 g of Li_2CO_3. When the reaction has subsided, the slurry obtained is heated for 1 h at 473 K; the temperature is then increased to 548 K and maintained for 5 h. The solid thus obtained is ground and added to 800 cm^3 of water. After 15 min of stirring, the insoluble material (probably long-chain lithium polyphosphate) is removed by filtration. The resulting solution contains both diphosphate and cyclohexaphosphate. This solution is then passed through a column of a strong acid ion-exchange resin in its protonated form. The solution is then neutralized with sodium carbonate until the pH is between 5 and 6, and 300 cm^3 of methyl alcohol are added. The first compound to crystallize is $Na_6P_6O_{18}\cdot 6H_2O$. If an excess of methyl alcohol is added, precipitation of the diphosphate occurs. The precipitated $Na_6P_6O_{18}\cdot 6H_2O$ is filtered and air dried. If the sample contains an excessive amount of the diphosphate, the purity can be improved by redissolving it in water and reprecipitating with methyl alcohol. Between 10 and 20 g of sodium cyclohexaphosphate are obtained from this procedure.

4.5.3.2. The Schülke–Kayser Process

The Griffith–Buxton process has recently been greatly improved and simplified by Schülke and Kayser.[1052] This process is based on the structure-controlled thermal dehydration–condensation of LiH_2PO_4 seeded by $Li_6P_6O_{18}$ crystals. The experimental procedure is as follows. A mixture of 104 g of LiH_2PO_4 and 21 g of $Li_6P_6O_{18}\cdot6H_2O$ is finely ground and heated at 623 K for 30 min. The resulting product is then ground and dissolved in 500 ml of water. Usually, three hours of mechanical stirring are necessary to dissolve the product. The insoluble part, mainly $LiPO_3$ long-chain polyphosphate, is removed by filtration, and 500 ml of methanol is then added to the solution, leading to a complete precipitation of $Li_6P_6O_{18}\cdot6H_2O$ within one day. The yield is almost quantitative. Schematically, the reaction is

$$6LiH_2PO_4 + 0.1Li_6P_6O_{18} \xrightarrow[30\ min]{623\ K} 1.1Li_6P_6O_{18} + 6H_2O$$

4.5.3.3. The Boullé's Process

Since the characterization of the silver salt, $Ag_6P_6O_{18}\cdot H_2O$, by Averbuch-Pouchot,[1053] a metathesis reaction derived from Boullé's process[805] has been commonly used for the preparation of water-soluble cyclohexaphosphates. The reaction is

$$Ag_6P_6O_{18}\cdot H_2O + 6MCl \rightarrow M_6P_6O_{18} + 6AgCl$$

Many cyclohexaphosphates were prepared using this process, including all the presently known organic cation cyclohexaphosphates.

4.5.3.4. The Use of Fluorides

The insolubility of LiF can be used to prepare most of the alkali cyclohexaphosphates according to the following reaction:

$$6MF + Li_6P_6O_{18} \rightarrow M_6P_6O_{18} + 6LiF$$

4.5.3.5. Ion-Exchange Resins

Production of cyclohexaphosphoric acid from $Na_6PO_{18}\cdot6H_2O$ or $Li_6PO_{18}\cdot6H_2O$ is a common process for the synthesis of the alkali or other cyclo-hexaphosphates.

4.5.3.6. Flux Methods

Flux methods have been successfully used for the production of some anhydrous cyclohexaphosphates (e.g., $Cu_2Li_2P_6O_{18}$, $Cr_2P_6O_{18}$). As has already been mentioned, the optimization of a flux method for the production of a given compound is a time-consuming and tedious undertaking.

4.5.3.7. Classical Methods

Most of the monovalent–divalent cation cyclohexaphosphates can be prepared as polycrystalline samples by very classical reactions such as

$$(NH_4)_6P_6O_{18} + 2Cd(NO_3)_2 \rightarrow Cd_2(NH_4)_2P_6O_{18} + 4NH_4NO_3$$

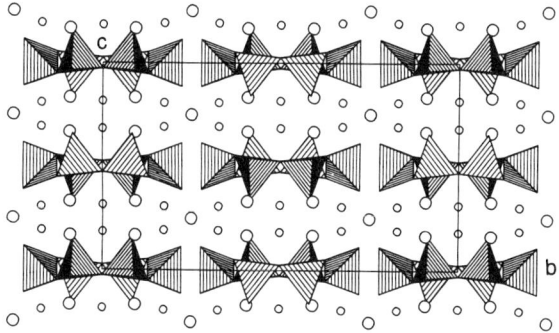

Figure 4.5.1. Projection, along the *a* direction, showing the layer organization of $Na_6P_6O_{18} \cdot 6H_2O$ structure. The smaller circles represent the sodium atoms, and the larger ones the water molecules.

However, the production of single crystals is achieved by rather unorthodox processes that we cannot report here in detail.

4.5.4. Some Atomic Arrangements of Cyclohexaphosphates

4.5.4.1. Sodium Cyclohexaphosphate Hexahydrate

$$Na_6P_6O_{18} \cdot 6H_2O,^{34,1047} \text{ orthorhombic, } Ccmb, Z = 4$$

$$a = 11.58(2), \quad b = 18.54(4), \quad c = 10.48(2) \text{ Å}$$

This structure is a typical layer organization. As shown by Fig. 4.5.1, a projection along the **a** direction, planes of ring anions in $z = 0$ and 1/2 alternate with layers containing the sodium atoms and the water molecules. A projection along the **c** axis (Fig. 4.5.2) shows the stacking of the ring anions, sodium atoms and water molecules in this very symmetrical arrangement. The ring anion has $2/m$ internal symmetry and is thus built by only two independent PO_4 tetrahedra.

One of the two crystallographically independent sodium atoms is located in a mirror plane, and the second one occupies a general position. Both are surrounded by distorted oxygen octahedra, with Na–O distances ranging from 2.311 to 2.507 Å*. The Na–Na distances (3.121 and 3.185 Å) are relatively short.

All the water molecules are involved in sodium coordination polyhedra.*

4.5.4.2. Ammonium Cyclohexaphosphate Monohydrate

$$(NH_4)_6P_6O_{18} \cdot H_2O,^{1064} \text{ trigonal (rhombohedral), } R\bar{3}$$

$$a = 15.445(10), \quad c = 7.553(7) \text{ Å}, \quad Z = 3 \text{ (hexagonal setting)}$$

---

*Values recalculated by the author.

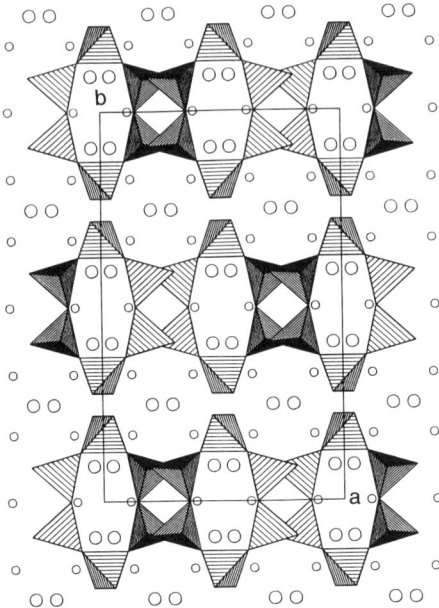

Figure 4.5.2. Projection, along the **c** direction, of the atomic arrangement in $Na_6P_6O_{18} \cdot 6H_2O$. The smaller circles represent the sodium atoms, and the larger ones the water molecules.

$$a = 9.266 \text{ Å}, \quad \alpha = 112.91°, \quad Z = 1 \text{ (rhombohedral setting)}$$

The atomic arrangement, depicted in Fig. 4.5.3 by a projection along the **c** axis, can be simply described by examining what occurs along a threefold axis. The P_6O_{18} ring anions located around this $\bar{3}$ axis alternate with pseudohexagonal rings composed of six ammonium groups. These $(NH_4)_6$ rings have at their center a water molecule located on the threefold axis. This water molecule is not involved in the associated-cation coordination. The P_6O_{18} group and the $(NH_4)_6$ ring are separated by a distance of $c/2$. Located close to $z = 0$, the phosphoric ring anion has $\bar{3}$ internal symmetry and is thus built by only one independent PO_4 tetrahedron.

Within a range of 3.5 Å, the ammonium group has eight neighbors with N–O distances ranging from 2.796 to 3.264 Å.

The corresponding silver salt, $Ag_6P_6O_{18} \cdot H_2O$, has a very similar arrangement, but, owing to slight differences in the orientations of some of the PO_4 groups, it is not possible to affirm its isotypy with the ammonium salt.

4.5.4.3. Potassium Cyclohexaphosphate Trihydrate

$$K_6P_6O_{18} \cdot 3H_2O,^{[1062]} \text{ monoclinic, } P2_1/m, Z = 2$$

$$a = 6.803(5), \quad b = 17.447(2), \quad c = 9.195(8) \text{ Å}, \quad \beta = 107.18(6)°$$

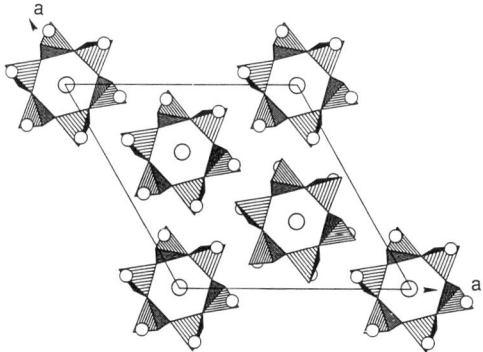

Figure 4.5.3. Projection, along the *c* axis, of the atomic arrangement in $(NH_4)_6P_6O_{18} \cdot H_2O$. The PO_4 groups are represented by hatched tetrahedra; the larger circles represent the water molecules, and the smaller ones the ammonium groups.

In this atomic arrangement the P_6O_{18} group has internal mirror symmetry, two bonding oxygen atoms (diametrical opposites) of the ring being located in a mirror plane. This compound represents the only known example of such an internal symmetry for a P_6O_{18} group.

As shown by Fig. 4.5.4, in the mirror planes ($y = 1/4$ and $3/4$) are also located two of the four crystallographically independent potassium atoms as well as one of the two independent water molecules. Two of the potassium atoms have eightfold oxygen coordination, and the other two sevenfold. Within these coordination polyhedra the K–O

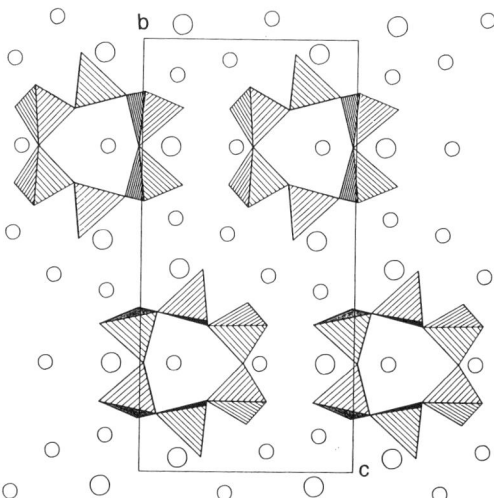

Figure 4.5.4. Projection of the $K_6P_6O_{18} \cdot 3H_2O$ structure along the *a* direction. The smaller circles represent the potassium atoms, and the larger ones the water molecules.

distances range from 2.704 to 3.158 Å. By sharing vertices, edges, and faces the KO_n polyhedra form corrugated layers perpendicular to the **a** direction. All the water molecules are involved in the potassium coordination polyhedra.

4.5.4.4. Lithium–Sodium Cyclohexaphosphate Dodecahydrate.

$$Li_3Na_3P_6O_{18}\cdot 12H_2O,^{[1068]} \text{ trigonal (rhombohedral), } R\bar{3}c$$

$$a = 10.474(8), \quad c = 41.68(5) \text{ Å}, \quad Z = 6 \text{ (hexagonal setting)}$$

$$a = 15.152 \text{ Å}, \quad \alpha = 40.44°, \quad Z = 2 \text{ (rhombohedral setting)}$$

In spite of its apparent complexity and of the large size of the unit cell, the atomic arrangement in this salt can be easily described as a succession of planes perpendicular to the threefold axis. A first family of planes contains the P_6O_{18} ring anions. These planes are located approximately in $z = nxc/6$ and thus are separated by a distance of about 7 Å. The second family of planes contains a two-dimensional network of corner sharing LiO_4 tetrahedra and NaO_6 octahedra, with large voids centered around the $\bar{3}$ axes. In addition, isolated $Na(H_2O)_6$ octahedra are located at the centers of these voids. These planes alternate with those containing the phosphoric anions, located approximately in $z = (2n+1)c/12$. The two independent P_6O_{18} ring anions have a $\bar{3}$ internal symmetry. Figure 4.5.5, a projection along the **b** axis, shows a group of planes of P_6O_{18} ring anions, while Fig. 4.5.6 gives a representation of the two-dimensional network of LiO_4 and NaO_6 octahedra situated in $z = c/12$.

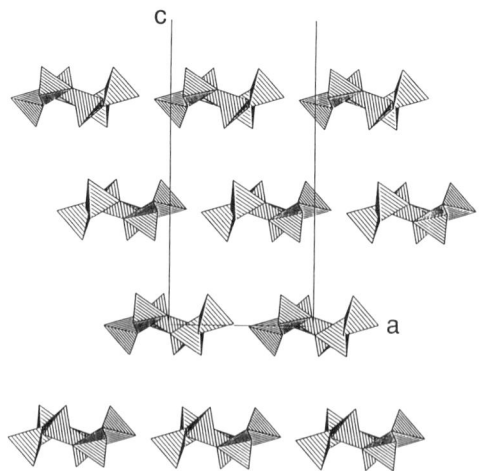

Figure 4.5.5. Projection, along the **b** axis, of a group of planes of P_6O_{18} ring anions in the $Li_3Na_3P_6O_{18}\cdot 12H_2O$ atomic arrangement.

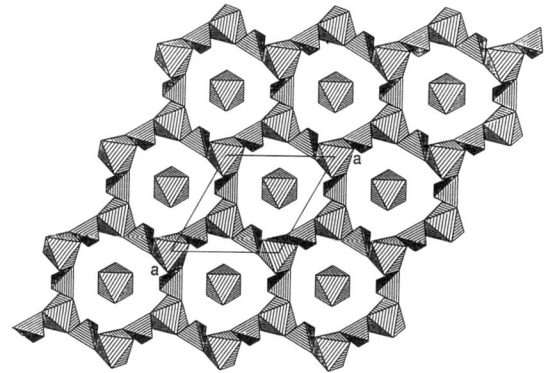

Figure 4.5.6. Projection, along the c axis, of the two-dimensional network of LiO_4 and NaO_6 octahedra situated in $z = c/12$ in the $Li_3Na_3P_6O_{18}\cdot12H_2O$ atomic arrangement. The isolated octahedra represent the $Na(H_2O)_6$ groups.

4.5.4.5. Lithium–Silver Cyclohexaphosphate Dihydrate

$$Li_2Ag_4P_6O_{18}\cdot2H_2O,^{1070}\text{triclinic, } P\overline{1}, Z = 1$$

$$a = 8.408(2), \quad b = 7.602(2), \quad c = 7.566(2) \text{ Å}$$

$$\alpha = 107.47(3), \quad \beta = 106.09(3), \quad \gamma = 72.64(3)°$$

This atomic arrangement is a typical layer organization in which planes of corner-sharing LiO_4 tetrahedra and P_6O_{18} ring anions alternate with planes of Ag atoms. The lithium atoms have tetrahedral coordination, involving three oxygen atoms and one water molecule. These tetrahedra share their three oxygen atoms with three different P_6O_{18} groups. The phosphoric groups located around the inversion center at 0, 0, 0 are centrosymmetric and share six of their external oxygen atoms with six adjacent $LiO_3(H_2O)$ tetrahedra. Such a network forms tetrahedral layers parallel to the (a, b) plane. One such

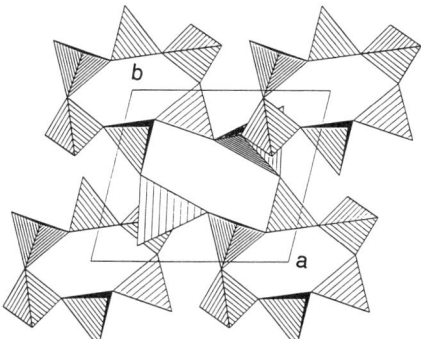

Figure 4.5.7. Projection, along the c direction, of a layer of P_6O_{18} and $LiO_3(H_2O)$ groups in the $Li_2Ag_4P_6O_{18}\cdot2H_2O$ atomic arrangement. The silver atoms are not represented. The projection is limited to $-0.45 < z < 0.45$.

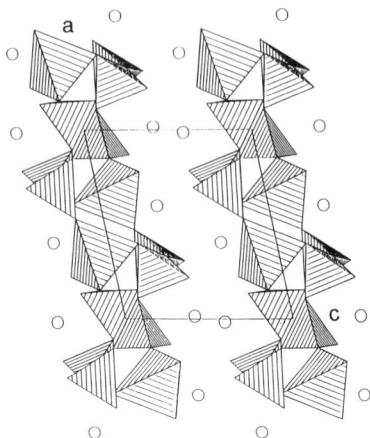

Figure 4.5.8. Projection, along the **b** direction, of the atomic arrangement in Li$_2$Ag$_4$P$_6$O$_{18}$·2H$_2$O. The open circles represent the silver atoms.

layer is shown in Fig. 4.5.7, a projection along the **c** direction. All Ag atoms are situated between these layers. Within a range of 3 Å, one of the two independent silver atoms is fivefold coordinated by five oxygen atoms, and the other one has sixfold coordination including one water molecule. Within the Ag atom layer, the shortest Ag–Ag distances are 3.140 and 3.304 Å. Figure 4.5.8 shows clearly this layer organization.

4.5.4.6. Silver–Ammonium Cyclohexaphosphate Monohydrate

$$Ag_3(NH_4)_3P_6O_{18} \cdot H_2O, ^{1069} \text{ trigonal, } R\bar{3}$$

$$a = 15.172(5), \quad c = 13.994(5) \text{ Å}, \quad Z = 6 \text{ (hexagonal description)}$$

$$a = 9.924 \text{ Å}, \quad \alpha = 99.71°, \quad Z = 2 \text{ (rhombohedral setting)}$$

The unit cell is closely related to those of two very similar phosphates, Ag$_6$P$_6$O$_{18}$·H$_2$O and (NH$_4$)$_6$P$_6$O$_{18}$·H$_2$O. The observed a value is approximately halfway between the values measured for the ammonium and silver salts, and the c value is close to the sum of those measured for these two phosphates. These observations suggest that the atomic arrangement of this salt must be very similar to that determined for Ag$_6$P$_6$O$_{18}$·H$_2$O and (NH$_4$)$_6$P$_6$O$_{18}$·H$_2$O, with, in addition, an order established between the associated cations to explain the doubling of the **c** axis. The determination of the atomic arrangement confirmed these assumptions. Here, two independent P$_6$O$_{18}$ ring anions coexist in the structure, both located around the $\bar{3}$ axis, the first one at $z = 0$ and the second one at $z = 1/2$, thus both of them have $\bar{3}$ internal symmetry. Halfway between these groups, in $z = 1/4$ and 3/4, are located pseudo-hexagonal rings of associated cations, made up three silver atoms and three ammonium groups. At the centers these rings of associated cations, built around the $\bar{3}$ axis, are the water molecules located on this axis. Thus, the two main differences between the atomic arrangement in this salt and that in Ag$_6$P$_6$O$_{18}$·H$_2$O and (NH$_4$)$_6$P$_6$O$_{18}$·H$_2$O are the existence of two crystallographically independent P$_6$O$_{18}$ groups

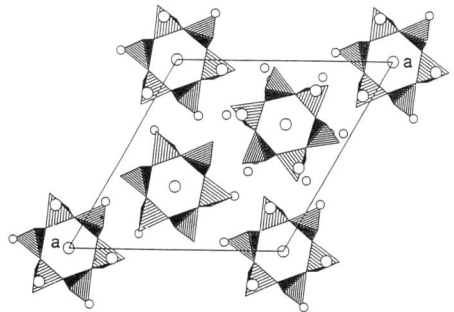

Figure 4.5.9. Projection, along the *c* axis, of $Ag_3(NH_4)_3P_6O_{18} \cdot H_2O$. The small open circles represent the silver atoms, and the larger ones the ammonium groups. The ring anions have water molecules at their centers.

and the order established in the hexagonal ring of associated cations. Figure 4.5.9 shows a projection of this structure along the *c* axis.

The corresponding lithium–potassium salt, $Li_3K_3P_6O_{18} \cdot H_2O$,[1068] is isotypic.

4.5.4.7. Copper Cyclohexaphosphate Tetradecahydrate

$$Cu_3P_6O_{18} \cdot 14H_2O,^{1077} \text{ triclinic, } P\bar{1}, Z = 1$$

$$a = 10.944(8), \quad b = 7.539(4), \quad c = 8.974(4) \text{ Å}$$

$$\alpha = 110.49(5), \quad \beta = 110.14(5), \quad \gamma = 77.82(5)°$$

There are two crystallographically independent copper atoms in this arrangement. The first one, located on the inversion centre at 0,0,0 is at the centre of a distorted octahedron built by four oxygen atoms and two water molecules. This octahedron shares its four oxygen atoms with two adjacent P_6O_{18} groups, building P_6O_{18}–$CuO_4(H_2O)_2$–P_6O_{18} infinite chains extending along the **c** axis. These chains are themselves interconnected by a centrosymmetric $Cu_2(H_2O)_8O_2$ cluster composed of two edge-sharing $CuO(H_2O)_5$ octahedra built around the second copper atom, located in a general position. In addition, it must be noted that two of the water molecules are not involved in the copper coordination. The phosphoric ring anion is centrosymmetric. Figure 4.5.10 gives a projection along **b** of this atomic arrangement.

4.5.4.8. Cadmium Cyclohexaphosphate Hexahydrate

$$Cd_3P_6O_{18} \cdot 6H_2O,^{1078} \text{ trigonal (rhombohedral)}$$

$$a = 15.056(8), \quad c = 16.080(8) \text{ Å}, \quad Z = 6 \text{ (hexagonal setting)}$$

$$a = 10.212 \text{ Å}, \quad \alpha = 94.98°, \quad Z = 2 \text{ (rhombohedral description)}$$

As shown in Fig. 4.5.11, the P_6O_{18} groups are located around the threefold axis. They are built by two independent PO_4 tetrahedra and have threefold internal symmetry. Along

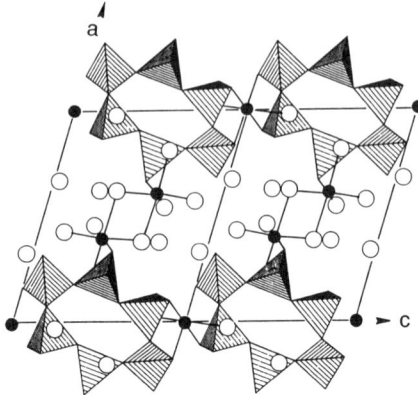

Figure 4.5.10. Projection of the atomic arrangement in $Cu_3P_6O_{18}\cdot14H_2O$ along the *b* axis. The closed circles represent the copper atoms and large empty circles figurate the water molecules. Hydrogen atoms have been omitted.

a threefold axis, these rings are separated by a distance of $c/2$ (8.04 Å) and are almost superimposed in projection. Also around the threefold axis, halfway between the phosphoric groups, are located rings composed of six water molecules. These water-molecule rings have threefold internal symmetry. Within such a ring, each water molecule is connected

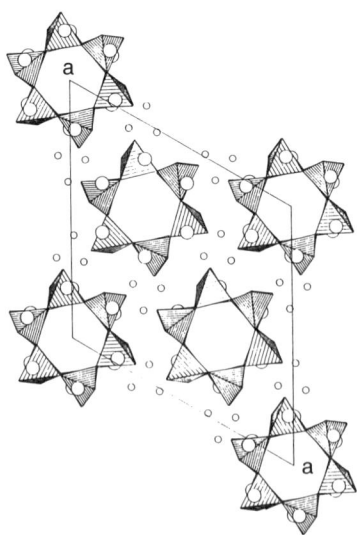

Figure 4.5.11. Projection of the atomic arrangement of $Cd_3P_6O_{18}.6H_2O$ along the *c* axis. Small circles are cadmium atoms, and the open circles represent the water molecules. The hydrogen atoms have been omitted.

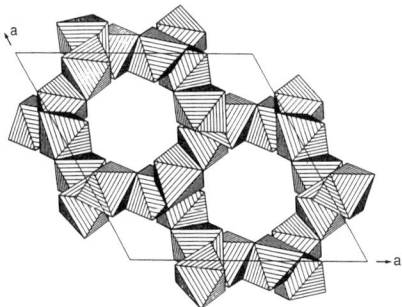

Figure 4.5.12. Projection, along the c axis, of the three-dimensional network of CdO_6 octahedra in the $Cd_3P_6O_{18}.6H_2O$ atomic arrangement. The large hexagonal channels contain the phosphoric ring anions.

to its neighbor by a hydrogen bond, the second hydrogen of the water molecule establishing an H bridge with an external oxygen atom of the phosphoric ring anion. The cadmium atom is coordinated by four oxygen atoms and two water molecules. As the cadmium atoms are located close to the helical axes, the CdO_6 octahedra form edge-sharing spirals around these axes. These spirals are themselves interconnected, thus forming a three-dimensional network of octahedra as shown in Fig. 4.5.12, a projection along the c direction. This network creates large hexagonal channels, parallel to the c axis, in which are located the P_6O_{18} anions.

The corresponding manganese salt is isotypic.[1079]

4.5.4.9. Barium Cyclohexaphosphate Octahydrate

$$Ba_3P_6O_{18}.8H_2O,^{[1082]} \text{ monoclinic, } C2/c, Z = 4$$

$$a = 20.98(2), \quad b = 7.227(3), \quad c = 17.44(3) \text{ Å}, \quad \beta = 119.56(3)°$$

In this arrangement, the P_6O_{18} group is located around a twofold axis. Contrary to the impression given by Fig. 4.5.13, a projection of the atomic arrangement along the **b** axis, the phosphoric ring anion is far from being flat. Projections along **a** and **c** for an isolated anion (Figure 4.5.14) show clearly an organization made by two levels of tetrahedra. Two crystallographically independent barium atoms are present in this arrangement, one in a general position, the other one on the twofold axis. Within a range of 3 Å, the first one has ten oxygen neighbors, including three water molecules, while the second one has a more regular eightfold coordination, including two water molecules. These barium polyhedra are located in two layers approximately centered by the planes $z = 1/4$ and $3/4$. Within these layers, through corner and edge sharing, they produce a tiling whose repetition unit is a ring of eight BaO_n polyhedra assembled so as to form a nonconvex polygon. Thus, the cohesion between the two layers of barium polyhedra is provided by the external oxygen atoms of the large phosphoric groups.

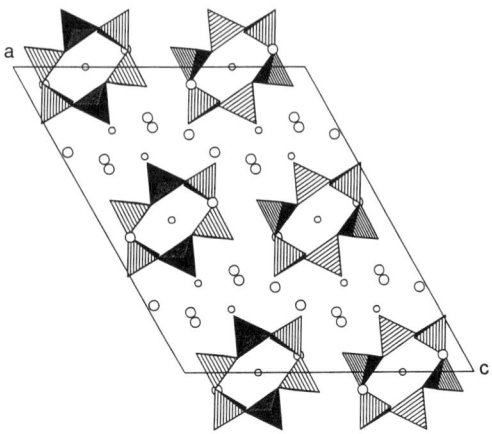

Figure 4.5.13. Projection of the atomic arrangement in $Ba_3P_6O_{18}\cdot8H_2O$ along the *b* axis. The small open circles represent the barium atoms, and the large ones the water molecules.

4.5.4.10. Copper–Ammonium Cyclohexaphosphate Octahydrate

$$Cu_2(NH_4)_2P_6O_{18}\cdot8H_2O,^{[1087]}\text{ triclinic, } P\bar{1}, Z = 1$$

$$a = 7.413(3), \quad b = 9.234(4), \quad c = 9.634(4) \text{ Å}$$

$$\alpha = 116.23(5), \quad \beta = 107.98(5), \quad \gamma = 83.10(5)°$$

The P_6O_{18} ring anion is centrosymmetric, located around the inversion center at 0, 0, 1/2. The two copper atoms are also located on inversion centers, the first one at 0, 0, 0, the second one at 1/2, 1/2, 0. A projection of the atomic arrangement along the **a** axis is

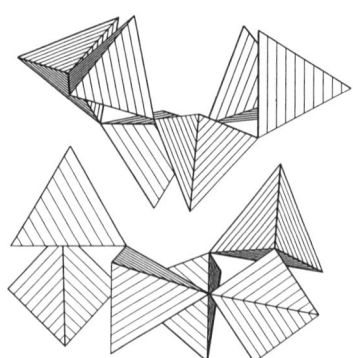

Figure 4.5.14. Projection along the *a* axis (top) and along the *c* axis (bottom) of the P_6O_{18} group in the $Ba_3P_6O_{18}\cdot8H_2O$ atomic arrangement.

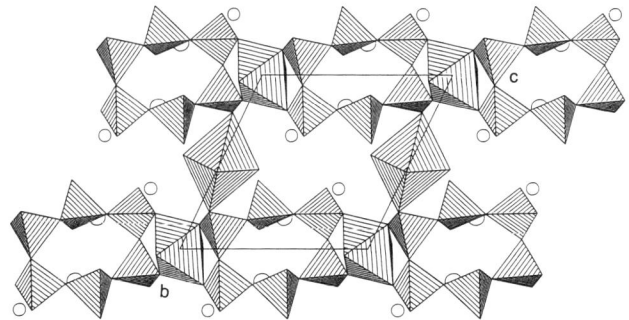

Figure 4.5.15. Projection, along the **a** axis, of the atomic arrangement in $Cu_2(NH_4)_2P_6O_{18} \cdot 8H_2O$. The open circles represent the ammonium groups.

presented in Fig. 4.5.15. The atomic arrangement can be easily described as a succession of ribbons built up by the P_6O_{18} phosphoric rings and one kind of CuO_6 octahedra and extending parallel to the **c** direction. These ribbons are themselves interconnected along the **b** direction by the second kind of CuO_6 octahedra so as to form a layer parallel to the (b, c) plane. The connections thus established by the CuO_6 octahedra, are not identical; one of the CuO_6 octahedron shares four of its oxygen atoms with the two adjacent P_6O_{18} groups, while the other one shares only two. The three-dimensional cohesion between these layers is established by the ammonium polyhedra and the hydrogen-bond network.

4.5.4.11. Chromium Cyclohexaphosphate

$$Cr_2P_6O_{18},^{1050} \text{ monoclinic, } P2_1/a, Z = 2.$$

$$a = 8.311(4), \quad b = 15.221(8), \quad c = 6.220(3) \text{ Å}, \quad \beta = 105.85(5) \text{ °}$$

As shown by Fig. 4.5.16, a projection along the **a** direction, the atomic arrangement can be considered as built by thick layers perpendicular to the **b** direction containing both chromium atoms and P_6O_{18} ring anions. The internal organization of such a layer is given by Fig. 4.5.17, a projection along the **b** axis of an isolated layer.

In addition, Fig. 4.5.18, a projection along **c**, the shortest axis, shows the respective locations of the chromium atoms and the phosphoric anions. The P_6O_{18} ring anion is centrosymmetric with a strong pseudo $2/m$ internal symmetry.

The chromium atoms have a rather regular octahedral coordination, with Cr–O distances ranging from 1.949 to 1.982 Å and O–Cr–O angles ranging from 88.9 to 91.0°. These CrO_6 octahedra do not share any oxygen atoms, and each of them is connected to four different P_6O_{18} groups, with Cr–O distances ranging between 3.15 and 3.28 Å.

The corresponding gallium[258] and aluminium[1094] salts are isotypic.

4.5.4.12. Chromium Cyclohexaphosphate Heneicosahydrate

$$Cr_2P_6O_{18} \cdot 21H_2O,^{1096} \text{ cubic, } P\bar{4}3n, Z = 8, a = 19.052(10) \text{ Å}$$

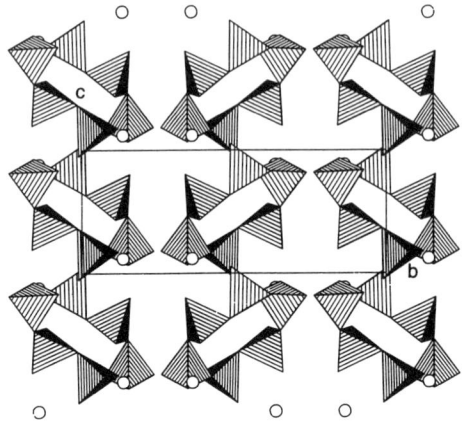

Figure 4.5.16. Projection of the structure of $Cr_2P_6O_{18}$ along the *a* direction, showing the layered arrangement of the structure. The open circles represent the chromium atoms.

This arrangement can be described as a stacking of P_6O_{18} ring anions, $Cr(H_2O)_6$ octahedra, and nonbonded water molecules. In such a framework, in which there is no common oxygen atom between the phosphoric anion and the associated-cation polyhedra, the three-dimensional cohesion is provided only by hydrogen bonds.

The unit cell contains eight P_6O_{18} rings located around the threefold axes and thus built by only two independent PO_4 tetrahedra. These rings are rather regular, with P–P–P angles ranging from 104.4 to 109.5 °.

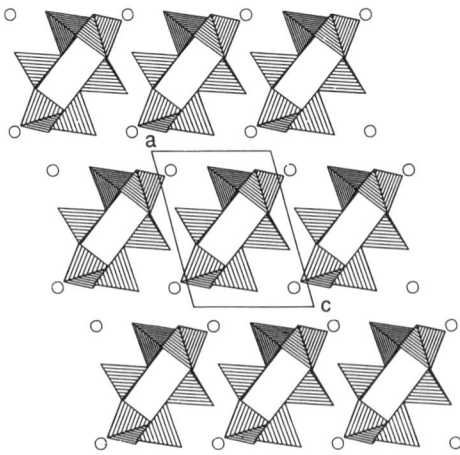

Figure 4.5.17. Projection along the *b* axis of an isolated layer $(0.25 < y < 0.75)$ in the $Cr_2P_6O_{18}$ structure.

Figure 4.5.18. Projection, along the *c* direction, of the atomic arrangement in $Cr_2P_6O_{18}$. The open circles represent the chromium atoms.

The $Cr(H_2O)_6$ octahedra do not share any corners or edges and adopt various internal symmetries (222, 3). Within these different $Cr(H_2O)_6$ octahedra the Cr–O distances range between 1.931 and 2.058 Å. In addition, it should be noted that 9 of the 21 water molecules are not involved in the associated-cation polyhedra.

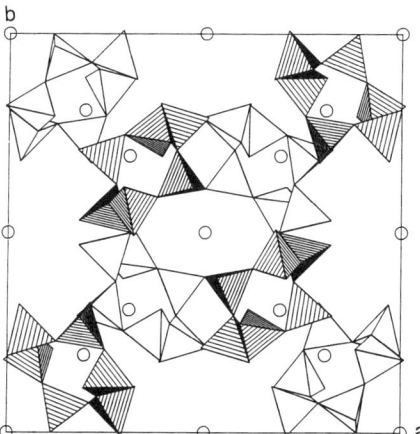

Figure 4.5.19. Projection, along the *c* axis, of the arrangement of P_6O_{18} ring anions and Cr atoms in $Cr_2P_6O_{18}\cdot21H_2O$. The water molecules have been omitted and some Cr atoms are superimposed in projection.

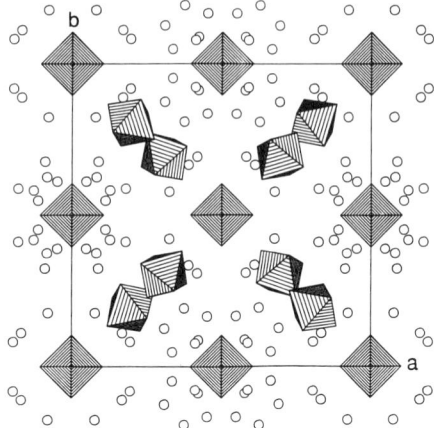

Figure 4.5.20. Projection along the **c** axis of $Cr(H_2O)_6$ octahedra and non bonded water molecules in $Cr_2P_6O_{18}\cdot21H_2O$. The ring anions have been omitted. Some of the $Cr(H_2O)_6$ groups are superimposed in projection.

A projection of this arrangement along the **c** direction is shown in Fig. 4.5.19, while Fig. 4.5.20 represents the respective locations of $Cr(H_2O)_6$ octahedra and nonbonded water molecules in a projection along the same direction.

4.5.4.13. Cerium Cyclohexaphosphate Decahydrate

$$Ce_2P_6O_{18}\cdot10H_2O,^{1098} \text{ orthorhombic, } P2_12_12, Z = 2$$

$$a = 13.522(5), \quad b = 13.105(9), \quad c = 6.938(3) \text{ Å}$$

The P_6O_{18} ring anion is located around the binary axis parallel to the **c** axis. This twofold symmetry is relatively rare for this kind of anion, the only other example being found in $Ca_2(NH_4)_2P_6O_{18}\cdot6H_2O$. It is interesting to note that the unit-cell parameters, space group and anionic distribution for the latter compound are similar to those for the cerium derivative that we describe here.

Figure 4.5.21 shows a projection along the **c** direction of this atomic arrangement. The cerium atoms are coordinated by six oxygen atoms of the phosphoric ring and three water molecules, forming a distorted three-capped trigonal prism, with Ce–O distances ranging from 2.441 to 2.556 Å and Ce–H_2O distances ranging from 2.602 to 2.645 Å. As shown by Fig. 4.5.22, the P_6O_{18} rings form corrugated layers perpendicular to the **c** axis.

Four of the ten water molecules in the formula do not belong to the cation coordination polyhedron and are located in the channels parallel to the **c** axis.

4.5.4.14. Ethylammonium Cyclohexaphosphate Tetrahydrate

$$(C_2H_5\cdot NH_3)_6\cdot P_6O_{18}\cdot4H_2O,^{1103} \text{ monoclinic, } P2_1/a, Z = 4$$

$$a = 16.804(10), \quad b = 23.883(10), \quad c = 10.623(8) \text{ Å}, \quad \beta = 109.66(2)°$$

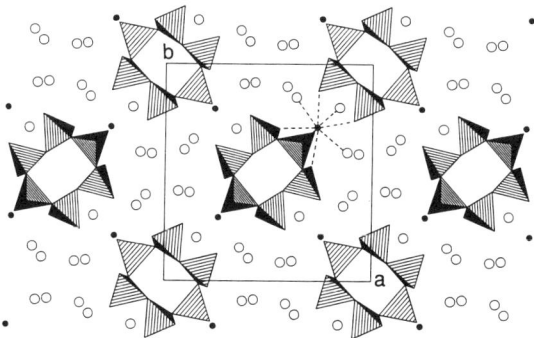

Figure 4.5.21. Projection along the c axis of the atomic arrangement in $Ce_2P_6O_{18} \cdot 10H_2O$. The closed circles represent the cerium atoms, and the open circles represent the water molecules.

This atomic arrangement has a very strong pseudo-hexagonal symmetry and can be simply described as being built up of arrays of ethylammonium groups and P_6O_{18} ring anions which are both parallel to the c axis. These arrays are arranged in an almost regular hexagonal way, as can be seen in Fig. 4.5.23. This pseudo-hexagonal stacking creates large channels parallel to the c direction; the water molecules are located in these channels. The phosphoric anions, with no internal symmetry, and the ethylammonium groups have their usual conformations.

It is worth noting that among the 18 H atoms belonging to the NH_3 radicals of the six independent organic entities, only one establishes a hydrogen bond with a water molecule; the remaining ones are connected to the external O atoms of the phosphoric rings. This can probably be attributed to the clustering of the water molecules described below.

The water molecules of this compound are not dispersed inside the arrangement but rather are assembled in centrosymmetric clusters of eight molecules. The central part of

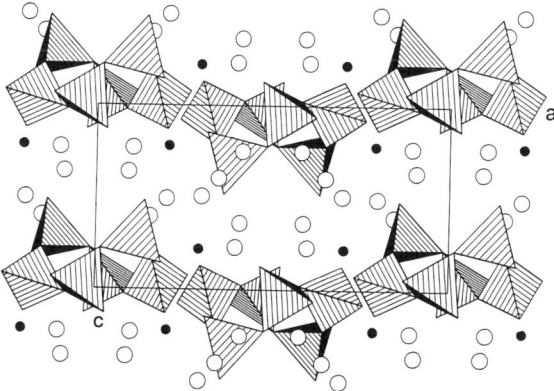

Figure 4.5.22. Projection along the a axis of the atomic arrangement in $Ce_2P_6O_{18} \cdot 10H_2O$. The closed circles represent the cerium atoms, and the open circles represent the water molecules.

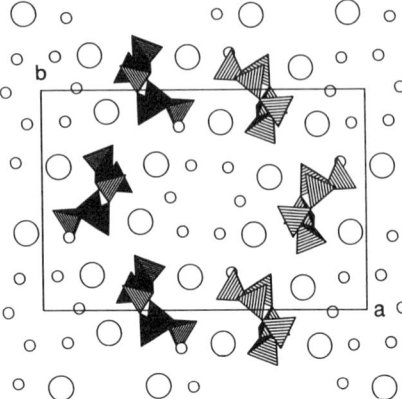

Figure 4.5.23. Projection along the *c* axis of the atomic arrangement in $(C_2H_5NH_3)_6P_6O_{18}\cdot4H_2O$. The small open circles represent the water molecules, and the large ones represent the channels containing the arrays of ethylammonium groups.

the cluster is an almost regular hexagon with H_2O–H_2O–H_2O angles ranging between 111.1 and 123.7° and H_2O–H_2O distances ranging between 2.751 and 2.775 Å. Inside this ring the water molecules are connected by strong H bonds. Two centrosymmetric branches complete the cluster. The water molecules building these branches, located at a distance of 2.917 Å from the hexagonal ring, are not connected by H bonds to the neighboring water molecules of the ring. Similar polygonal clusters of water molecules have been observed in other condensed phosphates: an almost regular pentagon in $Li_4P_4O_{12}\cdot5H_2O$ and a hexagon of symmetry $\bar{3}$ in $Cd_3P_6O_{18}\cdot6H_2O$.

4.5.4.15. Potassium Cyclohexaphosphate–Telluric Acid Trihydrate

$K_6P_6O_{18}\cdot2Te(OH)_6\cdot3H_2O$,[1106] trigonal (rhombohedral), $R\bar{3}c$

$a = 13.084(5),\quad c = 34.80(2)$ Å, $Z = 6$ (hexagonal setting)

$a = 13.843$ Å, $\alpha = 56.40°$, $Z = 2$ (rhombohedral setting)

The atomic arrangement can be described in terms of the alternation of two types of planes perpendicular to the threefold axis and separated by a distance of about $c/12$. Planes of the first type contain the P_6O_{18} groups, $Te(OH)_6$ octahedra, and one kind of potassium atoms. Figure 4.5.24 shows a projection of such a plane along the **c** axis. Planes of the second type contain the water molecules and the second type of potassium atoms. Figure 4.5.25 gives the respective locations of water molecules and potassium atoms in such a plane. The cohesion between these planes is established through the potassium polyhedra and the hydrogen bonds. The P_6O_{18} groups have $\bar{3}$ internal symmetry.

4.5.4.16. Rubidium Cyclohexaphosphate–Telluric Acid Trihydrate

$Rb_6P_6O_{18}\cdot3Te(OH)_6\cdot4H_2O$,[1106] triclinic, $P\bar{1}$, $Z = 1$

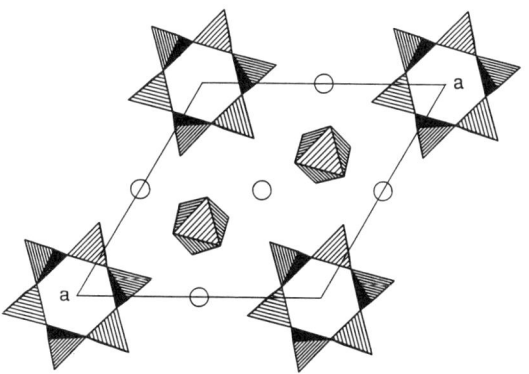

Figure 4.5.24. Projection, along the *c* axis, of a plane of the first type in $K_6P_6O_{18} \cdot 2Te(OH)_6 \cdot 3H_2O$. The open circles represent the potassium atoms, and the hatched octahedra represent $Te(OH)_6$ groups.

$$a = 11.222(8), \quad b = 8.077(6), \quad c = 11.731(9) \text{ Å}$$

$$\alpha = 111.11(2), \quad \beta = 104.66(2), \quad \gamma = 83.25(2)°$$

The atomic arrangement in this salt can be described as a very simple stacking of $Te(OH)_6$ and P_6O_{18} groups, which are both located around inversion centers of the triclinic cell and are interconnected by RbO_7 and RbO_9 polyhedra and water molecules. Within the various RbO_n polyhedra, the Rb–O distances range between 2.889 and 3.454 Å. Two of the three independent $Te(OH)_6$ groups aligned along the **c** axis are interconnected by hydrogen bonds whereas the third one is not hydrogen bonded to the other telluric groups. The projection of this very simple arrangement along the **b** direction is shown in Fig. 4.5.26.

The corresponding cesium salt, $Cs_6P_6O_{18} \cdot 3Te(OH)_6 \cdot 4H_2O$,[1106] is isotypic.

4.5.4.17. Manganese–Ammonium Cyclohexaphophate Oxalate

$$Mn_2(NH_4)_4P_6O_{18}(C_2O_4) \cdot 6H_2O,^{[1108]} \text{ triclinic, } P\bar{1}, Z = 1$$

$$a = 9.747(3), \quad b = 9.751(3), \quad c = 7.689(3) \text{ Å}$$

$$\alpha = 99.25(5), \quad \beta = 105.88(5), \quad \gamma = 100.08(5)°$$

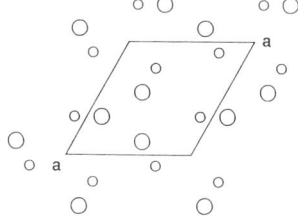

Figure 4.5.25. A plane of the second type in $K_6P_6O_{18} \cdot 2Te(OH)_6 \cdot 3H_2O$. The smaller open circles represent the potassium atoms, and the larger ones represent the water molecules.

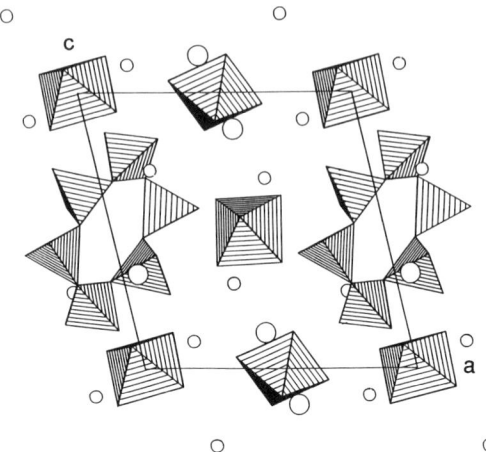

Figure 4.5.26. Projection, along the **b** direction, of the atomic arrangement in Rb$_6$P$_6$O$_{18}$·3Te(OH)$_6$·4H$_2$O. The smaller open circles represent the rubidium atoms, and the larger ones the water molecules.

The atomic arrangement as represented in Fig. 4.5.27 can be easily described as built by infinite ribbons of P$_6$O$_{18}$ ring anions and MnO$_6$ octahedra extending parallel to the **b** axis. These ribbons are themselves interconnected by the oxalic groups located between two MnO$_6$ octahedra belonging to two different ribbons.

The P$_6$O$_{18}$ ring anion is located around the inversion center located at 1/2,1/2,1/2 and the oxalic group around the inversion center situated at the origin.

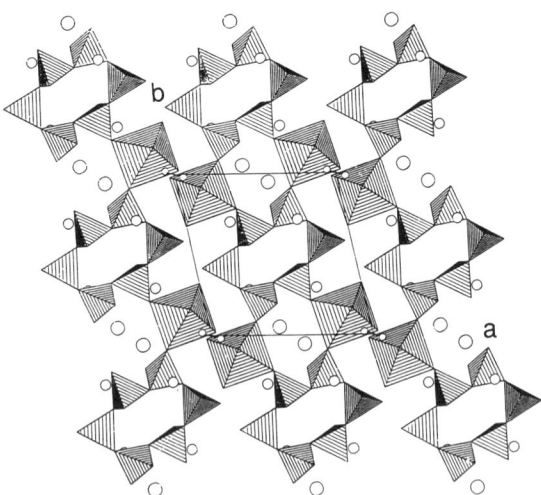

Figure 4.5.27. Projection, along the **c** direction, of the atomic arrangement in Mn$_2$(NH$_4$)$_4$P$_6$O$_{18}$(C$_2$O$_4$)6·H$_2$O. The hatched octahedra represent the MnO$_6$ groups; the largest open circles represent the water molecules, the intermediate ones the ammonium groups, and the smallest ones the carbon atoms. The hydrogen atoms have been omitted.

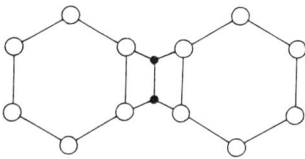

Figure 4.5.28. Schematic representation of the situation around the inversion center at (0,0,0) in the structure of $Mn_2(NH_4)_4P_6O_{18}(C_2O_4)\cdot 6H_2O$. The two related MnO_6 octahedra are represented by hexagons. The closed circles represent the carbon atoms of the oxalate group.

Owing to the connection of the ribbons by the oxalic groups, an unusual situation occurs around the inversion center situated at the origin of the unit cell. Two manganese octahedra are related by this inversion center, and a very short O–O distance (2.232 Å) is observed between these two polyhedra. This situation is depicted schematically in Fig. 4.5.28.

Within a range of 3.50 Å, the two independent ammonium groups are surrounded by nine and seven oxygen neighbors, with N–O distances ranging from 2.846 to 3.426 Å.

The MnO_6 octahedron is rather regular, with Mn–O distances ranging from 2.085 to 2.242 Å. In addition to two external oxygen atoms of the phosphoric ring, two water molecules and two oxygen atoms of the oxalic group take part in this coordination polyhedron.

$Mg_2(NH_4)_4P_6O_{18}(C_2O_4)\cdot 6H_2O$, the corresponding magnesium–ammonium compound, is isotypic.[1108]

4.5.4.18. *Potassium Cyclohexaphosphate–Nitrate Monohydrate*

$$K_6P_6O_{18}\cdot KNO_3\cdot H_2O,^{[1109]} \text{ triclinic, } P1, Z = 1$$

$$a = 9.700(7), \quad b = 9.749(8), \quad c = 6.543(5) \text{ Å}$$

$$\alpha = 107.23(6), \quad \beta = 106.37(6), \quad \gamma = 75.88(6)°$$

The highly pseudo-centrosymmetric stacking of the P_6O_{18} anions and of the potassium atoms creates channels parallel to the **c** axis in which are located the NO_3 groups and the water molecules, interconnected by hydrogen bonds so as to form—NO_3–H_2O–NO_3–H_2O chains. As shown by Fig. 4.5.29, the P_6O_{18} ring anion is located around the origin of the unit cell and is highly pseudocentrosymmetric.

The seven independent potassium atoms have, within a range of 3.50 Å for the K–O distances, various coordination polyhedra consisting of six to ten oxygen atoms or water molecules with K–O distances ranging from 2.624 to 3.500 Å. In a projection along **c**, as given by Fig. 4.5.29, the potassium atoms are in a pseudo-hexagonal arrangement. A set of six of them is located around the origin of the unit cell, forming a channel including the phosphoric groups. In addition, this pseudo-hexagonal channel of potassium atoms parallel to the **c** axis is centered around one of the potassium atoms located at 0,0,1/2, so that phosphoric groups and potassium atoms alternate along the **c** axis. Another pseudo-hex-

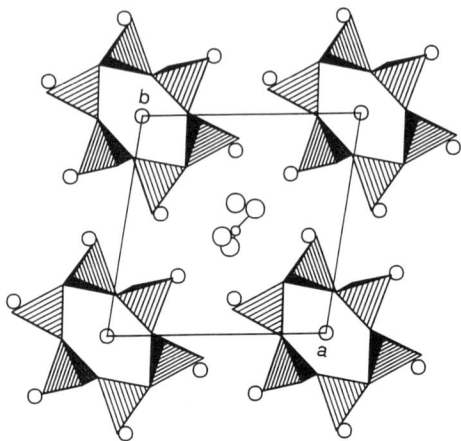

Figure 4.5.29. Projection, along the *c* axis, of the atomic arrangement in $K_6P_6O_{18} \cdot KNO_3 \cdot H_2O$. The smallest open circles represent the nitrogen atoms, the intermediate ones the potassium atoms, and the largest ones the oxygen atoms. The hydrogen atoms have been omitted.

agonal channel of potassium atoms develops, parallel to the **c** axis, at the center of the unit cell. In this last channel are situated the NO_3 groups and the water molecules.

Within the NO_3 group, the N–O distances range from 1.245 to 1.284 Å, and the O–N–O angles from 117.8 to 122.1°, and the O–O distances from 2.161 to 2.213 Å. From the three O–N–O angle values, this group can be considered planar within the experimental errors.

The two hydrogen bridges built by the water molecule connect it to two different NO_3 groups, thus establishing $-NO_3-H_2O-NO_3-H_2O-$ chains lying parallel to the **c** axis inside the second type of tunnel built by the potassium atoms. Figure 4.5.30, a projection along the **b** direction, shows clearly the organization of such a chain.

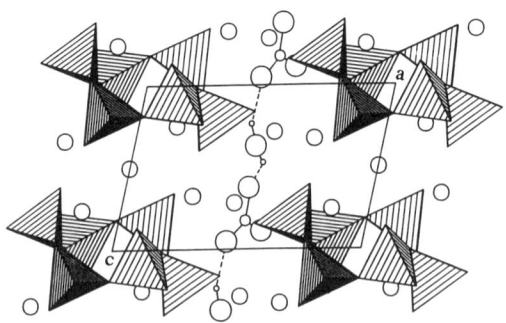

Figure 4.5.30. Projection, along the *b* axis, of the atomic arrangement in $K_6P_6O_{18} \cdot KNO_3 \cdot H_2O$. In order of decreasing size, the open circles represent the oxygen, the potassium, the nitrogen, and the hydrogen atoms.

4.5.5. Geometry of the P_6O_{18} Rings

To date more than 40 accurate determinations of cyclohexaphosphate atomic arrangements were performed, making it possible to compare of a good number of different P_6O_{18} ring anions.

Before we examine the bond distances and angles in these rings, it seems interesting to classify them according to their internal symmetry. Among all the presently known examples, 3 have no internal, 1 has m symmetry, 1 has $2/m$ symmetry, 2 have 3 symmetry, 7 have $\bar{3}$ and 18 have $\bar{1}$ symmetry.

Despite the small number of representatives, the examination of bond distances and angles in these rings seems worthwhile. We limit this examination to what we consider to be the main geometric features of a ring: the P–P distances and the P–O–P and P–P–P angles.

The observed P–P distances are quite comparable to those in other condensed phosphoric anions. The overall average P–P distance is 2.927 Å, and the spread of the observed values is small, the extreme values being 2.857 and 3.029 Å. The P–O–P angles also exhibit values typical for condensed phosphoric anions, with an average value of 132.4 and extreme values of 145.9 and 123.4°.

The situation is very different when one compares the observed P–P–P angles in cyclohexaphosphates with those in well-represented types of phosphoric rings such as cyclotri- and cyclotetraphosphates. In these two classes of cyclophosphates, the P–P–P angles never depart significantly from their ideal values, $60 \pm 2°$ for cyclotriphosphates and $90 \pm 4°$ for cyclotetraphosphates. In the case of the P_6O_{18} ring anions, if we examine first the most common ones ($\bar{1}$ internal symmetry) very large deviations from the ideal value (120°) are observed. The average angle is 111.9° with extrema of 85.9 and 142.8°. For higher symmetry rings (3 and $\bar{3}$), one could expect a more regular distribution of the P–P–P angles because of the higher ring strain, but here also the spread in the observed values is large. The average is 109.2°, with 87.5 and 131.7° as extreme values.

Most of the structural data available for cyclohexaphosphates are recent. Thus, for the numerical values given in Tables 4.5.9–4.5.11, the estimated standard deviations can be evaluated as within 0.001 Å for the P–P distances, between 0.01 and 0.05° for the P–P–P angles, and less than 0.1° for the P–O–P angles.

Table 4.5.9. Main Geometrical Features of Centrosymmetric P_6O_{18} Ring Anions[a]

Formula	P–P (Å)	P–O–P (°)	P–P–P (°)	Reference
$Li_6P_6O_{18} \cdot 5H_2O$	2.927	133.0	117.5	1054
	2.927	131.7	106.1	
	2.915	133.1	122.6	
$Li_6P_6O_{18} \cdot (5 + n)H_2O$	2.958	137.2	126.9	1055
	2.935	135.2	111.2	
	2.930	134.3	121.8	
$Cs_2(UO_2)_2P_6O_{18}$	2.866	131.7	107.9	1051
	2.918	129.6	99.1	
	2.857	129.4	105.6	

(continued)

Table 4.5.9. (continued)

Formula	P–P (Å)	P–O–P (°)	P–P–P (°)	Reference
$Cs_6P_6O_{18}\cdot6H_2O$	2.960	133.5	93.2	1066
	2.918	127.0	111.5	
	2.880	129.9	142.5	
$Ag_4Li_2P_6O_{18}\cdot2H_2O$	2.902	145.9	94.9	1070
	3.029	131.1	140.6	
	2.940	135.4	98.5	
$Na_4Rb_2P_6O_{18}\cdot6H_2O$	2.906	130.1	141.7	1073
	2.975	129.1	95.3	
	2.885	135.2	109.4	
$Na_2(NH_4)_4P_6O_{18}\cdot2H_2O$	2.882	127.5	132.6	1073
	2.939	133.1	128.4	
	2.927	131.7	98.7	
$Cu_3P_6O_{18}\cdot14H_2O$	2.920	132.4	139.9	1077
	2.904	130.3	96.3	
	2.883	128.6	114.1	
$Mn_2Li_2P_6O_{18}\cdot10H_2O$	2.964	136.5	111.5	1084
	2.960	135.8	90.6	
	2.938	132.8	114.1	
$Ca_2Li_2P_6O_{18}\cdot8H_2O$	2.867	125.9	133.3	1085
	2.906	130.3	97.1	
	2.877	127.0	129.3	
$Cd_2Na_2P_6O_{18}\cdot14H_2O$	2.906	132.4	94.2	1086
	2.937	130.5	142.8	
	2.896	129.4	110.1	
$Cu_2(NH_4)_2P_6O_{18}\cdot8H_2O$	2.945	134.6	138.9	1087
	2.965	136.4	97.4	
	2.922	131.4	95.5	
$Zn_2(NH_4)_2P_6O_{18}\cdot8H_2O$	2.932	133.6	117.8	1090
	2.933	127.9	94.2	
	2.874	133.7	135.1	
$Cr_2P_6O_{18}$	2.909	131.9	112.7	1051
	2.970	139.2	121.5	
	2.911	133.0	125.5	
$Cd_2(NH_4)_2P_6P_{18}\cdot9H_2O$	2.977	137.5	109.1	1091
	2.885	128.1	92.2	
	2.916	130.8	141.6	
$Nd_2P_6O_{18}\cdot12H_2O$	2.905	131.3	117.9	1099
	2.948	126.3	127.6	
	2.867	132.3	85.9	

(continued)

Table 4.5.9. (continued)

Formula	P–P (Å)	P–O–P (°)	P–P–P (°)	Reference
$(NH_3OH)_6P_6O_{18} \cdot 4H_2O$	2.873	127.7	138.9	1101
	2.932	123.4	112.5	
	2.935	133.4	93.5	
$(EDA)_3P_6O_{18} \cdot 2H_2O$	2.980	126.9	110.8	1102
	2.878	137.8	91.8	
	2.930	132.3	111.9	
$Cu(EDA)_2P_6O_{18} \cdot 6H_2O$	2.885	127.5	121.1	1102
	2.921	132.8	114.4	
	2.919	131.8	106.5	
$(EDA)_3P_6O_{18} \cdot 2Te(OH)_6 \cdot 2H_2O$	2.929	132.1	101.1	1107
	2.935	131.4	139.5	
	2.973	137.2	99.5	
$(NH_4)_6P_6O_{18} \cdot Te(OH)_6 \cdot 2H_2O$	2.876	127.3	114.3	1105
	2.960	135.3	115.2	
	2.933	132.4	96.1	
$Rb_6P_6O_{18} \cdot 3Te(OH)_6 \cdot 4H_2O$	2.940	133.3	96.7	1106
	2.911	127.3	144.9	
	2.884	131.1	102.2	
$Mn_2(NH_4)_4(P_6O_{18})(C_2O_4) \cdot 5H_2O$	2.928	132.0	93.1	1108
	2.942	129.2	103.6	
	2.891	134.3	116.1	
$(N_2H_5)_2(N_2H_6)_2P_6O_{18}$	2.857	124.7	108.5	1103
	2.929	133.4	110.3	
	2.934	133.1	94.1	

^{a}EDA, ethylenediammonium, $[NH_3-(CH_2)_2-NH_3]^{2+}$.

Table 4.5.10. Main Geometrical Features of P_6O_{18} Ring Anions with No Symmetry

Formula	P–P (Å)	P–O–P (°)	P–P–P (°)	Reference
$Cu_2Li_2P_6O_{18}$	2.892	131.3	113.5	1049
	2.931	135.0	134.3	
	2.992	141.4	111.1	
	2.926	134.4	113.9	
	2.950	136.3	133.6	
	2.989	141.1	112.4	
$(C_2H_5NH_3)_6P_6O_{18} \cdot 4H_2O$	2.925	131.2	104.5	1103
	2.963	129.3	105.5	
	2.902	134.0	115.7	
	2.954	131.4	99.6	
	2.917	136.0	111.1	
	2.974	135.7	108.5	
$K_6P_6O_{18} \cdot KNO_3 \cdot H_2O$	2.886	127.9	100.§	1110
	2.888	128.1	118.0	
	2.958	132.8	95.9	
	2.903	128.5	99.6	
	2.892	127.0	117.7	
	2.951	133.1	95.4	

Table 4.5.11. Main Geometrical Features of P_6O_{18} Ring Anions
with Higher Symmetries

Formula	P–P (Å)	P–O–P (°)	P–P–P (°)	Reference
Rings with 2 symmetry				
$Ca_2(NH_4)_2P_6O_{18} \cdot 6H_2O$	2.900	130.3	116.1	1089
	2.955	134.3	112.9	
	2.945	134.3	87.8	
$Ce_2P_6O_{18} \cdot 10H_2O$	2.906	131.1	87.5	1098
	2.922	133.8	114.4	
	2.944	134.4	115.6	
$Ba_3P_6O_{18} \cdot 8H_2O$	2.865	125.2	98.95	1094
	2.990	138.9	116.4	
	2.926	133.3	102.6	
Rings with 3 symmetry				
$K_6P_6O_{18}$	2.957	130.0	105.8	1060
	2.917	134.9	103.4	
$Cd_3P_6O_{18} \cdot 6H_2O$	2.934	133.8	116.4	1078
	2.933	133.7	115.8	
$Cr_2P_6O_{18} \cdot 21H_2O$	2.926	133.6	104.4	1096
	2.929	135.2	109.5	
Rings with $\bar{3}$ symmetry				
$(NH_4)_6P_6O_{18} \cdot H_2O$	2.954	134.3	108.5	1064
$Ag_6P_6O_{18} \cdot H_2O$	2.909	130.3	112.2	1053
$Li_3K_3P_6O_{18} \cdot 12H_2O^a$	2.921	130.6	111.9	1068
	2.937	133.9	113.9	
$Ag_3(NH_4)_3P_6O_{18} \cdot H_2O^a$	2.916	131.0	110.4	1069
	2.907	131.0	110.3	
$Li_3Na_3P_6O_{18} \cdot H_2O$	2.931	133.2	106.5	1068
$K_6P_6O_{18} \cdot 2Te(OH)_6 \cdot 3H_2O$	2.934	134.8	107.8	1106
Rings with m symmetry				
$K_6P_6O_{18} \cdot 3H_2O$	3.004	141.9	106.1	1054
	2.965	136.2	106.5	
	2.865	125.6	107.1	
	2.962	131.1		
Rings with 2/m symmetry				
$Na_6P_6O_{18} \cdot 6H_2O$	2.870	131.3	131.7	34
	2.935	124.9	96.6	

^{a}Two independent rings in the unit cell.

4.6. Cyclooctaphosphates

4.6.1. Introduction

Like other large phosphoric rings, P_8O_{24} anions were clearly characterized during paper-chromatography experiments as early as 1956 by Van Wazer and Karl-Kroupa.[1111] However, it was not until 1968 that Schülke[1112,1113] reported the first characterization of a crystalline cyclooctaphosphate, $Pb_4P_8O_{24}$. This author described the chemical preparation of this salt and used it to produce the corresponding sodium salt.[1112,1113] Schülke's process will be described in more detail in the next section. By using an ion-exchange

Table 4.6.1. Main Crystallographic Data for Cyclooctaphosphates

Formula	a (Å) α (°)	b (Å) β (°)	c (Å) γ (°)	S.G.	Z	Reference(s)
$Na_8P_8O_{24} \cdot 6H_2O$	6.622(2) 104.06(5)	10.031(4) 101.21(5)	11.250(4) 90.88(5)	$P\bar{1}$	1	1115
$(NH_4)_8P_8O_{24} \cdot 3H_2O$	26.27(1)	6.700(3) 112.06(6)	20.59(1)	Cc	4	1116
$Cs_8P_8O_{24} \cdot 8H_2O$	7.666(8) 100.2(1)	11.569(9) 106.5(2)	11.634(9) 92.2(1)	$P\bar{1}$	1	1117
$(NH_4)_2Cu_3P_8O_{24}$	9.846(2) 80.98(3)	7.962(2) 110.79(3)	7.261(2) 110.61(3)	$P\bar{1}$	1	37, 680
$Cu_3Rb_2P_8O_{24}$	9.797(4) 80.93(3)	8.035(3) 110.35(3)	7.256(3) 110.48(3)	$P\bar{1}$	1	595, 680
$Cu_3Cs_2P_8O_{24}$	9.913(2) 81.55(2)	7.998(1) 109.20(2)	7.298(1) 109.14(2)	$P\bar{1}$	1	674, 680
$Cu_3Tl_2P_8O_{24}$	9.862(2) 81.62(3)	7.922(2) 110.43(3)	7.273(2) 109.88(3)	$P\bar{1}$	1	680
$Ga_2K_2P_8O_{24}$	5.138(3)	12.290(5) 101.04(5)	16.802(13)	$A2/m$	2	1124
$V_2K_2P_8O_{24}$	5.223(2)	12.277(4) 101.19(5)	16.867(4)	$A2/m$	2	41
$Al_2K_2P_8O_{24}$	5.070	12.266 97.5	16.510	$A2/m$	2	1123
$Fe_2K_2P_8O_{24}$	5.351	12.290 110.97	17.735	$A2/m$	2	1123

resin, Schülke and Chudinova[1114] prepared the other alkali cyclooctaphosphates. However despite the possibility of producing good starting materials for the investigation of this class of compounds, the first structural characterization of a P_8O_{24} ring anion did not occur before 1975, when Laügt and Guitel[37] determined the crystal structure of $Cu_3(NH_4)_2P_8O_{24}$. It should also be noted that, until recently, most structural investigations of cyclooctaphosphates were performed on compounds characterized either during the elaboration of phase-equilibrium diagrams or during investigations of various systems by flux methods, crystalline samples of these compounds usually being prepared by flux methods at relatively high temperatures.

Nevertheless, during the past three years, there has been renewed interest, at least from a structural perspective, in this family of cyclophosphates.

The main crystallographic features of cyclooctaphosphates and various compounds containing P_8O_{24} ring anions are given in Table 4.6.1.

4.6.2. Present State of the Cyclooctaphosphate Chemistry

4.6.2.1. Alkali and Monovalent Cation Cyclooctaphosphates

Soon after Schülke reported the preparation of the sodium salt,[1112,1113] Schülke and Chudinova[1114] described the chemical preparation and investigated the thermal behavior and solubility of five new hydrates, $M_8P_8O_{24} \cdot 6H_2O$, with M = Li, K, Rb, Cs, and NH_4. All five were prepared from the sodium salt through the use of ion-exchange resins. According to these authors, the solubility of the sodium salt is relatively low in comparison with that of the other alkali or the ammonium derivatives. It is only very recently that structural investigations of these salts have begun. Except in the case of the sodium salt, all the crystalline samples obtained for these structural investigations have a state of hydration different from that originally reported by Schülke and Chudinova.

$Na_8P_8O_{24}\cdot6H_2O$. A crystal structure determination for this salt was performed by Schülke et al.[1115] The very accurate atomic arrangement determined by these authors cannot account for the very low solubility of this salt.

$(NH_4)_8P_8O_{24}\cdot3H_2O$. Recently, the same authors[1116] obtained the ammonium salt as a trihydrate, $(NH_4)_8P_8O_{24}\cdot3H_2O$, and performed its crystal structure determination. The crystals used for this last investigation were grown at room temperature from a material prepared by ion exchange from the sodium salt. The study of the thermal behavior of this salt shows that its stability is very limited. As low as 343 K it loses its crystallization water, and at 413 K it transforms into form I of long-chain ammonium polyphosphate via an exothermic reaction:

$$(NH_4)_8P_8O_{24}\cdot3H_2O \xrightarrow{343\,K} (NH_4)_8P_8O_{24} + 3H_2O \xrightarrow{413\,K} 8NH_4PO_3$$

$Cs_8P_8O_{24}\cdot8H_2O$. Very recently, Brühne and Jansen[1117] prepared crystals of $Cs_8P_8O_{24}\cdot8H_2O$ by using a gel technique and performed its crystal structure determination.

4.6.2.2. Divalent Cation Cyclooctaphosphates

$Pb_4P_8O_{24}$. During a very careful study of the thermal behavior of lead cyclotetraphosphate tetrahydrate, Schülke[1112,1113] observed that this salt can be converted into anhydrous lead cyclooctaphosphate under well-defined conditions. According to this author, the best yields are obtained when $Pb_2P_4O_{12}\cdot4H_2O$ is first heated at 383 K for 30 min and then at 623 K again for 30 min. This transformation can be represented schematically by the following two steps:

$$2Pb_2P_4O_{12}\cdot4H_2O \rightarrow 2Pb_2H_4(PO_4)(P_3O_{10}) + 4H_2O \rightarrow Pb_4P_8O_{24} + 8H_2O$$

In fact, the process is not so simple and must be separated into four distinct steps:

(a) Formation of crystalline $Pb_2P_4O_{12}\cdot2H_2O$ by partial dehydration
(b) Hydrolysis of the cyclotetraphosphate anion by the residual crystal water, leading to a mixture of mono-, di-, and triphosphates
(c) Condensation of these various anions into 30% lead polyphosphate, $Pb(PO_3)_2$, and 70% anhydrous lead cyclotetraphosphate, $Pb_2P_4O_{12}$
(d) Conversion of $Pb_2P_4O_{12}$ into $Pb_4P_8O_{24}$

According to Schülke, steps c and d occur simultaneously.

Then, sodium cyclooctaphosphate hexahydrate can be obtained from the lead salt by the action of sodium sulfide:

$$Pb_4P_8O_{24} + 4Na_2S \rightarrow Na_8P_8O_{24} + 4PbS$$

$Ca_4P_8O_{24} \cdot 16H_2O$. During the same study[1112,1113] Schülke also characterized this salt.

$Ni_4P_8O_{24} \cdot 19H_2O$, $Mn_4P_8O_{24} \cdot 17H_2O$ and $Cd_4P_8O_{24} \cdot 12H_2O$. Lavrov *et al.*[989] prepared and characterized these three hydrates. They were obtained as crystalline compounds by the reaction of $H_8P_8O_{24}$ with the corresponding divalent cation perchlorates. No structural information is available for these salts.

4.6.2.3. Divalent–Monovalent Cation Cyclooctaphosphates

$Cu_3M_2P_8O_{24}$ (M = Rb, Cs, Tl, NH$_4$). A series of four isotypic cyclooctaphosphates of general formula $Cu_3M_2P_8O_{24}$ (M = Rb, Cs, Tl, and NH$_4$) was characterized by Laügt[1118] during a systematic study of the MPO_3–$Cu_2P_4O_{12}$ phase-equilibrium diagrams (M = Rb, Cs, and Tl) and a careful investigation of the P_2O_5–$(NH_4)_2O$–Cu_2O–H_2O system. Two of these compounds, $Cu_3Rb_2P_8O_{24}$ and $Cu_3Tl_2P_8O_{24}$, were not observed in the first versions of the $RbPO_3$–$Cu_2P_4O_{12}$ and $TlPO_3$–$Cu_2P_4O_{12}$ phase-equilibrium diagrams reported by Laügt *et al.*,[668] but appear as incongruent melting compounds in the revised versions elaborated by Laügt[595,680] (Figs. 4.6.1 and 4.6.2). $Cu_3Rb_2P_8O_{24}$ and $Cu_3Tl_2P_8O_{24}$ decompose at 893 and 909 K respectively. In the $CsPO_3$–$Cu_2P_4O_{12}$ phase-equilibrium diagram elaborated by Laügt and Martin[674] $Cu_3Cs_2P_8O_{24}$ appears as an incongruent melting compound decomposing at 939 K. It is isotypic with the corresponding rubidium and

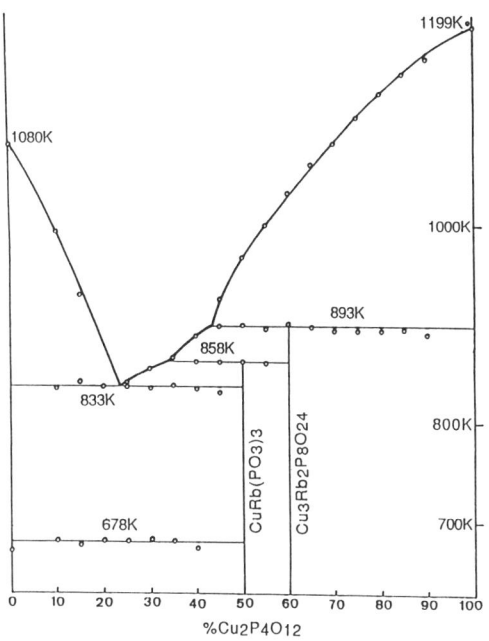

Figure 4.6.1. The revised $RbPO_3$–$Cu_2P_4O_{12}$ phase-equilibrium diagram. The phase transformation of $RbPO_3$ appears at 678 K on the left-hand side of the diagram.

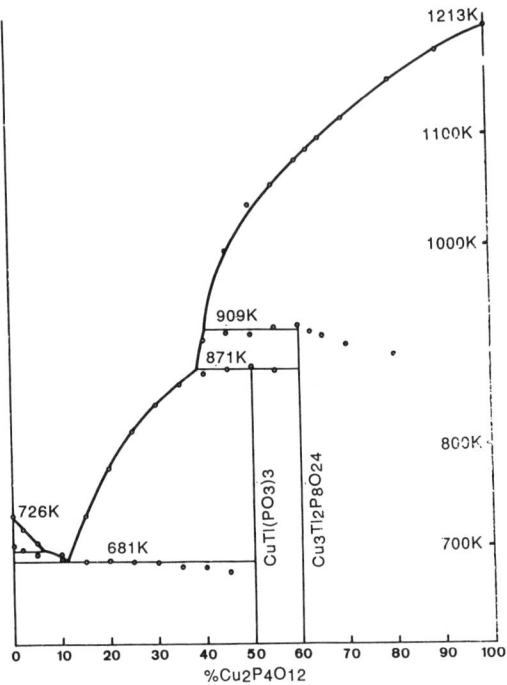

Figure 4.6.2. The revised TlPO₃–Cu₂P₄O₁₂ phase-equilibrium diagram.

thallium salts. $Cu_3(NH_4)_2P_8O_{24}$ was prepared by Laügt and Guitel[37] and recognized as an isotype of the alkali derivatives. The accurate structure determination that was performed by these authors will be described in Section 4.6.3.

4.6.2.4. Trivalent and Trivalent–Monovalent Cation Cyclooctaphosphates

$[Ga(OH)_2]_8P_8O_{24} \cdot 16H_2O$, $GaNa_5P_8O_{24}$, and $Ga_8(P_8O_{24})_3$. The reaction of gallium chloride with sodium cyclooctaphosphate was investigated by Lazarevski *et al.*[1119] These authors reported the existence of three compounds. Over a wide pH range (3–8) an amorphous basic gallium salt, $[Ga(OH)_2]_8P_8O_{24} \cdot 16H_2O$, is precipitated. When equimolar amounts of the reactants are employed, the readily soluble $GaNa_5P_8O_{24}$ is obtained, and in a more acidic medium (pH 2.8) the neutral $Ga_8(P_8O_{24})_3$ is obtained.

$Y_8(P_8O_{24})_3 \cdot nH_2O$ (n ~ 30) and $YK_5P_8O_{24} \cdot 10H_2O$. During an investigation of the YCl_3–$K_8P_8O_{24}$–H_2O system Lazarevski *et al.*[1120] characterized $Y_8(P_8O_{24})_3 \cdot nH_2O$ with n ~ 30, and $YK_5P_8O_{24} \cdot 10H_2O$. These authors did not report any crystal data.

$A_2B_2P_8O_{24}$ compounds. A number of compounds of general formula $A_2B_2P_8O_{24}$ with A = Al, Ga, Cr, Fe or V and B = K, Rb, or NH_4 have been described in the chemical literature. In spite of some deficiencies in the crystallographic characterizations of some of them, it can be assumed that they are isotypic. All these compounds were characterized during investigations of various P_2O_5–A_2O_3–B_2O–H_2O systems using flux methods. $Ga_2K_2P_8O_{24}$ was characterized during several investigations of the P_2O_5–K_2O–Ga_2O_3–H_2O system. In the first of these performed by Chudinova *et al.*[1121] it appears as the only phase crystallizing between 573 and 623 K in mixtures corresponding to an initial P:K:Ga molar ratio of 15:5:1. It was originally identified by the authors as a long-chain polyphosphate. Another study of the same system between 423 and 773 K performed by Chudinova *et al.*[1122] showed the existence of additional compounds and confirmed the existence of $Ga_2K_2P_8O_{24}$, still described as a long-chain polyphosphate. During this investigation it was observed that $Ga_2K_2P_8O_{24}$ can be obtained by firing $KGa(H_2P_2O_7)_2$, a double diphosphate characterized in the same system. The reaction is

$$2KGa(H_2P_2O_7)_2 \xrightarrow{603\ K} Ga_2K_2P_8O_{24} + 4H_2O$$

The melting point of this cyclooctaphosphate was measured as 1003 K.

The P_2O_5–Rb_2O–Ga_2O_3–H_2O system was investigated by Chudinova, *et al.*[260] over the same range of temperatures. These authors reported the existence of $Ga_2Rb_2P_8O_{24}$ which they identified as a cyclooctaphosphate isotypic with the corresponding potassium salt.

During a very similar investigation of ammonium–gallium phosphates between 423 and 623 K performed by Chudinova *et al.*[258] $Ga_2(NH_4)_2P_8O_{24}$ was characterized as a product of the thermal transformation of $GaNH_4HP_3O_{10}$. This cyclooctaphosphate is stable up to 623 K. Above this temperature, it decomposes according to the following scheme:

$$Ga_2(NH_4)_2P_8O_{24} \xrightarrow{T > 623\ K} 2NH_3 + 2HPO_3 + 2Ga(PO_3)_3(C)$$

$Cr_2K_2P_8O_{24}$, $Cr_2(NH_4)_2P_8O_{24}$, and $Cr_2Rb_2P_8O_{24}$ were discovered by Grunze and Chudinova[445] during investigations of the same type of systems between 473 and 673 K and identified as isotypes of $Ga_2K_2P_8O_{24}$.

$V_2K_2P_8O_{24}$ was obtained by Lavrov *et al.*[41] during an investigation of several V_2O_3–M_2O–P_2O_5–H_2O systems. This compound is stable in argon up to 933 K. Crystals are pale yellow, acicular prisms. According to the authors, $V_2K_2P_8O_{24}$ can be prepared in very high yield by heating well-crystallized $VK(H_2P_2O_7)_2$ at 620 K for 10–12 h.

The chemical preparation of $Al_2K_2P_8O_{24}$, $Ga_2K_2P_8O_{24}$, $Fe_2K_2P_8O_{24}$ and $Al_2Rb_2P_8O_{24}$ was described by Grunze and co-workers.[275,242,1123] These salts are obtained by thermal decomposition of the corresponding $M^{III}M^I(H_2P_2O_7)_2$ diphosphates according the scheme already described for the formation of the gallium–potassium cyclooctaphosphate. The temperatures used are similar—663 K for the aluminum salts and at 648 K in the case of gallium.

The atomic arrangement common to all these compounds was determined by Palkina *et al.*[1124] with the gallium–potassium salt, $Ga_2K_2P_8O_{24}$. It will be described in Section 4.6.3, devoted to the structural aspect of cyclooctaphosphates.

$Sn_2P_8O_{24} \cdot 5H_2O$. The chemical preparation of this salt was described by Nariai *et al.*[1125] These authors investigated the effect of water vapor on the thermal decomposition of this salt.

Table 4.6.2. Main Crystallographic Data for Organic Cation Cyclooctaphosphates and Various Adducts Containing P_8O_{24} Ring Anions

Formula[a]	a (Å) α (°)	b (Å) β (°)	c (Å) γ (°)	S.G.	Z	Reference(s)
$(Gua)_8P_8O_{24}\cdot2H_2O$	12.621(4)	20.41(1) 110.45(4)	9.365(6)	$P2_1/n$	2	1126
$(EDA)_4P_8O_{24}\cdot6H_2O$	11.833(2)	21.844(3) 98.49(3)	7.467(2)	$P2_1/n$		1127
$Ag_9NaP_8O_{24}\cdot(NO_3)_2\cdot4H_2O$	17.254(5)	7.543(1)	23.465(5)	$Cmcm$	4	1128
$K_8P_8O_{24}\cdot Te(OH)_6\cdot2H_2O$	11.315(9) 108.72(5)	10.67(1) 100.30(2)	7.547(3) 66.80(5)	$P\bar{1}$	1	1129
$(NH_4)_8P_8O_{24}\cdot Te(OH)_6\cdot2H_2O$	15.146(6) 117.15(4)	11.049(6) 109.72(4)	12.189(6) 90.54(4)	$P\bar{1}$	2	1130
$(EDA)_8P_8O_{24}\cdot Te(OH)_6\cdot2H_2O$	11.526(3) 106.12(3)	12.723(3)	13.581(3)	$P2_1/c$	2	1131

[a]Abbreviations: Gua, guanidinium, $C(NH_2)_3^+$; EDA, ethylenediammonium, $[NH_3-(CH_2)_2-NH_3]^{2+}$.

4.6.2.5. *Organic Cation Cyclooctaphosphates*

Very recently, two organic cation cyclooctaphosphates were synthesized: guanidinium cyclooctaphosphate dihydrate and ethylenediammonium cyclooctaphosphate hexahydrate. The guanidinium derivative, $[C(NH_2)_3]_8P_8O_{24}\cdot2H_2O$, was synthesized by Averbuch-Pouchot et al.[1126] by the action of $H_8P_8O_{24}$ on guanidinium carbonate. The ethylenediammonium salt, $[NH_3(CH_2)_2NH_3]_4P_8O_{24}\cdot6H_2O$, investigated by the same authors,[1127] was prepared by the reaction between $H_8P_8O_{24}$ and ethylenediamine. In both cases, the crystal structures were accurately determined.

4.6.2.6. *Other Compounds Containing Cyclooctaphosphate Anions*

Cyclooctaphosphate anions have also been observed in two kinds of adducts, cyclooctaphosphate-nitrates and cyclooctaphosphate-tellurates.

$Ag_8P_8O_{24}\cdot AgNO_3\cdot(5-6)H_2O$. The existence of this adduct was reported by Schülke,[1112] but no crystal data have been reported for this salt.

$Ag_9NaP_8O_{24}\cdot(NO_3)_2\cdot4H_2O$. This silver–sodium cyclooctophosphate nitrate was characterized by Averbuch-Pouchot and Durif[1128] during attempts to prepare $Ag_8P_8O_{24}\cdot xH_2O$ in order to investigate the possibility of using Boullé's metathesis reaction[805] for the synthesis of water-soluble cyclooctaphosphates. The authors reported a complete description of the structure. The atomic arrangement of this first example of a condensed phosphate nitrate is described in Section 4.6.3.5.

$K_8P_8O_{24}\cdot Te(OH)_6\cdot2H_2O$ and $(NH_4)_8P_8O_{24}\cdot Te(OH)_6\cdot2H_2O$. The potassium compound prepared and investigated by Averbuch-Pouchot and Durif[1129] provided the first example of an adduct between telluric acid and an alkali cyclooctaphosphate. It was simply prepared by slow evaporation at room temperature of an aqueous solution of the two components in a stoichiometric ratio. Its atomic arrangement is described in Section 4.6.3.6. Later, the corresponding ammonium salt, prepared under the same conditions, was described by Averbuch-Pouchot and Durif.[1130] The two salts are not isotypic.

$[NH_3(CH_2)_2NH_3]_4P_8O_{24}\cdot Te(OH)_6\cdot 2H_2O$. Ethylenediammonium cyclooctaphosphate tellurate dihydrate was very recently prepared and characterized by Schülke and Averbuch-Pouchot[1131] who report a complete description of the atomic arrangement.

4.6.3. *Atomic Arrangements of Cyclooctaphosphates*

For this relatively new field of crystal chemistry, we report all the atomic arrangements that have been presently determined thus far.

4.6.3.1. *Sodium Cyclooctaphosphate Hexahydrate*

$$Na_8P_8O_{24}\cdot 6H_2O,^{1115} \text{ triclinic, } P\bar{1}, Z = 1$$

$$a = 6.622(2), \quad b = 10.031(4), \quad c = 11.250(4) \text{ Å}$$

$$\alpha = 104.06(5), \quad \beta = 101.21(5), \quad \gamma = 90.88(5)°$$

This very simple atomic arrangement is represented in Fig. 4.6.3 by a projection along the **a** direction. The P_8O_{24} ring anion is located around the inversion center at 0, 0, 0. It thus has $\bar{1}$ internal symmetry and is consequently built by four independent PO_4 tetrahedra. As can be seen in Fig. 4.6.3, this ring has in projection a strong pseudo-$2/m$ symmetry, the pseudo-mirror including two centrosymmetric bonding oxygen atoms, while the pseudo-twofold axis almost perpendicular to the mirror includes two centrosymmetric oxygen atoms of the same kind.

One of the four independent sodium atoms has fivefold coordination, involving external oxygen atoms of the phosphoric ring. The three remaining ones have sixfold

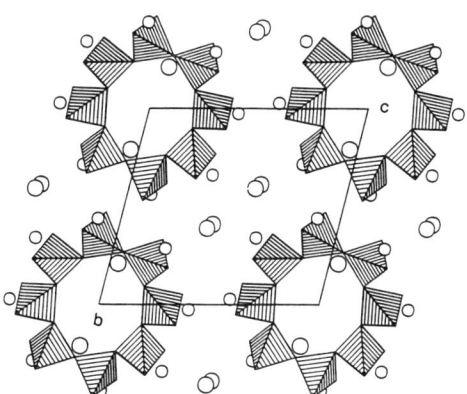

Figure 4.6.3. Projection, along the *a* direction, of the atomic arrangement in $Na_8P_8O_{24}\cdot 6H_2O$. The large open circles represent water molecules, and the smaller ones the sodium atoms.

coordination involving both water molecules and oxygen atoms in a moderately distorted octahedral arrangement. The Na–O distances range from 2.318 to 2.635 Å. In the three-dimensional hydrogen-bond network, the hydrogen atoms belonging to two of the water molecules are involved, as can be expected, in bonds connecting these water molecules either to external oxygen atoms of the phosphoric ring or to other water molecules. The bonding of the two hydrogen atoms of one of the water molecules is different; both are connected to bonding oxygen atoms belonging to two different phosphoric rings. This feature is rather unusual in condensed phosphate chemistry, where bonding oxygen atoms are almost never involved in associated-cation coordination or in hydrogen bonds.

4.6.3.2. *Ammonium Cyclooctaphosphate Trihydrate*

$$(NH_4)_8P_8O_{24}\cdot 3H_2O,^{[1116]} \text{ monoclinic, } Cc, Z = 4$$

$$a = 24.27(1), \quad b = 6.700(3), \quad c = 20.59(1) \text{ Å}, \quad \beta = 112.06(6)°$$

For the most part, the present atomic arrangement, built by the stacking of the P_8O_{24} and NH_4 groups, is highly pseudo-centrosymmetric. Fig. 4.6.4 gives a projection of this structure along the **b** axis. The only clear departure from centrosymmetry is observed for the water molecules. For the description of this atomic arrangement, the arbitrary atomic coordinates were fixed so as to locate the ring anion around the origin of the unit cell. The phosphoric ring anion has a strongly pseudo-centrosymmetric configuration.

The eight independent ammonium groups have various types of oxygen coordination. Within a range of 3.50 Å, three of them have eight neighbors, three have seven, one has

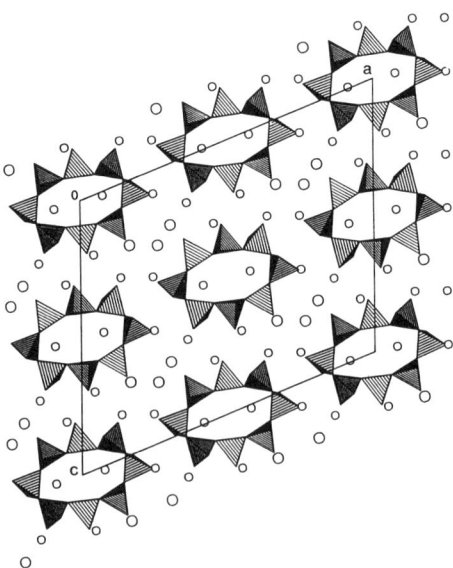

Figure 4.6.4. Projection along the **b** direction of the atomic arrangement in $(NH_4)_8P_8O_{24}\cdot 3H_2O$. The large open circles represent the water molecules, and the smaller ones the ammonium groups.

six, and the last one has only five. Within these various NO_n polyhedra, the N–O distances range between 2.772 and 3.483 Å .

The water molecules are not, as is most frequently observed, dispersed inside the arrangement. They are assembled in groups of three, building an irregular triangle O(W1)–O(W2)–O(W3) with the following edge lengths and angles:

O(W1)–O(W2) 3.479(8) Å		O(W2)–O(W1)–O(W3) 49.8(2)°	
O(W1)–O(W3) 3.205(6)		O(W1)–O(W2)–O(W3) 60.1(2)	
O(W2)–O(W3) 2.824(9)		O(W1)–O(W3)–O(W2) 70.2(2)	

Inside this triangle the only hydrogen bond is established between O(W2) and O(W3). All the other hydrogen atoms of these water molecules are involved in H bonds with external oxygen atoms of the adjacent phosphoric groups.

4.6.3.3. Copper–Ammonium Cyclooctaphosphate

$$Cu_3(NH_4)_2P_8O_{24},^{37,680} \text{triclinic}, P\bar{1}, Z = 1$$

$$a = 9.846(2), \quad b = 7.962(2), \quad c = 7.261(2) \text{ Å}$$

$$\alpha = 80.98(3), \quad \beta = 110.79(3), \quad \gamma = 110.61(3)°$$

A projection of this atomic arrangement along the **b** axis is shown in Fig. 4.6.5. The centrosymmetric P_8O_{24} ring anion is located around the inversion center at 1/2,1/2,1/2. Its main geometrical features are given in Table 4.6.3. One of the copper atoms occupies a special position at 0,1/2,0, and the second one is in a general position; both have distorted octahedral coordination, with Cu–O distances ranging between 1.927 and 2.873 Å. The ammonium group has ninefold coordination, with N–O distances ranging from 2.902 to 3.372 Å. The associated-cation polyhedra build up a three-dimensional network in which chains of corner-sharing CuO_6 octahedra connect planes built by CuO_6 and NO_9 polyhedra.

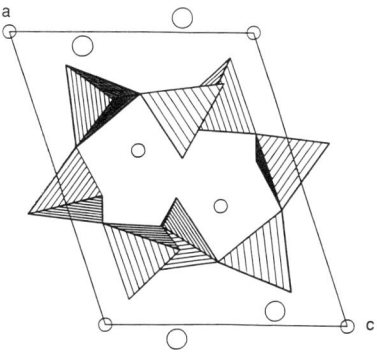

Figure 4.6.5. Projection of the atomic arrangement in $Cu_3(NH_4)_2P_8O_{24}$ along the **b** axis. The smaller open circles represent the copper atoms, and the larger ones the ammonium groups.

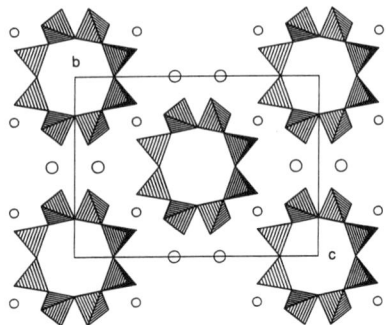

Figure 4.6.6. Projection of the atomic arrangement in $Ga_2K_2P_8O_{24}$ along the *a* axis. The small open circles represent the gallium atoms, and the larger ones the potassium atoms.

The corresponding rubidium, cesium, and thallium salts, $Cu_3Rb_2P_8O_{24}$, $Cu_3Cs_2P_8O_{24}$, and $Cu_3Tl_2P_8O_{24}$, are isotypic (Table 4.6.1).

4.6.3.4. Gallium–Potassium Cyclooctaphosphate

$$K_2Ga_2P_8O_{24},^{1124} \text{ monoclinic, } A2/m, Z = 2$$

$$a = 5.138(3), \quad b = 12.290(5), \quad c = 16.802(13) \text{ Å}, \quad \beta = 101.04(5)°$$

Figure 4.6.6 shows a projection of the atomic arrangement of $Ga_2K_2P_8O_{24}$ along the **a** axis. As can be seen from this figure, the respective locations of the P_8O_{24} rings and of the associated cations are such that wide, infinite, empty channels (5.2 x 5.4 Å), parallel to the **a** axis, pass through the rings. The P_8O_{24} groups are interconnected by slightly distorted centrosymmetric GaO_6 octahedra and KO_6 polyhedra. The KO_6 polyhedra are distorted trigonal prisms with mirror symmetry. They are arranged in space as isolated pairs with a common edge. Within these pairs, the K–O distances range from 2.754 to 2.925 Å. In the octahedral GaO_6 group the Ga–O distances range from 1.924 to 1.978 Å. The P_8O_{24} ring anion has $2/m$ internal symmetry.

The corresponding vanadium, iron, and aluminum compounds are isotypic (Table 4.6.1).

4.6.3.5. Silver–Sodium Cyclooctaphosphate-Nitrate Tetrahydrate

$$Ag_9NaP_8O_{24}(NO_3)_2 \cdot 4H_2O,^{1128} \text{ orthorhombic, } Cmcm, Z = 4$$

$$a = 17.254(5), \quad b = 7.543(1), \quad c = 23.465(5) \text{ Å}$$

In this arrangement the P_8O_{24} ring anion has $2/m$ internal symmetry. Figure 4.6.7 gives two different projections of this anion. As shown by Fig. 4.6.8, a projection of the entire atomic arrangement along the **b** axis, all the P_8O_{24} entities lie perpendicular to this direction, forming arrays extending around the planes $z = 0$ and $1/2$. The central nitrogen atom and one oxygen atom of the NO_3 group are located in a mirror plane. The values measured for the O–N–O angles (121.0 and 119.3°) show this group to be, as usual, almost

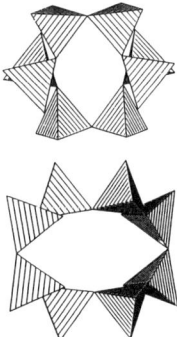

Figure 4.6.7. Projection of the phosphoric ring in $Ag_9NaP_8O_{24} \cdot (NO_3)_2 \cdot 4H_2O$ along the b direction (top) and along the c direction (bottom).

planar within the experimental error. All the NO_3 entities are located in the planes $z = 1/4$ and $3/4$ thus alternating with the layers of phosphoric groups.

One of the four associated-cation crystallographic sites is statistically occupied by both sodium and silver atoms. A careful determination of the Ag/P atomic ratio by energy dispersion spectrometry confirmed the formula reported above. Within a range of 3 Å, three of the four silver atoms have sixfold coordination with various internal symmetries (m or 2), while the fourth one has seven neighbors.

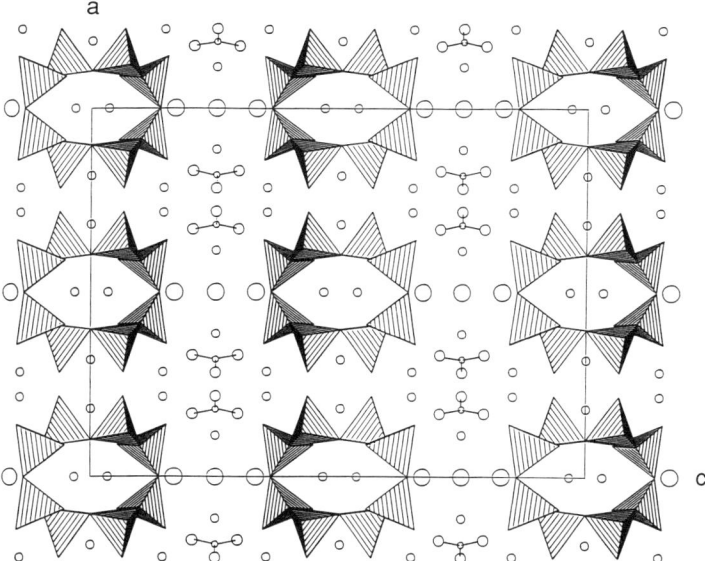

Figure 4.6.8. Projection, along the b direction, of the atomic arrangement in $Ag_9N/aP_8O_{24}(NO_3)_2 \cdot 4H_2O$. In order of decreasing size, the open circles represent the water molecules, the oxygen atoms of the NO_3 groups, the silver atoms and the nitrogen atoms.

The three independent water molecules are all situated in special positions (m or mm). All of them are thus located in the planes $x = 0$ and $1/2$. In addition, it should be noted that one of them is not involved in the associated-cation coordination polyhedra and, within a range of 3 Å, has only two water molecules as neighbors, at distances of 2.842 and 2.855 Å, corresponding probably to hydrogen bonds.

4.6.3.6. Potassium Cyclooctaphosphate-Telluric Acid Dihydrate

$$K_8P_8O_{24} \cdot Te(OH)_6 \cdot 2H_2O,^{[1129]} \text{ triclinic, } P\bar{1}, Z = 1$$

$$a = 11.315(9), \quad b = 10.67(1), \quad c = 7.547(3) \text{ Å}$$

$$\alpha = 108.72(5), \quad \beta = 100.30(2), \quad \gamma = 66.80(5)°$$

As always observed in adducts between monovalent cation phosphates and telluric acid, the phosphoric anion and the $Te(OH)_6$ group are independent entities and thus do not share any oxygen atoms. The $Te(OH)_6$ group is located around the inversion center at $0,0,0$. The six oxygen atoms build an almost regular octahedron around the central tellurium atom, with three Te–O distances ranging from 1.910 to 1.922 Å and O–Te–O angles ranging between 87.61 and 89.18°. The three observed Te–O–H angles—105, 106, and 120°—are in accordance with all the values previously observed in such groups.

Four independent PO_4 tetrahedra build the centrosymmetric P_8O_{24} ring anion located around the inversion center at $1/2,1/2,1/2$. Within this ring, the geometrical features of the four independent PO_4 tetrahedra are quite comparable to all those previously reported for condensed phosphoric anions.

Figure 4.6.9, a projection along the c direction, shows the respective locations of the main components of this atomic arrangement. This projection may give the impression

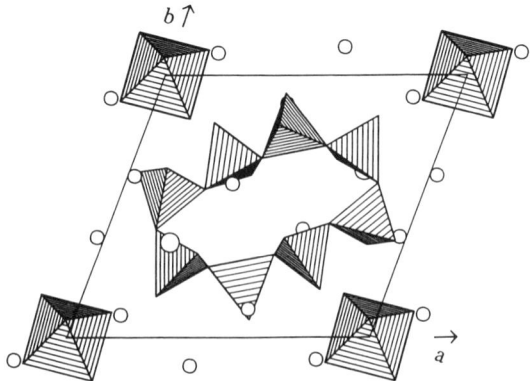

Figure 4.6.9. Projection along c of the atomic arrangement in $K_8P_8O_{24} \cdot Te(OH)_6 \cdot 2H_2O$. The small open circles represent the potassium atoms, and the larger ones the water molecules.

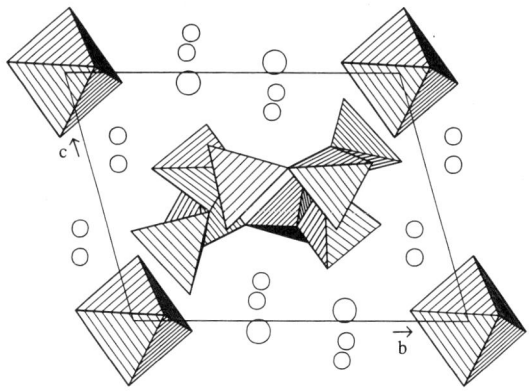

Figure 4.6.10. Projection, along the *a* direction of the atomic arrangement in $K_8P_8O_{24}\cdot Te(OH)_6\cdot 2H_2O$. The smaller open circles represent the potassium atoms, and the larger ones the water molecules.

that the P_8O_{24} ring is almost flat, but a projection along **a** (Fig. 4.6.10) providing a profile view of this anion, shows clearly its compact configuration.

The telluric groups are not interconnected by hydrogen bonds; all their hydrogen atoms establish H bonds with external oxygen atoms of the adjacent phosphoric rings. Thus, four of them establish H bridges connecting $Te(OH)_6$ and P_8O_{24} entities along the **c** direction, while the other two establish the same type of connections in the (*a*, *b*) plane. The two hydrogen atoms of the water molecule connect phosphoric rings along the **c** direction. It is noteworthy that one of the two hydrogen bonds involving the water molecule is established with a bonding oxygen atom of the phosphoric ring, as this type of oxygen is rarely observed to act as an H-bond acceptor in condensed phosphate crystal chemistry.

The four independent potassium atoms have, within a range of 3.50 Å, sevenfold and eightfold coordination. Within these various polyhedra, the K–O distances range from 2.582 to 3.472 Å. The water molecule is involved in two potassium coordination polyhedra.

4.6.3.7. *Ammonium Cyclooctaphosphate–Telluric Acid Dihydrate*

$(NH_4)_8P_8O_{24}\cdot Te(OH)_6\cdot 2H_2O,$[1130] triclinic, $P\bar{1}$, $Z = 2$

$$a = 15.146(6), \quad b = 11.049(6), \quad c = 12.189(6) \text{ Å}$$

$$\alpha = 117.15(4), \quad \beta = 109.72(4), \quad \gamma = 90.54(4)°$$

As shown by Fig. 4.6.11, a projection along the **b** direction, this atomic arrangement is a layer organization. Layers containing the phosphoric ring anions and the $Te(OH)_6$ groups alternate perpendicular to the [101] direction with corrugated layers including the ammonium groups and the water molecules. Two crystallographically independent P_8O_{24}

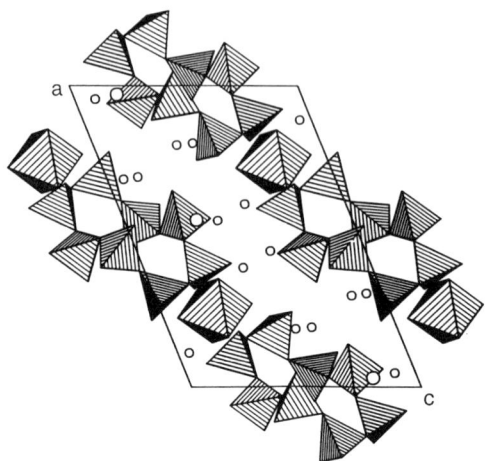

Figure 4.6.11. Projection, along the *b* direction, of the atomic arrangement in $(NH_4)_8P_8O_{24} \cdot Te(OH)_6 \cdot 2H_2O$. The hatched octahedra represent the $Te(OH)_6$ groups; the large open circles represent the water molecules, and the smaller ones the ammonium groups.

ring anions coexist inside this arrangement. Both are centrosymmetric and are located around inversion centers at 0,0,1/2 and 1/2,1/2,0.

The ammonium groups build a three-dimensional network of various NO_6, NO_7, and NO_8 polyhedra, with N–N distances ranging from 3.424 to 4.852 Å. Within these various NO_n polyhedra, the N–O distances range between 2.729 to 3.453 Å.

The $Te(OH)_6$ group is, as usual, an almost regular TeO_6 octahedron, with Te–O distances ranging from 1.905 to 1.917 Å and O–Te–O angles from 88.4 to 89.8°. The Te–O–H angles range between 109 and 117°. No H bonds exist between the telluric groups.

4.6.4. Geometries of the Cyclooctaphosphate Anion

In Table 4.6.3 what we consider to be the main geometrical features of a ring framework—the P–P distances and P–O–P and P–P–P angles—are tabulated for the nine P_8O_{24} rings presently known. With so few examples, the kind of discussion that was possible for other smaller rings, P_6O_{18} for instance,[1144] would be fruitless here. Nevertheless, it may be noted that among the nine rings presently investigated, two have 2/*m* internal symmetry, six are centrosymmetric, and one has no internal symmetry. In so far as the numerical data presented in Table 4.6.3 are concerned, it may be noted that the P–P distances, ranging from 2.818 to 3.018 Å, and the P–O–P angles ranging from 123.1 to 146.3°, are within the ranges commonly observed in cyclophosphate crystal chemistry. The spread in the values reported for the P–P–P angles (92.1–147.8°) may appear to be very wide, but is in fact quite comparable to the range of values observed in cyclohexaphosphates (85.9–142.8°).[1132]

Table 4.6.3. Main Geometrical Features of P_8O_{24} Ring Anions

Formula	P–P–P (°)	P–O–P (°)	P–P (Å)	Symmetry	Reference
$Na_8P_8O_{24}\cdot6H_2O$	123.9	138.4	3.011	$\bar{1}$	1115
	121.0	126.3	2.880		
	147.8	127.3	2.891		
	146.7	128.5	2.902		
$(NH_4)_8P_8O_{24}\cdot3H_2O$	128.5	126.7	2.887	None	1116
	118.2	130.5	2.915		
	100.6	129.7	2.910		
	109.9	127.8	2.876		
	123.1	128.2	2.887		
	121.7	128.7	2.890		
	100.4	131.5	2.930		
	112.1	126.2	2.874		
$Cs_8P_8O_{24}\cdot8H_2O$			2.997(4)	$\bar{1}$	1117
			2.900		
			2.882		
			2.886		
$Cu_3(NH_4)_2P_8O_{24}$	119.9	129.1	2.888	$P\bar{1}$	37
	92.1	134.8	2.928		
	112.2	146.3	2.930		
	123.3	134.9	3.018		
$Ga_2K_2P_8O_{24}$	131.4	123.1	2.818	$2/m$	1123
	138.0	134.4	2.933		
		136.0	2.947		
$Ag_9NaP_8O_{24}\cdot(NO_3)_2\cdot4H_2O$	102.5	133.9	2.958	$2/m$	1128
	108.7	129.5	2.889		
		128.8	2.900		
$(NH_4)_8P_8O_{24}\cdot Te(OH)_6\cdot2H_2O$	110.3	130.0	2.884	$\bar{1}$	1130
	118.7	137.0	2.936		
	105.7	134.6	2.946		
	106.8	136.4	2.965		
	106.8	136.5	2.960	$\bar{1}$	
	103.7	134.9	2.952		
	114.6	126.3	2.868		
	119.9	131.0	2.908		
$K_8P_8O_{24}\cdot Te(OH)_6\cdot2H_2O$	107.4	132.4	2.946	$\bar{1}$	1129
	105.1	132.6	2.939		
	117.5	132.2	2.925		
	101.1	133.3	2.953		
$(EDA)_4P_8O_{24}\cdot2Te(OH)_6\cdot2H_2O^a$	109.1	133.1	2.941	$\bar{1}$	1131
	118.8	127.5	2.875		
	118.4	128.4	2.895		
	123.8	127.3	2.871		

[a]EDA, ethylenediammonium, $[NH_3(CH_2)_2NH_3]^{2+}$.

4.7. Cyclodecaphosphates

4.7.1. Introduction

Cyclodecaphosphates are very rare compounds, and, until recently, $Ba_2Zn_3P_{10}O_{30}$ was the only one prepared in a crystalline state. It was characterized during the investigation of the BaO–ZnO–P_2O_5 system by Bagieu-Beucher and El-Horr.[1133] During experiments to optimize a flux method in order to prepare single crystals of $Ba_2Zn(PO_3)_6$, a long-chain polyphosphate appearing in the $Zn(PO_3)_2$–$Ba(PO_3)_2$ phase-equilibrium diagram, these authors obtained a new species, which they identified as probably being a cyclodecaphosphate, $Ba_2Zn_3P_{10}O_{30}$. The geometrical feature of the anion in the crystal structure of this compound, which was determined by Bagieu-Beucher et al.,[40,1134] confirmed its identification as a cyclodecaphosphate. It should be noted that this compound does not appear in the phase diagram.

All attempts to elaborate from this compound a starting material for the convenient preparation of other cyclodecaphosphates failed until very recently, when Schülke[1135] succeeded in producing alkali salts in a reproducible way. Since the discovery of this process, several new structural investigations were performed, providing a better knowledge of the geometry of the $P_{10}O_{30}$ ring anions.

4.7.2. Present State of Cyclodecaphosphate Chemistry

$K_{10}P_{10}O_{30} \cdot 4H_2O$. Potassium cyclodecaphosphate tetrahydrate was prepared for the first time by Schülke[1135]; the procedure employed has not yet been published in detail, but it was outlined by Schülke et al.[1136] in a publication reporting a crystal structure determination for the compound.

This salt is very soluble in water and very resistant to hydrolytic ring cleavage. It is converted at very low temperature (433–443 K) to crystalline long-chain potassium polyphosphate, KPO_3, in an exothermic reaction.

$(NH_4)_{10}P_{10}O_{30}$. The anhydrous ammonium salt was prepared from the potassium salt by Schülke,[1135] using an ion-exchange resin. This salt is also very soluble in water.

$Ag_{10}P_{10}O_{30}$. The existence of this salt was reported by Schülke et al.,[1136] but no crystal data are available.

$Ag_4K_6P_{10}O_{30} \cdot 10H_2O$. During attempts to prepare the silver salt in order to synthesize water-soluble cyclodecaphosphates by using Boullé's metathesis reaction, Averbuch-Pouchot[1137] observed the formation of several mixed silver–potassium salts. Among them, $Ag_4K_6P_{10}O_{30} \cdot 10H_2O$ was clearly characterized and its atomic arrangement determined by Averbuch-Pouchot et al.[1138]

$Mn_4K_2P_{10}O_{30} \cdot 18H_2O$. This salt prepared by Schülke and Averbuch-Pouchot[1140] crystallizes upon evaporation, at room temperature, of an aqueous solution of $K_{10}P_{10}O_{30} \cdot 4H_2O$ to which an aqueous solution of manganese(II) chloride has been added. The initial K/Mn ratio used by the authors was 2. A complete description of the atomic arrangement was given.

Table 4.7.1. Main Crystallographic Data for Cyclodecaphosphates

Formula	a (Å) α (°)	b (Å) β (°)	c (Å) γ (°)	S.G.	Z	Reference(s)
$K_{10}P_{10}O_{30} \cdot 4H_2O$	15.342(5)	11.846(5)	19.264(5)	$C2/c$	4	1136
$Ag_4K_6P_{10}O_{30} \cdot 10H_2O$	14.267(7)	7.305(1)	10.319(4)	$P\bar{1}$	1	1138
	105.38(5)	101.03(5)	87.51(5)			
$Mn_4K_2P_{10}O_{30} \cdot 18H_2O$	14.546(10)	15.211(10)	9.860(6)	$P2_1/a$	2	1140
		105.12(4)				
$(Gua)_{10}P_{10}O_{30} \cdot 4H_2O^a$	12.192(8)	14.083(9)	9.317(6)	$P\bar{1}$	1	1141
	91.25(3)	103.61(3)	71.22(2)			

^{a}Gua, guanidinium, $[C(NH_2)_3]^+$.

$Ba_2Zn_3P_{10}O_{30}$. The characterization of $Ba_2Zn_3P_{10}O_{30}$ was described in Section 4.7.1. This salt is quite insoluble in neutral aqueous solutions. In strongly acidic or alkaline solutions, its dissolution is accompanied by hydrolytic degradation of the ring anion. Infrared and Raman spectra of polycrystalline $Ba_2Zn_3P_{10}O_{30}$ were recorded by Cabello and Baran.[1139]

$[C(NH_2)_3]_{10}P_{10}O_{30} \cdot 4H_2O$. Very recently, the first organic-cation derivative, guanidinium cyclodeca-phosphate tetrahydrate, $[C(NH_2)_3]_{10}P_{10}O_{30} \cdot 4H_2O$, was prepared by Averbuch-Pouchot and Schülke.[1141] A detailed description of the atomic arrangement was given by the authors.

Crystallographic Data. Table 4.7.1 contains the main crystallographic data for cyclodecaphosphates.

4.7.3. Chemical Preparation of Cyclodecaphosphates

Because this area of cyclophosphate chemistry is very new, it seems worthwhile to provide detailed descriptions of the syntheses of cyclodecaphosphates.

$Ba_2Zn_3P_{10}O_{30}$. The preparation of single crystals of $Ba_2Zn_3P_{10}O_{30}$ by the use of a flux method was first reported by Bagieu-Beucher *et al.*[1134] These authors dissolved $BaCO_3$ and $ZnCO_3$ in stoichiometric amounts in an excess of H_3PO_4 and heated the mixture for several days at 673 K. Polycrystalline samples were obtained by heating a stoichiometric mixture of $BaCO_3$, $ZnCO_3$, and $(NH_4)_2HPO_4$ for several days at 873–923 K.

Recently, Schülke[1135,1136] considerably improved these procedures. When a mixture of $BaCO_3$, $ZnCO_3$, and H_3PO_4 in the stoichiometric ratio was seeded with a few crystals of $Ba_2Zn_3P_{10}O_{30}$ and heated at 773 K in a platinum dish for 2 h, $Ba_2Zn_3P_{10}O_{30}$ was obtained in 100% yield. Crystals used as seeds in this procedure are prepared by melting a stoichiometric mixture of $BaCO_3$, $ZnCO_3$, and H_3PO_4 at 1023 K for 10 min. The melt is then tempered at 773 K for 12 h. This procedure is sometimes unsuccessful and leads to a mixture of various condensed phosphates.

$K_{10}P_{10}O_{30} \cdot 4H_2O$. We report here in detail the procedure employed by Schülke[1135,1136] for the preparation of the potassium salt. Before giving the experimental details, let us say first that this process can be represented schematically by three steps, corresponding to the following reactions:

$$Ba_2Zn_3P_{10}O_{30} + 2K_2SO_4 \rightarrow 2BaSO_4 + K_4Zn_3P_{10}O_{30} \text{ (amorphous)}$$

$$K_4Zn_3P_{10}O_{30} + 3Na_2S \rightarrow 3ZnS + K_4Na_6P_{10}O_{30} \text{ (solution)}$$

$$K_4Na_6P_{10}O_{30} + K^+ \text{ cation exchanger} \rightarrow K_{10}P_{10}O_{30} \text{ (solution)}$$

The experimental procedure is as follows. A mixture of 18.9 g of $Ba_2Zn_3P_{10}O_{30}$, 5.2 g of K_2SO_4, and 40 ml of H_2O is subjected intensive grinding in an agate swing mill for 8 h. A solution of 10.5 g of $Na_2S \cdot 9H_2O$ in 400 ml of H_2O is then added to the suspension with vigorous stirring. The ZnS and $BaSO_4$ formed during the first two steps are then removed by filtration, and an aqueous solution of $K_4Na_6P_{10}O_{30}$ is thus obtained. The latter salt can be precipitated by addition of 400 ml of methanol.

The pure potassium salt, $K_{10}P_{10}O_{30} \cdot 4H_2O$, is then prepared from $K_4Na_6P_{10}O_{30}$ (aq.) by ion exchange with a strongly acidic cationic exchanger in the K^+ form or by precipitation of the insoluble silver salt, $Ag_{10}P_{10}O_{30}$, and reaction of this salt with KCl in aqueous solution. The yield is about 70%.

4.7.4. Atomic Arrangements of Cyclodecaphosphates

4.7.4.1. Potassium Cyclodecaphosphate Tetrahydrate

$$K_{10}P_{10}O_{30} \cdot 4H_2O,^{1136} \text{ monoclinic}, C2/c, Z = 4$$

$$a = 15.342(5), \quad b = 11.846(5), \quad c = 19.264(5) \text{ Å}, \quad \beta = 91.27(3)°$$

As shown in Fig. 4.7.1, a projection of the atomic arrangement along the **b** axis, the ring anions have a layer organization in planes parallel to the [101] direction. In this figure,

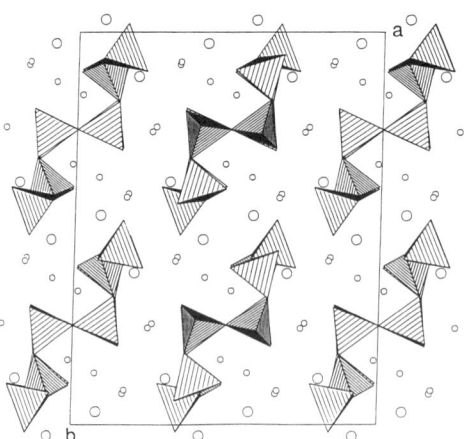

Figure 4.7.1. Projection, along the **b** axis, of the atomic arrangement in $K_{10}P_{10}O_{30} \cdot 4H_2O$. The small empty circles represent the potassium atoms, and the larger ones the water molecules.

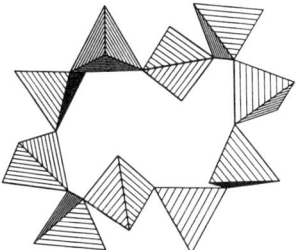

Figure 4.7.2. Projection of the $P_{10}O_{30}$ anion along the [1 1 1] direction.

which is presented only to give a general idea of the atomic arrangement, the profile projection of the ring anion does not allow for an understanding of its geometry as some tetrahedra of the ring are superimposed. This phosphoric group has a binary internal symmetry induced by a twofold axis passing through two bonding oxygen atoms. A projection of this ring along the [111] direction is shown in Fig. 4.7.2. This projection may give the impression that the ring is rather planar, but the two additional projections in Figs. 4.7.3 and 4.7.4 show clearly the two-stage organization of the anion induced by its twofold internal symmetry. The main geometrical features of this ring are reported and discussed in the Section 4.5.5.

Five independent potassium atoms, all located in general positions, coexist in this atomic arrangement. Within a range of 3.50 Å, three of them have eightfold coordination and two have sevenfold coordination. Within these various polyhedra, the K–O distances range between 2.643 and 3.423 Å.

The water molecules are not dispersed inside the atomic arrangement but are located in pairs close to the inversion centers, forming finite networks of hydrogen bonds in association with the neighboring external oxygen atoms of the ring. The H_2O–O distances observed in the H-bond network range from 2.830 to 3.029 Å.

4.7.4.2. Silver–Potassium Cyclodecaphosphate Decahydrate

$$Ag_4K_6P_{10}O_{30} \cdot 10H_2O,[1138] \text{ triclinic, } P\bar{1}, Z = 1$$

Figure 4.7.3. Projection of the $P_{10}O_{30}$ ring anion along the c axis.

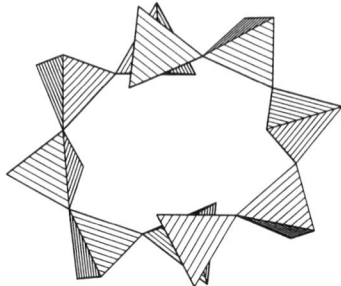

Figure 4.7.4. Projection of the $P_{10}O_{30}$ ring anion along the a axis.

$$a = 14.267(7), \quad b = 7.305(1), \quad c = 10.319(4) \text{ Å}$$

$$\alpha = 105.38(5), \quad \beta = 101.03(5), \quad \gamma = 87.51(5)°$$

Figure 4.7.5 shows a projection of this structure along the **b** direction. The centrosymmetric $P_{10}O_{30}$ ring anion is situated around the inversion center at 0,0,1/2. Its configuration is discussed in Section 4.7.5. The three independent potassium atoms have different coordination polyhedra. Within a range of 3.50 Å one of them has sevenfold coordination, the second one has eightfold coordination, and the third one has sixfold coordination. Within these various KO_n polyhedra, the K–O distances range between 2.727 and 3.262 Å. Some bonding oxygen atoms of the phosphoric ring are involved in the potassium coordination polyhedra.

The silver atoms are not dispersed inside the atomic arrangement but assemble around the inversion centers in such a manner as to form clusters containing four silver atoms. Within such a cluster, the Ag–Ag distances are 3.265 and 3.321 Å. Each silver atom has four oxygen neighbors, assembled in an unusual configuration. Within an AgO_4 entity, the

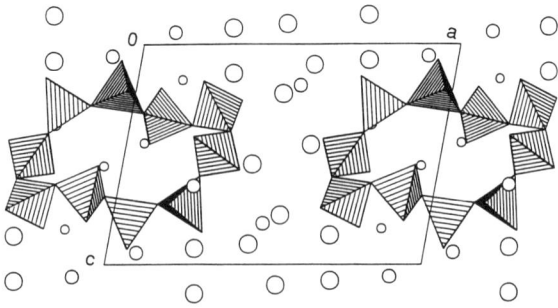

Figure 4.7.5. Projection, along the **b** direction, of the atomic arrangement in $Ag_4K_6P_{10}O_{30}\cdot10H_2O$. The small open circles represent the silver atoms, the intermediate ones the potassium atoms, and the largest ones the water molecules.

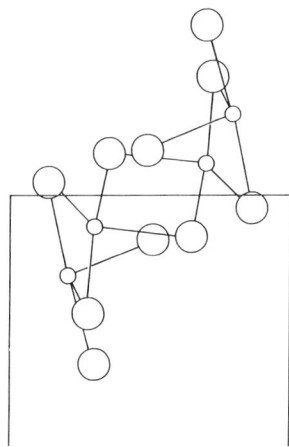

Figure 4.7.6. Projection, along the *c* direction, of the Ag_4O_{10} cluster situated around the $(0,1/2,1/2)$ inversion center in the structure of $Ag_4K_6P_{10}O_{30}\cdot10H_2O$. The smaller circles represent the silver atoms.

silver atom and three of the oxygen atoms are almost coplanar; with O–Ag–O angles close to 90°, while the fourth oxygen atom establishes an Ag-O bond almost perpendicular to this plane. These AgO_4 groups are connected by edge sharing, thus building a centrosymmetric cluster of formula Ag_4O_{10}. Such an assembly is represented in Fig. 4.7.6, a projection along the **c** axis. Within such a cluster, the Ag–O distances range from 2.269 to 2.618 Å. As far as we know, such a configuration has not yet been observed in silver salts.

4.7.4.3. Zinc–Barium Cyclodecaphosphate

$$Ba_2Zn_3P_{10}O_{30}, [40] \text{ monoclinic, P2/n, Z = 2}$$

$$a = 21.738(15), \quad b = 5.356(5), \quad c = 10.748(8) \text{ Å}, \quad \beta = 99.65(3)°$$

This atomic arrangement can be simply described as a three-dimensional framework of ZnO_6 octahedra and BaO_9 polyhedra organized in such a way as to build large channels in which are located the phosphoric groups and the ZnO_4 tetrahedra. The phosphoric anions form rows parallel to the **b** axis. In these rows, the $P_{10}O_{30}$ groups are interconnected by the ZnO_4 tetrahedra. Such a row is depicted in Fig. 4.7.7, a projection along the **c** axis. BaO_9 polyhedra and ZnO_6 octahedra are interconnected in a three-dimensional array building wide channels parallel to the **b** axis. Inside these channels are located the $\cdots P_{10}O_{30}$–ZnO_4–$P_{10}O_{30}\cdots$ rows. Figure 4.7.8, a projection along the **b** axis, depicts the organization of such a channel. The phosphoric ring anion has twofold internal symmetry, and its main geometrical features are reported in Section 4.7.5.

Within the BaO_9 polyhedron, the Ba–O distances range from 2.744 to 3.237 Å. In the two crystallographically independent ZnO_6 octahedra, the Zn–O distances range from 2.029 to 2.260 Å, whereas in the $ZnO4$ tetrahedra they range from 1.904 to 1.911 Å. In addition, it may be noted that the Zn–P distances are relatively short ranging between 3.163 and 3.334 Å.

Figure 4.7.7. Representation of a row of $P_{10}O_{30}$ ring anions interconnected by ZnO_4 tetrahedra (shaded tetrahedra) in the structure of $Ba_2Zn_3P_{10}O_{30}$.

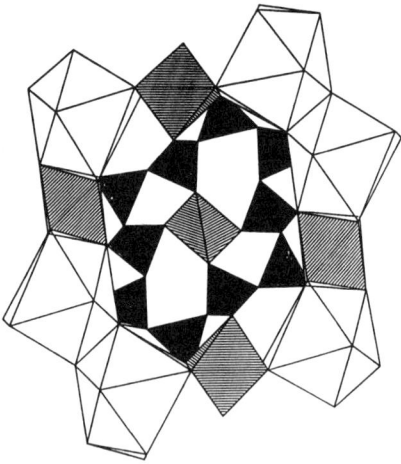

Figure 4.7.8. Details of the organization of a channel of ZnO_6 and BaO_9 polyhedra around a phosphoric ring, viewed along the *b* axis, in the structure of $Ba_2Zn_3P_{10}O_{30}$. The central ZnO_4 tetrahedron interconnecting the $P_{10}O_{30}$ groups along the b direction is also shown.

4.7.5. Geometries of the $P_{10}O_{30}$ Ring Anion

As mentioned above, $P_{10}O_{30}$ ring anions are very rare; in fact, only five examples have been reported to date. With so few data available, general discussion of the geometries of $P_{10}O_{30}$ ring anions would be fruitless. Nevertheless, we present in Table 4.7.2 what we

Table 4.7.2. Main Geometrical Data for $P_{10}O_{30}$ Ring Anions

Formula	P–P–P (°)	P–O–P (°)	P–P (Å)	Symmetry	Reference
$Ba_2Zn_3P_{10}O_{30}$	112.9	130.0	2.915	2	40,1134
	102.1	133.8	2.915		
	131.9	144.3	3.002		
	99.7	123.3	2.838		
	126.6	128.6	2.870		
$K_{10}P_{10}O_{30}\cdot4H_2O$	86.6	137.1	3.011	2	1136
	111.5	133.1	2.950		
	133.3	128.9	2.911		
	107.1	131.3	2.936		
	90.5	134.9	2.963		
		138.2	3.006		
$Ag_4K_6P_{10}O_{30}\cdot10H_2O$	99.8	131.5	2.986	$\bar{1}$	1138
	96.0	130.1	2.925		
	136.9	127.9	2.911		
	116.3	134.2	2.891		
	99.4	137.1	2.945		
$Mn_4K_2P_{10}O_{30}\cdot18H_2O$	114.6	135.9	2.938	$\bar{1}$	1140
	138.8	129.4	2.928		
	85.5	133.5	2.890		
	87.9	134.3	2.944		
	98.0	127.1	2.869		
$(Gua)_{10}P_{10}O_{30}\cdot4H_2O^a$	138.7	130.9	2.925	$\bar{1}$	1141
	144.1	125.6	2.867		
	138.8	135.5	2.958		
	121.4	127.8	2.881		
	107.6	129.2	2.904		

^{a}Gua, guanidinium, $[C(NH_2)_3]^+$.

consider to be the main geometrical features of a ring framework—the P–P distances and P–O–P and P–P–P angles—for the five rings presently known. The P–P distances, ranging from 2.870 to 3.011 Å, and the P–O–P angles, ranging between 123.3 and 144.3°, are within the ranges commonly observed in cyclophosphate crystal chemistry. The wild range of values the P–P–P angles (86.6–144.1°) is quite comparable to the range of values observed in cyclohexaphosphates (85.9–142.8°).[1132]

Two of the three rings included in Table 4.7.2 have twofold internal symmetry, but they in fact have fundamentally different geometries: in $Ba_2Zn_3P_{10}O_{30}$ the binary axis is perpendicular to the mean plane formed by the phosphorus atoms, whereas in the case of $K_{10}P_{10}O_{30}\cdot4H_2O$ the binary axis is parallel to this mean plane, as it passes through two centrosymmetric bonding oxygen atoms of the ring.

4.8. Cyclododecaphosphates

4.8.1. Present State of Cyclododecaphosphate Chemistry

Cyclododecaphosphates are very rare compounds. The first example of cyclododecaphosphate, $Cs_3V_3P_{12}O_{36}$, was recently characterized by Lavrov *et al.*[41] during an investigation of the V_2O_3-M_2O-P_2O_5-H_2O systems. The crystalline compound appears as greenish-yellow pentagonal dodecahedra belonging to the cubic system. This salt is stable in argon up to 1173 K. At this temperature, it decomposes into $V(PO_3)_3$ (C form) and $CsPO_3$:

$$Cs_3V_3P_{12}O_{36} \rightarrow 3CsPO_3 + 3V(PO_3)_3$$

The corresponding iron–casium compound is isotypic. A crystal structure determination for $Cs_3V_3P_{12}O_{36}$ was performed by Lavrov et al.[1142]

Grunze et al.[278] investigated the thermal behavior of various gallium–casium phosphates. They obtained $Cs_3Ga_3P_{12}O_{36}$ by thermal decomposition of $CsGaHP_3O_{10}$ according to the following scheme:

$$6CsGaHP_3O_{10} \rightarrow Cs_3Ga_3P_{12}O_{36} + 3CsGaP_2O_7 + 3H_2O$$

As $CsGaHP_3O_{10}$ is trimorphic, the temperature at which the above decomposition occurs is a function of the form used and varies from 823 to 853 K. The same authors carried out a crystal structure determination for $Cs_3Ga_3P_{12}O_{36}$ and reported that this compound is isotypic with the corresponding vanadium compound. During an investigation of the Cr_2O_3-M_2O-P_2O_5-H_2O systems between 473 and 673 K, Grunze and Chudinova[445] characterized three compounds—$(NH_4)_3Cr_3P_{12}O_{36}$, $Rb_3Cr_3P_{12}O_{36}$, and $Cs_3Cr_3P_{12}O_{36}$— but did not report their unit-cell dimensions. Nevertheless, on the basis of the unindexed X-ray powder patterns given by the authors, they are probably isotypic with the cesium–vanadium salt.

Recently, we have been informed that U. Schülke successfully performed the synthesis of alkali cyclododecaphosphates.

4.8.2. Crystal Chemistry of Cyclododecaphosphates

All the compounds described above are isotypic and crystallize with cubic unit cells close to 14.5 Å in length. The unit cell dimensions of only two of these compounds have been accurately measured:

$$V_3Cs_3P_{12}O_{36}, \qquad a = 14.543 \text{ Å}^{41,445}$$

$$Ga_3Cs_3P_{12}O_{36}, \qquad a = 14.374 \text{ Å}^{278}$$

The common space group is $Pa3$ with $Z = 4$. We describe here the structure determined with the vanadium salt. This highly symmetrical atomic arrangement is rather difficult to describe. There are two crystallographically independent vanadium atoms, both located on ternary axes and having almost regular octahedral coordination. The two independent cesium atoms have 9- and 12-fold coordination. The $P_{12}O_{36}$ ring anions built around the ternary axes have in fact $\overline{3}$ internal symmetry. A vanadium atom is situated at the center of the ring and is coordinated by six of the external atoms of the ring in an octahedral arrangement. This organization, represented by a projection along a threefold axis in Fig. 4.8.1, is the only known example of a phosphoric ring having an associated-cation polyhedron at its center. Along the threefold axis, the CsO_9 polyhedra and the other type

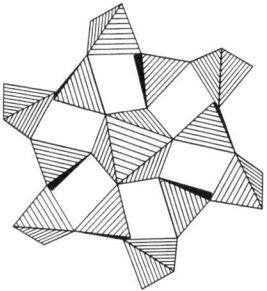

Figure 4.8.1. Projection, along the ternary axis, of the $P_{12}O_{36}$ ring anion. A VO_6 octahedron is situated at the center of the ring.

of VO_6 octahedra form columns by sharing faces. The geometrical features of the phosphoric ring as observed in the vanadium salt are reported below:

P–P–P (°)	P–O–P (°)	P–P (Å)
85.8	133.8	2.945
111.3	136.3	2.952

These values are not significantly different from those reported for other large phosphoric rings.

4.9. Thiocyclophosphates

4.9.1. Introduction

Wolf and Meisel[1143] described the synthesis of several phosphorus oxide–sulfides $P_4O_{(10-n)}S_n$ by reaction of P_4S_{10} with P_4O_{10}. Depending as the P_4S_{10}/P_4O_{10} ratio in the starting mixture, various phosphorus oxide–sulfides are formed. In addition to the well-known $P_4O_6S_4$, the following compounds were obtained for the first time by these authors: $P_4O_5S_5, P_4O_4S_6, P_4O_3S_7, P_4O_2S_8$, and P_4OS_9. They were separated by fractional distillation or crystallization. A detailed study of these phosphorus oxide–sulfides was reported by Meisel.[1144]

One must also mention the existence of $P_4O_3S_6$, first prepared by Wolf and Meisel.[1145] In this hygroscopic compound one of the four phosphorus atoms is trivalent. Its crystal structure was determined by Palkina *et al.*[1146] According to Wolf and Meisel,[1147] at room temperature this compound slowly decomposes to H_3PO_4 and H_2S.

Some of these compounds were used for the preparation of thiocyclophosphates by a procedure similar to that described for the production of cyclotetraphosphoric acid by the hydrolysis of P_4O_{10} at low temperature. In the case of $P_4O_6S_4$, the reaction is

$$P_4O_6S_4 + 2H_2O \rightarrow H_4P_4O_8S_4$$

leading to the formation of tetrathiocyclotetraphosphoric acid.

In the thiocyclotetraphosphate anions that have been obtained, one of the two external oxygen atoms of each tetrahedron in the corresponding cyclophosphate anion is replaced by a sulfur atom, except in the case of $(NH_4)_3P_3S_9$, in which all the oxygen atoms appear to be substituted.

4.9.2. Present State of Thiocyclophosphate Chemistry

4.9.2.1. Thiocyclotriphosphates

$(NH_4)_3P_3S_9$. The reaction of P_4S_{10} or P_4S_9 in liquid ammonia below 240 K, investigated by Wolf and Meisel,[1148] yields to ammonium nonathiocyclotriphosphate, $(NH_4)_3P_3S_9$, which crystallizes on cooling to 195 K.

$$P_4S_9 + 6NH_3 \rightarrow (NH_4)_3P_3S_9 + P(NH_2)_3$$

$$P_4S_{10} + 6NH_3 \rightarrow (NH_4)_3P_3S_9 + SP(NH_2)_3$$

$(NH_4)_3P_3S_9$ reacts with PCl_3, rebuilding the adamantane-like structure of P_4S_9:

$$(NH_4)_3P_3S_9 + PCl_3 \rightarrow P_4S_9 + 3NH_4$$

Heating $(NH_4)_3P_3S_9$ at 513 K for 100 h, in vacuum, leads to the formation of PNS:

$$(NH_4)_3P_3S_9 \rightarrow 3PNS + 6H_2S$$

No structural study has been performed to confirm the ring nature of the anion suggested by the chemical formula of this salt.

$(NH_4)_3P_3O_6S_3$. This salt is obtained by the action of ammonium fluoride on $P_4O_6S_4$ in glacial acetic acid.[1149] Finely ground $P_4O_6S_4$ is slowly added to an ice-cooled saturated solution of ammonium fluoride in acetic acid. Crystals of $(NH_4)_3P_3O_6S_3$ precipitate after two hours of stirring. The reaction scheme is

$$3NH_4F + P_4O_6S_4 + 2AcOH \rightarrow (NH_4)_3P_3O_6S_3 + H(POSF_2) + Ac_2O + HF$$

The crystal structure determination was performed by Meisel et al.[1149]

$M_3P_3O_3S_6$ (M = Na, Tl, K, Cs). $Na_3P_3O_3S_6$ was prepared by Wolf and Meisel[1147] according to the following scheme:

$$P_4O_3S_6 + 4NaHCO_3 \rightarrow Na_3P_3O_3S_6 + NaH(PO_3H) + 4CO_2 + H_2O$$

The reaction is carried out at 283 K. The thallium derivative was also prepared by the authors in a good crystalline state. Recently, crystal structures of the potassium and cesium salts were performed by Palkina et al.[1150] The main crystallographic data for these compounds are included in Table 4.9.1.

4.9.2.2. Thiocyclotetraphosphates

The chemical preparation and the main properties of salts of tetrathiocyclotetraphosphoric acid with monovalent cations and some organic cations were described by Kuvshi-

Table 4.9.1. Main Crystallographic Data for Thiocyclophosphates

Formula	a (Å) α (°)	b (Å) β (°)	c (Å) γ (°)	S.G.	Z	Reference(s)
$(NH_4)_3P_3O_6S_3$	12.450(8)	12.755(8)	8.154(6)	$Pnma$	4	1149
$K_3P_3O_6S_3$	8.013(1)	11.088(1)	13.627(2)	$Pnma$	4	1150
$Cs_3P_3O_6S_3$	13.013(3)	8.337(2) 91.54(2)	13.108(3)	$P2_1/a$	4	1150
$Na_4P_4O_8S_4 \cdot 6H_2O$	8.815(1) 78.33(4)	9.313(4) 95.07(2)	14.259(3) 119.26(2)	$P\bar{1}$	2	1152, 1154
$Cs_4P_4O_8S_4$	7.784(3) 95.55(3)	7.889(2) 114.67(4)	8.610(5) 75.46(3)	$P\bar{1}$	1	1153, 1154
$Ba_2P_4O_8S_4 \cdot 10H_2O$	8.600(2) 88.69(1)	8.904(2) 73.20(1)	9.234(1) 61.77(1)	$P\bar{1}$	1	1155
$(Gua)_4P_4O_8S_4$ [a]	10.894(8)	13.126(8) 105.74(5)	8.919(6)	$P2_1/n$	2	1159

[a]Gua, guanidinium, $[C(NH_2)_3]^+$.

nova *et al.*[1151] The existence of $Li_4P_4O_8S_4 \cdot 9H_2O$, $Na_4P_4O_8S_4 \cdot 6H_2O$, $K_4P_4O_8S_4 \cdot 2H_2O$, $Rb_4P_4O_8S_4 \cdot 2H_2O$, $Cs_4P_4O_8S_4$, $Tl_4P_4O_8S_4$, $(NH_4)_4P_4O_8S_4 \cdot 2H_2O$, $(Eta)_4P_4O_8S_4 \cdot 2H_2O$ (Eta = ethyleneammonium = $C_2H_5NH_3+$), $(Gua)_4P_4O_8S_4$ (Gua = guanidinium = $C(NH_2)_3+$), $(Pyr)_4P_4O_8S_4$ (Pyr = pyridinium = $C_5H_5NH^+$), and $(Qui)_4P_4O_8S_4$ (Qui = quinolinium = $C_9H_7NH^+$) was reported by these authors.

The thermal behavior of the alkali derivatives was investigated by Kuvshinova *et al.*[1151] After several steps of dehydration with the successive formation of the octa-, tetra-, and dihydrate, the lithium salt transforms first into $Li_4P_4O_{12}$ (at about 723 K) and then into a mixture of $LiPO_3$ (HT) and $Li_6P_6O_{18}$ (at about 823 K). The tetrahydrate and the dihydrate of the sodium derivative are observed between room temperature and 423 K. Formation of the final product, $Na_3P_3O_9$, is observed above 700 K. Anhydrous $K_4P_4O_8S_4$ and $Rb_4P_4O_8S_4$ are observed above 423 K and then transform into long-chain potassium and rubidium polyphosphate. The cesium salt melts at about 650 K; above this temperature, it transforms into $CsPO_3$.

During these investigations, some unidentified crystalline phases also appeared.

Some of these cyclotetraphosphates have been prepared as single crystals and investigated from a structural point of view. These studies are discussed below. The main crystallographic data for these compounds are included in Table 4.9.1.

$Na_4P_4O_8S_4 \cdot 6H_2O$. This salt was obtained by Ilyukhin *et al.*[1152] by slowly adding ground $P_4O_6S_4$ to a cooled aqueous solution of $NaHCO_3$ (about a 20% excess of carbonate with respect to the stoichiometry of the reaction was used). At the end of the reaction, the excess of carbonate was destroyed by acetic acid, and ethanol was added to precipitate the sodium salt. $Na_4P_4O_8S_4 \cdot 6H_2O$ is very difficult to crystallize and is not very stable in air. Nevertheless, the authors were able to perform a crystal structure determination.

$Cs_4P_4O_8S_4$. This cesium salt was prepared by Ilyukhin *et al.*[1153] by slowly adding pulverized $P_4O_6S_4$ to a cold aqueous solution of cesium carbonate. This solution must contain about a 15% excess of carbonate relative to the stoichiometry of the reaction. Upon completion of the reaction, the excess of carbonate is removed by acetic acid, and $Cs_4P_4O_8S_4$ is precipitated by ethanol. The crystals obtained are lamellar. The authors performed the determination of the atomic arrangement. Ilyukhin *et al.*[1154] subsequently reported a comparative study of the atomic arrangements of $Cs_4P_4O_8S_4$ and $Na_4P_4O_8S_4 \cdot 6H_2O$.

$Ba_2P_4O_8S_4 \cdot 10H_2O$. Nikolaev and Kuvshinova[1155] reported the chemical preparation and the crystal structure of $Ba_2P_4O_8S_4 \cdot 10H_2O$. This salt was synthesized by adding $P_4O_6S_4$ to a suspension of $BaCO_3$ in water. Large, prismatic, colorless crystals were obtained by recrystallization from acetic acid. They are stable for years in the mother liquor, but in air they disintegrate into the hexahydrate within minutes.

The thermal transformations of Ba and Sr tetrathiocyclotetraphosphate hexahydrates were studied in vacuum by Prodan *et al.*[1156] and in air by Kubshinova *et al.*[1157] The influence of water vapor on their topochemical transformations was investigated by Petrovskaya *et al.*[1158]

$[C(NH_2)_3]_4P_4O_8S_4$. Guanidinium tetrathiocyclotetraphosphate was prepared by slowly adding guanidinium carbonate to a cooled solution of tetrathiocyclotetraphosphoric acid, obtained by careful dissolution of the phosphorus oxide–sulfide $P_4O_6S_4$ in water at 273–278 K.[1151] The reactions are

$$P_4O_6S_4 + 2H_2O \rightarrow H_4P_4O_8S_4$$

$$H_4P_4O_8S_4 + 2(Gua)_2CO_3 \rightarrow (Gua)_4P_4O_8S_4 + 2CO_2 + 2H_2O$$

The salt has relatively low solubility in water and is stable at room temperature. Crystals up to 1 mm long were obtained as colorless prisms by evaporation at room temperature. The atomic arrangement in this salt was determined by Meisel *et al.*[1159]

4.9.3. Some Atomic Arrangements of Thiocyclophosphates

4.9.3.1. Ammonium Trithiocyclotriphosphate

$(NH_4)_3P_3O_6S_3$,[1149] orthorhombic, *Pnma*, $Z = 4$

$a = 12.450(8)$, $b = 12.755(8)$, $c = 8.154(6)$ Å

Figure 4.9.1 shows the projection along the **a** direction of this atomic arrangement. The ring anion has internal mirror symmetry, with one phosphorus atom, one sulfur atom, and two oxygen atoms located in the mirror plane. One of the two crystallographically independent ammonium group is also located in this mirror plane. This ammonium groups and the $P_3O_6S_3$ groups build up layers centered by the mirror planes ($y = 1/4$ and 3/4). The ammonium groups in general positions form corrugated layers between the mirror planes. The $P_3O_6S_3$ ring anion is built by two independent strongly distorted tetrahedra in which the P–S bond is longer than 1.9 Å and the three P–O distances are normal. The ammonium

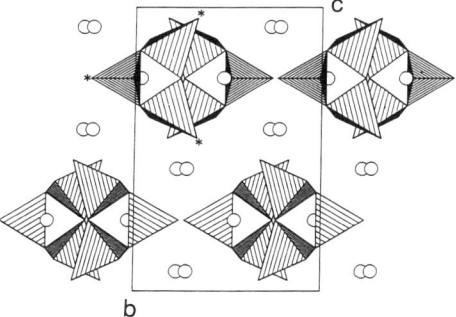

Figure 4.9.1. Projection, along the *a* axis, of the atomic arrangement in $(NH_4)_3P_3O_6S_3$. The starred corners of the tetrahedra indicate the locations of the sulfur atoms. The open circles represent the ammonium groups.

group located in the mirror plane is coordinated by six oxygen atoms, while the one in a general position has sevenfold coordination including three sulfur atoms.

Within these two polyhedra, the N–O distances range between 2.744 and 3.273 Å, and the N–S distances between 3.395 and 3.497 Å.

4.9.3.2. Cesium Tetrathiocyclotetraphosphate

$$Cs_4P_4O_8S_4,^{[1149]} \text{ triclinic, } P\bar{1}, Z = 1$$

$$a = 7.784(3), \quad b = 7.889(2), \quad c = 8.610(5) \text{ Å}$$

$$\alpha = 95.55(3), \quad \beta = 114.67(4), \quad \gamma = 75.46(3)°$$

The centrosymmetric $P_4O_8S_4$ ring anion is located around the center of inversion at 1/2,0,0. As can be expected, the two independent PO_3S tetrahedra are very distorted, with three normal P–O distances and P–S distances of 1.926 Å in the first tetrahedron and 1.968 Å in the second one. Figure 4.9.2 shows clearly the layer organization of this arrangement built by planes of ring anions and cesium atoms parallel to the [$\bar{1}01$] direction. Within a

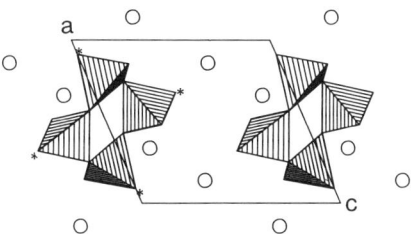

Figure 4.9.2. Projection, along the *b* axis, of the atomic arrangement in $Cs_4P_4O_8S_4$. The open circles represent the cesium atoms. The starred corners of the tetrahedra indicate the locations of the sulfur atoms.

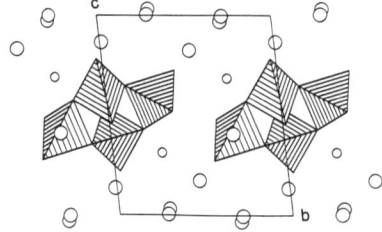

Figure 4.9.3. Projection, along the *a* axis, of the atomic arrangement in Ba₂P₄O₈S₄. The small open circles represent the barium atoms, and the large ones the water molecules. The starred corners of the tetrahedra indicate the locations of the sulfur atoms.

range of 3.80 Å, the two independent cesium atoms have eightfold coordination involving four oxygen and four sulfur atoms. Within these two polyhedra, the Cs–O distances range from 2.971 to 3.945 Å, and the Cs–S distances from 3.637 to 3.757 Å.

4.9.3.3. Barium Tetrathiocyclotetraphosphate Decahydrate

$$Ba_2P_4O_8S_4.10H_2O,^{1155} \text{ triclinic, } P\bar{1}, Z = 1$$

$$a = 8.600(2), \quad b = 8.904(2), \quad c = 9.234(1) \text{ Å}$$

$$\alpha = 88.69(1), \quad \beta = 73.20(1), \quad \gamma = 61.77(1)°$$

The $P_4O_8S_4$ ring anion, built by two independent PO_3S tetrahedra, is located around the inversion center at 1/2,0,1/2. Figure 4.9.3 shows the projection of this atomic arrangement along the **a** direction. Within these two distorted tetrahedra, the P–S distances are 1.950 and 1.953 Å while the P–O distances range between 1.488 and 1.608 Å.These latter values thus fall within the range normally observed in all condensed phosphoric anions.

The barium coordination polyhedron is built, within a range of 3.50 Å, by three oxygen atoms, four water molecules, and two sulfur atoms, with Ba–O distances ranging from 2.731 to 2.849 Å and Ba–S distances ranging from 3.392 to 3.438 Å. These $BaO_3(H_2O)_4S_2$ polyhedra assemble in pairs through a common O–O edge so as to form centrosymmetric $Ba_2O_{12}S_4$ clusters around the inversion center at 0,1/2,1/2.

4.9.3.4. Guanidinium Tetrathiocyclotetraphosphate

$$[C(NH_2)_3]_4P_3O_8S_4,^{1159} \text{ monoclinic, } P2_1/n, Z = 2$$

$$a = 10.894(8), \quad b = 13.126(8), \quad c = 8.919(6) \text{ Å}, \quad \beta = 105.74(5)°$$

As in many previously reported examples, the $P_4O_8S_4$ ring anion is centrosymmetric and thus built by two independent PO_3S tetrahedra. Within these tetrahedra, the P–S distances, 1.943 and 1.944 Å, are comparable to those observed in the other thiocyclophosphoric anions. A net of hydrogen bonds between the NH_2 groups and the oxygen or sulfur atoms of the ring anions establishes the three–dimensional cohesion of this arrangement. Figure 4.9.4 shows a projection along the **b** direction of this atomic arrangement. Among the hydrogen bonds involving oxygen atoms, the N–O distances range between 2.810 and

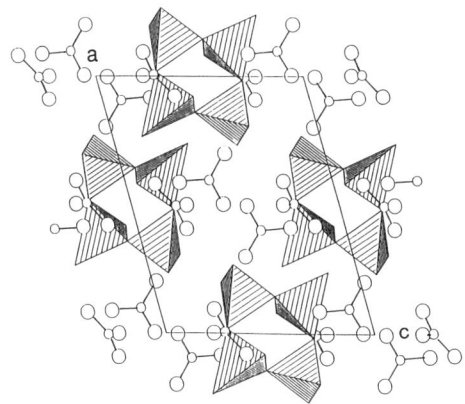

Figure 4.9.4. Projection, along the *b* direction, of the atomic arrangement in $[C(NH_2)_3]P_3O_6S_3$. The small open circles represent the central carbon atoms of the guanadinium groups, and the larger ones represent the nitrogen atoms. The hydrogen atoms have been omitted.

3.243 Å, with N–H–O angles from 151 to 178°. Those involving the sulfur atoms are significantly longer, ranging from 3.408 to 3.533 Å, with N–H–S angles from 137 to 171°.

The guanidinium group, almost coplanar within the experimental error, is as usual very regular with C–N distances ranging from 1.313 to 1.321 Å.

4.9.4. The Thiocyclophosphate Ring

The structural data for the thiocyclotriphosphate rings in the five thiocyclotriphosphates that have been examined to date are tabulated in Table 4.9.2. Clearly, the number

Table 4.9.2. Main Geometrical Features of $P_3O_6S_3$ and $P_4O_8S_4$ Anions

Formula	P–P (Å)	P–O–P (°)	P–P–P (°)	Reference(s)
Ring with m symmetry				
$(NH_4)_3P_3O_6S_3$	2.938	131.3	2×60.4	1149
	2×2.977	2×136.0	59.1	
Rings with $\bar{1}$ symmetry				
$Na_4P_4O_8S_4 \cdot 6H_2O^a$	2.951	132.0	97.7	1152, 1154
	2.993	133.0	82.3	
	2.957	133.0	83.7	
	3.015	141.5	96.3	
$Cs_4P_4O_8S_4$	2.973	133.7	82.8	1153, 1154
	2.981	134.2	97.2	
$Ba_2P_4O_8S_4 \cdot 10H_2O$	2.935	131.1	82.6	1155
	2.953	132.0	97.4	
$(Gua)_4P_4O_8S_4^{\ b}$	2.981	135.3	95.6	1159
	2.977	135.1	84.4	

[a]Two independent rings in the unit cell.

[b]Gua, guanidinium, $[C(NH_2)_3]^+$.

Table 4.9.3. Main Geometrical Features of a PO_3S
Tetrahedron in the $P_4O_8S_4$ Ring Anion Observed in
$[C(NH_2)_3]_4P_4O_8S_4{}^a$

P	S	O(E)	O(L1)	O(L2)
S	1.9427(8)	2.976(2)	2.876(2)	2.926(2)
O(E)	119.37(8)	1.495(2)	2.560(2)	2.478(2)
O(L1)	107.84(6)	111.17(9)	1.608(2)	2.504(2)
O(L2)	110.04(7)	105.29(8)	101.70(8)	1.621(1)

^{a}The P–O and P–S distances are given along the diagonal, angles below the diagonal,
and the O–O(S) distances above the diagonals. O(E) is an external oxygen of the ring,
while O(L) atoms belong to the ring. Distances are in angstroms and angles are in
decimal degrees.

of examples is not sufficient to permit any valuable comparisons. In fact, only one
trithiocyclotriphosphate has been well investigated. Nevertheless, one may note that in the
five $P_4O_8S_4$ rings, all centrosymmetric, the P–P–P angles depart more significantly from
the ideal value than was observed in the case of the P_4O_{12} rings. Thus, for these five $P_4O_8S_4$
rings the P–P–P angles range from 97.7 to 95.6° (Table 4.9.2).

The most important difference between P_3O_9 or P_4O_{12} phosphoric rings and the
thiorings considered here is the strong deformation of the tetrahedral configuration around
the phosphorus atom, owing to the presence of the sulfur atom. The geometry of such a
PO_3S tetrahedron, as observed in $[C(NH_2)_3]_4P_4O_8S_4$, is presented in Table 4.9.3.

The main differences between this PO_3S tetrahedron and the normal PO_4 tetrahedra
observed in other types of condensed phosphoric anions, arise from the length of the P–S
bond. Because the P–S distance significantly longer than the three P–O distances, the three
edges of the tetrahedron corresponding to the S–O distances are much longer (2.876 < S–O
< 2.976 Å) than the three edges corresponding to the O–O distances (2.478 < O–O < 2.560
Å). In contrast, the value of 119.37° measured for the S–P–O(E) angle is quite comparable
to the O(E)–P–O(E) values observed in other phosphoric rings, and the average value of
the O–P–O(S) angle in this tetrahedron (109.23°) is similar to what is commonly measured
in any kind of condensed phosphoric anions.

5

Mixed-Anion Phosphates

5.1. Introduction

Today more than 20 phosphates containing two kinds of phosphoric anions having different degrees of condensation are known. Except in the case of the very few of them that have been prepared by classical reactions of aqueous chemistry, such as $K_2H_5(PO_4)(P_2O_7)$ and $K_2H_8(PO_4)_2(P_2O_7)$, there are no rules for the chemical preparation of these mixed-anion phosphates. The vast majority of them have been obtained in various flux-method experiments as unexpected products that could not be identified until their crystal structures had been solved. The main crystallographic data for monophosphate–diphosphates are presented in Table 5.1, and those for mixed-anion phosphates containing anions with higher degrees of condensation are given in Table 5.2.

Table 5.1. Main Crystallographic Data for Monophosphate–Diphosphates[a]

Formula	a (Å) α (°)	b (Å) β (°)	c (Å) γ (°)	S.G.	Z	Reference
$Na_5H_2(PO_4)(P_2O_7)$	10.862(2) 128.27(4)	8.426(3) 99.21(4)	7.001(2) 88.84(3)	$P\bar{1}$	2	1173
$K_2H_5(PO_4)(P_2O_7)$	31.272(7)	7.428(1) 99.85(1)	9.253(1)	$C2/c$	8	1174
$K_2H_8(PO_4)_2(P_2O_7)$	9.364(2)	7.458(2)	19.560(2)	$Pca2_1$	4	1177
$Na_7Al_4(PO_4)(P_2O_7)_4$	14.046(3)	14.046(3)	6.169(2)	$P42_1c$	2	1175
$Na_7Cr_4(PO_4)(P_2O_7)_4$	14.061(3)	14.061(3)	6.306(2)	$P42_1c$	2	1175
$Na_7Fe_4(PO_4)(P_2O_7)_4$	14.105(3)	14.105(3)	6.378(2)	$P42_1c$	2	1175
$Na_3Th(PO_4)(P_2O_7)$	8.734(2) 93.33(3)	8.931(2) 108.29(4)	6.468(1) 110.10(4)	$P\bar{1}$	2	1176
$K_2Ni_4(PO_4)_2(P_2O_7)$	10.304(2)	13.682(3) 102.91(2)	18.139(3)	$C2/c$	8	1179
$Cs(MoO)_2(PO_4)(P_2O_7)$	6.342(1) 83.083(7)	9.676(1) 97.42(1)	10.035(1) 108.30(1)	$P\bar{1}$	2	1180
$Cs_3Mo_2(Mo_2O)_2(PO_4)_2(P_2O_7)_4$	9.511(3) 91.15(5)	14.232(4) 105.92(5)	6.437(5) 90.04(2)	$P\bar{1}$	1	1183
$K(MoO)_2(PO_4)(P_2O_7)$	10.743(2)	14.084(1) 126.42(1)	8.852(1)	$C2/c$	8	1182
$K(MoO)_2(PO_4)(P_2O_7)$	8.846(8) 56.49(1)	8.846(9) 55.59(1)	10.01(1) 68.87(1)	$P\bar{1}$	1	1181
$KMoWO(PO_4)(P_2O_7)$	8.818(1)	9.157(1)	12.384(1)	$Pbcm$	4	1184

[a]The geometrical configurations of the P_2O_7 in these compounds are described in Chapter 2 (Section 2.2.5).

Table 5.2. Main Crystallographic Data for Mixed-Anion Phosphates Containing Anions Having Higher Degrees of Condensation[a]

Formula	a (Å) α (°)	b (Å) β (°)	c (Å) γ (°)	S.G.	Z	Reference
$CsTa_2(PO_4)_2(P_3O_{10})$	5.1355(6) 82.54(1)	10.900(1) 87.13(1)	14.392(1) 85.05(1)	$P\bar{1}$	2	1185
$Ta_2Rb_2H(PO_4)_2(P_5O_{16})$	5.173(1)	18.458(5) 95.63(2)	10.803(4)	$P2_1/m$	2	509
$(NH_4)Cd_6(P_2O_7)_2(P_3O_{10})$	6.785(3)	5.494(2) 107.28(4)	27.199(5)	Pm	4	1186
$Cs_2Mo_5O_2(P_2O_7)_3(P_3O_{10})$	14.395(1) 98.316(6)	15.600(1) 90.824(7)	6.4571(6) 90.073(6)	$P\bar{1}$	2	1187
$CaNb_2O(P_2O_7)(P_4O_{13})$	13.264(9)	10.577(8) 96.09(1)	12.393(9)	$C2/m$	4	506
$KTa(P_2O_7)(PO_3)_2$	7.045(1)	8.402(1)	17.494(4)	$P2_12_12_1$	4	1188
$KAl_2(H_2P_3O_{10})(P_4O_{12})$	11.864(3)	8.332(3) 99.67(2)	17.317(4)	$C2/c$	4	1190
$Pb_2Cs_3(P_4O_{12})(PO_3)_3$	6.808(5) 86.23(1)	7.875(6) 96.96(1)	22.12(1) 113.98(1)	$P\bar{1}$	2	688
$Sr_2Cs_3(P_4O_{12})(PO_3)_3$	6.922(5) 86.91(5)	8.055(5) 97.99(5)	21.97(5) 115.15(5)	$P\bar{1}$	2	688

[a]The main geometrical features of the various condensed anions in these compounds are given in the sections devoted to these anions in Chapters 2–4.

5.2. Present State of Mixed-Anion Phosphate Chemistry

5.2.1. Monophosphate–Diphosphates

In this class of compounds, isolated PO_4 groups and P_2O_7 diphosphate groups coexist as independent units in the atomic arrangements.

$Na_5H_2(PO_4)(P_2O_7)$. By heating at 623 K mixtures of Na_2O and $Na_4P_4O_{12}.H_2O$ in sealed gold tubes for 30 days, Wiench and Jansen[1160] obtained crystals of $Na_5H_2(PO_4)(P_2O_7)$. A detailed description of the crystal structure was reported by these authors. Some doubts remain concerning the proposed hydrogen-bond scheme.

$K_2H_5(PO_4)(P_2O_7)$. By evaporation under vacuum of aqueous solutions of K_2HPO_4 and $H_4P_2O_7$ in a stoichiometric ratio, Larbot *et al.*[1161] obtained crystals of $K_2H_5(PO_4)(P_2O_7)$. They reported a complete description of the atomic arrangement.

$Na_7M_4(PO_4)(P_2O_7)_4$. (M = Fe, Al, Cr). These three compounds were first described by De la Rochère *et al.*[1162] The authors reported detailed preparative procedures and crystal data for these three isomorphous salts. The atomic arrangement was determined by using the iron–sodium compound. This class of compounds seems to have good cation-transport properties.

$Na_3Th(PO_4)(P_2O_7)$. The preparation of crystalline $Na_3Th(PO_4)(P_2O_7)$ by a flux method was described by Kojic-Prodic *et al.*[1163] The same authors reported a detailed determination of the crystal structure.

$K_2H_8(PO_4)_2(P_2O_7)$. This salt was prepared from an aqueous solution of KH_2PO_4 and $H_4P_2O_7$ in a 2:1 ratio by Larbot *et al.*[1164] These authors reported a complete description of the atomic arrangement.

$K_2Ni_4(PO_4)_2(P_2O_7)$. The existence of this salt was first reported by Ladvig and Worzala,[1165] who described its chemical preparation and its main properties. Later, Palkina and Maksimova[1166] determined the crystal structure of this salt.

$M(NH_4)_3H(PO_4)(P_2O_7)$ (M = Al, Fe) and $Fe(NH_4)_3H(PO_4)(P_2O_7)\cdot H_2O$ Detailed procedures for the chemical preparation of these three insoluble compounds were reported by Frazier *et al.*[183] and Lehr *et al.*[64] The optical data given by the authors indicate that these salts are orthorhombic.

$Cs(MoO)_2(PO_4)(P_2O_7)$. This compound of pentavalent molybdenum was described by Chen *et al.*[1167] The detailed determination of its atomic arrangement by these authors showed the coexistence of PO_4 and P_2O_7 groups.

$K(MoO)_2(PO_4)(P_2O_7)$. Four forms are presently known for this salt (see Ref. 1168). This mixed-anion phosphate of pentavalent molybdenum was prepared by Leclaire *et al.*[1169] These authors reported a detailed description of the atomic arrangement. Form IV was also prepared and its structure determination performed by Leclaire *et al.*[1168]

$Cs_3Mo_2(Mo_2O)_2(PO_4)_2(P_2O_7)_4$. Chemical preparation of this salt was described by Lii and Wang.[1170] These authors report a detailed description of the atomic arrangement and discussed the relationship between this compound and $Cs_4Mo_8P_{12}O_{52}$.

$KMoWO(PO_4)(P_2O_7)$. This compound was described by Benmoussa *et al.*[1171] These authors reported a detailed determination of the crystal structure.

5.2.2. Monophosphate–Polyphosphates

$CsTa_2(PO_4)_2(P_3O_{10})$. This salt was characterized by Sadikov *et al.*[1172] during their investigation of the P_2O_5–Cs_2O–Ta_2O_5–H_2O system in the temperature range 423–673 K. These authors reported a description of the atomic structure.

$Rb_2Ta_2H(PO_4)_2(P_5O_{16})$. This compound was characterized by Sadikov *et al.*[502] during an investigation by flux methods of the P_2O_5–Rb_2O–Ta_2O_5–H_2O system. These authors reported a complete description of the crystal structure. The observation of the pentapolyphosphoric anion, P_5O_{16}, in this arrangement is the second piece of structural evidence for this very rare type of condensed anion (see section 2.5).

5.2.3. Diphosphate–Triphosphates

$NH_4Cd_6(P_2O_7)_2(P_3O_{10})$. Crystals of this phosphate were discovered during investigations of the products of hydrothermal crystallization in the CdO–$(NH_4)_2HPO_4$–$NaCl$–H_2O system at 723 K and pressures of 800 to 1400 atm by Ivanov *et al.*[1173] These authors reported a complete crystal structure determination.

$Cs_2Mo_5O_2(P_2O_7)_3(P_3O_{10})$. This compound containing both trivalent and tetravalent molybdenum was reported by Haushalter and Lai.[1174] These authors reported a detailed determination of the atomic arrangement.

5.2.4. Diphosphate–Tetraphosphates

$CaNb_2O(P_2O_7)(P_4O_{13})$. This compound was characterized during a study of the P_2O_5–CaO–Nb_2O_5-system by Averbuch-Pouchot.[499] The author reported a detailed chemical preparation of this salt and a complete description of its atomic arrangement.

5.2.5. Diphosphate–Polyphosphates

$KTa(P_2O_7)(PO_3)_2$. $KTa(P_2O_7)(PO_3)_2$ was obtained as single crystals during an investigation of the P_2O_5–K_2O–Ta_2O_5–H_2O system in the temperature range 573–673 K by Nikolaev et al.[1175] These authors reported the chemical preparation of this salt and a detailed structural analysis.

5.2.6. Triphosphate–Cyclotetraphosphates

$MN_2(H_2P_3O_{10})(P_4O_{12})$. The existence of eight compounds corresponding to the above general formula, with M = K, NH_4, Rb, and Cs and N = Al and Fe was reported by Grunze and Grunze.[242] They are prepared by heating various mixtures of MH_2PO_4, $Al(OH)_3$ or $Fe(NO_3)_3.9H_2O$, and H_3PO_4 at temperatures ranging from 443 to 573 K. Grunze et al.[1176] showed that the four aluminum salts are isotypic. The atomic arrangement in compounds of this type was determined by Grunze et al.[1177] using the aluminum–potassium salt. These authors produced single crystals during their reinvestigation of the crystallization products obtained by heating various mixtures of KH_2PO_4, $Al(OH)_3$, and H_3PO_4. They reported a detailed procedure for the chemical preparation of this salt and an accurate crystal structure showing the coexistence of P_4O_{12} ring anions and P_3O_{10} groups in the atomic arrangement.

5.2.7. Long-Chain Polyphosphate-Cyclotetraphosphates

$Pb_2Cs_3(P_4O_{12})(PO_3)_3$ and $Sr_2Cs_3(P_4O_{12})(PO_3)_3$. These two salts were identified by Averbuch-Pouchot[679] as isotypic compounds during investigations of the P_2O_5–Cs_2O–PbO- and P_2O_5–Cs_2O–SrO systems. A detailed crystal structure determination carried out by the author, with the lead salt, shows the coexistence of P_4O_{12} ring anions and $(PO_3)_n$ infinite phosphoric chains in the atomic arrangement.

5.3. Atomic Arrangements of Some Mixed-Anion Phosphates

5.3.1. Tantalum–Rubidium–Hydrogen Monophosphate–Pentaphosphate

$$Ta_2Rb_2H(PO_4)_2(P_5O_{16}),^{502} \text{ monoclinic, } P2_1/m, Z = 2$$

$$a = 5.173(1), b = 18.458(5), c = 10.803(4) \text{ Å}, \beta = 95.63(2)°$$

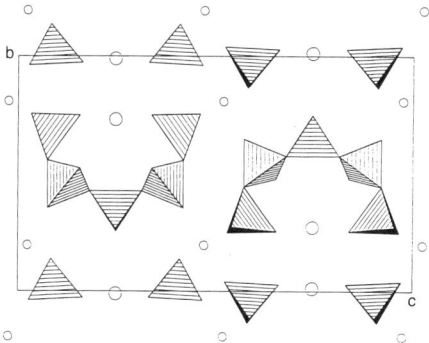

Figure 5.1. Projection of the structure of $Ta_2Rb_2H(PO_4)_2(P_5O_{16})$ structure along the **a** direction. The small open circles represent the tantalum atoms, and the larger ones the rubidium atoms.

A projection of this very simple arrangement along the **a** axis is shown in Figure 5.1. The pentaphosphate group has the mirror symmetry, its central phosphorus atom being located on a mirror plane. The main geometrical features of this group have been given in Chapter 2 (Table 2.5.2). The two independent rubidium atoms are also located on the mirror plane. Each pentaphosphate group is connected to six TaO_6 octahedra. The two end tetrahedra are linked to two TaO_6 octahedra, and the adjacent ones to one TaO_6 octahedron. The TaO_6 octahedra are relatively little distorted, with O-Ta-O angles ranging from 88.1 to 93.1° and Ta-O distances ranging from 1.93 to 1.99 Å. The two independent rubidium atoms are situated in wide oblong channels parallel to the **a** direction in which they form chains. One of them has eightfold oxygen coordination, and the other one sevenfold. Within these two polyhedra, the Rb-O distances range from 2.845 to 3.138 Å.

5.3.2. Calcium–Niobium Diphosphate–Tetraphosphate

$$CaNb_2O(P_2O_7)(P_4O_{13}),^{499} \text{ monoclinic, } C2/m, Z = 4$$

$$a = 13.264(9), b = 10.577(8), c = 12.393(9) \text{ Å}, \beta = 96.09(1)°$$

The structure of this mixed-anion phosphate is represented in Fig. 5.2 by a projection on the (a,c) plane. The two anionic groups alternate in rows parallel to the **a** axis. The P_4O_{13} group, built up by only two crystallographically independent PO_4 tetrahedra, has twofold symmetry around its central oxygen atom, located on a twofold axis. The main geometrical features of this anion have been presented in Chapter 2 (Table 2.4.2). They are quite similar to those commonly observed in this type of anion. On the other hand, the geometry of the P_2O_7 group observed in this compound is quite unusual. Five atoms of this group—the two phosphorus atoms and three oxygen atoms—are located in a mirror plane. In addition, the average P-O distances observed in the two PO_4 tetrahedra building this anion are very short (1.522 and 1.516 Å).

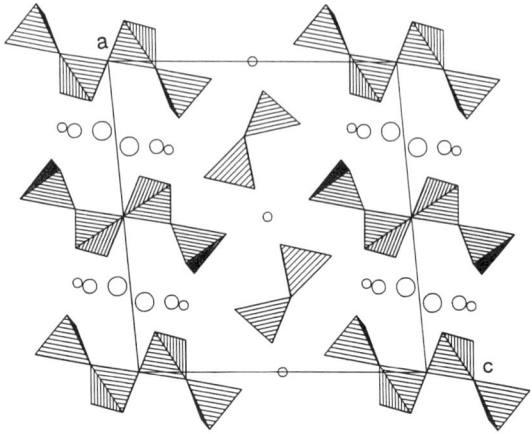

Figure 5.2. Projection, along the **b** direction, of the atomic arrangement in CaNb$_2$O(P$_2$O$_7$)(P$_4$O$_{13}$). The largest open circles represent the oxygen atoms belonging to the coordination polyhedron of one of the niobium atoms, the intermediate ones represent the niobium atoms and the smallest ones the calcium atoms.

The two crystallographically independent niobium atoms of this arrangement have very different situations. The first one is located on a twofold axis, and its six oxygen neighbors build an almost regular octahedron, with Nb-O distances ranging from 1.912 to 1.981 Å. This polyhedron shares its corners with four P$_2$O$_7$ and two P$_4$O$_{13}$ groups. The second niobium atom is located in the mirror plane and has a less regular sixfold coordination, with five normal Nb-O distances ($1.975 < $ Nb-O $ < 2.195$ Å) and a short one (1.698 Å). This situation is comparable to that observed in numerous oxoniobium salts.

The calcium atom is located in a mirror plane and has classical sixfold coordination, with Ca-O distances ranging from 2.279 to 2.449 Å.

5.3.3. Potassium–Aluminum Triphosphate–Cyclotetraphosphate

$$KAl_2(H_2P_3O_{10})(P_4O_{12}),^{1177} \text{ monoclinic, } C2/c, Z = 4$$

$$a = 11.864(3), b = 8.332(3), c = 17.317(4) \text{ Å}, \beta = 99.67(2)°$$

The atomic arrangement in this salt, as shown in Fig. 5.3, a projection along the **b** axis, is a typical layer organization. The planes $z = 0$ and $1/2$ contain the P$_4$O$_{12}$ ring anions, whereas the planes $z = 1/4$ and $3/4$ contain the potassium atoms and the P$_3$O$_{10}$ triphosphate groups. Between these two kinds of layers, the aluminum atoms are intercalated in planes $z = (2n+1)/8$. The P$_4$O$_{12}$ ring anions are centrosymmetric, and the P$_3$O$_{10}$ groups have twofold internal symmetry. According to Grunze *et al.*,[1177] the hydrogen atoms belonging to terminal oxygen atoms of the P$_3$O$_{10}$ group establish a set of strong hydrogen bonds connecting the phosphoric anions along the **c** direction.

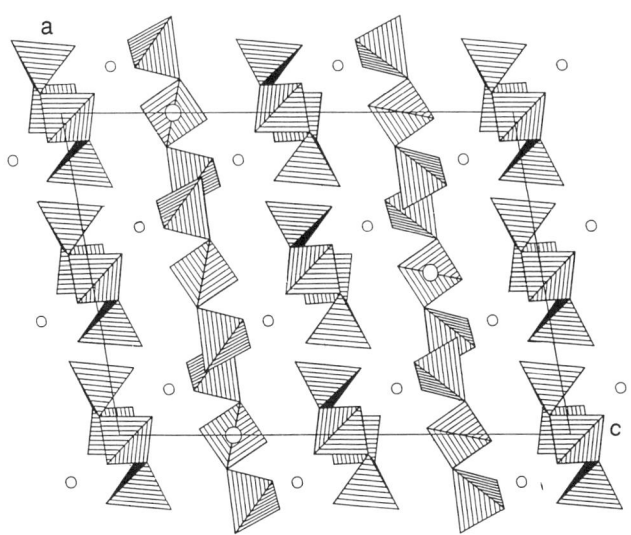

Figure 5.3. Projection, along the **b** axis, of the atomic arrangement in $KAl_2(H_2P_3O_{10})(P_4O_{12})$. The large open circles represent the potassium atoms, and the smaller ones the aluminum atoms.

5.3.4. Lead–Cesium Cyclotetraphosphate–Long-Chain Polyphosphate

$$Pb_2Cs_3(P_4O_{12})(PO_3)_3,^{679} \text{ triclinic, } P\bar{1}, Z = 2$$

$$a = 6.808(5), b = 7.875(6), c = 22.12(1) \text{ Å}$$

$$\alpha = 86.23(1), \beta = 96.96(1), \gamma = 113.98(1)°$$

Figure 5.4. Projection of the structure of $Pb_2Cs_3(P_4O_{12})(PO_3)_3$ along the **b** axis. The large open circles represent the cesium atoms, and the smaller ones represent the lead atoms.

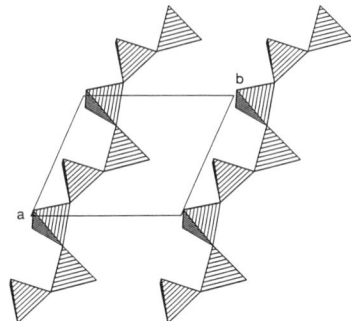

Figure 5.5. The infinite phosphoric chain as observed in a projection of the structure of $Pb_2Cs_3(P_4O_{12})(PO_3)_3$ along the **c** axis in a layer centered in $z = 0.25$.

If one considers only the stacking of the anions, this atomic arrangement can be considered as a layer organization. As can be seen in Fig. 5.4, all the P_4O_{12} anions are located in the planes $z = 0$ and 1/2, while the infinite $(PO_3)_n$ chains spread in the planes $z = 1/4$ and 3/4. Two crystallographically independent ring anions, both centrosymmetric, coexist in the structure. Their main geometrical features have already been reported in Table 4.3.13. The infinite $(PO_3)_n$ anion has a period of three tetrahedra. The general aspect of such a chain as observed in a layer centered on the plane $z = 0.25$ is shown in Fig. 5.5. The two independent lead atoms both have sevenfold coordination, while within a range of 3.5 Å two of the cesium atoms have eight neighbors and the third one nine.

The strontium salt, $Sr_2Cs_3(P_4O_{12})(PO_3)_3$, is isotypic.

6

Ultraphosphates

6.1. Introduction

Ultraphosphates comprise a group of P_2O_5-rich condensed phosphates that have not yet been fully investigated. In our discussion of the classification and nomenclature of condensed phosphates in Chapter 1, we gave a chemical definition of ultraphosphates and briefly described their geometries. Let us simply repeat that the general formula for ultraphosphate anions is

$$[P_{(n+2)}O_{(3n+5)}]^{n-}$$

Such anions have been well characterized for $n = 2, 3, 4,$ and 6.

Among condensed phosphates, ultraphosphates were for a long time the poor relations in the family. In fact, erroneous beliefs about their properties that were commonly held until recently discouraged the study of these compounds and contributed greatly to their bad reputation. Ultraphosphates were said to hydrolyze very rapidly in contact with water vapor present in air owing to the instability of the triply linked PO_4 tetrahedra present in the anionic framework. The "antibranching rule" put forward by Van Wazer and Holst in 1950[1178] summarizes well the commonly accepted opinion at that time:

> In any environment in which reactions involving the degradation of condensed phosphates are possible, it is to be expected that assemblies in which three of the four oxygens of a PO_4 tetrahedron are shared with other PO_4 tetrahedra will be exceedingly unstable and will degrade much more rapidly as compared to those in which one or two oxygens are shared.

Van Wazer, who remains unsurpassed as a reviewer of phosphorus chemistry, add a great and probably prophetic interest in this kind of compounds (he devoted many pages to ultraphosphates and to their possible anionic framework geometry), but nevertheless only two examples of crystalline ultraphosphates, CaP_4O_{11} and $Ca_2P_6O_{17}$, were mentioned in his 1966 book.[27]

A deeper investigation of the chemical literature, going back to last century, shows that, as early as 1885, Clève[1179] described the chemical preparation of the samarium ultraphosphate, SmP_5O_{14}, and reported its main chemical properties. In 1889, Johnsson,[1180] while investigating some P_2O_5–M_2O_3 systems, prepared lanthanum, cerium, and samarium ultraphosphates. He also reported some of their chemical properties as well as very detailed optical morphological measurements of a monoclinic LaP_5O_{14} crystal. Both Clève and Johnsson called these P_2O_5-rich compounds "anhydro-metaphosphates." Twenty-five years later, Kroll,[1181–1183] found evidence for the existence of some other P_2O_5-rich compounds during investigations of various systems and became convinced of the existence of salts deriving from the hypothetical HP_3O_8, $H_2P_4O_{11}$, and $H_4P_6O_{17}$ acids. He was, in fact, the first to characterize $Ca_2P_6O_{17}$. Apparently unaware of the work of the previous investigators, he named these compounds "ultraphosphates" and was for a long period of time credited with their discovery. For more than thirty years after Kroll published his results, the literature about ultraphosphates was purely speculative, and it was only in 1944 that Hill et al.[47,568] characterized clearly two calcium ultraphosphates, CaP_4O_{11} and $Ca_2P_6O_{17}$, during their investigation of a portion of the P_2O_5–CaO system.

The rapid development of X-ray facilities played an important rôle in the improvement of knowledge about ultraphosphates, all the more so because the application of paper chromatography had proved fruitless in the investigation of these compounds.

Beucher[1018] prepared all the LnP_5O_{14} ultraphosphates and performed a detailed crystallographic investigation of these compounds, showing the existence of three structure types in this class of compounds. Almost simultaneously, Jaulmes[1184] reported crystal data for the lanthanum salt. These compounds were soon recognized as efficient laser materials, and during the 1970s there was an explosive development of the scientific and technical literature devoted to these ultraphosphates (mainly the Nd salt) or any kind of rare-earth-containing phosphates. This upsurge of interest was beneficial with regard to the development of the crystal chemistry of rare-earth condensed phosphates, which had been rather neglected before.

The demonstration that ultraphosphates can be stable compounds provoked renewed interest in these salts, mainly in the Soviet Union, where, during the past 20 years, a good number of ultraphosphates have been synthesized and clearly characterized.

6.2. Present State of Ultraphosphate Chemistry

To date, no example of a crystalline monovalent cation ultraphosphate has been reported in the chemical literature. Nevertheless, the behavior of some phosphoric fluxes currently used in crystal growth experiments has provided evidence exists for the existence of ultraphosphate anions in the presence of alkali metals.

6.2.1. Divalent Cation Ultraphosphates

MgP_4O_{11} and ZnP_4O_{11}. These two isotypic compounds were very recently described by Stachel et al.[1185] and Baez-Doelle et al.[1186] respectively. Crystal structure determinations were performed for the two compounds.

A second form of MgP_4O_{11} was very recently described by Yakubovich et al.[1187]

CaP_4O_{11} and $Ca_2P_6O_{17}$. These two compounds have been clearly characterized by Hill et al.[47,568] during their investigation of the P_2O_5–$Ca(PO_3)_2$ phase-equilibrium diagram. The first one appears as a congruent-melting compound (mp = 1083 K) while the second one decomposes at 1047 K. This phase-equilibrium diagram is given in Fig. 6.1. Crystal data for CaP_4O_{11} were first reported by Beucher,[1188] and its atomic arrangement was later determined by Tordjman et al.[1189] Crystal structure determination for $Ca_2P_6O_{17}$ was recently performed by Stachel et al.[1190] This last salt is isotypic with the corresponding cadmium salt.

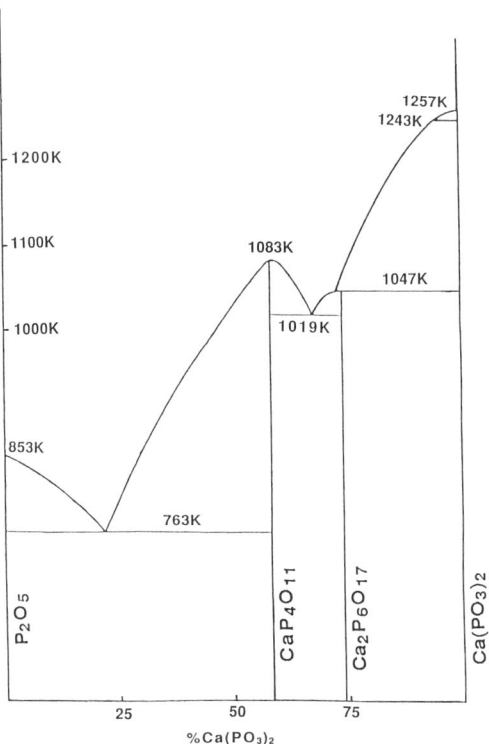

Figure 6.1. The P_2O_5–$Ca(PO_3)_2$ phase-equilibrium diagram as elaborated by Hill *et al.*[44, 568] The $\beta \to \alpha$ transformation of $Ca(PO_3)_2$ appears at 1243 K on the right-hand side of the diagram.

MnP_4O_{11}. The atomic arrangement in MnP_4O_{11} was determined by Minacheva *et al.*[1191] This ultraphosphate is isotypic with the corresponding calcium salt.

NiP_4O_{11} and CoP_4O_{11}. Lavrov *et al.*[1192] prepared this nickel ultraphosphate from a mixture of NiO and $(NH_4)_2HPO_4$ heated at 893–933 K for 30–40 h and investigated its thermal stability. When heated in air, it decomposes into the cyclotetraphosphate:

$$2NiP_4O_{11} \to Ni_2P_4O_{12} + 2P_2O_5$$

Very recently Olbertz *et al.*[1193] prepared the isotypic cobalt salt and determined its crystal structure.

$Cd_2P_6O_{17}$ and $Sr_2P_6O_{17}$. Antsyshkina and co-workers determined the atomic arrangement in the cadmium salt[1194] and in the strontium salt[1195] and discussed the relationship between the atomic arrangements in these two salts.[1196] A survey of MP_4O_{11} ultraphosphate crystal chemistry was presented by Lavrov *et al.*[1197]

Crystallographic Data. The main crystallographic data for divalent cation ultraphosphates are included in Table 6.1.

Table 6.1. Main Crystallographic Data for Divalent Cation and Various Other Ultraphosphates

Formula	a (Å) α (°)	b (Å) β (°)	c (Å) γ (°)	S.G.	Z	Reference(s)
MgP_4O_{11}	5.343(1)	22.228(5) 110.89(1)	7.451(2)	$P21/c$	4	1185
MgP_4O_{11}	9.670(5)	35.40(4)	14.521(8)	$Pmc2_1$	24	1187
ZnP_4O_{11}	5.302(1)	22.242(4) 110.13(1)	7.412(1)	$P21/c$	4	1186
CaP_4O_{11}	8.856(4)	12.72(1) 92.48(5)	8.707(4)	$P2_1/n$	4	1188–1189
MnP_4O_{11}	8.608(2)	8.597(4) 97.30(4)	12.464(5)	$P2_1/n$	4	1191
$Ca_2P_6O_{17}$	5.753(2)	18.265(5) 111.12(1)	7.625(2)	$P2_1/c$	2	1190
$Sr_2P_6O_{17}$	7.158(2)	13.02(2) 105.26(2)	7.172(1)	$P2_1$	2	1195
$Cd_2P_6O_{17}$	7.566(4)	5.486(4) 111.22(4)	18.082(5)	$P2_1/n$	2	1194
$Na_3FeP_8O_{23}$	11.915(7)	11.915(7)	11.915(7)	$P4_132$	4	1201
$Na_3GaP_8O_{23}$	11.884(3)	11.884(3)	11.884(3)	$P4_132$	4	1202
$Na_3AlP_8O_{23}$	11.870(2)	11.870(2)	11.870(2)	$P4_132$	4	1202
$(UO_2)_2P_6O_{17}$	8.653(1)	11.092(4) 106.12(2)	17.453(4)	Cc	4	1200

6.2.2. Various Other Ultraphosphates

Various other types of ultraphosphates have also been described.

$(TaO_2)_4P_6O_{17}$. Chernorukov *et al.*[1198] observed the formation of an ultraphosphate during the thermal decomposition of $TaH(PO_4)_2$. According to these authors, the reaction is:

$$4TaH(PO_4)_2 \xrightarrow{723-773 \text{ K}} (TaO_2)_4P_6O_{17} + 2H_2O + P_2O_5$$

$(UO_2)_2P_6O_{17}$. This salt was obtained in a good crystalline state by Lavrov[1199] during a study of the $P_2O_5-UO_3-H_2O$ system. According to this author, this ultraphosphate can also be prepared as a polycrystalline material by roasting $(UO_2)H(PO_3)_3$ at 973 K. Its crystal structure was determined by Gorbunova *et al.*[200]

$Na_3FeP_8O_{23}$, $Na_3AlP_8O_{23}$ and $Na_3GaP_8O_{23}$. The iron derivative was for a time the only example of a mixed-cation ultraphosphate. It was characterized by Chudinova *et al.*[1201] during an investigation of the $P_2O_5-Fe_2O_3-Na_2O$ system. These authors reported a complete description of the atomic arrangement.

Very recently Palkina *et al.*[1202] prepared the isotypic aluminum and gallium derivatives and refined their crystal structures.

Crystallographic Data. The main crystallographic data reported for the ultraphosphates considered in this section are given in Table 6.1.

6.2.3. Yttrium, Rare-Earth, and Bismuth Ultraphosphates

In this section, we do not attempt to exhaustively review the large amount of literature concerning yttrium, rare-earth, and bismuth ultraphosphates that have been published since 1969. We have instead selected studies that have revealed some new feature of the structural chemistry or chemical behavior of these salts. In spite of this narrow focus, this review will probably appear to be rather repetitious, because a number of crystal structure determinations for a given type of compounds, have been performed of the same type of compounds and many papers dealing with the same subjects have been published almost simultaneously. In addition, in many of these papers, the technological and scientific parts are intertwined.

All the compounds described in this section have the common formula TP_5O_{14}, (T = Y, Bi or Ln) and belong to four structure types. We have already mentioned that some of them were characterized at the end of last century by Clève[1179] and Johnsson,[1180] but the fundamental pioneering work was performed by Beucher[1018] who prepared single crystals of YP_5O_{14} and all the LnP_5O_{14} ultraphosphates and performed a careful crystallographic investigation. She found that these compounds belonged to three different structure types and that some of them were polymorphic.

Crystal structures of the following representatives of these three structure types were determined by Bagieu-Beucher and co-workers:

Table 6.2. Main Crystallographic Data for Bismuth and Rare-Earth Ultraphosphates Belonging to the First Structure Type[a]

Formula	a (Å) α (°)	b (Å) β (°)	c (Å) γ (°)	Reference(s)
BiP_5O_{14}	13.06(1)	9.02(1) 90.6(1)	8.77(1)	1206
LaP_5O_{14}	13.18(1)	9.112(3) 90.38(5)	8.820(3)	1018
CeP_5O_{14}	13.11(1)	9.063(3) 90.45(5)	8.790(3)	1018
PrP_5O_{14}	13.08(1)	9.041(3) 90.42(5)	8.787(3)	1018
NdP_5O_{14}	13.03(1)	9.001(3) 90.48(5)	8.768(3)	705, 1018, 1211
SmP_5O_{14}	12.99(1)	8.944(3) 90.41(5)	8.757(3)	1018, 1203
EuP_5O_{14}	12.93(1)	8.930(3) 90.45(5)	8.751(3)	1018, 1215
GdP_5O_{14}	12.93(1)	8.904(3) 90.49(5)	8.743(3)	1018
HoP_5O_{14} [b]	12.855	8.822 90.54	8.695	1018
TbP_5O_{14}	12.91(1)	8.887(3) 90.49(5)	8.728(3)	1018

[a]The common space group is $P2_1/a$ and $Z = 4$.
[b]HP: High-pressure form.

Table 6.3. Main Crystallographic Data for Yttrium and Rare-Earth Ultraphosphates Belonging to the Second Structure Type[a]

Formula	a (Å) α (°)	b (Å) β (°)	c (Å) γ (°)	Reference(s)
TbP_5O_{14}	12.91(1)	12.80(1) 91.31(5)	12.48(1)	1018
DyP_5O_{14}	12.90(1)	12.79(1) 91.30(5)	12.46(1)	1018
HoP_5O_{14}	12.88(1)	12.77(1) 91.34(5)	12.42(1)	1018, 1204
ErP_5O_{14}	12.85(1)	12.74(1) 91.28(5)	12.40(1)	1018, 1216
TmP_5O_{14}	12.84(1)	12.73(1) 91.26(5)	12.38(1)	35, 1018
YbP_5O_{14}	12.84(1)	12.72(1) 91.25(5)	12.37(1)	1018, 1218
LuP_5O_{14}	12.81(1)	12.71(1) 91.24(5)	12.34(1)	1018
YP_5O_{14}	12.87(1)	12.76(1) 91.28(5)	12.43(1)	1018

[a]The common space group is $C2/c$; $Z = 8$.

- Type I: SmP_5O_{14} (Ref. 1203)
- Type II: HoP_5O_{14} (Ref. 1204)
- Type III: HoP_5O_{14} (Ref. 1205)

BiP_5O_{14} was soon after recognized as being isotypic with the lanthanum salt by Chudinova and Jost.[1206] The main crystallographic features of the compounds belonging to these three structure types are presented in Tables 6.2–6.4.

Later, a fourth structure type was described by Rzaigui *et al.*[1207] for the cerium salt. The unit-cell dimensions of the form of CeP_5O_{14} corresponding to this fourth structure type are:

Table 6.4. Main Crystallographic Data for Yttrium, Bismuth, and Rare-Earth Ultraphosphates Belonging to the Third Structure Type[a]

Formula	a (Å) α (°)	b (Å) β (°)	c (Å) γ (°)	Reference
DyP_5O_{14}	8.726(3)	12.75(1)	8.950(3)	1018
HoP_5O_{14}	8.720(3)	12.71(1)	8.926(3)	1018
ErP_5O_{14}	8.712(3)	12.68(1)	8.919(3)	1018
YP_5O_{14}	8.718(3)	12.73(1)	8.939(3)	1018
BiP_5O_{14} (HT)	8.726(12)	12.950(20)	8.958(18)	726

[a]The common space group is *Pnma*; $Z = 4$.
[b]HT: High-temperature form.

$$a = 9.227(5), b = 8.890(5), c = 7.219(4)\text{Å}$$

$$\alpha = 110.12(5), \beta = 102.68(5), \gamma = 82.13(5)°$$

The space group is $P1$, and $Z = 2$. This compound is the only known representative of this structure type.

All these compounds have been prepared as single crystals by various flux methods, usually involving a large excess of phosphoric acid. They are very stable under normal conditions, but at high temperatures they lose P_2O_5 and transform into the very stable monophosphates according to the following scheme:

$$LnP_5O_{14} \rightarrow LnPO_4 + 2P_2O_5$$

The mechanism of this decomposition was investigated for rare-earth and bismuth ultraphosphates by Dudko *et al.*[1208]

The cation coordination in the various forms of the TP_5O_{14} compounds was carefully examined by Bagieu.[686]

Some brief reviews dealing with these TP_5O_{14} compounds have been published by Bagieu-Beucher and Tranqui,[35] Bagieu,[686] Durif,[1209] and Bondar *et al.*[1210] All the remaining literature may be roughly classified into four main categories: investigations of various systems, crystal structure determinations, reports on crystal growth, and studies of physical properties. Selected papers on these various topics are listed in Tables 6.5–6.8.

Table 6.5. Investigations of Various Systems Involving the Formation of Rare-Earth Ultraphosphates

Formula	Nature of the investigation	Reference
LaP_5O_{14}	Study of the P_2O_5–La_2O_3- by Park and Kreidler showing the existence of six intermediate compounds: among them, LaP_5O_{14}, melting congruently at 1368 K	490
	Investigation between 373 and 773 K of the P_2O_5–La_2O_3–H_2O system by Chudinova *et al.* and determination of crystallization regions for several La phosphates including LaP_5O_{14}	701
CeP_5O_{14}	Study between 293 and 1173 K of the reaction between CeO_2 and $NH_4H_2PO_4$ by Vaivada and Konstant and observation of the formation of CeP_5O_{14} for P/Ce starting ratio greater than 5	704
	Investigation by Szczygiel and Znamierowska of the P_2O_5-rich portion of the P_2O_5–Ce_2O_3–Na_2O-system and determination of the $Ce(PO_3)_3$–CeP_5O_{14} and CeP_5O_{14}–$NaCe(PO_3)_4$ phase equilibrium diagrams (melting point of CeP_5O_{14}, 1337 K)	1212
NdP_5O_{14}	Reinvestigation at 773 K under 100 MPa pressure of the P_2O_5–Nd_2O_3–H_2O system by Yoshimura *et al.* in order to elaborate a method of crystal growth for NdP_5O_{14}	1213
YbP_5O_{14}	Determination of the crystallization region of YbP_5O_{14} in the P_2O_5–Yb_2O_3–H_2O system between 373 and 673 K by Chudinova *et al.*	709

Table 6.6. Crystal Structure Determinations and Crystallographic Investigations of
Rare-Earth and Bismuth Ultraphosphates

Formula	Nature of the investigation	Reference
NdP_5O_{14}	Crystal structure determination by Hong	705
	Crystal structure determination by Albrand *et al.*	1211
	Observation by Schulz *et al.* of a reversible transformation at 388 K: orthorhombic space group *Pncm* for the high-temperature form	1214
EuP_5O_{14}	Crystal structure determination by Parrot *et al.*	1215
HoP_5O_{14}	Observation by Bagieu of the transformation of form II into form I at 873 K under 20-kbar pressure	695
ErP_5O_{14}	Crystal structure determination by Jezowska-Trzebiatowska *et al.* [(1216)]	1216
TmP_5O_{14}	Crystal structure determination by Kangjing *et al.*	1217
YbP_5O_{14}	Crystal structure determination by Hong and Pierce	1218
BiP_5O_{14}	Transformation from form I to form III observed at 373 K by Hilmer *et al.*	716

Table 6.7. Crystal Growth Investigations in the Field of Rare-Earth Ultraphosphates

Formula	Nature of the investigation	Reference
PrP_5O_{14}	Chemical synthesis and crystal growth of laser quality PrP_5O_{14} by Borkowski *et al.*	1219
NdP_5O_{14}	Investigation of crystal growth conditions and optical properties of NdP_5O_{14} by Tofield *et al.*	1220
	Study by Danielmayer *et al.* of the growth of NdP_5O_{14} crystals, optimization of the growth parameters for the production of laser quality and laser size crystals, and observation of the effect of addition of various impurities on the emission lifetime	1221
	Study of the morphological control of NdP_5O_{14} single crystals grown from polyphosphoric acids by Kasano and Furuhata	1222
	Crystal growth of NdP_5O_{14} and $Nd_xLa_{(1-x)}P_5O_{14}$ by Marais *et al.*	1223
	Development of a method for growing large laser quality $Nd_xLn_{(1-x)}P_5O_{14}$ crystal for Ln = La and Y by Plattner *et al.*	1224
	Investigation and optimization of crystal growth of Nd and La ultraphosphates by Miller *et al.* Crystals up to 14 mm long obtained	1225
	Crystal growth of (Nd, Y)P_5O_{14} thin layers from tin melt by Kasano and Furuhata	1226
	Elaboration of a process for the production of single-crystal fibers of (Nd, La)P_5O_{14} by Tofield *et al.* growth in fiber form by controlling the water content of the atmosphere above the crucible during crystal growth; fibers up to 2 mm long	1227
	Crystal growth of NdP_5O_{14} and $Nd_xY_{(1-x)}P_5O_{14}$ by Ito and Ashida and discussion of the defects in the crystals obtained	1228

Table 6.8. Investigations of Various Physical Properties of Rare-Earth and Bismuth
Ultraphosphates

Formula	Nature of the investigation	Reference
LnP_5O_{14}	Investigation of the ferroelastic behavior and of the monoclinic–orthorhombic phase transition in LaP_5O_{14}, PrP_5O_{14}, NdP_5O_{14} and TbP_5O_{14} by Weber *et al.* Transition temperatures reported at 391, 413, 419, and 447 K respectively	1229
	Investigation of the ferroelasticity in LnP_5O_{14} (Ln = La to Tb) by DSC, Weissenberg technique, and polarizing microscope by Kobayashi *et al.*	1230
CeP_5O_{14}	Chemical preparation and investigation of the optical properties of the triclinic form of CeP_5O_{14} by Rzaigui and Kbir Ariguib	1231
NdP_5O_{14}	Investigation of the influence of local site symmetry on fluorescence lifetime in high-Nd concentration laser materials including NdP_5O_{14} by Hong and Chinn	1232
	Measurements by DSC of the specific heats of NdP_5O_{14} and PrP_5O_{14} near their ferroelastic phase transitions in the temperature range 385–440 K by Loicano *et al.* Transition temperatures reported at 426.5 and 416.5 K respectively	1233
GdP_5O_{14}	Electron paramagnetic resonance study of gadolinium in GdP_5O_{14} by Parrot *et al.*	1215
BiP_5O_{14}	Study of X-ray photo electron spectrum of BiP_5O_{14} by Chudinova	720
	Study of NMR spectrum of BiP_5O_{14} by Grimmer and Chudinova	1234

6.3. Atomic Arrangements of Ultraphosphates

Owing to the great complexity of most of ultraphosphate atomic arrangements, we will limit this review to the main geometrical features of the anionic frameworks. These frameworks can adopt various geometries: finite groups, infinite ribbons, infinite layers, or three-dimensional networks.

The only ultraphosphate *finite group*, $P_8O_{23}^{6-}$, was first observed in $Na_3FeP_8O_{23}$ and later in the corresponding aluminum and gallium derivatives. This anion has ternary internal symmetry and can be simply described as being built by two tetrahedra located on the ternary axis interconnected by three P_2O_7 groups. Another way to describe this anion is to consider it as a ring of six PO_4 tetrahedra in which two opposite tetrahedra, corresponding to those located on the ternary axis, are linked by a P_2O_7 group. Figure 6.2

Figure 6.2. Schematic representation of the P_8O_{23} ultraphosphate anion. Only the phosphorus atoms are represented.

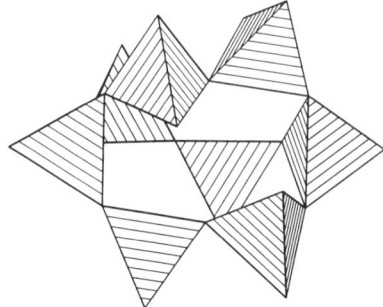

Figure 6.3. Perspective view of the eight tetrahedra constituting the P_8O_{23} anion. The six-membered ring appears at the front of the projection. Behind the ring, a group of two tetrahedra interconnect the two ternary tetrahedra.

is a skeletal representation of this anion in which only the phosphorus atoms are represented, while Fig. 6.3 is a perspective view of the eight tetrahedra building the anion.

Infinite ribbon anions have been observed in two forms (I and III) of the rare-earth ultraphosphates. The infinite ultraphosphate anion observed in form I can be described in several ways. One can consider it as built by two infinite $(PO_3)_n$ chains lying parallel to the **a** axis and interconnected by some PO_4 groups. A projection of this anion, as observed in SmP_5O_{14}, along the **c** axis is presented in Fig. 6.4. Alternatively, this anion may be described as being constructed from a succession of centrosymmetric eight-membered rings interconnected through the ternary tetrahedra.

With some slight differences in the orientation of the tetrahedra owing to the different symmetry elements, the anion observed in form III of the rare-earth ultraphosphates can be described in almost the same terms. In form III the eight-membered rings building the ribbon have mirror symmetry. A projection of this anion, as observed in HoP_5O_{14}, along the **c** axis is shown in Fig. 6.5. The profile views of these two anions presented in Fig. 6.6 show clearly their very similar organization. In both cases the repetition unit is $P_{10}O_{28}$.

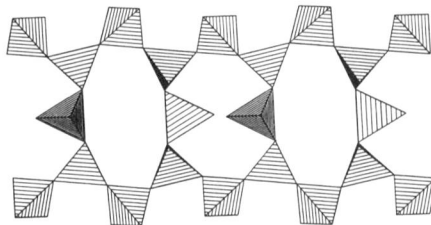

Figure 6.4. Projection, along the **c** axis, of the infinite ultraphosphate ribbon observed in SmP_5O_{14}.

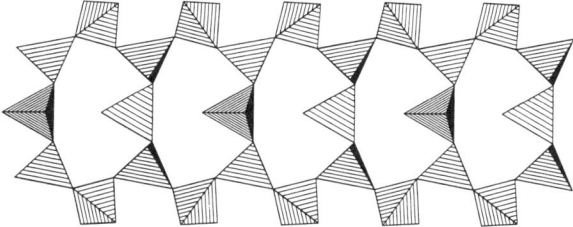

Figure 6.5. Projection, along the **c** axis, of the infinite ultraphosphate ribbon observed in HoP_5O_{14}.

Infinite two-dimensional anions have been found in CaP_4O_{11}, in form IV of rare-earth ultraphosphates, in $Sr_2P_6O_{17}$, and in $Cd_2P_6O_{17}$. These types of infinite anions are difficult to describe clearly.

The layered organization of the atomic arrangement in CaP_4O_{11} is well illustrated by Fig. 6.7, a projection along the **b** axis. Flat planes formed by the PO_4 tetrahedra alternate with layers of associated cations. The internal organization of the phosphoric layer is depicted in Fig. 6.8, a projection along the **c** axis. Here also, as in the previous examples, this network may be described in several ways. One can consider it as built up by a set of infinite $(PO_3)_n$ chains interconnected by PO_4 tetrahedra or as a tiling of P_8O_{24} and $P_{12}O_{36}$ rings.

In $Cd_2P_6O_{17}$ the atomic arrangement can be described as a succession of two kinds of layers parallel to the (110) planes. The first type of layer contains the infinite phosphoric anion and alternates with planes of cadmium atoms. A projection of this phosphoric layer along the **a** axis is presented in Fig. 6.9, while Fig. 6.10 shows its projection along the **b** axis. This phosphoric anion can be depicted as built by centrosymmetric rings of 14 tetrahedra interconnected in a two-dimensional way through their ternary PO_4 tetrahedra. Each ring is so connected to six of its neighbors.

As shown in Fig. 6.11, the atomic arrangement in $Sr_2P_6O_{17}$ is also a layer organization. Two kinds of very corrugated layers alternate perpendicular to the **a** axis. The first type of layer contains the strontium atoms, and the second one is formed by the phosphoric anion.

Figure 6.6. Profile views of the infinite ultraphosphate ribbons observed in form I (top) and in form III (bottom) of rare-earth ultraphosphates.

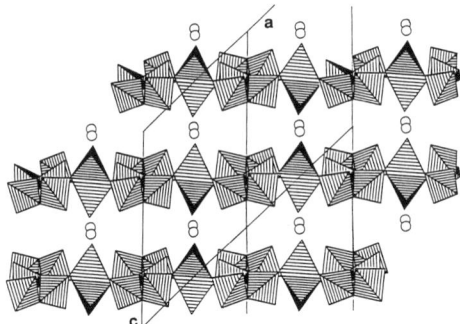

Figure 6.7. Projection, along the **b** axis, of the atomic arrangement in CaP$_4$O$_{11}$, showing the layer organization of this structure. The open circles represent the calcium atoms. The unit cell used for this drawing corresponds to the $P2_1/c$ description with $a = 8.856$, $b = 12.72$, and $c = 12.148$ Å and $\beta = 134.26°$.

Figure 6.8. Projection, along the axis of the infinite P$_4$O$_{11}$ ultraphosphate anion **a** as observed in CaP$_4$O$_{11}$. See the caption of Fig 6.7 for the choice of the unit-cell description.

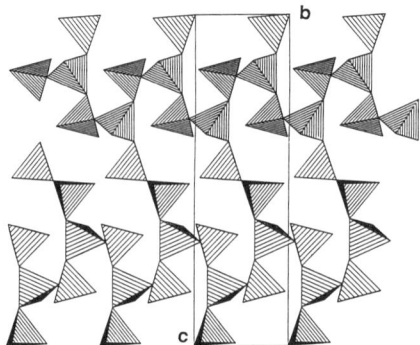

Figure 6.9. Projection, along the **a** axis, of the P$_6$O$_{17}$ infinite ultraphosphate anion as observed in Cd$_2$P$_6$O$_{17}$.

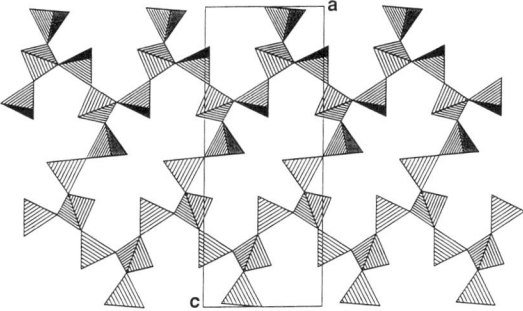

Figure 6.10. Projection, along the **b** axis, of the P_6O_{17} infinite ultraphosphate anion as observed in $Cd_2P_6O_{17}$.

Details of the anion framework are given in Fig. 6.12. As in the case of the cadmium ultraphosphate, the infinite anion can be described as a tiling of 14-membered rings interconnected through their ternary tetrahedra.

Three-dimensional anionic networks have been observed in form II of rare-earth ultraphosphates and in the uranyl ultraphosphate.

In $(UO_2)_2P_6O_{17}$ the framework of PO_4 tetrahedra building the anion can be described in several ways. It seems to us that the most appropriate is to consider it as built by a three-dimensional network of corrugated $P_{14}O_{42}$ rings interconnected by some common PO_4 groups. An isolated $P_{14}O_{42}$ ring is represented in Fig. 6.13, while the whole phosphoric network is shown in Fig. 6.14, a projection along the **a** axis.

In LnP_5O_{14} (form II) the anionic framework can be described as built up by three different types of interconnected thick layers. Layers of the first type are centered approximately by the planes $x = 1/4$ and $3/4$ and contain zigzag infinite $(PO_3)_n$ chains extending along the **b** axis. Figure 6.15 shows the projection of such a layer along the **a** axis.

A second, very different type of layer is perpendicular to the **b** axis. Two such layers cross the unit cell, being approximately centered by the planes $y = 1/4$ and $3/4$. This type of layer is constituted by a tiling of cyclic groups of eight tetrahedra, quite comparable to

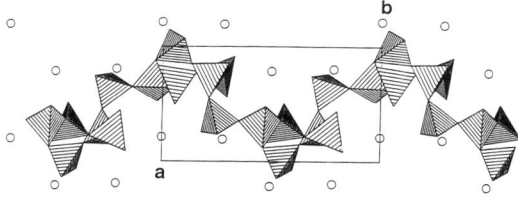

Figure 6.11. Projection, along the **c** axis, of the atomic arrangement in $Sr_2P_6O_{17}$, showing the corrugated layers of associated cations alternating with the infinite phosphoric anion. The open circles represent the strontium atoms.

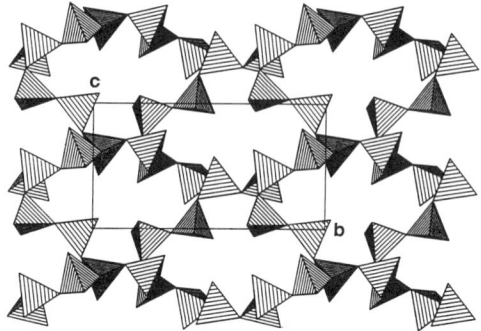

Figure 6.12. Projection, along the **a** axis, of the infinite ultraphosphate anion observed in $Sr_2P_6O_{17}$.

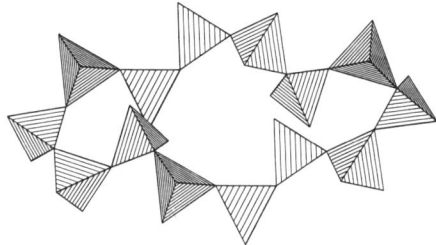

Figure 6.13. Representation of an isolated $P_{14}O_{42}$ ring as observed in uranyl ultraphosphate.

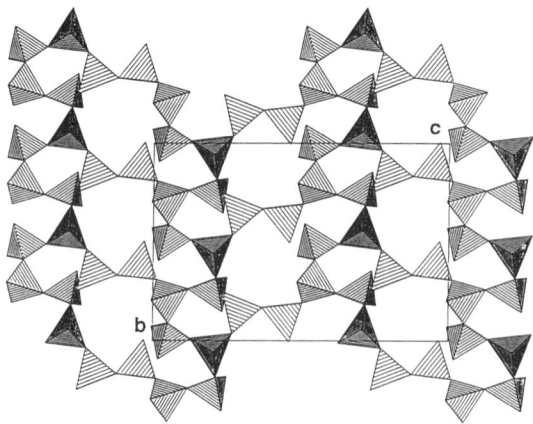

Figure 6.14. Projection, along the **a** axis, of the P_6O_{17} anionic framework of uranyl ultraphosphate.

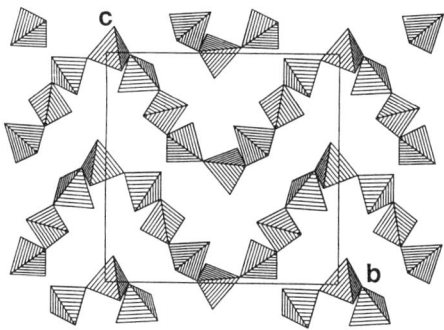

Figure 6.15. Projection, along the **a** axis, of a layer of the first type in HoP_5O_{14} (II).

the P_8O_{24} ring anions already observed in cyclooctaphosphates. Within the layer these rings do not share any oxygen atoms. A representation of such a layer is given in Fig. 6.16.

Layers of the third type are perpendicular to the **c** axis and are centered by planes $z = 0$ and $1/2$. Like those of the first type, they contain $(PO_3)_n$ infinite chains lying along the (110) direction, but these chains are much less corrugated than the chains observed in the first type of layer. A layer of this type is depicted in Fig. 6.17. A general view of this anionic framework is presented in Fig. 6.18, showing how the intersections of the three types of layers create a series of channels in which are located the associated cations, which have been omitted in the various drawings that we have used to illustrate the atomic arrangement in this very complex anion. All these drawings have been performed using the data reported for HoP_5O_{14}.[1204]

We have deliberately limited this survey to the external geometries of ultraphosphate anions, but much work remains to be done in comparing their internal geometries with those of other condensed phosphoric anions found in poly- or cyclophosphates. For

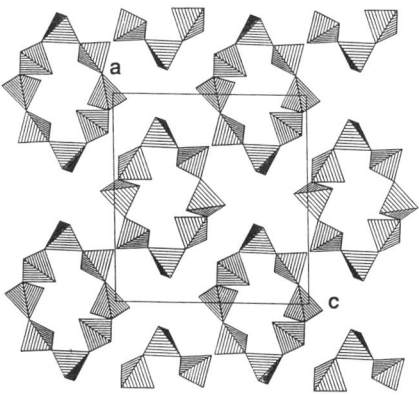

Figure 6.16. Projection, along the **b** axis, of a layer of the second type in HoP_5O_{14} (II).

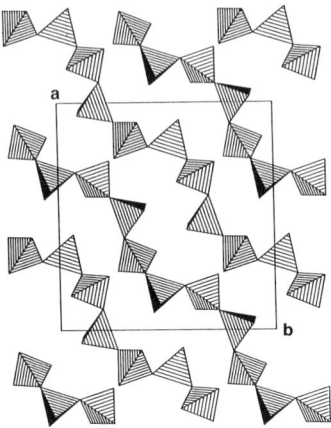

Figure 6.17. Projection, along the **c** axis, of a layer of the third type in HoP$_5$O$_{14}$ (II).

instance, it has not yet been ascertained whether the P–P–P and P–O–P angles and P–P distances measured in ultraphosphates, particularly in the vicinity of the ternary phosphorus, are comparable to or significantly different from those commonly observed in classical condensed phosphates. Most of these numerical values were recalculated from the original data, some of which are unfortunately not accurate enough to permit meaningful comparisons. Bagieu-Beucher[686] undertook a survey of the geometrical features of the PO$_4$ tetrahedron in this class of compounds. She did not observe significant differences between the characteristics of tetrahedra corresponding to a ternary phosphorus and those of normal ones; however, because of the small amount of available data and their poor accuracy (with estimated standard deviations of 0.5 to 1.0° for angles and more than 0.01 Å for distances), no firm conclusions could be drawn from this survey. The established additional data reported since then are not enough, to change this situation significantly. Thus, in this area of phosphate chemistry, accurate new data are needed.

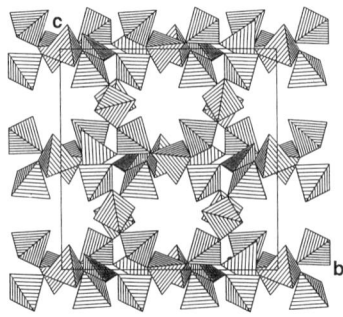

Figure 6.18. Projection, along the **a** axis, of the anionic framework in HoP$_5$O$_{14}$ (II).

7

The PO$_4$ Tetrahedron in Condensed Phosphates

In various sections of this book, we have discussed the geometries of the condensed anions that we have described, but nothing has been said concerning the geometry of the basic unit of such anions, the PO$_4$ tetrahedron. A feature common to all PO$_4$ tetrahedra involved in the constitution of phosphoric condensed anion is that they share one or two of their oxygen atoms with the adjacent tetrahedra. Thus, a primary distinction can be made between two kinds of oxygen atoms inside the PO$_4$ tetrahedron of a given condensed anion. One or two oxygen atoms are involved in the P–O–P bonds and are generally denoted O(L), whereas the others are usually called external oxygens and denoted O(E). In the vast majority of condensed phosphates, only the O(E) oxygen atoms are involved in the associated-cation coordination. In some rare cases, O(L) atoms are considered as being shared with an associated cation, but such cases generally involve large cations, whose coordination is never well determined.

The P–O(E) distances are the shortest inside the tetrahedron. They normally range from 1.470 to 1.502 Å with an average value of 1.481 Å. The P–O(L) bonds are significantly longer (1.595–1.628 Å) with an average P–O(L) distance of 1.609 Å. However, the most remarkable feature of the P–O distances is that their average value is almost constant in a given tetrahedron. This value never departs significantly from 1.545 Å, with nevertheless some small, but significant, variations in tetrahedra having internal symmetry (m, 2, 3). The rather large difference between P–O(E) and P–O(L) distances has been discussed by Cruickshank[1235] in terms of $d\pi$–$p\pi$ orbital overlap and later by Pakhomov.[1236]

Among the six O–P–O angles of the tetrahedron, one can distinguish three categories—the O(E)–P–O(E), the O(L)–P–O(L), and the O(E)–P–O(L) angles. Angles of the first category never depart very significantly from the average value of 119.6° and range between 118.4 and 121.3°. The O(L)–P–O(L) angles range between 99.3 and 102.7° with an average value of 101.3°, and the O(E)–P–O(L) between 104.8 and 112.0° with an average of 108.7°. The overall average value for the O–P–O angles is 109.3°, close to the value in a regular tetrahedron.

The numerical values reported here have been obtained from data for 60 tetrahedra, extracted from very accurately determined atomic arrangements. A more extensive review involving the analysis of several hundred PO$_4$ tetrahedra is planned.

References

1. D. E. C. Corbridge, *The Structural Chemistry of Phosphorus*, Elsevier, Amsterdam (1966).
2. M. T. Averbuch-Pouchot and A. Durif, *J. Solid State Chem. 46*, 193–196 (1983).
3. J. Berzelius, *Ann. Physik 54*, 31–52 (1816).
4. T. Clark, *Edinburgh J. Sci. 7*, 298–314 (1827).
5. T. Graham, *Phil. Trans. Roy. Soc. 123A*, 253–284 (1833).
6. R. Maddrell, *Liebigs Ann. 61*, 53–63 (1847).
7. T. Fleitmann and W. Henneberg, *Liebigs Ann. Chem. 65*, 304–334, 387–390 (1848).
8. T. Fleitmann, *Ann. Physik und Chem. 78*, 233–260, 338–366 (1849).
9. *Ann. Chemi. 72*, 228–248 (1849).*
10. F. Schwarz, *Z. anorg. Chem. 9*, 249–266 (1895).
11. E. Thilo and R. Ratz, *Z. anorg. Chem. 260*, 255–266 (1949).
12. C. G. Lindbom, *Berichte. 8*, 122–124 (1875).
13. A. Glatzel, Inaugural Dissertation, Würzburg University, Würzburg Germany (1880).
14. P. Sabatier, *C. R. Acad. Sci. 106*, 63–66 (1888).
15. P. Sabatier, *C. R. Acad. Sci. 108*, 738–741 (1889).
16. G. Tammann, *Z. Phys. Chem. 6*, 122–140 (1890).
17. G. Tammann, *J. Prakt. Chem. 45*, 417–474 (1892).
18. M. Stange, *Z. anorg. Chem. 12*, 444–463 (1898).
19. P. Groth, *Chemische Krystallographie*, W. Engelman Verlag, Leipzig (1906).
20. G. Knorre, *Z. anorg. Chem. 24*, 369–401 (1900).
21. F. Warschauer, *Z. anorg. Chem. 36*, 137–200 (1903).
22. P. Pascal, *Bull. Soc. Chim. 33*, 1611–1627 (1923).
23. A. Boullé, *C. R. Acad. Sci. 206*, 517–518 (1938).
24. P. Bonneman-Bémia, *Anal. Chim. 16*, 395–476 (1941).
25. A. E. R. Westman and A. E. Scott, *Nature (London) 168*, 740 (1951).
26. J. P. Ebel, *Bull. Soc. Chim. Fr.* (1968), 1663–1670.
27. J. R. Van Wazer, *Phosphorus and Its Compounds*, Interscience, New York (1966).
28. G. R. Levi and G. Peyronel, *Z. Kristallogr. 92*, 190–209 (1935).
29. L. Pauling and J. S. Sherman, *Z. Kristallogr. 96*, 481–487 (1937).
30. V. Cagliotti, G. Giacomello, and E. Bianchi, *Rendi. Conti. Atti Accad. Italia 3*, 761–769 (1942).
31. D. E. C. Corbridge, *Acta Crystallogr. 9*, 308–314 (1956).
32. D. R. Davies and D. E. C. Corbridge, *Acta Crystallogr. 11*, 315–319 (1958).
33. E. D. Eanes and H. M. Ondik, *Acta Crystallogr. 15*, 1280–1285 (1962).
34. K. H. Jost, *Acta Crystallogr. 19*, 555–560 (1965).
35. M. Bagieu-Beucher and D. Tranqui, *Bull. Soc. fr. Minéral. Cristallogr. 93*, 505–508 (1970).
36. K. H. Jost, *Acta Crystallogr., Sect. B, 28*, 732–738 (1972).
37. M. Laügt and J. C. Guitel, *Z. Kristallogr. 141*, 203–216 (1975).

*Although this reference is very often attributed to Th. Fleitmann and W. Henneberg, it is in fact an anonymous report on their works.

38. A. Durif, M. T. Averbuch-Pouchot, and J. C. Guitel, *Acta Crystallogr., Sect. B, 32*, 2957–2960 (1976).
39. Yu. I. Smolin, Yu. F. Shepelev, A. I. Domanskii, and J. Majling, *Sov. Phys. Crystallogr. 23*, 715–717 (1978).
40. M. Bagieu-Beucher, A. Durif, and J. C. Guitel, *J. Solid State Chem. 45*, 159–163 (1982).
41. A. V. Lavrov, M. Ya. Voitenko, and E. G. Tselebrovskaya, *Izv. Akad. Nauk SSSR, Neorg. Mater. 17*, 99–103 (1981).
42. B. Topley, *Quart. Rev. 3*, 345–368 (1949).
43. E. Thilo, *Adv. Inorg. Chem. Radiochem. 4*, 1–75 (1962).
44. S. Y. Kalliney, in: *Topics in Phosphorus Chemistry*, Vol. 7, (E. J. Griffith and M. Grayson, eds.), pp. 255–309, Interscience, New York (1972).
45. R. X. Fischer, *J. Appl. Crystallogr. 18*, 258–262 (1985).
46. E. J. Griffith and R. L. Buxton, *J. Am. Chem. Soc. 89*, 2884–2890 (1967).
47. W. L. Hill, G. T. Faust, and D. S. Reynolds, *Am. J. Sci. 242*, 457–477 (1944).
48. J. R. Van Wazer and S. Ohashi, *J. Am. Chem. Soc. 80*, 1010–1013 (1958).
49. S. Ohashi and J. R. Van Wazer, *J. Am. Chem. Soc. 81*, 830–832 (1959).
50. W. Wieker, A. R. Grimmer, and E. Thilo, *Z. anorg. allg. Chem. 330*, 78–90 (1964).
51. J. Nakano, T. Yamada, and S. Miyazawa, *J. Am. Ceram. Soc. 62*, 465–467 (1979).
52. T. Y. Tien and F. A. Hummel, *J. Am. Ceram. Soc. 44*, 206–208 (1961).
53. V. B. Lazarev, I. D. Sokolova, G. A. Sharpataya, I. S. Shaplygin, and I. B. Markina, *Thermochim. Acta 86*, 243–249 (1985).
54. K. Y. Leung and C. Calvo, *Can. J. Chem. 50*, 2519–2526 (1972).
55. R. V. G. Sundara Rao and N. S. Nampoothiri, *Acta Crystallogr. 8*, 850 (1955).
56. D. E. C. Corbridge, *Acta Crystallogr. 10*, 85 (1957).
57. D. M. MacArthur and C. A. Beevers, *Acta Crystallogr. 10*, 428–432 (1957).
58. D. W. J. Cruickshank, *Acta Crystallogr. 17*, 672–673 (1964).
59. W. S. McDonald and D. W. J. Cruickshank, *Acta Crystallogr. 22*, 43–48 (1967).
60. R. K. Osterheld and T. J. Moser, *J. Inorg. Nucl. Chem. 35*, 3463–3465 (1973).
61. T. Yamada and H. Koizumi, *J. Cryst. Growth 64*, 558–562 (1983).
62. E. Thilo and K. Dostal, *Z. anorg. allg. Chem. 298*, 100–115 (1959).
63. R. K. Osterheld and L. F. Audrieth, *J. Phys. Chem. 56*, 38–42 (1952).
64. J. R. Lehr, E. H. Brown, A. W. Frazier, J. P. Smith, and R. D. Thrasher, Crystallographic Properties of Fertilizer Compounds, *Chemical Engineering Bulletin*, No 6 (1967).
65. L. Brun, *Rev. Chim. Minér. 4*, 839–897 (1967).
66. A. W. Frazier, R. M. Scheib, and J. R. Lehr, *J. Agric. Food Chem. 20*, 146–150 (1972).
67. A. W. Frazier, J. P. Smith, and J. R. Lehr, *J. Agric. Food Chem. 13*, 316–322 (1965).
68. K. Dostal, V. Kocman, and V. Ehrenbergrova, *Z. anorg. allg. Chem. 367*, 80–91 (1969).
69. D. S. Emmerson and D. E. C. Corbridge, *Phosphorus 3*, 75 (1973).
70. E. Ingerson and G. W. Morey, *Am. Mineral 28*, 488–498 (1943).
71. J. M. Verdier, C. Dorémieux-Morin, and A. Boullé, *C. R. Acad. Sci. 257*, 917–919 (1963).
72. Y. Dumas, J. L. Galigne, and J. Falgueirettes, *Acta Crystallogr., Sect. B 29*, 1623–1630 (1973).
73. M. T. Averbuch-Pouchot and A. Durif, *Eur. J. Solid State Inorg. Chem. 29*, 993–999 (1992).
74. R. L. Collin and M. Willis, *Acta Crystallogr., Sect. B 27*, 291–302 (1971).
75. J. D. Lee, *J. Chem. Soc. A, 1968*, 2881–2882.
76. A. Larbot, Dissertation, Montpellier University, Montpellier France (1971).
77. A. Larbot, J. Durand, A. Norbert, and L. Cot, *Acta Crystallogr., Sect. C 39*, 6–8 (1983).
78. D. S. Emmerson and D. E. C. Corbridge, *Phosphorus 2*, 159–160 (1972).
79. Y. Dumas, J. L. Galigné, and J. Falgueirettes, *Acta Crystallogr., Sect. B 29*, 2913–2918 (1973).
80. M. T. Averbuch-Pouchot and A. Durif, *Eur. J. Solid State Inorg. Chem. 29*, 191–198 (1992).
81. A. Larbot, A. Norbert, and L. Cot, *C. R. Acad. Sci. 289C*, 185–187 (1979).
82. M. T. Averbuch-Pouchot and A. Durif, *C. R. Acad. Sci., Ser. 2 316*, 469–476 (1993).
83. M. T. Averbuch-Pouchot and A. Durif, *C. R. Acad. Sci., Ser. 316*, 41–46 (1993).
84. M. T. Averbuch-Pouchot and A. Durif, unpublished results.

85. H. Giran, *Ann. Chim. Phys. 30* 247–261 (1903).
86. J. R. Partington and H. E. Wallsom, *Chem. News 136*, 97–103 (1928).
87. A. Norbert and C. Dautel, *C. R. Acad. Sci. 262*, 1534–1536 (1966).
88. A. Larbot, J. Durand, and A. Norbert, *Rev. Chim. Minér. 17*, 548–554 (1980).
89. L. Brun, *C. R. Acad. Sci. 266C*, 918–920 (1968).
90. G. Brun, Dissertation, Montpellier University, Montpellier, France (1967).
91. Y. Dumas, *Acta Crystallogr., Sect. B 34*, 3514–3519 (1978).
92. Y. Dumas and J. Lapasset, *Acta Crystallogr., Sect. B 35*, 1977–1982 (1979).
93. Y. Dumas, J. Lapasset, and J. Vicat, *Acta Crystallogr., Sect. B 36,* 2754–2757 (1980).
94. Y. Dumas, A. Escande, and J. L. Galigné, *Acta Crystallogr., Sect. B 34*, 710–715 (1978).
95. A. W. Frazier, E. F. Dillard, R. D. Thrasher, and K. R. Waerstad, *J. Agric. Food Chem. 21*, 700–704 (1973).
96. K. R. Waerstad and A. W. Frazier, *J. Appl. Crystallogr. 13*, 614–615 (1980).
97. W. H. Zachariasen, *Z. Kristallogr. 73*, 1–6 (1930).
98. A. G. Nord and P. Kierkegaard, *Chemi. Scri. 15*, 27–39 (1980).
99. I. D. Brown and C. Calvo, *J. Solid State Chem. 1*, 173–179 (1970).
100. S. Jaulmes, Thesis, University of Paris (1964).
101. A. de Sallier-Dupin, *C. R. Acad. Sci. 253*, 2984–2985 (1961).
102. A. de Sallier-Dupin, Dissertation, Paris University, France (1966).
103. A. Boullé and A. de Sallier-Dupin, *C. R. Acad. Sci. 254*, 122–124 (1961).
104. B. Bleyer and B. Müller, *Z. anorg. allg. Chem. 79*, 263–276 (1913).
105. J. Berak, *Rocz. Chem. 32*, 17–22 (1958).
106. F. L. Katnack and F. A. Hummel, *J. Electrochem. 105*, 125–133 (1958).
107. K. Lukaszewicz, *Bull. Acad. Pol. Sci.* 15 (1967).
108. C. Calvo, *Acta Crystallogr. 23*, 289–295 (1967).
109. K. Lukaszewicz, *Rocz. Chem. 35*, 31–35 (1961).
110. C. Calvo, *Can. J. Chem. 43*, 1139–1146 (1965).
111. Z. A. Konstant and A. I. Dimante, *Izv. Akad. Nauk SSSR, Neorg. Mater. 13*, 99–103 (1977).
112. T. Stefanidis and A. G. Nord, *Acta Crystallogr., Sect. C 40*, 1995–1999 (1984).
113. P. Royen and J. Korinth, *Z. anorg. allg. Chem. 291*, 227–238 (1957).
114. T. Stefanidis and A. G. Nord, *Z. Kristallogr. 159*, 255–264 (1982).
115. J. T. Hoggins, J. S. Swinnea, and H. Steinfink, *J. Solid State Chem. 50*, 278–283 (1983).
116. J. F. Sarver, *Trans. Br. Ceram. Soc. 65*, 191–198 (1966).
117. N. Krishnamachari and C. Calvo, *Acta Crystallogr., Sect. B 28*, 2883–2885 (1972).
118. B. M. Wanklyn, F. R. Wondre, W. Davison, and R. Salmon, *J. Mater. Sc. Lett. 2*, 511–515 (1983).
119. K. Lukaszewicz, *Bull. Acad. Pol. Sci., Ser. Sci. Chim. 15*, 47–51 (1967).
120. A. Pietraszko and K. Lukaszewicz, *Bull. Acad. Polon. Sci., Ser. Sci. Chim. 16*, 183–187 (1968).
121 R. Masse, J. C. Guitel, and A. Durif, *Mater. Res. Bull. 14*, 337–341 (1979).
122. T. Ericsson and A. G. Nord, *Acta Chem. Scand. 44*, 990–993 (1990).
123. K. Lukaszewicz and E. Nagler, *Rocz. Chem 35*, 1167–1168 (1961).
124. B. E. Robertson and C. Calvo, *Acta Crystallogr. 22*, 665–672 (1967).
125. B. E. Robertson and C. Calvo, *Can. J. Chem. 46*, 605–612 (1968).
126. M. C. Ball, *J. Chem. Soc. A* (1968) 1113–1115.
127. V. B. Lazarev, I. D. Sokolova, G. A. Sharpataya, I. S. Shaplygin, and I. B. Markina, *Thermochim. Acta 86*, 243–249 (1985).
128. J. J. Brown and F. A. Hummel, *Trans. Br. Ceram. Soc. 64*, 387–396 (1965).
129. B. E. Robertson and C. Calvo, *J. Solid State Chem. 1*, 120–133 (1970).
130. C. Calvo, *Can. J. Chem. 43*, 1147–1153 (1965).
131. M. T. Averbuch-Pouchot, *J. Appl. Crystallogr. 7*, 511–512 (1974).
132. C. Calvo, *Inorg. Chem. 7*, 1347–1351 (1968).
133. N. C. Webb, *Acta Crystallogr. 21*, 942–948 (1966).
134. E. H. Brown, J. R. Lehr, J. P. Smith, and A. W. Frazier, *J. Agric. Food Chem. 11*, 214–222 (1963).

135. W. L. Hill, S. B. Hendricks, E. J. Fox, and J. G. Cady, *Ind. Eng. Chem. 39*, 1667–1672 (1947).

136. J. R. Lehr, A. W. Frazier, and J. P. Smith, *J. Agric. Food Chem. 14*, 27–33 (1966).

137. E. H. Brown, J. R. Lehr, J. P. Smith, W. E. Brown, and A. W. Frazier, *J. Phys. Chem. 61*, 1669 (1957).

138. E. H. Brown and J. R. Lehr, *J. Agric. Food Chem. 12*, 201–204 (1964).

139. E. H. Brown, J. R. Lehr, A. W. Frazier, and J. P. Smith, *J. Agric. Food Chem. 12*, 70–73 (1964).

140. R. C. Ropp, M. A. Aia, C. W. W. Hoffman, T. J. Veleker, and R. W. Mooney, *Anal. Chem. 31*, 1164–1167 (1959).

141. E. R. Kreidler and F. A. Hummel, *Inorg. Chem. 6*, 884–891 (1967).

142. C. W. W. Hoffman and R. W. Mooney, *J. Electrochem. Soc. 107*, 854–855 (1960).

143. L. O. Hagman, I. Jansson, and C. Magneli, *Acta Chem. Scand. 22*, 1419–1429 (1968).

144. J. C. Grenier and R. Masse, *Bull. Soc. fr. Minéral. Cristallogr. 90*, 285–292 (1967).

145. J. C. Grenier and R. Masse, *Bull. Soc. fr. Minéral. Cristallogr. 92*, 91–92 (1969).

146. R. Klement, *Chem. Ber. 93*, 2314–2316 (1960).

147. R. A. McCauley and F. A. Hummel, *Trans. Br. Ceram. Soc. 67*, 619–628 (1968).

148. P. W. Ranby, D. H. Mash, and S. T. Henderson, *Br. J. Appl. Phys. Suppl. 4*, 18–25 (1955).

149. S. I. Lopatiyn and G. A. Semenov, *Izv. Akad. Nauk SSSR, Neorg. Mater. 25,* 1404–1406 (1989).

150. C. Calvo and P. K. L. Au, *Can. J. Chem. 47*, 3409–3416 (1969).

151. L. H. Brixner, P. E. Bierstedt, and C. M. Foris, *J. Solid State Chem. 7*, 430–432 (1973).

152. A. Bruckner and H. Worzala, *Z. anorg. allg. Chem. 584*, 173–177 (1990).

153. D. F. Mullica, H. O. Perkins, D. A. Grossie, L. A. Boatner, and B. C. Sales, *J. Solid State Chem. 62*, 371–376 (1986).

154. D. Riou and B. Raveau, *Acta Crystallogr., Sect. C 47*, 1708–1709 (1991).

155. D. Riou and M. Goreaud, *Acta Crystallogr., Sect. C 46*, 1191–1193 (1990).

156. A. Boukhari, A. Moqine, and S. Flandrois, *J. Solid State Chem. 87*, 251–256 (1990).

157. A. Alaoui, E. Belghiti, A. Boukhai, and E. M. Holt, *Acta Crystallogr., Sect. C 47*, 473–477 (1991).

158. D. Riou, P. Labbé, and M. Goreaud, *C. R. Acad. Sci., Ser. 2 307*, 903–907 (1988).

159. A. Moqine, A. Boukhari, and E. M. Holt, *Acta Crystallogr., Sect. C 47*, 2294–2297 (1991).

160. E. V. Murashova, Yu. A. Velikodnyi, and V. K. Trunov, *Russ. J. Inorg. Chem. 36*, 479–481 (1991).

161. E. V. Murashova, Yu. A. Velikodnyi, and V. K. Trunov, *Russ. J. Inorg. Chem. 36*, 481–483 (1991).

162. A. P. Dindune, V. V. Krasnikov, and Z. A. Konstant, *Izv. Akad. Nauk SSSR, Neorg. Mater. 20*, 1553–1556 (1984).

163. V. V. Krasnikov, Z. A. Konstant, and V. K. Bel'skii, *Izv. Akad. Nauk SSSR, Neorg. Mater. 21*, 1560–1563 (1985).

164. C. Calvo, *J. Electrochem. Soc. 115*, 1095–1096 (1968).

165. R. A. Terpstra, F. C. M. Driessens, and R. M. H. Verbeeck, *Z. anorg. allg. Chem. 515*, 213–224 (1984).

166. J. J. Brown and F. A. Hummel, *J. Electrochem. Soc. 111*, 1052–1057 (1964).

167. J. Oka and A. Kawahara, *Acta Crystallogr., Sect. B 38*, 3–5 (1982).

168. V. V. Kokhanovskii, Z. N. Zemtsova, and S. G. Tereshkova, *Russ. J. Inorg. Chem. 25*, 1305–1308 (1980).

169. A. V. Lavrov and T. A. Bykanova, *Izv. Akad. Nauk SSSR, Neorg. Mater. 18*, 111–116 (1982).

170. O. T. Quimby and H. W. McCune, *Anal. Chem. 29*, 248–253 (1957).

171. N. M. Selivanova, N. Yu. Morozova, L. Kh. Kravchenko, and T. I. Khozhainova, *Russ. J. Inorg. Chem. 20*, 327–329 (1975).

172. S. Schneider and R. Collin, *Inorg. Chem. 12*, 2136–2139 (1973).

173. M. V. Goloshchapov and B. V. Martynenko, *Izv. Akad. Nauk SSSR, Neorg. Mater. 12*, 485–490 (1976).

174. H. Effenberger and F. Pertlik, *Z. Kristallogr., Suppl. Issue No.5*, 54–54 (1992).

175. W. L. Lindsay, A. W. Frazier, and H. F. Stephenson, *Soil Sci. Soc. Am. Proc. 26*, 446–452 (1962).

176. W. L. Lindsay and A. W. Taylor, *Transactions of the 7th International Congress on Soil Science,* Vol. 3, Elsevier, Amsterdam (1961).

177. M. S. Mandel, *Acta Crystallogr., Sect. B 31*, 1730–1734 (1975).

178. R. Klement, *Chem. Ber. 93*, 2314–2316 (1960).

179. J. Majling and F. Hanic, *J. Solid State Chem.* 7, 370–373 (1973).
180. F. Hanic and Z. Zak, *J. Solid State Chem.* 10, 12–19 (1974).
181. V. V. Kokhanovskii, Z. N. Zemtsova, and S. G. Tereshkova, *Russ. J. Inorg. Chem.* 25, 1305–1308 (1980).
182. L. Zh. Atstinya, A. P. Dindune, and Z. A. Konstant, *Izv. Akad. Nauk SSSR, Neorg. Mater.* 20, 1548–1552 (1984).
183. A. W. Frazier, J. P. Smith, and J. R. Lehr, *J. Agric. Food Chem.* 14, 522–529 (1966).
184. P. T. Cheng, K. P. H. Pritzker, and S. C. Nyburg, *Acta Crystallogr., Sect. B 36*, 921–924 (1980).
185. P. T. Cheng, S. C. Nyburg, M. E. Adams, and K. P. H. Pritzker, *Cryst. Struct. Commun.* 8, 313–317 (1979).
186. E. H. Brown, W. E. Brown, J. R. Lehr, J. P. Smith, and A. W. Frazier, *J. Phys. Chem.* 62, 366 (1958).
187. J. R. Lehr, O. P. Engelstad, and E. H. Brown, *Soil Sci. Soc. Am. Proc.* 28, 396–400 (1964).
188. V. A. Lyutsko, A. S. Lyakhov, and P. L. Frenkel, *Russ. J. Inorg. Chem.* 34, 1245–1247 (1989).
189. V. K. Trunov, Yu. V. Oboznenko, S. P. Sirotinkin, and N. B. Tskhelashvili, *Izv. Akad. Nauk SSSR, Neorg. Mater.* 27, 1993–1994 (1991).
190. R. Faggiani and C. Calvo, *Can. J. Chem.* 54, 3319–3324 (1976).
191. V. K. Trunov, Yu. V. Oboznenko, S. P. Sirotinkin, and N. B. Tskhelashvili, *Izv. Akad. Nauk SSSR, Neorg. Mater.* 27, 2370–2374 (1991).
192. J. Liebertz and S. Stähr, *Z. Kristallogr.* 162, 313–314 (1983).
193. M. V. Goloshchapov and B. V. Martynenko, *Izv. Akad. Nauk SSSR, Neorg. Mater.* 12, 485–490 (1976).
194. M. V. Goloshchapov and B. V. Martynenko, *Russ. J. Inorg. Chem.* 19, 611–612 (1974).
195. V. V. Kokhanovskii and E. A. Prodan, *Russ. J. Inorg. Chem.* 34, 1415–1419 (1989).
196. V. V. Kokhanovskii, *Russ. J. Inorg. Chem.* 35, 1720–1724 (1990).
197. M. Gabelica-Robert, *C. R. Acad. Sci., Ser. 2 293*, 497–499 (1981).
198. P. Lightfoot, A. K. Cheetham, and A. W. Sleight, *J. Solid State Chem.* 85, 275–282 (1990).
199. V. V. Kokhanovskii and E. A. Prodan, *Russ. J. Inorg. Chem.* 33, 428–431 (1988).
200. Z. N. Zemtsova, *Izv. Akad. Nauk SSSR, Neorg. Mater.* 25, 103–106 (1989).
201. M. T. Averbuch-Pouchot, *Z. Kristallogr.* 171, 113–119 (1985).
202. T. A. Bykanova and A. V. Lavrov, *Izv. Akad. Nauk SSSR, Neorg. Mater.* 14, 2044–2048 (1978).
203. I. D. Sokolova, I. S. Shaplygin, G. A. Sharpataya, and I. B. Markina, *Russ. J. Inorg. Chem.* 28, 18–21 (1983).
204. P. S. Yuen and R. L. Collin, *Acta Crystallogr., Sect. B 30*, 2513–2516 (1974).
205. I. A. Spintse, A. P. Dindune, and Z. A. Konstant, *Izv. Akad. Nauk SSSR, Neorg. Mater.* 25, 1356–1359 (1989).
206. E. L. Krivoviazov, N. K. Voskresenskaya, and K. K. Palkina, *Izv. Akad. Nauk SSSR, Inorg. Mater.* 6, 1057–1061 (1969).
207. V. B. Lazarev, I. D. Sokolova, G. A. Sharpataya, I. S. Shaplygin, and I. B. Markina, *Thermochim. Acta 86*, 243–249 (1985).
208. N. Yu. Morozova and N. M. Selivanova, *Russ. J. Inorg. Chem.* 21, 878–879 (1976).
209. G. A. Selivanova, N. T. Krudryavtsev, and Yu. M. Khozhainov, *Russ. J. Inorg. Chem.* 17, 806–807 (1972).
210. M. T. Averbuch-Pouchot, *Bull. Soc. fr. Minéral. Cristallogr 95*, 513–515 (1972).
211. G. A. Selivanova and N. T. Kudryavtsev, *Russ. J. Inorg. Chem.* 15, 1097–1099 (1970).
212. G. A. Selivanova and N. T. Kudryavtsev, *Russ. J. Inorg. Chem.* 19, 1067–1069 (1974).
213. A. Rosenheim, *Z. anorg. allg. Chem.* 153, 126–132 (1926).
214. A. Rulmont, P. Tarte, and J. M. Winand, *Eur. J. Solid State Inorg. Chem.* 28, 1021–1034 (1991).
215. I. D. Sokolova, I. B. Markina, and I. S. Shaplygin, *Russ. J. Inorg. Chem.* 27, 172–173 (1982).
216. Y. Laligant, *Eur. J. Solid State Inorg. Chem.* 29, 83–94 (1992).
217. Y. Laligant, *Eur. J. Solid State Inorg. Chem.* 29, 239–247 (1992).
218. M. T. Averbuch-Pouchot and J. C. Guitel, *Acta Crystallogr., Sect. B 33*, 3460–3462 (1977).
219. A. F. Selevich and V. D. Lyutsko, *Russ. J. Inorg. Chem.* 29, 364–369 (1984).
220. F. d' Yvoire, *Bull. Soc. Chim. Fr. 1962*, 1224–1236.

221. A. Durif and M. T. Averbuch-Pouchot, *Acta Crystallogr., Sect. B 38*, 2883–2885 (1982).
222. N. N. Chudinova, G. M. Balagina, and L. P. Shklover, *Izv. Akad. Nauk SSSR, Inorg. Mater. 13*, 2075–2082 (1977).
223. A. A. Medvedev, A. V. Lavrov, N. N. Chudinova, and I. V. Tananaev, *Izv. Akad. Nauk SSSR, Neorg. Mater. 6*, 1650–1656 (1970).
224. I. V. Tananaev, V. G. Kuznetsov, and V. P. Vasil'eva, *Izv. Akad. Nauk SSSR, Inorg. Mater. 3*, 107–113 (1967).
225. L. A. Tezikova, N. N. Chudinova, P. M. Fedorov, and A. V. Lavrov, *Izv. Akad. Nauk SSSR, Neorg. Mater. 10*, 2057–2063 (1974).
226. N. N. Chudinova, *Izv. Akad. Nauk SSSR, Neorg. Mater. 11*, 1662–1666 (1975).
227. P. Rémy and A. Boullé, *C. R. Acad. Sci. 258*, 927–929 (1964).
228. V. A. Leitsin and S. D. Grekov, *Russ. J. Inorg. Chem. 11*, 152–154 (1966).
229. K. Schlesinger, B. Ziemer, E. W. Hanke, and G. Ladwig, *Z. anorg. allg. Chem. 500*, 104–116 (1983).
230. K. K. Palkina, S. I. Maksimova, N. T. Chibiskova, K. Schlesinger, and G. Ladwig, *Z. anorg. allg. Chem. 529*, 89–96 (1985).
231. I. V. Tananaev and V. P. Vasil'eva, *Russ. J. Inorg. Chem. 9*, 1141–1144 (1964).
232. V. G. Kuznetsov and V. P. Vasil'eva, *Izv. Akad. Nauk SSSR, Inorg. Mater. 3*, 361–367 (1967).
233. I. V. Tananaev and V. P. Vasil'eva, *Russ. J. Inorg. Chem. 9*, 1237–1239 (1964).
234. V. G. Kuznetsov and V. P. Vasil'eva, *Izv. Akad. Nauk SSSR, Neorg. Mater. 3*, 401–402 (1967).
235. M. Kizilyalli, *J. Inorg. Nucl. Chem. 38*, 483–486 (1976).
236. I. V. Tananaev, V. G. Kuznetsov, and S. M. Petushkova, *Russ. J. Inorg. Chem. 11*, 157–160 (1966).
237. I. V. Tananaev and S. M. Petushko, *Russ. J. Inorg. Chem. 11*, 154–156 (1966).
238. T. P. Haromy, C. F. Linck, W. W. Cleland, and M. Sundaralingam, *Acta Crystallogr., Sect. C 46*, 951–957 (1990).
239. T. P. Haromy, W. B. Knight, D. Dunaway-Mariano, and M. Sundaraligam, *Acta Crystallogr., Sect. C 40*, 223–226 (1984).
240. E. A. Merritt and M. Sundaraligam, *Acta Crystallogr., Sect. B 36*, 2576–2584 (1980).
241. E. A. Genkina, B. A. Maksimov, V. A. Timofeeva, A. B. Bykov, and O. K. Mel'Nikov, *Dokl. Akad. Nauk SSSR. 284*, 864–867 (1985).
242. I. Grunze and H. Grunze, *Z. anorg. allg. Chem. 512*, 39–47 (1984).
243. D. Tranqui, S. Hamdoune, and Y. Le Page, *Acta Crystallogr., Sect. C 43*, 201–202 (1987).
244. J. P. Gamondes, F. d'Yvoire, and A. Boullé, *C. R. Acad. Sci. 272*, 49–52 (1971).
245. Y. P. Wang, K. H. Lii, and S. L. Wang, *Acta Crystallogr., Sect. C 45*, 1417–1418 (1989).
246. L. Bohaty, J. Liebertz, and R. Frohlich, *Z. Kristallogr. 161*, 53–59 (1982).
247. M. Gabelica-Robert, *Proceedings of the 2nd European Conference on Solid State Chemistry*, Elsevier, Amsterdam, pp. 475–478 (1983).
248. F. Gabelica-Robert, M. Goreaud, P. Labbé, and B. Raveau, *J. Solid State Chem. 45*, 389–395 (1982).
249. T. Moya-Pizarro, R. Salmon, S. L. Fournes, G. Le Flem, B. S. Wanklin, and P. Hagenmuller, *J. Solid State Chem. 53*, 387–397 (1984).
250. A. Leclaire, M. M. Borel, A. Grandin, and B. Raveau, *J. Solid State Chem. 76*, 131–135 (1988).
251. M. A. Avaliani, N. N. Chudinova, and I. V. Tananaev, *Izv. Akad. Nauk SSSR, Neorg. Mater. 20*, 282–286 (1984).
252. N. Yu. Anisimova, V. K. Trunov, and N. N. Chudinova, *Izv. Akad. Nauk SSSR, Neorg. Mater. 24*, 268–272 (1988).
253. H. N. Ng and C. Calvo, *Can. J. Chem. 51*, 2613–2620 (1973).
254. L. S. Guzeeva, A. V. Lavrov, and I. V. Tananaev, *Izv. Akad. Nauk SSSR, Neorg. Mater. 18*, 1850–1855 (1982).
255. D. Riou, P. Labbé, and M. Goreau, *Eur. J. Solid. State Inorg. Chem. 25*, 215–229 (1988).
256. M. A. Avaliani, N. N. Chudinova, and I. V. Tananaev, *Izv. Akad. Nauk SSSR, Neorg. Mater. 15*, 1688–1689 (1979).
257. A. Leclaire, M. M. Borel, A. Grandin, and B. Raveau, *J. Solid State Chem. 78*, 220–226 (1989).

258. N. N. Chudinova, I. Grunze, and L. S. Guzeeva, *Izv. Akad. Nauk SSSR, Neorg. Mater.* 23, 616–621 (1987).
259. J. Belkouch, L. Monceaux, F. Oudet, E. Bordes, and P. Courtine, *Mater. Res. Bull.* 25, 1099–1107 (1990).
260. N. N. Chudinova, M. A. Avaliani, L. S. Guzeeva, and I. V. Tananaev, *Izv. Akad. Nauk SSSR, Neorg. Mater.* 15, 2176–2179 (1979).
261. U. Florke, *Z. Kristallogr. New Crystal Structures,* 1990, 5–6.
262. L. S. Guzeeva and I. V. Tananaev, *Izv. Akad. Nauk SSSR, Neorg. Mater.* 24, 646–650 (1988).
263. A. Akrim, D. Zambon, J. Metin, and J. C. Cousseins, *Eur. J. Solid State Inorg. Chem.* 30, 483–495 (1993).
264. J. M. Millet and B. F. Mentzen, *Eur. J. Solid State Inorg. Chem.* 28, 493–504 (1991).
265. E. Dvoncova and K.-H. Lii, *J. Solid State Chem.* 105, 279–286 (1993).
266. Y. P. Wang and K. H. Lii, *Acta Crystallogr., Sect. C* 45, 1210–1211 (1989).
267. S. A. Linde and Yu. E. Gorbunova, *Izv. Akad. Nauk SSSR, Neorg. Mater.* 18, 464–467 (1982).
268. K. H. Lii., and R. C. Haushalter, *Acta Crystallogr., Sect. C* 43, 2036–2038.
269. M. Jansen, G. Q. Wu, and K. Koenigstein, *Z. Kristallogr.* 197, 245–261 (1991).
270. A. A. Medvedev, A. V. Lavrov, N. N. Chudinova, and I. V. Tananaev, *Izv. Akad. Nauk SSSR, Neorg. Mater.* 6, 1650–1656 (1970).
271. I. V. Tananaev and V. P. Vasil'eva, *Russ. J. Inorg. Chem.* 9, 1141–1144 (1964).
272. V. G. Kuznetsov and V. P. Vasil'eva, *Izv. Akad. Nauk SSSR, Inorg. Mater.* 3, 361–367 (1967).
273. I. V. Tananaev, V. G. Kuznetsov, and S. M. Petushkova, *Russ. J. Inorg. Chem.* 11, 157–160 (1966).
274. I. V. Tananaev and S. M. Petushkova, *Russ. J. Inorg. Chem.* 11, 154–156 (1966).
275. J. Grunze, W. Hilmer, N. N. Chudinova, and H. Grunze, *Izv. Akad. Nauk SSSR, Neorg. Mater.* 20, 287–291 (1984).
276. O. V. Yakubovich, M. S. Dadashov, and B. N. Litvin, *Sov. Phys. Crystallogr.* 33, 16–19 (1988).
277. L. S. Guzeeva and I. V. Tananaev, *Izv. Akad. Nauk SSSR, Neorg. Mater.* 24, 651–654 (1988).
278. I. Grunze, K. K. Palkina, N. N. Chudinova, L. S. Guzeeva, M. A. Avaliani, and S. I. Maksimova, *Izv. Akad. Nauk SSSR, Neorg. Mater.* 23, 610–615 (1987).
279. I. Grunze, S. I. Maksimova, K. K. Palkina, N. T. Chibiskova, and N. N. Chudinova, *Izv. Akad. Nauk SSSR, Neorg. Mater.* 24, 264–267 (1988).
280. S. I. Berul and N. V. Sizova, *Russ. J. Inorg. Chem.* 17, 126–128 (1972).
281. S. I. Berul and N. V. Sizova, *Russ. J. Inorg. Chem.* 22, 13–16 (1977).
282. S. I. Berul and N. V. Sizova, *Russ. J. Inorg. Chem.* 19, 1843–1846 (1974).
283. C. Masquelier, F. d'Yvoire, and N. Rodier, *Acta Crystallogr., Sect. C* 46, 1584–1588 (1990).
284. C. Masquelier, F. d'Yvoire, and N. Rodier, *J. Solid State Chem.* 95, 156–167 (1990).
285. M. Ijjaali, G. Venturini, R. Gerardin, B. Malaman, and C. Gleitzer, *Eur. J. Solid State Inorg. Chem.* 28, 983–998 (1991).
286. G. Venturini, M. Ijjaali, B. Malaman, and C. Gleitzer, *Eur. J. Solid State Inorg. Chem.* 29, 1189–1204 (1992).
287. B. Malaman, M. Ijjaali, R. Gerardin, G. Venturini, and C. Gleitzer, *Eur. J. Solid State Inorg. Chem.* 29, 1269–1284 (1992).
288. S. Wang and S. J. Hwu, *J. Solid State Chem.* 90, 31–41 (1991).
289. A. Leclaire, J. Chardon, M. M. Borel, A. Grandin, and B. Raveau, *Z. anorg. allg. Chem.* 617, 127–130 (1992).
290. L. Benhamada, A. Grandin, M. M. Borel, A. Leclaire, and B. Raveau, *Acta Crystallogr.* C47, 2437–2438 (1991).
291. F. Liebau, G. Bissert, and N. Koppen, *Z. anorg. allg. Chem.* 359, 113–134 (1968).
292. F. Liebau and G. Bissert, *Bull. Soc. Chim. Fr.,* 1968, 1742–1744.
293. G. Bissert and F. Liebau, *Acta Crystallogr., Sect. B* 26, 233–240 (1970).
294. F. Liebau and K. H. Hesse, *Naturwissenschaften* 56, 634–635 (1969).
295. K. F. Hesse, *Acta Crystallogr., Sect. B* 35, 724–725 (1979).
296. E. Tillmanns, W. Gebert, and W. H. Baur, *J. Solid State Chem.* 7, 69–84 (1973).

297. V. V. Pechkovskii, E. D. Dzyuba, G. I. Salonets, V. N. Yaglov, and A. I. Volkov, *Russ. J. Inorg. Chem. 20*, 329–332 (1975).

298. A. E. Mal'shikov, O. V. Egorova, and I. A. Bondar, *Russ. J. Inorg. Chem. 33*, 722–726 (1988).

299. K. A. Avduevskaya and I. V. Tananaev, *Russ. J. Inorg. Chem. 10*, 197–200 (1965).

300. A. E. Mal'shikov and I. A. Bondar, *Russ. J. Inorg. Chem. 34*, 723–727 (1989).

301. R. Sacks, Y. Avigal, and E. Banks, *J. Electrochem. Soc. 129*, 726–729 (1982).

302. A. E. Mal'shikov and I. A. Bondar, *Russ. J. Inorg. Chem. 34*, 1207–1209 (1989).

303. M. Tsuhako, S. Ikeuchi, T. Matsuo, I. Motooka, and T. Kobayashi, *Chem. Lett.*, 1977, 195–198.

304. S. A. Merkusheva, N. A. Skorik, V. N. Kumok, and V. V. Serebrennikov, *Russ. J. Inorg. Chem. 12*, 1793–1794 (1967).

305. B. Wellman and F. Liebau, *J. Less-Commmon. Met. 77*, 31–39 (1981).

306. A. Leclaire, M. M. Borel, A. Grandin, and B. Raveau, *Eur. J. Solid State Inorg. Chem. 25*, 323–328 (1988).

307. E. Banks and S. R. Sacks, *Mater. Res. Bull. 17*, 1053–1055 (1982).

308. R. M. Douglass and E. Staritzky, *Anal. Chem. 28*, 1211–1212 (1956).

309. H. P. Kirchner, K. M. Merz, and W. R. Brown, *J. Am. Ceram. Soc. 46*, 137–141 (1963).

310. A. Burdese and M. L. Bordela, *Ann. Chim. 53*, 333–343 (1963).

311. M. L. Bordela and A. Burdese, *Atti Accad. Sci; Torino 1959*, 89–95.

312. A. Burdese and M. L. Bordela, *Ann. Chim. 53*, 344–355.

313. E. Steger and G. Leukroth, *Z. anorg. allg. Chem. 303*, 169–176 (1960).

314. H. Vollenkle, A. Wittman, and H. Nowotny, *Monatsh. Chem. 94*, 956–963 (1963).

315. J. Lecomte, A. Boullé, C. Dorémieux-Morin, and B. Lelong, *C. R. Acad. Sci. 255*, 621–624 (1962).

316. J. Lecomte, A. Boullé, C. Dorémieux-Morin, and B. Lelong, *C. R. Acad. Sci. 258*, 131–134 (1964).

317. J. Lecomte, A. Boullé, C. Dorémieux-Morin, and B. Lelong, *C. R. Acad. Sci. 258*, 1447–1451 (1964).

318. M. T. Averbuch-Pouchot and A. Durif, *Z. Kristallogr. 180*, 195–202 (1987).

319. M. T. Averbuch-Pouchot and A. Durif, *Acta Crystallogr., Sect. C 43*, 1861–1863 (1987).

320. A. Verbaere, S. Oyetola, D. Guyomard, and Y. Piffard, *J. Solid State Chem. 75*, 217–224 (1988).

321. Yu. E. Gorbunova, S. A. Linde, A. V. Lavrov, and I. V. Tananaev, *Dokl. Akad. Nauk SSSR. 250*, 350–353 (1980).

322. K. H. Lii and S. L. Wang, *J. Solid State Chem. 82*, 239–246 (1989).

323. A. Leclaire, H. Chahboun, D. Groult, and B. Raveau, *J. Solid State Chem. 77*, 170–179 (1988).

324. K. H. Lii, Y. P. Wang, and S. L. Wang, *J. Solid State Chem. 80*, 127–132 (1989).

325. C. S. Lee and K. H. Lii, *J. Solid State Chem. 92*, 362–369 (1991).

326. J. Protas, B. Menaert, G. Marnier, and B. Boulanger, *Acta Crystallogr., Sect. C 47*, 698–701 (1991).

327. A. Leclaire, M. M. Borel, A. Grandin, and B. Raveau, *Z. Kristallogr. 184*, 247–255 (1989).

328. A. Leclaire, M. M. Borel, A. Grandin, and B. Raveau, *J. Solid State Chem. 78*, 220–226 (1989).

329. S. A. Linde, Yu. E. Gorbunova, A. V. Lavrov, and I. V. Tananaev, *Dokl. Akad. Nauk SSSR 250*, 96–99 (1980).

330. C. Gueho, M. M. Borel, A. Grandin, A. Leclaire, and B. Raveau, *Z. anorg. allg. Chem. 615*, 104–108 (1992).

331. V. P. Nikolaev, G. G. Sadikov, A. V. Lavrov, and M. A. Porai-Koshits, *Dokl. Akad. Nauk SSSR 264*, 859–862 (1982).

332. M. T. Averbuch-Pouchot, *Acta Crystallogr., Sect. C 44*, 2046–2048 (1988).

333. M. T. Averbuch-Pouchot. *J. Solid State Chem. 79*, 296–299 (1989).

334. H. Barten and E. H. P. Cordfunken, *J. Inorg. Nucl. Chem. 42*, 75–78 (1980).

335. H. Barten and E. H. P. Cordfunken, *Thermochim. Acta 40*, 357–365 (1980).

336. A. V. Lavrov, A. B. Pobenida, and I. A. Rozanov, *Russ. J. Inorg. Chem. 25*, 583–587 (1980).

337. S. A. Linde, Yu. U. Gorbunova, A. V. Lavrov, and A. B. Pobedina, *Russ. J. Inorg. Chem. 29*, 879–881 (1984).

338. A. V. Lavrov, A. B. Pobedina, I. A. Rozanov, and I. V. Tananaev, *Russ. J. Inorg. Chem. 28*, 1070–1072 (1983).

339. A. V. Lavrov, A. B. Pobedina, and I. A. Rozanov, *Russ. J. Inorg. Chem. 25*, 583–587 (1980).

340. S. A. Linde, Yu. E. Gorbunova, A. V. Lavrov, and A. B. Pobedina, *Izv. Akad. Nauk SSSR, Neorg. Mater. 17*, 1062–1066 (1981).
341. J. M. Adams and V. Ramdas, *Acta Crystallogr., Sect. B 32*, 3224–3227 (1976).
342. J. M. Adams and V. Ramdas, *Acta Crystallogr., Sect. B 33*, 3654–3657 (1977).
343. J. M. Adams and V. Ramdas, *Acta Crystallogr., Sect. B 34*, 2150–2156 (1978).
344. S. Kamoun, A. Jouini, and A. Daoud, *C. R. Acad. Sci., Ser. 2 308*, 923–925.
345. S. Kamoun, A. Jouini, and A. Daoud, *J. Solid State Chem. 99*, 18–28 (1992).
346. M. T. Averbuch-Pouchot and A. Durif, *Eur. J. Solid State Inorg. Chem., 29*, 411–418 (1992).
347. M. T. Averbuch-Pouchot and A. Durif, *C. R. Acad. Sci., Ser. 2 316*, 187–192 (1993).
348. A. Gharbi, A. Jouini, M. T. Averbuch-Pouchot, and A. Durif, *J. Solid State Chem., 111*, 330-337 (1994).
349. M. T. Averbuch-Pouchot and A. Durif, *Acta Crystallogr., Sect. B 39*, 27–28 (1983).
350. M. T. Averbuch-Pouchot and A. Durif, *Acta Crystallogr., C48*, 973–975 (1992).
351. M. T. Averbuch-Pouchot and A. Durif, *Acta Crystallogr., C48*, 1912 (1992).
352. M. T. Averbuch-Pouchot and A. Durif, *Eur. J. Solid State Inorg. Chem. 30*, 1153–1162 (1993).
353. S. Gali, K. Byrappa, and G. S. Gopalakrishna, *Acta Crystallogr., Sect. C 45*, 1667–1669 (1989).
354. R. E. Marsh, *Acta Crystallogr., Sect. C 46*, 2497–2499 (1990).
355. J. Durand, H. Falius, J. L. Galigné, and L. Cot, *J. Solid State Chem. 24*, 345–349 (1978).
356. J. D. Garrett, J. E. Greedan, R. Faggiani, S. Carbotte, and I. D. Brown, *J. Solid State Chem. 42*, 183–190 (1982).
357. M. Trojan, D. Brandova, J. Fabry, J. Hybler, K. Jurek, and V. Petricek, *Acta Crystallogr., Sect. C 43*, 2038–2040 (1987).
358. G. W. Morey and E. Ingerson, *Am. J. Sci. 242*, 1 (1944).
359. E. Ingerson and G. W. Morey, *Am. Mineral. 28*, 48 (1943).
360. G. W. Morey, *J. Am. Chem. Soc. 76*, 4724 (1954).
361. V. A. Sotnikova-Yuzhik, G. V. Pelsyak and E. A. Prodan, *Russ. J. Inorg. Chem. 32*, 1505–1507 (1987).
362. J. J. Dymon and A. J. King, *Acta Crystallogr. 4*, 378–379 (1951).
363. D. E. C. Corbridge, *Acta Crystallogr. 13*, 263–269 (1960).
364. P. Bonneman and M. Bassiere, *C. R. Acad. Sci. 206*, 1379–1380 (1938).
365. D. R. Dyroff, Thesis, California Institute of Technology, Pasadena (1965).
366. D. M. Wiench, M. Jansen, and R. Hoppe, *Z. anorg. allg. Chem. 488*, 80–86 (1982).
367. J. D. Lee, *J. Chem. Soc. A, 12*, 2881–2882 (1968).
368. A. Lamotte and J. C. Merlin, *Bull. Soc. Chim. Fr.* 1968, 4311–4312.
369. T. D. Farr, J. D. Fleming, and J. D. Hatfield, *J. Chem. Eng. Data. 12*, 141–142 (1967).
370. K. R. Waerstad and G. H. McClellan, *J. Appl. Crystallogr. 7*, 404–405 (1974).
371. V. A. Sotnikova-Yuzhik and E. A. Prodan, *Russ. J. Inorg. Chem. 26*, 848–849 (1981).
372. I. L. Shashkova, V. A. Lyutsko, and E. A. Prodan, *Izv. Akad. Nauk SSSR, Neorg. Mater. 23*, 986–991 (1987).
373. V. A. Lyutsko, I. L. Shashkova, and E. A. Prodan, *Zh. Prikl. Spektrosk. 3*, 415–420 (1982).
374. E. A. Prodan and I. L. Shashkova, *Russ. J. Inorg. Chem. 28*, 180–183 (1983).
375. E. A. Prodan and N. V. Bulavkina, *Izv. Akad. Nauk SSSR, Neorg. Mater. 17*, 1662–1667 (1981).
376. E. A. Prodan, I. L. Shashkova, and L. A. Lesnikovich, *Russ. J. Inorg. Chem. 24*, 1454–1458 (1979).
377. I. L. Shashkova, E. A. Prodan, and V. A. Lyutsko, *Izv. Akad. Nauk SSSR, Neorg. Mater. 24*, 259–263 (1988).
378. V. Lyutsko and G. Johansson, *Acta Chem. Scand. A38*, 663–669 (1988).
379. E. L. Krivoviazov, V. P. Volkova, and N. K. Vostresenskaya, *Izv. Akad. Nauk SSSR, Neorg. Mater. 9*, 761–765 (1970).
380. V. A. Sotnikova-Yuzhik, *Izv. Akad. Nauk. SSSR, Neorg. Mater. 27*, 1011–1013 (1991).
381. N. V. Bulavkina, E. A. Prodan, and L. I. Petrovskaya, *Russ. J. Inorg. Chem. 29*, 1104–1106 (1984).
382. E. J. Griffith and R. L. Buxton, *J. Chem. Eng. Data 13*, 145–148 (1968).
383. E. A. Prodan, L. I. Petrovskaya, and V. N. Korzhuev, *Russ. J. Inorg. Chem. 25*, 1013–1017 (1980).

384. E. A. Prodan, B. M. Galogadja, P. N. Petruskaia, and B. H. Kordjev, *Dokl. Akad. Nauk. SSSR 25*, 163–165 (1981).

385. M. T. Averbuch-Pouchot and A. Durif, *Acta Crystallogr., Sect. C 41*, 1553–1555 (1985).

386. M. T. Averbuch-Pouchot and J. C. Guitel, *Acta Crystallogr., Sect. B 32*, 1670–1673 (1976).

387. M. T. Averbuch-Pouchot and A. Durif, *J. Appl. Crystallogr. 8*, 564 (1975).

388. M. T. Averbuch-Pouchot, A. Durif, and J. C. Guitel, *Acta Crystallogr., Sect. B 31*, 2482–2486 (1975).

389. H. Worzala and K. H. Jost, *Z. anorg. allg. Chem. 445*, 36–46 (1978).

390. M. Bagieu-Beucher, A. Durif, and M. T. Averbuch-Pouchot, *J. Appl. Crystallogr. 9*, 52 (1976).

391. M. T. Averbuch-Pouchot, A. Durif, J. Coing-Boyat, and J. C. Guitel, *Acta Crystallogr., Sect. B 33*, 203–205 (1977).

392. A. S. Lyakhov, V. A. Lyutsko, L. I. Prodan, and K. K. Palkina, *Izv. Akad. Nauk SSSR, Neorg. Mater. 27*, 1014–1018 (1991).

393. H. Huber, *Angew. Chem. 50*, 323–330 (1937).

394. D. E. C. Corbridge and F. R. Tromans, *Anal Chem. 30*, 1101–1110 (1968).

395. M. T. Averbuch-Pouchot and J. C. Guitel, *Acta Crystallogr., Sect. B 33*, 1427–1431 (1977).

396. E. A. Prodan and Yu. G. Zonov, *Russ. J. Inorg. Chem. 17*, 635–638 (1972).

397. P. Bonneman-Bémia, *C. R. Acad. Sci. 207*, 214–216 (1939).

398. E. Rakotomahanina, M. T. Averbuch-Pouchot, and A. Durif, *Bull. Soc. fr. Minéral. Cristallogr. 95*, 516–520 (1972).

399. M. Herceg, 2nd European Meeting of Crystallography, Keszthely, Hungary (1974).

400. A. Jouini and A. Durif, *C. R. Acad. Sci., Ser. 2 297*, 573–575 (1983).

401. A. Jouini, M. Dabbabi, M. T. Averbuch-Pouchot, A. Durif, and J. C. Guitel, *Acta Crystallogr., Sect. C 40*, 728–730 (1984).

402. V. Lyutsko and G. Johansson, *Acta. Chem. Scand. A38*, 415–417 (1984).

403. W. Dewald, *Z. anorg. allg. Chem. 298*, 279–284 (1959).

404. M. T. Averbuch-Pouchot and J. C. Guitel, *Acta Crystallogr., Sect. B 32*, 2270–2274 (1976).

405. E. A. Prodan and O. P. Ol'Shevskaya, *Izv. Akad. Nauk SSSR, Neorg. Mater. 15*, 1997–2002 (1979).

406. O. P. Ol'Shevskaya and E. A. Prodan, *Russ. J. Inorg. Chem. 31*, 348–351 (1986).

407. A. S. Lyakhov, V. A. Lyutsko, T. N. Galkova, and K. K. Palkina, *Russ. J. Inorg. Chem. 38*, 1174–1178 (1993).

408. O. P. Ol'shevskaya, T. N. Galkova, and E. A. Prodan, *Russ. J. Inorg. Chem. 27*, 947–949 (1982).

409. T. N. Galkova and E. A. Prodan, *Russ. J. Inorg. Chem. 30*, 798–801 (1985).

410. E. A. Prodan and T. N. Galkova, *Russ. J. Inorg. Chem. 25*, 663–666 (1980).

411. T. N. Galkova and E. A. Prodan, *Russ. J. Inorg. Chem. 32*, 675–678 (1987).

412. E. A. Prodan, T. N. Galkova, M. M. Pavlyuchenko, and N. N. Sadovnikova, *Russ. J. Inorg. Chem. 24*, 1190–1193 (1979).

413. Z. A. Konstant, I. Sikach, and A. P. Dindune, *Izv. Akad. Nauk SSSR, Neorg. Mater. 20*, 1893–1897 (1984).

414. A. S. Lyakhov, V. A. Lyutsko, T. N. Galkova, and K. K. Palkina, *Russ. J. Inorg. Chem. 36*, 1715–1718 (1991).

415. O. P. Ol'Shevskaya and E. A. Prodan, *Russ. J. Inorg. Chem. 33*, 1427–1428 (1988).

416. V. A. Lyutsko, A. S. Lyakhov, G. K. Tuchkovskii, and K. K. Palkina, *Russ. J. Inorg. Chem. 36*, 662–664 (1981).

417. A. F. Selevich and V. D. Lyutsko, *Russ. J. Inorg. Chem. 29*, 364–369 (1984).

418. N. N. Chudinova, M. A. Avaliani, and L. S. Guzeeva, *Izv. Akad. Nauk SSSR, Neorg. Mater. 13*, 2229–2233 (1977).

419. P. P. Mel'Nikov, V. A. Efremov, A. K. Stepanov, T. S. Romanova, and L. N. Komissarova, *Russ. J. Inorg. Chem. 21*, 26–28 (1976).

420. Z. Ya. Kanepe and Z. A. Konstant, *Izv. Akad. Nauk SSSR, Neorg. Mater. 19*, 969–971 (1983).

421. K. K. Palkina, S. I. Maksimova, and V. G. Kuznetsov, *Izv. Akad. Nauk SSSR, Neorg. Mater. 15*, 2168–2170 (1979).

422. A. S. Lyakhov, K. K. Palkina, V. A. Lyutsko, S. I. Maksimova, and N. T. Chibiskova, *Izv. Akad. Nauk SSSR, Neorg. Mater. 26*, 1064–1068 (1990).

423. M. T. Averbuch-Pouchot and J. C. Guitel, *Acta Crystallogr., Sect. B 33*, 1613–1615 (1977).

424. E. A. Genkina, Yu. A. Gorbunov, B. A. Maksimov, A. A. Shternberg, and O. K. Mel'Kinov, *Sov. Phys. Crystallogr. 29*, 128–130 (1984).

425. V. A. Lyutsko, M. Nikanovich, and K. Lapko, *Izv. Akad. Nauk SSSR, Neorg. Mater. 16*, 1613–1617 (1980).

426. V. A. Lyutsko, A. V. Tuchkovskaya, and A. L. Shifrina, *Russ. J . Inorg. Chem. 32*, 1541–1544 (1987).

427. V. A. Lyutsko, A. V. Tuchkovskaya, and A. L. Shifrina, *Russ. J. Inorg. Chem. 33*, 1431–1433 (1988).

428. G. V. Rodicheva, E. N. Deichman, I. V. Tananaev, and V. V. Klimov, *Russ. J. Inorg. Chem. 17*, 1199–1201 (1972).

429. E. Giesbrecht and E. B. Melardi, *An. Acad. Bras. Cienc. 35*, 527–532 (1963).

430. G. V. Rodicheva, I. V. Tananaev, and N. M. Romanova, *Russ. J. Inorg. Chem. 33*, 274–277 (1988).

431. S. M. Petushkova, I. V. Tananaev, and S. O. Samilova, *Russ. J. Inorg. Chem. 16*, 61–63 (1971).

432. V. G. Kuznetsov, S. M. Petushkova, and I. V. Tananaev, *Russ. J. Inorg. Chem. 17*, 663–665 (1972).

433. G. V. Rodicheva, I. V. Tananaev, and N. M. Romanova, *Izv. Akad. Nauk SSSR, Neorg. Mater. 17*, 126–130 (1981).

434. G. V. Rodicheva, I. V. Tananaev, and N. M. Romanova, *Izv. Akad. Nauk. SSSR, Neorg. Mater. 16*, 1458–1461 (1980).

435. G. V. Rodicheva and N. M. Romanova, *Izv. Akad. Nauk SSSR, Neorg. Mater. 15*, 963–968 (1979).

436. S. M. Petushkova, V. G. Kuznetsov, I. V. Tananaev, and S. O. Samoilova, *Russ. J. Inorg. Chem. 17*, 1234–1236 (1972).

437. M. T. Averbuch-Pouchot, A. Durif, and J. C. Guitel, *Acta. Crystallogr. B33*, 1436–1438 (1977).

438. V. A. Lyutsko and O. G. Pap, *Russ. J. Inorg. Chem. 34*, 665–668 (1989).

439. V. A. Lyutsko, O. G. Pap, and N. M. Ksenofontova, *Izv. Akad. Nauk SSSR, Neorg. Mater. 22*, 1773–1777 (1986).

440. M. A. Avaliani, *Izv. Akad. Nauk SSSR, Neorg. Mater. 26*, 2647–2648 (1990).

441. B. Klinkert and M. Jansen, *Z. anorg. allg. Chem. 567*, 77–86 (1988).

442. A. I. Teterevkov and G. K. Mikhailovskaya, *Russ. J. Inorg. Chem. 25*, 781–782 (1980).

443. V. V. Krasnikov, Z. A. Konstant, and V. S. Fundamenskii, *Izv. Akad. Nauk SSSR, Neorg. Mater. 19*, 1373–1378 (1983).

444. M. A. Vaivada, Z. A. Konstant, and V. V. Krasnikov, *Izv. Akad. Nauk. SSSR, Neorg. Mater. 21*, 1555–1559 (1985).

445. I. Grunze and N. N. Chudinova, *Izv. Akad. Nauk SSSR, Neorg. Mater. 24*, 988–993 (1988).

446. N. N. Chudinova, I. V. Tananaev, and M. A. Avaliani, *Izv. Akad. Nauk SSSR, Neorg. Mater. 13*, 2234–2235 (1977).

447. N. N. Chudinova, M. A. Avaliani, L. S. Guzeeva, and I. V. Tananaev, *Izv. Akad. Nauk SSSR, Neorg. Mater. 14*, 2054–2060 (1978).

448. N. N. Chudinova, I. Grunze, L. S. Guzeeva, and M. A. Avaliani, *Izv. Akad. Nauk SSSR, Neorg. Mater. 23*, 604–609 (1987).

449. V. A. Lyutsko, O. G. Pap, and T. I. Prisedskaya, *Izv. Akad. Nauk SSSR, Neorg. Mater. 24*, 693–694 (1988).

450. M. A. Avaliani, N. N. Chudinova, and I. V. Tananaev, *Izv. Akad. Nauk SSSR, Neorg. Mater. 15*, 1688–1689 (1979).

451. N. V. Vinogradova and N. N. Chudinova, *Izv. Akad. Nauk SSSR, Neorg. Mater. 19*, 116–119 (1983).

452. N. N. Chudinova, L. P. Shklover, L. I. Shkol'Nikova, A. E. Balanevskaya, and G. M. Balagina, *Izv. Akad. Nauk SSSR, Neorg. Mater. 14*, 1324–1328 (1978).

453. M. A. Vaivada and Z. A. Konstant, *Izv. Akad. Nauk SSSR, Neorg. Mater. 22*, 2026–2028 (1986).

454. N. N. Chudinova, N. V. Vinogradova, G. M. Balagina, and K. K. Palkina, *Izv. Akad. Nauk SSSR, Neorg. Mater. 13*, 1494–1499 (1977).

455. W. Hilmer, N. N. Chudinova, K. H. Jost, and N. V. Vinogradova, *Izv. Akad. Nauk SSSR, Neorg. Mater. 15*, 332–334 (1979).

456. W. Hilmer, N. N. Chudinova, K. H. Jost, and N. V. Vinogradova, *Izv. Akad. Nauk SSSR, Neorg. Mater.* 15, 1123–1125 (1979).

457. M. T. Averbuch-Pouchot and M. Bagieu-Beucher, *Z. anorg. allg. Chem.* 552, 171–180 (1987).

458. V. A. Lyutsko, M. V. Nikanovich, K. N. Lapko, and V. F. Tikavyi, *Russ. J. Inorg. Chem.* 28, 1105–1109 (1983).

459. S. I. Berul and N. K. Voskresenskaya, *Izv. Akad. Nauk SSSR, Neorg. Mater.* 3, 534–538 (1967).

460. T. N. Galkova and E. A. Prodan, *Russ. J. Inorg. Chem.* 32, 1318–1321 (1987).

461. S. M. Petushkova, I. V. Tananaev, and S. O. Samoilova, *Russ. J. Inorg. Chem.* 16, 61–63 (1971).

462. G. V. Rodicheva, I. V. Tananaev, and N. M. Romanova, *Russ. J. Inorg. Chem.* 33, 274–277 (1988).

463. G. V. Rodicheva and N. M. Romanova, *Izv. Akad. Nauk SSSR, Neorg. Mater.* 15, 963–968 (1979).

464. G. V. Rodicheva, I. V. Tananaev, and N. M. Romanova, *Izv. Akad. Nauk SSSR, Neorg. Mater.* 17, 126–130 (1981).

465. I. V. Tananaev, A. V. Lavrov, N. N. Chudinova, and V. G. Kuznetsov, *Izv. Akad. Nauk SSSR, Neorg. Mater.* 4, 1966–1971 (1968).

466. E. Giesbrecht and G. Vicentini, *An. Assoc. Bras. Quim.* 19, 61–72 (1960).

467. Z. Ruzic-Toros, B. Kojic-Prodic, R. Liminga, and S. Popovic, *Inorg. Chim. Acta* 8, 273–278 (1974).

468. R. K. Osterheld and R. P. Langguth, *J. Phys. Chem.* 59, 76–80 (1955).

469. R. O. Langguth, R. K. Osterheld, and E. Karl-Kroupa, *J. Phys. Chem.* 60, 1335–1336 (1956).

470. O. T. Quimby, *J. Phys. Chem.* 58, 615–624 (1954).

471. J. I. Watters, P. E. Sturrock, and R. E. Simonaitis, *Inorg. Chem.* 2, 765–767 (1963).

472. F. Schulz and M. Jansen, *Z. anorg. allg. Chem.* 543, 152–160 (1986).

473. E. J. Griffith, *J. Inorg. Nucl. Chem.* 26, 1381–1383 (1964).

474. M. T. Averbuch-Pouchot and A. Durif, *J. Solid State Chem.* 18, 391–393 (1976).

475. K. R. Waerstad and G. H. McClellan, *J. Appl. Crystallogr.* 7, 404–405 (1974).

476. T. D. Farr, J. W. Williard, and J. D. Hatfield, *J. Chem. Eng. Data.* 17, 313–317 (1972).

477. A. H. Mc Keag and E. G. Steward, *Br. J. Appl. Phys.*, suppl. 4, p. S26-S31 (1955).

478. R. A. McCauley and F. A. Hummel, *Trans. Br. Ceram. Soc.* 67, 619–625 (1968).

479. *Powder Diffraction 1*, 80 (1986).

480. J. M. Millet, H. S. Parker, and R. S. Roth, *J. Am. Ceram. Soc.* 69C, 103–105 (1986).

481. B. M. Gatehouse, S. N. Platts, and R. S. Roth, *Acta Crystallogr., Sect. C* 47, 2285–2287 (1991).

482. M. T. Averbuch-Pouchot and A. Durif, *C. R. Acad. Sci., Ser.* 2 303, 543–545 (1986).

483. M. T. Averbuch-Pouchot and A. Durif, *Acta Crystallogr., Sect. C* 43, 631–632 (1987).

484. I. Schulz, *Z. anorg. allg. Chem.* 257, 106–112 (1956).

485. J. F. Argyle and F. A. Hummel, *J. Am. Ceram. Soc.* 43, 542–547 (1960).

486. K. H. Lii, Y. B. Chen, C. C. Su, and S. L. Wang, *J. Solid State Chem.* 82, 156–160 (1989).

487. N. Hilmer, N. N. Chudinova, and K. H. Jost, *Izv. Akad. Nauk SSSR, Neorg. Mater.* 8, 1507–1515 (1978).

488. M. Bagieu-Beucher and M. T. Averbuch-Pouchot, *Z. Kristallogr.* 180, 165–170 (1987).

489. D. Agrawal and F. A. Hummel, *J. Electrochem. Soc.* 1550–1554 (1980).

490. H. D. Park and E. R. Kreidler, *J. Am. Ceram. Soc.* 67, 23–26 (1984).

491. G. V. Rodicheva, E. N. Deichman, I. V. Tananaev, and Zh. K. Shaidarbekova, *Russ. J. Inorg. Chem.* 22, 1647–1650 (1977).

492. G. V. Rodicheva, E. N. Deichman, I. V. Tananaev, and Zh. K. Shaidarbekova, *Russ. J. Inorg. Chem.* 20, 1316–1318 (1975).

493. G. V. Rodicheva, E. N. Deichman, I. V. Tananaev, and Zh. K. Shaidarbekova, *Russ. J. Inorg. Chem.* 19, 814–817 (1974).

494. G. V. Rodicheva, E. N. Deichman, I. V. Tananaev, and Zh. K. Shaidarbekova, *Russ. J. Inorg. Chem.* 19, 1467–1470 (1974).

495. M. T. Averbuch-Pouchot, *J. Appl. Crystallogr.* 10, 200 (1977).

496. V. P. Nikolaev, G. G. Sadikov, A. V. Lavrov, and M. A. Porai-Koshits, *Izv. Akad. Nauk SSSR, Neorg. Mater.* 22, 1364–1368 (1986).

497. L. Kh. Minacheva, A. S. Antsyshkina, A. V. Lavrov, V. G. Sakhrova, V. P. Nikolaev, and M. A. Porai-Koshits, *Russ. J. Inorg. Chem.* 24, 51–53 (1979).
498. A. V. Lavrov, M. Ya. Voitenko, and L. A. Tezikova, *Izv. Akad. Nauk SSSR, Neorg. Mater.* 14, 2073–2077 (1978).
499. M. T. Averbuch-Pouchot, *Z. anorg. allg. Chem.* 545, 118–124 (1987).
500. J. Majling and F. Hanic, *J. Appl. Crystallogr.* 12, 244 (1979).
501. B. Klinkert and M. Jansen, *Z. anorg. allg. Chem.* 567, 87–94 (1988).
502. G. G. Sadikov, V. P. Nikolaev, A. V. Lavrov, and M. A. Porai-Koshits, *Dokl. Akad. Nauk SSSR* 266, 354–358 (1982).
503. F. Liebau, *Structural Chemistry of Silicates*, Springer-Verlag, Berlin, (1985).
504. B. M. Beglov, *Izv. Akad. Nauk SSSR, Neorg. Mater.* 13, 2236–2241 (1977).
505. U. Schülke and R. Kayser, *Z. anorg. allg. Chem.* 531, 167–176 (1985).
506. E. Thilo and H. Grunze, *Z. anorg. allg. Chem.* 281, 262–283 (1955).
507. J. C. Grenier and A. Durif, *Z. Kristallogr.* 137, 10–16 (1973).
508. J. C. Guitel and I. Tordjman, *Acta Crystallogr., Sect. B* 32, 2960–2966 (1976).
509. R. ßenkhoucha and B. Wunderlich, *Acta. Crystallogr. B*35, 265–267 (1979).
510. V. G. Zakzhevskii, A. I. Boldyrev, and O. P. Charkin, *Russ. J. Inorg. Chem.* 25, 1443–1446 (1980).
511. D. E. C. Corbridge, *Acta Crystallogr.* 8, 520 (1955).
512. K. H. Jost, *Acta Crystallogr.* 14, 844–847 (1961).
513. K. H. Jost, *Acta Crystallogr.* 16, 640–642 (1963).
514. A. McAdam, K. H. Jost, and B. Beagley, *Acta Crystallogr., Sect. B* 24, 1621–1622 (1968).
515. A. Immirzi and W. Porzio, *Acta Crystallogr., Sect. B* 38, 2788–2792 (1982).
516. K. R. Andress and K. Fischer, *Z. anorg. allg. Chem.* 273, 193–199 (1953).
517. K. H. Jost, *Acta Crystallogr.* 16, 623–626 (1963).
518. K. H. Jost and H. J. Schülze, *Acta Crystallogr., Sect. B* 25, 1110–1118 (1969).
519. K. H. Jost and H. J. Schülze, *Acta Crystallogr., Sect. B* 27, 1345–1353 (1971).
520. A. B. Bekturov, D. Z. Serazetdinov, E. V. Poletaev, and S. M. Divnenko, *Russ. J. Inorg. Chem.* 13, 20–23 (1968).
521. N. M. Dombrovskii and V. A. Koval, *Izv. Akad. Nauk SSSR, Neorg. Mater.* 12, 738–741 (1976).
522. E. P. Egan and Z. T. Wakefield, *J. Phys. Chem.* 64, 1955 (1960).
523. K. H. Jost, *Acta Crystallogr.* 14, 779–784 (1961).
524. H. N. Terem and S. Akalan, *C. R. Acad. Sci.* 228, 1437–1441 (1949).
525. E. V. Margulis, L. J. Beisekeeva, and M. A. Fishman, *J. Appl. Chem. (USSR)*, 39, 2216–2220 (1966).
526. S. J. Kiehl and T. H. Hill, *J. Am. Chem. Soc.* 49, 123–127 (1927).
527. V. A. Kopilevich and L. N. Shegrov, *Russ. J. Inorg. Chem.* 32, 1261–1264 (1987).
528. C. Y. Shen, N. E. Stahlheber, and D. R. Dyroff, *J. Am. Chem. Soc.* 91, 62–67 (1969).
529. D. W. J. Cruickshank, *Acta Crystallogr.* 17, 681–682 (1964).
530. N. N. Chudinova, L. A. Borodina, U. Schülke, and K. H. Jost, *Izv. Akad. Nauk SSSR, Neorg. Mater.* 25, 459–465 (1989).
531. K. Dostal, V. Kocman, and V. Ehrenbergrova, *Z. anorg. allg. Chem.* 367, 80–91 (1969).
532. N. El-Horr, *J. Solid State Chem.* 90, 386–387 (1991).
533. A. S. Alikhanyan, A. V. Steblevskii, I. D. Sokolova, and V. I. Gorgoraki, *Russ. J. Inorg. Chem.* 22, 335–338 (1977).
534. I. V. Mardirosova and G. A. Bukhalova, *Russ. J. Inorg. Chem.* 11, 1275–1277 (1966).
535. C. Cavero-Ghersi, Thesis, Grenoble University, Grenoble France (1975).
536. N. El-Horr, C. Cavero-Ghersi, and M. Bagieu-Beucher, *C. R. Acad. Sci., Ser. 2* 297 479–482 (1983).
537. N. El-Horr, M. Bagieu, and I. Tordjman, *Acta Crystallogr., Sect. C* 39, 1597–1599 (1983).
538. N. El-Horr and M. Bagieu, *Acta Crystallogr., Sect. C* 41, 1157–1159 (1985).
539. M. T. Averbuch-Pouchot, A. Durif, and J. C. Guitel, *Acta Crystallogr., Sect. B* 32, 2440–2443 (1976).
540. N. El Horr and M. Bagieu, *C. R. Acad. Sci., Ser. 2* 312, 373–375 (1991).
541. N. El-Horr and M. Bagieu, *Acta Crystallogr., Sect. C* 43, 603–605 (1987).

542. J. Majling and F. Hanic, in: *Topics in Phosphorus Chemistry*, Vol. 10, Grayson and Griffith editors, 341–502, Interscience New York (1980).

543. M. T. Averbuch-Pouchot, *J. Solid State Chem. 102*, 93–99 (1993).

544. M. A. Savenkova, I. V. Mardirosova, and E. V. Poletaev, *Russ. J. Inorg. Chem. 20*, 1374–1376 (1975).

545. M. A. Savenkova, L. V. Kubasova, I. V. Mardirosova, and E. V. Poletaev, *Izv. Akad. Nauk SSSR, Neorg. Mater. 11*, 2200–2202 (1975).

546. I. D. Sokolova, *Russ. J. Inorg. Chem. 11*, 502–503 (1966).

547. I. D. Sokolova, E. L. Krivovyazov, and N. K. Voskresenskaya, *Russ. J. Inorg. Chem. 8*, 1375–1378 (1963).

548. A. V. Steblevskii, A. S. Alikhanyan, I. D. Sokolova, and V. I. Gorgoraki, *Russ. J. Inorg. Chem. 22*, 11–13 (1977).

549. S. Ohashi and K. Yamagishi, *J. Jpn. Chem. Soc. 33*, 1431–1435 (1960).

550. N. S. Slobodyanik, P. G. Nagornyi, and T. I. Zhunkovskaya, *Russ. J. Inorg. Chem. 26*, 838–839 (1981).

551. I. B. Markina and N. K. Voskresenskaya, *Russ. J. Inorg. Chem. 12*, 407–411 (1967).

552. V. P. Kochergin, Z. A. Shevrina, L. V. Paderova, and A. N. Kruglov, *Russ. J. Inorg. Chem. 22*, 22–23 (1977).

553. G. A. Bukhalova and I. V. Mardirosova, *Russ. J. Inorg. Chem. 11*, 85–87 (1966).

554. G. A. Bukhalova and I. V. Mardirosova, *Russ. J. Inorg. Chem. 12*, 1158–1161 (1967).

555. M. Amadori, *Atti Accad. Naz. Lincei, Sci. Fis. Mat. Nat. 21*, 588–599 (1913).

556. E. N. Kovarskaya and Yu. I. Rodionov, *Izv. Akad. Nauk SSSR, Neorg. Mater. 24*, 642–645 (1988).

557. E. N. Kovarskaya, V. S. Mityakhina, Yu. I. Rodionov, and M. Yo. Silin, *Izv. Akad. Nauk SSSR, Neorg. Mater. 24,* 655–660 (1988).

558. E. Thilo and I. Grunze, *Z. anorg. allg. Chem.*, 290, 209–222 (1957).

559. S. Jaulmes, *Rev. Chim. Minér. 1*, 617–671 (1964).

560. M. Bagieu-Beucher and A. Durif, *Bull. Soc. fr. Minéral. Crist. 93*, 129–130 (1970).

561. M. T. Averbuch-Pouchot, A. Durif, and I. Tordjman, *Acta Crystallogr., Sect. B 33*, 3462–3464 (1977).

562. E. Schultz, Dissertation, University of Kiel, Kiel Germany (1974).

563. E. Schultz and F. Liebau, *Naturwiss.*, 429–430 (1973).

564. E. Schultz and F. Liebau, *Z. Kristallogr. 154*, 115–126 (1981).

565. M. Bagieu-Beucher, M. Gondrand, and M. Perroux, *J. Solid. State Chem. 19*, 353–357 (1976).

566. M. T. Averbuch-Pouchot, A. Durif, and M. Bagieu-Beucher, *Acta Crystallogr., Sect. C 39*, 25–26 (1983).

567. E. L. Krivovyazov, I. D. Sokolova, and N. K. Voskresenskaya, *Izv. Akad. Nauk SSSR, Neorg. Mater. 3*, 530–533 (1967).

568. W. L. Hill, G. T. Faust, and D. S. Reynolds, *Am. J. Sci. 242*, 542–562 (1944).

569. G. Trömel, *Stahl Eisen 63*, 21–30 (1943).

570. G. Trömel, H. J. Harkort, and W. Hotop, *Z. anorg. Chem. 256*, 253–272 (1948).

571. P. E. Stone, E. P. Egan, and J. R. Lehr, *J. Am. Ceram. Soc. 39,* 89–98 (1956).

572. E. Thilo and I. Grunze, *Z. anorg. Chem. 290*, 223–237 (1957).

573. C. Morin, *Bull. Soc. Chim. Fr.* (1961), 1726–1734.

574. M. Schneider, K. H. Jost, and P. Leibnitz, *Z. anorg. allg. Chem. 527*, 99–104 (1985).

575. W. Rothammel, F. H. Burzlaff, and R. Specht, *Acta Crystallogr., Sect. C 45*, 551–553 (1989).

576. E. P. Egan and Z. T. Wakefield, *J. Am. Chem. Soc. 78*, 4245–4249 (1956).

577. E. H. Brown, J. R. Lehr, J. P. Smith, W. E. Brown, and A. W. Frazier, *J. Phys. Chem. 61*, 1669–1670 (1957).

578. E. O. Huffman and J. D. Fleming, *J. Phys. Chem. 64*, 240–244 (1960).

579. J. J. Brown and F. A. Hummel, *J. Electrochem. Soc. 111*, 660–665 (1964).

580. M. Beucher and I. Tordjman, *Bull. Soc. fr. Minéral. Cristallogr. 91*, 207–209 (1968).

581. I. Tordjman, M. Beucher, J. C. Guitel, and G. Bassi, *Bull. Soc. fr. Minéral. Cristallogr. 91*, 344–349 (1968).

582. M. Bagieu-Beucher, J. C. Guitel, I. Tordjman, and A. Durif, *Bull. Soc. fr. Minéral. Cristallogr. 97*, 481–484 (1974).

583. M. Laügt, M. Bagieu-Beucher, and J. C. Grenier, *C. R. Acad. Sci. 275C*, 1283–1285 (1972).

584. M. Bagieu-Beucher, M. Brunel-Laügt, and J. C. Guitel, *Acta Crystallogr., Sect. B 35*, 292–295 (1979).

585. R. C. Ropp and M. A. Aia, *Anal. Chem. 34*, 1288–1291 (1962).

586. A. Durif, M. Bagieu-Beucher, C. Martin, and J. C. Grenier, *Bull. Soc. fr. Minéral. Cristallogr. 95*, 146–148 (1972).

587. J. C. Grenier and C. Martin, *Bull. Soc. fr. Minéral. Cristallogr. 98*, 107–110 (1975).

588. J. C. Grenier, C. Martin, A. Durif, Tranqui D., and J. C. Guitel, *Bull. Soc. fr. Minéral. Cristallogr. 90*, 24–31 (1967).

589. J. Coing-Boyat, M. T. Averbuch-Pouchot, and J. C. Guitel, *Acta Crystallogr., Sect. B 34*, 2689–2692 (1978).

590. M. I. Kuz'menkov, V. V. Pechkovskii, and S. V. Plyshevskii, *Russ. J. Inorg. Chem. 17*, 985–987 (1972).

591. S. I. Lopatin and G. A. Semenov, *Izv. Akad. Nauk SSSR, Neorg. Mater. 25*, 1404–1406 (1989).

592. K. H. Jost, *Acta Crystallogr. 17*, 1539–1544 (1964).

593. M. Beucher, Dissertation, University of Grenoble, Grenoble France (1968).

594. K. K. Palkina, S. I. Maksimova, A. V. Lavrov, and N. A. Chalisova, *Sov. Phys. Dokl. 23*, 691–692 (1978).

595. M. Laügt, *C. R. Acad. Sci. 275C*, 1197–1200 (1972).

596. M. Laügt and J. C. Guitel, *Acta Crystallogr., Sect. B 31*, 1148–1153 (1975).

597. I. V. Mardirosova, V. A. Matrosova, M. A. Savenkova, and G. A. Bukhalova, *Izv. Akad. Nauk SSSR, Neorg. Mater. 15*, 2079–2081 (1979).

598. M. Bagieu-Beucher and N. El-Horr, unpublished results.

599. G. A. Bukhalova, I. A. Tokman, and V. M. Shpakova, *Russ. J. Inorg. Chem. 15*, 865–866 (1970).

600. M. T. Averbuch-Pouchot, *J. Appl. Crystallogr. 8*, 389–390 (1975).

601. M. T. Averbuch-Pouchot, A. Durif, and J. C. Guitel, *Acta Crystallogr., Sect. B 31*, 2453–2456 (1975).

602. I. A. Tokman and E. V. Poletaev, *Izv. Akad. Nauk SSSR, Neorg. Mater. 12*, 735–737 (1976).

603. J. F. Sarver and F. A. Hummel, *J. Electrochem. Soc. 106*, 500–504 (1959).

604. I. A. Tokman and G. A. Bukhalova, *Russ. J. Inorg. Chem. 22*, 578–580 (1977).

605. G. A. Bukhalova and I. A. Tokman, *Russ. J. Inorg. Chem. 22*, 1051–1052 (1977).

606. I. A. Tokman and G. A. Bukhalova, *Izv. Akad. Nauk SSSR, Neorg. Mater. 13*, 1104–1105 (1977).

607. M. I. Kuzmenkov, S. V. Plysevskii, and V. V. Peckovskii, *Zh. Neorg. Chim. 19*, 1621–1624 (1974).

608. M. Laügt, *C. R. Acad. Sci. 269C*, 1122–1124 (1969).

609. M. Laügt, I. Tordjman, J. C. Guitel, and M. Roudaut, *Acta Crystallogr., Sect. B 28*, 2352–2358 (1972).

610. E. A. Genkina, B. A. Maksimov, Yu. K. Kabalov, and O. K. Mel'Nikov, *Dokl. Akad. Nauk SSSR 270*, 1113–1116 (1983).

611. E. A. Genkina, N. S. Triodina, O. K. Mel'nikov, and B. A. Maksimov, *Izv. Akad. Nauk SSSR, Neorg. Mater. 24*, 1158–1162 (1988).

612. M. T. Averbuch-Pouchot and A. Durif, *J. Appl. Crystallogr. 5*, 307–308 (1972).

613. G. A. Bukhalova, I. G. Rabkina, I. V. Mardirosova, and V. N. Mirnyl, *Ukr. Khim. Zh. 41*, 1144–1147 (1975).

614. M. T. Averbuch-Pouchot and E. Rakotomahanina-Rolaisoa, *Bull. Soc. fr. Minéral. Cristallogr. 93*, 394–396 (1970).

615. P. de Pontcharra and A. Durif, *C. R. Acad. Sci. 278C*, 175–178 (1974).

616. E. Rakotomahanina-Rolaisoa, Dissertation, University of Grenoble, Grenoble, France (1972).

617. Y. Henry and A. Durif, *C. R. Acad. Sci. 270C*, 423–425 (1970).

618. C. Martin and A. Durif, *Bull. Soc. fr. Minéral. Cristallogr. 92*, 489–490 (1969).

619. J. C. Grenier and I. Mahama, *C. R. Acad. Sci. 274C*, 1063–1065 (1972).

620. J. C. Guitel and M. Brunel-Laügt, *Acta Crystallogr., Sect. B 33*, 2713–2716 (1977).

621. M. T. Averbuch-Pouchot and A. Durif, *Mater. Res. Bull. 4*, 859–868 (1969).

622. M. T. Averbuch-Pouchot, I. Tordjman, and J. C. Guitel, *Acta Crystallogr., Sect. B 32*, 2953–2956 (1976).
623. C. Raholison C. and M. T. Averbuch-Pouchot, *C. R. Acad. Sci. 274C*, 1066–1068 (1972).
624. N. El-Horr and M. Bagieu-Beucher, *Acta Crystallogr., Sect. C 42*, 647–651 (1986).
625. C. Martin, Dissertation, University of Grenoble, Grenoble France (1972).
626. N. El-Horr, Dissertation, University of Grenoble, Grenoble France (1988).
627. N. El-Horr, M. Bagieu, J. C. Guitel, and I. Tordjman, *Z. Kristallogr. 169*, 73–82 (1984).
628. B. Thonnérieux, J. C. Grenier, A. Durif, and C. Martin, *C. R. Acad. Sci. 267C*, 968–970 (1968).
629. E. L. Krivovyasov, K. K. Palkina, and N. K. Voskresenskaya, *Dokl. Akad. Nauk SSSR 174*, 610–613 (1967).
630. M. T. Averbuch-Pouchot and A. Durif, *Mater. Res. Bull. 4*, 859–868 (1969).
631. M. T. Averbuch-Pouchot, E. Rakotomahanina-Rolaisoa, and A. Durif, *Bull. Soc. fr. Minéral. Cristallogr. 93*, 282–286 (1970).
632. Yu. F. Shepelev, Yu. I. Smolin, A. I. Domanskii, and A. V. Lavrov, *Dokl. Akad. Nauk SSSR 272*, 610–614 (1983).
633. G. W. Morey, *J. Am. Chem. Soc. 74*, 5783–5786 (1952).
634. J. C. Grenier, C. Martin, and A. Durif, *Bull. Soc. fr. Minéral. Cristallogr. 93*, 52–55 (1970).
635. A. Durif, *C. R. Acad. Sci. 275C*, 1379–1382 (1972).
636. C. Martin and A. Durif, *Bull. Soc. fr. Minéral. Cristallogr. 95*, 149–153 (1972).
637. J. L. Prisset, Dissertation, University of Grenoble, Grenoble France (1982).
638. M. Laügt, A. Durif, and C. Martin, *C. R. Acad. Sci. 266C*, 1700–1702 (1968).
639. M. Laügt, I. Tordjman, J. C. Guitel, and G. Bassi, *Acta Cystallogr. Sect.B 28*, 2721–2725 (1972).
640. M. Laügt, I. Tordjman, J. C. Guitel, and G. Bassi, *Cryst. Struct. Commun. 1*, 279–282 (1972).
641. G. A. Bukhalova, M. A. Savenkova, V. A. Matrosova, I. V. Mardirosova, and I. G. Rabkina, *Zh. Prikl. Khim. 53*, 1266–1269 (1980).
642. M. A. Savenkova, G. A. Bukhalova, and O. V. Tyumeneva, *Russ. J. Inorg. Chem. 21*, 468–469 (1976).
643. Ph. de Pontcharra, Dissertation, University of Grenoble, Grenoble France (1972).
644. M. T. Averbuch-Pouchot and A. Durif, *J. Solid State Chem. 49*, 341–352 (1983).
645. Y. Henry and A. Durif, *C. R. Acad. Sci. 270C*, 1984–1986 (1970).
646. M. Laügt, *C. R. Acad. Sci. 267C*, 1489–1491 (1968).
647. K. Omezzine and N. Kbir-Ariguib, *J. Soc. Chim. Tunis.* 37–44 (1982).
648. A. Durif, M. T. Averbuch-Pouchot, and J. C. Guitel, *Z. Kristallogr. 177*, 165–170 (1986).
649. I. Tordjman, D. Tranqui, A. Durif, and M. T. Averbuch-Pouchot, *Bull. Soc. fr. Minéral. Cristallogr. 91*, 242–246 (1968).
650. R. Andrieu and R. Diament. *C. R. Acad. Sci. 259*, 4708–4710 (1964).
651. M. T. Averbuch-Pouchot, C. Martin, E. Rakotomahanina-Rolaisoa, and A. Durif, *Bull. Soc. fr. Minéral. Cristallogr. 93*, 282–286 (1970).
652. B. Thonnérieux, D. Tranqui, A. Durif, and M. T. Averbuch-Pouchot, *C. R. Acad. Sci. 266C*, 208–210 (1968).
653. A. Durif, J. C. Grenier, M. T. Pouchot, and D. TranQui, *Bull. Soc. fr. Minér. Cristallogr. 89*, 273–274 (1966).
654. I. Tordjman, D. Tranqui, and M. Laügt, *Bull. Soc. fr. Minéral. Cristallogr. 93*, 160–165 (1970).
655. M. Laügt, I. Tordjman, G. Bassi, and J. C. Guitel, *Acta Crystallogr., Sect. B 30*, 1100–1104 (1974).
656. M. Laügt, J. C. Guitel, A. Durif, and C. Martin, *C. R. Acad. Sci. 265C*, 741–743 (1967).
657. A. Mermet, M. T. Averbuch-Pouchot, and A. Durif, *Bull. Soc. fr. Minéral. Cristallogr. 92*, 87–90 (1969).
658. C. Martin, I. Tordjman, and A. Durif, *Z. Kristallogr. 141*, 403–411 (1975).
659. C. Martin, I. Tordjman, and A. Mitschler, *Cryst. Struct. Commun. 1*, 349–352 (1972).
660. I. Mahama, M. Brunel-Laügt, and M. T. Averbuch-Pouchot, *C. R. Acad. Sci. 284C*, 681–684 (1977).
661. H. Grunze and F. Möwius, *Z. anorg. allg. Chem. 458*, 125–129 (1979).
662. F. Möwius and H. Grunze, *Z. anorg. allg. Chem. 494*, 43–48 (1982).
663. M. Brunel-Laügt and J. C. Guitel, *Acta Crystallogr., Sect. B 33*, 937–939 (1977).

664. M. T. Averbuch-Pouchot and D. Tranqui, *Bull. Soc. fr. Minéral. Cristallogr.* 92, 311–312 (1969).
665. D. Tranqui, J. C. Grenier, A. Durif, and J. C. Guitel, *Bull. Soc. fr. Minéral. Cristallogr.* 90, 252–256 (1967).
666. D. Tranqui, M. Laügt, and J. C. Guitel, *Bull. Soc. fr. Minéral. Cristallogr.* 92, 329–334 (1969).
667. M. T. Averbuch-Pouchot and A. Durif, *Mater. Res. Bull.* 4, 397–402 (1969).
668. M. Laügt, M. Scory, and A. Durif, *Mater. Res. Bull.* 3, 963–970 (1968).
669. M. T. Averbuch-Pouchot, *Bull. Soc. fr. Minéral. Cristallogr.* 95, 558–564 (1972).
670. E. Rakotomahanina-Rolaisoa, *Bull. Soc. fr. Minéral. Cristallogr.* 95, 143–145 (1972).
671. Y. Henry and A. Durif, *Bull. Soc. fr. Minéral. Cristallogr.* 92, 484–486 (1969).
672. M. T. Averbuch-Pouchot, *C. R. Acad. Sci.* 269C, 26–29 (1969).
673. B. Klinkert and M. Jansen, *Z. anorg. allg. Chem.* 570, 102–108 (1989).
674. M. Laügt and C. Martin, *Mater. Res. Bull.* 7, 1525–1534 (1972).
675. I. A. Tokman and G. A. Bukhalova, *Russ. J. Inorg. Chem.* 17, 84–87 (1972).
676. R. Masse and M. T. Averbuch-Pouchot, *Mater. Res. Bull.* 12, 13–16 (1977).
677. M. T. Averbuch-Pouchot and A. Durif, *Acta Crystallogr., Sect. C* 42, 928–930 (1986).
678. V. M. Shpakova, I. V. Mardirosova, and G. A. Bukhalova, *Izv. Akad. Nauk SSSR, Neorg. Mater.* 10, 2184–2186 (1974).
679. M. T. Averbuch-Pouchot, *Z. anorg. allg. Chem.* 529, 143–150 (1985).
680. M. Laügt, *C. R. Acad. Sci.* 278C, 1497–1500 (1974).
681. E. Rakotomahanina-Rolaisoa, Y. Henry, A. Durif, and C. Raholison, *Bull. Soc. fr. Minéral. Cristallogr.* 93, 43–51 (1970).
682. M. T. Averbuch-Pouchot, *C. R. Acad. Sci.* 286C, 1253–1255 (1969).
683. K. R. Johnson, *Ber. Dtsch. Chem. Ges.* 22, 976–980 (1889).
684. F. d'Yvoire, *Bull. Soc. Chim.* 1962, 1237–1243.
685. P. Rémy and A. Boullé, *Bull. Soc. Chim.* 1972, 2215–2221.
686. M. Bagieu, Dissertation, University of Grenoble, Grenoble France (1980).
687. M. Bagieu-Beucher, *Acta Crystallogr., Sect. B* 34, 1443–1446 (1978).
688. H. Van Der Meer, *Acta Crystallogr., Sect. B* 32, 2423–2426 (1976).
689. N. Middlemiss, F. Hawthorne, and C. Calvo, *Can. J. Chem.* 55, 1673–1679 (1977).
690. S. A. Linde, Yu. E. Gorbunova, and A. V. Lavrov, *Russ. J. Inorg. Chem.* 28, 16–18 (1983).
691. F. Liebau and H. P. Williams, *Angew. Chem.* 76, 303–304 (1964).
692. A. Colani, *C. R.. Acad. Sci.* 158, 499–501 (1914).
693. R. M. Douglass and E. Staritzky, *Anal. Chem.* 29, 985 (1957).
694. M. Bagieu-Beucher, *J. Appl. Crystallogr.* 9, 368–369 (1976).
695. Yu. I. Smolin, Yu. F. Shepelev, and A. I. Domanskii, *Izv. Akad. Nauk SSSR, Neorg. Mater.* 20, 1220–1226 (1984).
696. E. N. Deichman, I. V. Tananaev, Zh. A. Ezhova, and K. K. Palkina, *Izv. Akad. Nauk SSSR, Neorg. Mater.* 6, 1645–1649 (1970).
697. N. N. Chudinova, M. A. Avaliani, and L. S. Guzeeva, *Izv. Akad. Nauk SSSR, Neorg. Mater.* 13, 2229–2233 (1977).
698. P. P. Mel'nikov, V. A. Efremov, A. K. Stepanov, T. S. Romanova, and L. N. Komissarova, *Russ. J. Inorg. Chem.* 21, 26–28 (1976).
699. J. Matuszewski, J. Kropiwnicka, and T. Znamierowska, *J. Solid State Chem.* 75, 285–290 (1988).
700. N. N. Chudinova, L. P. Shklover, A. E. Balanevskaya, L. M. Shkol'nikova, A. E. Obodovskaya, and G. M. Balagina, *Izv. Akad. Nauk SSSR, Inorg. Mater.* 14, 727–733 (1978).
701. N. N. Chudinova, L. P. Shklover, and G. M. Balagina, *Izv. Akad. Nauk SSSR, Neorg. Mater.* 11, 686–689 (1975).
702. G. A. Bukhalova, I. V. Mardirosova, and M. M. Alí, *Russ. J. Inorg. Chem.* 33, 1438–1441 (1988).
703. M. Tsuhako, S. Ikeuchi, T. Matsuo, I. Motooka, and M. Kobayashi, *Chem. Lett.* 1977 195–198.
704. M. A. Vaivada and Z. A. Konstant, *Izv. Akad. Nauk SSSR, Neorg. Mater.* 16, 1810–1814 (1980).
705. H. Y-P. Hong, *Acta Crystallogr., Sect. B* 30, 468–474 (1974).

706. N. N. Chudinova, L. P. Shklover, L. I. Shkol'nikova, A. E. Balanevskaya, and G. M. Balagina, *Izv. Akad. Nauk SSSR, Neorg. Mater. 14*, 1324–1328 (1978).

707. G. I. Dorokhova and O. G. Karpov, *Sov. Phys. Crystallogr. 29*, 400–402 (1984).

708. O. S. Tarasenkova, G. I. Dorokhova N. N. Chudinova, B. N. Litvin, and N. V. Vinogradova, *Izv. Akad. Nauk SSSR, Neorg. Mater. 21*, 452–458 (1985).

709. N. N. Chudinova, G. M. Balagina, and L. P. Shklover, *Izv. Akad. Nauk SSSR, Neorg. Mater. 13*, 2075–2082 (1977).

710. H. Y-P. Hong, *Acta Crystallogr., Sect. B 30*, 1857–1861 (1974).

711. M. Rzaigui and N. Kbir-Ariguib, *Bull. Soc. Chim. Belg. 94*, 619–620 (1985).

712. N. N. Chudinova, A. V. Lavrov, and I. V. Tananaev, *Izv. Akad. Nauk SSSR, Neorg. Mater. 8*, 1971–1976 (1972).

713. K. K. Palkina and K. H. Jost, *Acta Crystallogr., Sect. B 31*, 2281–2285 (1975).

714. K. K. Palkina and K. H. Jost, *Acta Crystallogr., Sect. B 31*, 2285–2290 (1975).

715. N. N. Chudinova, *Izv. Akad. Nauk SSSR, Neorg. Mater. 11*, 1662–1666 (1975).

716. N. Hilmer, N. N. Chudinova, and K. H. Jost, *Izv. Akad. Nauk SSSR, Neorg. Mater. 14*, 1507–1515 (1978).

717. K. K. Palkina, N. N. Chudinova, G. M. Balagina, S. I. Maksimova, and N. T. Chibiskova, *Izv. Akad. Nauk SSSR, Neorg. Mater. 18*, 1561–1566 (1982).

718. P. P. Mel'nikov, L. N. Komissarova, and T. A. Butuzova, *Izv. Akad. Nauk SSSR, Neorg. Mater. 17*, 2110–2112 (1981).

719. G. M. Balagina, A. F. Banishev, Yu. K. Voron'ko, V. V. Osiko, A. A. Sobol, and N. N. Chudinova, *Izv. Akad. Nauk SSSR, Neorg. Mater. 21*, 712–720 (1985).

720. N. N. Chudinova, *Izv. Akad. Nauk SSSR, Neorg. Mater. 15*, 833–837 (1979).

721. A. Ya. Valtere, *Izv. Akad. Nauk SSSR, Neorg. Mater. 23*, 288–291 (1987).

722. R. M. Gupta and H. N. Bhargava, *Colloid Polym. Sci. 258*, 1226–1230 (1980).

723. Zh. A. Ezhova, I. V. Tananaev, and E. M. Koval', *Russ. J. Inorg. Chem. 23*, 1657–1662 (1978).

724. Zh. A. Ezhova, I. V. Tananaev, and E. M. Koval', *Russ. J. Inorg. Chem. 33*, 298–299 (1988).

725. Zh. A. Ezhova, I. V. Tananaev, L. N. Zorina, E. M. Koval', and N. P. Soshchin, *Izv. Akad. Nauk SSSR, Neorg. Mater. 14*, 2067–2072 (1978).

726. K. K. Palkina, N. N. Chudinova, B. N. Litvin, and N. V. Vinogradova, *Izv. Akad. Nauk SSSR, Neorg. Mater. 17*, 1501–1503 (1981).

727. S. V. Plyshevskii, M. I. Kuz'menkov, and V. V. Pechkovskii, *Russ. J. Inorg. Chem. 22*, 475–476 (1977).

728. G. A. Bukhalova, I. V. Mardirosova, N. P. Vassel, and M. A. Savenkova, *Izv. Akad. Nauk SSSR, Neorg. Mater. 27*, 828–831 (1991).

729. N. N. Chudinova, M. A. Avaliani, and I. V. Tananaev, *Izv. Akad. Nauk SSSR, Ser. Khim. 5*, 373–375 (1979).

730. K. K. Palkina, S. I. Maksimova, and N. T. Chibiskova, *Izv. Akad. Nauk SSSR, Neorg. Mater. 17*, 95–98. (1981).

731. N. N. Chudinova and N. V. Vinogradova, *Izv. Akad. Nauk SSSR, Neorg. Mater. 15*, 2171–2175. (1979).

732. M. F. Moktar, N. Kbir-Ariguib, and M. Trabelsi, *J. Solid State Chem. 38*, 133–137 (1981).

733. H. Koizumi, *Acta Crystallogr., Sect. B 32*, 266–268 (1976).

734. H. Y-P. Hong, *Mater. Res. Bull. 10*, 635–640 (1975).

735. J. Nakano, S. Miyazawa, and T. Yamada, *Mater. Res. Bull. 14*, 21–26 (1979).

736. L. N. Zorina, V. A. Bol'shukhin, O. A. Kruchnova, A. V. Lavrov, and N. P. Soshchin, *Izv. Akad. Nauk SSSR, Neorg. Mater. 16*, 126–130 (1980).

737. E. N. Fedorova, L. K. Shmatok, I. I. Kozhina, and T. R. Barabanova, *Izv. Akad. Nauk SSSR, Neorg. Mater. 22*, 480–484 (1986).

738. M. Rzaigui, M. Trabelsi, and N. Kbir-Ariguib, *C. R. Acad. Sci., Ser. 2 292*, 505–508 (1981).

739. H. Koizumi, *Acta Crystallogr., Sect. B 32*, 2254–2256 (1976).

740. S. I. Maksimova, V. A. Masloboev, K. K. Palkina, A. A. Sazhenkov, and N. T. Chibiskova, *Russ. J. Inorg. Chem. 33*, 1434–1435 (1988).
741. G. A. Bukhalova, R. S. Faustova, and M. A. Savenkova, *Ukr. Khim. Zh. 42*, 1152–1154 (1976).
742. M. Bagieu-Beucher, private communication.
743. Zh. Ezhova, I. V. Tananaev, and E. M. Koval, *Russ. J. Inorg. Chem. 23*, 1657–1662 (1978).
744. V. K. Trunov, N. Yu. Anisimova, N. B. Karmanovskaya, and N. N. Chudinova, *Izv. Akad. Nauk SSSR, Neorg. Mater. 26*, 1288–1290 (1990).
745. G. A. Bukhalova, I. V. Mardirosova, and M. M. Ali, *Dokl. Akad. Nauk SSSR, Neorg. Mater. 20*, 120–122 (1984).
746. N. N. Chudinova, N. V. Vinogradova, and K. K. Palkina, *Izv. Akad. Nauk SSSR, Neorg. Mater. 14*, 2049–2053 (1978).
747. K. K. Palkina, V. G. Kuznetsov, N. N. Chudinova, and N. T. Chibiskova, *Izv. Akad. Nauk SSSR, Neorg. Mater. 12*, 730–734 (1976).
748. M. Rzaigui, M. Dabbabi, and N. Kbir-Ariguib, *J. Chim. Phys. 78*, 563–566 (1981).
749. S. A. Linde, Yu. E. Gorbunova, and A. V. Lavrov, *Russ. J. Inorg. Chem. 28*, 804–807 (1983).
750. B. N. Litvin, G. I. Dorokhova, and O. S. Filipenko, *Sov. Phys. Dokl. 26*, 717–720 (1981).
751. K. K. Palkina, V. G. Kuznetsov, N. N. Chudinova, and N. I. Chibiskova, *Dokl. Akad. Nauk SSSR 226*, 357–360 (1976).
752. H. Y. P Hong, *Mater. Res. Bull. 10*, 1105–1110 (1975).
753. S. Miyazawa, H. Koizumi, K. Kubodera, and H. Iwasaki, *J. Crystal. Growth 47*, 351–356 (1979).
754. K. K. Palkina, A. Yu. Sazhenkov, S. I. Maksimova, N. T. Chibiskova, and V. A. Masloboev, *Russ. J. Inorg. Chem. 34*, 662–665 (1989).
755. N. N. Chudinova, N. V. Vinogradova, G. M. Balagina, and K. K. Palkina, *Izv. Akad. Nauk SSSR, Neorg. Mater. 13*, 1494–1499 (1977).
756. O. S. Tarasenkova, G. I. Dorokhova, N. N. Chudinova, B. N. Litvin, and N. V. Vinogradova, *Izv. Akad. Nauk SSSR, Neorg. Mater. 21*, 452–458 (1985).
757. V. M. Krutik, D. Yu. Pushcharovskii, E. A. Pobedimskaya, and N. V. Belov, *Sov. Phys. Dokl. 25*, 329–331 (1980).
758. A. M. Dago, D. Yu. Pushcharovskii, E. A. Pobedimskaya, and N. V. Belov, *Sov. Phys. Dokl. 25*, 231–233 (1980).
759. O. G. Karpov and G. I. Dorokhova, *Sov. Phys. Crystallogr. 34*, 607–608 (1989).
760. K. K. Palkina, S. I. Maksimova, N. N. Chudinova, N. V. VIinogradova, and N. T. Chibiskova, *Izv. Akad. Nauk SSSR, Neorg. Mater. 17*, 110–115 (1981).
761. G. A. Bukhalova, R. S. Faustova, and M. A. Savenkova, *Zh. Prikl. Khim. 50*, 171–173 (1977).
762. M. Bagieu-Beucher and J. C. Guitel, *Z. anorg. allg. Chem. 559*, 123–130 (1988).
763. M. Rzaigui and N. K. Ariguib, *J. Solid. State Chem. 49*, 391–398 (1983).
764. M. A. Vaivada and Z. A. Konstant, *Dokl. Akad. Nauk SSSR, Neorg. Mater. 15*, 824–827 (1979).
765. K. K. Palkina, V. V. Krasnikov, and Z. A. Konstant, *Izv. Akad. Nauk SSSR, Neorg. Mater. 17*, 1243–1247 (1981).
766. M. Rzaigui, K. Ariguib, M. T. Averbuch-Pouchot, and A. Durif, *J. Solid State Chem. 50*, 240–246 (1983).
767. M. T. Averbuch-Pouchot and M. Bagieu-Beucher, *Z. anorg. allg. Chem. 552*, 171–180 (1987).
768. S. I. Maksimova, K. K. Palkina, V. B. Loshchenov, and V. G. Kuznetsov, *Russ. J. Inorg. Chem. 23*, 1643–1646 (1978).
769. S. I. Maksimova, K. K. Palkina, and N. T. Chibiskova, *Izv. Akad. Nauk SSSR, Neorg. Mater. 18*, 653–659 (1982).
770. K. Byrappa and B. N. Litvin, *J. Mater. Sci. 18*, 2056–2062 (1983).
771. G. A. Bukhalova, I. V. Mardirosova, and M. M. Ali, *Izv. Akad. Nauk SSSR, Neorg. Mater. 20*, 1405–1408 (1984).
772. K. K. Palkina, S. I. Maksimova, and V. G. Kuznetsov, *Izv. Akad. Nauk SSSR, Neorg. Mater. 14*, 284–287 (1978).
773. H. Koizumi and J. Nakano, *Acta Crystallogr., Sect. B 34*, 3320–3323 (1978).

774. S. I. Maksimova, K. K. Palkina, V. B. Loshchenov, and V. G. Kuznetsov, *Izv. Akad. Nauk SSSR, Neorg. Mater.* 15, 969–974 (1979).

775. S. I. Maksimova, K. K. Palkina, and V. V. Loshchenov, *Izv. Akad. Nauk SSSR, Neorg. Mater.* 17, 116–120 (1981).

776. K. K. Palkina, V. G. Kuznetsov, S. I. Maksimova, N. N. Chudinova, and N. T. Chibiskova, *Koord. Khim.* 3, 275–276 (1977).

777. K. K. Palkina, S. I. Maksimova, V. G. Kuznetsov, and N. T. Chibiskova, *Koord. Khim.* 4, 1092–1095 (1978).

778. G. A. Bukhalova, R. S. Faustova, and M. A. Savenkova, *Russ. J. Inorg. Chem.* 2, 778–779 (1977).

779. N. V. Vinogradova and N. N. Chudinova, *Izv. Akad. Nauk SSSR, Neorg. Mater.* 17, 492–495 (1981).

780. K. Byrappa, I. I. Plyusnina, and G. I. Dorokhova, *J. Mater. Sci.* 17, 1847–1855 (1982).

781. K. K. Palkina, V. Z. Saifuddinov, V. G. Kuznetsov, and N. N. Chudinova, *Sov. Phys. Dokl.* 22, 698–700 (1977).

782. M. Rzaigui, K. Ariguib, and M. Bagieu-Beucher, *Mater. Chem. Phys.* 11, 49–64 (1984).

783. A. F. Banishev, N. V. Vinogradova, Yu. K. Voron'ko, A. A. Sobol, and N. N. Chudinova, *Izv. Akad. Nauk SSSR, Neorg. Mater.* 23, 292–297 (1987).

784. M. I. Kuz'menkov, S. V. Plyshevskii, and V. V. Pechkovskii, *Russ. J. Inorg. Chem.* 19, 881–883 (1974).

785. A. E. Mal'shikov and I. A. Bondar, *Russ. J. Inorg. Chem.* 34, 723–727 (1989).

786. A. E. Mal'shikov and I. A. Bondar, *Russ. J. Inorg. Chem.* 34, 1207–1209 (1989).

787. Yu. E. Gorbunova, V. V. Ilyukhin, V. G. Kuznetsov, A. V. Lavrov, and S. A. Linde, *Sov. Phys. Dokl.* 21, 309–310 (1976).

788. R. Masse and J. C. Grenier, *Bull. Soc. fr. Minéral. Cristallogr.* 95, 136–142 (1972).

789. M. Tsuhako, S. Ikeuchi, T. Matsuo, I. Motooka, and M. Kobayashi, *Chem. Lett.* 1977, 195–198.

790. A. Burdese and M. L. Borlera, *Ric. Sci.* 1960, 103–108.

791. A. Burdese and M. L. Borlera, *Ann. Chim.* 53, 344–345 (1963).

792. S. A. Linde, Yu. E. Gorbunova, V. V. Ilyukhin, A. V. Lavrov, and V. G. Kuznetsov, *Russ. J. Inorg. Chem.* 24, 989–991 (1979).

793. Y. Baskin, *J. Inorg. Nucl. Chem.* 29, 383–391 (1967).

794. R. M. Douglass, *Acta Crystallogr.* 15, 505–506 (1962).

795. S. A. Linde, Yu. E. Gorbunova, A. V. Lavrov, and V. G. Kusnetsov, *Dokl. Akad. Nauk SSSR 230*, 1376–1379 (1976).

796. V. A. Sarin, S. A. Linde, L. E. Fykin, V. Ya. Dudarev, and Yu. E. Gorbunova, *Russ. J. Inorg. Chem.* 28, 866–868 (1983).

797. A. V. Lavrov, *Izv. Akad. Nauk SSSR, Neorg. Mater.* 15, 942–946 (1979).

798. S. A. Linde, Yu. E. Gorbunova, A. V. Lavrov, and V. G. Kuznetsov, *Dokl. Akad. Nauk SSSR 235*, 394–397 (1977).

799. W. H. Zachariasen, *Z. Kristallogr. 74*, 139–146 (1930).

800. K. Dostal and V. Kocman, *Z. anorg. allg. Chem. 367*, 92–101 (1969).

801. M. Jansen and G. Brachtel, *Monatsh. Chem. 111*, 377–384 (1980).

802. C. L. Christ and J. R. Clark, *Acta Crystallogr. 9*, 830 (1956).

803. J. R. Clark, *Acta Crystallogr. 12*, 162–170 (1959).

804. I. Haiduc and D. B. Sowerby, *The Chemistry of Inorganic Homo-, and Heterocycles* (2 vols.), Academic Press, London (1987).

805. A. Boullé, *C. R. Acad. Sci. 206*, 517–519 (1938).

806. E. Thilo and H. Grunze, *Z. anorg. allg. Chem. 281*, 263–283 (1955).

807. E. D. Eanes, National Bureau of Standards Monograph, *25*, Section 2, p 20.

808. R. Masse, J. C. Grenier, G. Bassi, and I. Tordjman, *Z. Kristallogr. 137*, 17–23 (1973).

809. R. Masse, J. C. Grenier, G. Bassi, and I. Tordjman, *Crystallogr. Struct. Commun. 1*, 239–241 (1972).

810. V. A. Sotnikova-Yuzhik, G. V. Peslyak, and E. A. Prodan, *Russ. J. Inorg. Chem. 32*, 1505–1507 (1987).

811. E. J. Griffith, *J. Am. Chem. Soc. 78*, 3867–3870 (1956).

812. M. T. Averbuch-Pouchot, J. C. Guitel, and A. Durif, *Acta Crystallogr., Sect. C 39*, 809–810 (1983).
813. H. M. Ondik and J. W. Gryder, *J. Inorg. Nucl. Chem.* 1960, 240–246.
814. H. M. Ondik, *Acta Crystallogr. 18*, 226–232 (1965).
815. I. Tordjman and J. C. Guitel, *Acta Crystallogr., Sect. B 32*, 1871–1874 (1976).
816. E. A. Prodan and S. I. Pytlev, *Izv. Akad. Nauk SSSR, Neorg. Mater. 19*, 639–643 (1983).
817. E. D. Eanes, private communication, cited in *Crystal Data* (Joint Commitee Powder Diffraction Standard), Vol. 2 (1973).
818. J. C. Grenier, *Bull. Soc. fr. Minéral. Cristallogr. 96*, 171–178 (1973).
819. M. Bagieu-Beucher, A. Durif, and J. C. Guitel, *Acta Crystallogr., Sect. B 31*, 2264–2267 (1975).
820. J. C. Grenier and A. Durif, *Rev. Chim. Minér. 9*, 351–355 (1972).
821. M. Bagieu-Beucher, I. Tordjman, A. Durif, and J. C. Guitel, *Acta Crystallogr., Sect. B 32*, 1427–1430 (1976).
822. N. M. Dombrovskii and V. A. Koval, *Izv. Akad. Nauk SSSR, Neorg. Mater. 12*, 738–741 (1976).
823. I. Tordjman, R. Masse, and J. C. Guitel, *Acta Crystallogr., Sect. B 33*, 585–586 (1977).
824. N. Boudjada, Thesis, University of Grenoble, Grenoble France (1985).
825. G. Tammann and A. Ruppelt, *Z. anorg. allg. Chem. 197*, 65–89 (1931).
826. G. Morey, *J. Am. Chem. Soc. 76*, 4724–4726 (1954).
827. E. J. Griffith and J. R. Van Wazer, *J. Am. Chem. Soc. 77*, 4222 (1955).
828. G. A. Bukhalova and I. V. Mardirosova, *Russ. J. Inorg. Chem. 11*, 495–497 (1966).
829. C. Cavero-Ghersi and A. Durif, *C. R. Acad. Sci. 278C*, 459–461 (1974).
830. I. Tordjman, A. Durif, and C. Cavero-Ghersi, *Acta Crystallogr., Sect. B 30*, 2701–2704 (1974).
831. C. Cavero-Ghersi and A. Durif, *C. R. Acad. Sci. 280C*, 579–581 (1975).
832. C. Cavero-Ghersi, Thesis, Univiversity of Grenoble, Grenoble, France (1975).
833. M. T. Averbuch-Pouchot and A. Durif, *Eur. J. Solid State Inorg. Chem. 30*, 1075–1082 (1993).
834. M. T. Averbuch-Pouchot and A. Durif, *Z. Kristallogr. 135*, 318–319 (1972).
835. M. T. Averbuch-Pouchot, A. Durif, and I. Tordjman, *Crystallogr. Struct. Commun. 2*, 89–90 (1973).
836. M. T. Averbuch-Pouchot, A. Durif, and J. C. Guitel, *Acta Crystallogr., Sect. B 32*, 1533–1535 (1976).
837. D. Michot, Thesis, Univiversity of Dijon, Dijon France (1975).
838. M. H. Simonot-Grange, *J. Solid State Chem. 46*, 76–86 (1983).
839. M. H. Simonot-Grange and D. Michot, *Phosphorus 6*, 103–105 (1976).
840. M. H. Simonot-Grange and D. Michot, *Phosphorus Sulfur 4*, 35–38 (1978).
841. M. T. Averbuch-Pouchot, A. Durif, and J. C. Guitel, *Acta Crystallogr., Sect. B 32*, 1894–1896 (1976).
842. N. El-Horr and A. Durif, *C. R. Acad. Sci., Ser. 2 296*, 1185–1187 (1983).
843. A. Durif, unpublished results.
844. A. Durif, M. Bagieu-Beucher, C. Martin, and J. C. Grenier, *Bull. Soc. fr. Minéral. Cristallogr. 95*, 146–148 (1972).
845. I. Tordjman, A. Durif, and J. C. Guitel, *Acta Crystallogr., Sect. B 32*, 205–208 (1976).
846. J. C. Grenier and C. Martin, *Bull. Soc. fr. Minéral. Cristallogr. 98*, 107–110 (1975).
847. R. Masse, J. C. Guitel, and A. Durif, *Acta Crystallogr., Sect. B 32*, 1892–1894 (1976).
848. M. T. Averbuch-Pouchot and A. Durif, *Z. Kristallogr. 174*, 219–224 (1986).
849. D. Tacquenet, Thesis, University of Dijon, Dijon France (1978).
850. A. Thrierr-Sorel, D. Tacquenet, and M. H. Simonot-Grange, *Phosphorus Sulfur 8*, 73–78 (1980).
851. A. Durif and M. Brunel-Laügt, *J. Appl. Crystallogr. 9*, 154 (1976).
852. M. Brunel-Laügt, I. Tordjman, and A. Durif, *Acta Crystallogr., Sect. B 32*, 3246–3249 (1976).
853. M. I. Kuz'menkov, V. N. Makatun, and N. V. Semenova, *Russ. J. Inorg. Chem. 24*, 1151–1154 (1979).
854. A. Durif, M. T. Averbuch-Pouchot, and J. C. Guitel, *Acta Crystallogr., Sect. B 31*, 2680–2682 (1975).
855. E. J. Griffith, *Inorg. Chem. 2*, 962 (1962).
856. J. C. Grenier, C. Martin, and A. Durif, *Bull. Soc. fr. Minéral. Cristalogr. 93*, 52–55 (1970).
857. G. W. Morey, *J. Am. Chem. Soc. 74*, 5783–5784 (1952).
858. M. T. Averbuch-Pouchot and A. Durif, *Mater. Res. Bull. 4*, 859–868 (1969).
859. I. Mahama, M. T. Averbuch-Pouchot, and J. C. Grenier, *C. R. Acad. Sci. 280C*, 1105–1107 (1975).
860. M. T. Averbuch-Pouchot and A. Durif, *Z. Kristallogr. 164*, 307–313 (1983).

861. C. Martin and A. Durif, *Bull. Soc. fr. Minéral. Cristallogr. 95*, 149–153 (1972).
862. C. Martin and A. Mitschler, *Acta Crystallogr., Sect. B 28*, 2348–2352 (1972).
863. M. T. Pouchot, I. Tordjman, and A. Durif, *Bull. Soc. fr. Minéral. Cristallogr. 89*, 405–406 (1966).
864. M. T. Averbuch-Pouchot, *C. R. Acad. Sci. 268C*, 1253–1255 (1969).
865. A. Durif and M. T. Averbuch-Pouchot, *J. Appl. Crystallogr. 9*, 247–248 (1976).
866. M. A. Savenkova, I. V. Mardirosova, and V. A. Matrosova, *Izv. Akad. Nauk SSSR, Neorg. Mater. 12*, 1324–1325 (1976).
867. R. Andrieu and R. Diament, *C. R. Acad. Sci. 259*, 4708–4711 (1964).
868. R. Andrieu, R. Diament, A. Durif, M. T. Pouchot, and D. Tranqui, *C. R. Acad. Sci. 262B*, 718–721 (1966).
869. M. T. Averbuch-Pouchot, C. Martin, E. Rakotomahanina-Rolaisoa, and A. Durif, *Bull. Soc. fr. Minéral. Cristallogr. 93*, 282–286 (1970).
870. R. Masse, J. C. Grenier, M. T. Averbuch-Pouchot, D. Tranqui, and A. Durif, *Bull. Soc. fr. Minéral. Cristallogr. 90*, 158–161 (1967).
871. R. Masse, A. Durif, and J. C. Guitel, *Z. Kristallogr. 141*, 113–125 (1975).
872. E. Rakotomahanina-Rolaisoa, Thesis, University of Grenoble, Grenoble, France (1972).
873. A. Mermet, M. T. Averbuch-Pouchot, and A. Durif, *Bull. Soc. fr. Minéral. Cristallogr. 92*, 87–90 (1969).
874. M. T. Averbuch-Pouchot and A. Durif, *Acta Crystallogr., Sect. C 42*, 930–931 (1986).
875. J. C. Grenier and R. Masse, *Bull. Soc. fr. Minéral. Cristallogr. 91*, 428–439 (1968).
876. E. Rakotomahanina-Ralaisoa, *Bull. Soc. fr. Minéral. Cristallogr. 95*, 143–145 (1972).
877. Y. Henry and A. Durif, *Bull. Soc. fr. Minéral. Cristallogr. 92*, 484–486 (1969).
878. M. T. Averbuch-Pouchot, *C. R. Acad. Sci. 269C*, 26–29 (1969).
879. M. T. Averbuch-Pouchot and A. Durif, *Acta Crystallogr., Sect. B 33*, 3114–3116 (1977).
880. I. A. Tokman and G. A. Bukhalova, *Izv. Akad. Nauk SSSR, Neorg. Mater. 11*, 1654–1656 (1975).
881. E. Rakotomahanina-Rolaisoa, Y. Henry, A. Durif, and C. Raholison, *Bull. Soc. fr. Minéral. Cristallogr. 93*, 43–51 (1970).
882. M. T. Averbuch-Pouchot, *Bull. Soc. fr. Minéral. Cristallogr. 95*, 558–564 (1972).
883. M. T. Averbuch-Pouchot, Thesis, University of Grenoble, Grenoble, France (1974).
884. M. T. Averbuch-Pouchot and A. Durif, *Acta Crystallogr., Sect. C 42*, 932–933 (1986).
885. A. Durif, *C. R. Acad. Sci. 275C*, 1379–1382 (1972).
886. R. Zilber, I. Tordjman, A. Durif, and J. C. Guitel, *Z. Kristallogr. 140*, 350–359 (1974).
887. M. H. Simonot-Grange and P. Jamet, *Phosphorus Sulfur 3*, 197–202 (1977).
888. M. T. Averbuch-Pouchot and A. Durif, *Acta Crystallogr., Sect. C 43*, 390–392 (1987).
889. D. Seethanen, A. Durif, and J. C. Guitel, *Acta Crystallogr., Sect. B 32*, 2716–2719 (1977).
890. D. Seethanen and A. Durif, *Acta Crystallogr., Sect. B 34*, 1091–1093 (1978).
891. W. Feldman and I. Grunze, *Z. anorg. allg. Chem. 360*, 225–230 (1968).
892. A. Takenaka, I. Motooka, and H. Nariai, *Bull. Chem. Soc. Jpn. 62*, 2819–2823 (1989).
893. A. Durif, C. Martin, and G. Bassi, *Bull. Soc. fr. Minéral. Cristallogr. 98*, 19–24 (1975).
894. A. Jouini and M. Dabbabi, *C. R. Acad. Sci., Ser. 2 301*, 1347–1349 (1985).
895. A. Jouini and M. Dabbabi, *Rev. Chim. Minér. 23*, 776–781 (1986).
896. A. Jouini and M. Dabbabi, *Acta Crystallogr., Sect. C 42*, 268–270 (1986).
897. A. Durif and M. T. Averbuch-Pouchot, *Z. anorg. allg. Chem. 514*, 85–91 (1984).
898. D. Seethanen, I. Tordjman, and M. T. Averbuch-Pouchot, *Acta Crystallogr., Sect. B 34*, 2387–2390 (1978).
899. A. Durif and M. T. Averbuch-Pouchot, *Acta Crystallogr., Sect. C 43*, 819–821 (1987).
900. A. Jouini and M. Dabbabi, *Bull. Soc. Chim. Tunis. 2*, 29–34 (1987).
901. M. S. Belkhiria, M. Ben-Amara, and M. Dabbabi, *Acta Crystallogr., Sect. C 43*, 609–610 (1987).
902. M. T. Averbuch-Pouchot, *Acta Crystallogr., Sect. B 34*, 20–22 (1978).
903. D. Seethanen, A. Durif, and M. T. Averbuch-Pouchot, *Acta Crystallogr., Sect. B 34*, 14–17 (1978).
904. M. S. Belkhiria, M. Dabbabi, and M. Ben-Amara, *Acta Crystallogr., Sect. C 43*, 2270–2272 (1987).
905. O. A. Serra and E. Giesbrecht, *J. Inorg. Nucl. Chem. 30*, 793–799 (1968).

906. M. Bagieu-Beucher and A. Durif, *Bull. Soc. fr. Minéral. Cristallogr. 94*, 440–441 (1971).
907. M. Bagieu-Beucher, I. Tordjman, and A. Durif, *Rev. Chim. Minér. 8*, 753–760 (1971).
908. P. Birke and G. Kempe, *Z. Chem. 13*, 151–152 (1973).
909. P. Birke and G. Kempe, *Z. Chem. 3*, 65–66 (1973).
910. P. Birke and G. Kempe, *Z. Chem. 13*, 110–111 (1973).
911. D. Gobled, Thesis, University of Dijon, Dijon, France (1973).
912. M. Bagieu-Beucher, Thesis, University of Grenoble, Grenoble, France (1980).
913. M. Bagieu-Beucher and A. Durif, *Z. Kristallogr. 178*, 239–247 (1987)
914. M. Bagieu-Beucher, *C. R. Acad. Sci., Ser. 2 308*, 377–379 (1989).
915. M. T. Averbuch-Pouchot, A. Durif, and J. C. Guitel, *Acta Crystallogr., Sect. C 45*, 1320–1322 (1989).
916. M. T. Averbuch-Pouchot, A. Durif, and J. C. Guitel, *Acta Crystallogr., Sect. C 44*, 1907–1909 (1988).
917. M. T. Averbuch-Pouchot and A. Durif, *Acta Crystallogr., Sect. C 44*, 1909–1911 (1988).
918. M. T. Averbuch-Pouchot, A. Durif, and J. C. Guitel, *Acta Crystallogr., Sect. C 44*, 97–98 (1988).
919. M. T. Averbuch-Pouchot, A. Durif, and J. C. Guitel, *Acta Crystallogr., Sect. C 44*, 99–102 (1988).
920. M. T. Averbuch-Pouchot and A. Durif, *Eur. J. Solid State Inorg. Chem. 30*, 471–482 (1993).
921. N. Boudjada, *Z. anorg. allg. Chem. 477*, 225–228 (1981).
922. N. Boudjada, M. T. Averbuch-Pouchot, and A. Durif, *Acta Crystallogr., Sect. B 37*, 647–649 (1981).
923. N. Boudjada, M. T. Averbuch-Pouchot, and A. Durif, *Acta Crystallogr., Sect. B 37*, 645–647 (1981).
924. N. Boudjada, A. Boudjada, and J. C. Guitel, *Acta Crystallogr., Sect. C 39*, 656–658 (1983).
925. N. Boudjada and A. Durif, *Acta Crystallogr., Sect. B 38*, 595–597 (1982).
926. M. T. Averbuch-Pouchot, *Acta Crystallogr., Sect. C 44*, 1166–1168 (1988).
927. M. T. Averbuch-Pouchot and A. Durif, *Acta Crystallogr., Sect. C 43*, 1653–1655 (1987).
928. R. E. Marsh, *Acta Crystallogr., Sect. C 44*, 774 (1988).
929. A. Travers and Y. K. Chu, *C. R. Acad. Sci. 198*, 2169–2171 (1934).
930. B. Raistrick, *Discuss. Faraday Soc. 5*, 234–237 (1949).
931. R. N. Bell, L. F. Audrieth, and O. F. Hill, *Ind. Eng. Chem. 44*, 568–572 (1952).
932. E. Thilo and W. Wicker, *Z. anorg. allg. Chem. 277*, 27–36 (1952).
933. H. Grunze and E. Thilo, *Z. anorg. allg. Chem. 281*, 284–292 (1955).
934. H. Grunze, *Angew. Chem. 67*, 408 (1955).
935. M. T. Averbuch-Pouchot and A. Durif, *Acta Crystallogr., Sect. C 42*, 129–131 (1986).
936. D. M. Wiench and M. Jansen, *Monatsh. Chem. 114*, 699–709 (1983).
937. P. Bonneman, *C. R. Acad. Sci. 204*, 865–868 (1937).
938. K. A. Andress, W. Gehring, and K. Fischer, *Z. anorg. allg. Chem. 260*, 331–336 (1949).
939. D. L. Barney and J. W. Gryder, *J. Am. Chem. Soc. 77*, 3195–3198 (1955).
940. H. M. Ondik, S. Block, and C. H. Mac Gillavry, *Acta Crystallogr. 14*, 555–561 (1961).
941. H. M. Ondik, *Acta Crystallogr. 17*, 1139–1145 (1964).
942. M. T. Averbuch-Pouchot and A. Durif, *J. Solid State Chem. 58*, 119–132 (1985).
943. R. J. Gross, Dissertation, Johns Hopkins University, Baltimore, Maryland (1955).
944. E. J. Griffith, *Pure Appl. Chem. 44*, 173–200 (1975).
945. H. Nariai, I. Motooka, and M. Tsuhako, *Bull. Chem. Soc. Jpn. 64*, 3205–3206 (1991).
946. E. G. Griffith, *J. Am. Chem. Soc. 76*, 5862 (1954).
947. E. G. Griffith, *J. Am. Chem. Soc. 78*, 3867–3870 (1956).
948. G. W. Gryder, G. Donnay, and H. M. Ondik, *Acta Crystallogr. 11* (1958) 38–40.
949. O. H. Jarchow, *Acta Crystallogr. 17*, 1253–1262 (1964).
950. M. T. Averbuch-Pouchot and A. Durif, *Acta Crystallogr., Sect. C 41*, 1564–1566 (1985).
951. K. R. Waerstad, G. H. McClellan, A. W. Frazier, and R. C. Sheridan, *J. Appl. Crystallogr. 2*, 306–307 (1969).
952. C. Romers, J. A. A. Ketelaar, and C. H. Mac Gillavry, *Nature 164*, 960–961 (1949).
953. K. A. Andress and K. Fischer, *Acta Crystallogr. 3*, 399–400 (1950).
954. C. Romers, J. A. A. Ketelaar, and C. H. MacGillavry, *Acta Crystallogr. 4*, 114–120 (1951).
955. D. W. J. Cruickshank, *Acta Crystallogr. 17*, 675–676 (1964).
956. D. A. Koster and A. J. Wagner, *J. Chem. Soc. A* 1970, 435–441.

957. A. Takenaka, I. Motooka, and H. Nariai, *Bull. Chem. Soc. Jpn. 60*, 4299–4303 (1987).
958. M. T. Averbuch-Pouchot and A. Durif, *Acta Crystallogr., Sect. C 42*, 131–133 (1986).
959. M. T. Averbuch-Pouchot and A. Durif, *Acta Crystallogr., Sect. C 41*, 665–667 (1985).
960. K. Dostal, V. Kocman, and V. Ehrenbergrova, *Z. anorg. allg. Chemie 367*, 80–91 (1969).
961. K. Dostal and V. Kocman, *Z. Chem. 5*, 344 (1965).
962. J. K. Fawcett, V. Kocman, S. C. Nyburg, and R. J. O'Brien, *Chem. Commun. 18*, 1213 (1970).
963. J. K. Fawcett, V. Kocman, and S. C. Nyburg, *Acta Crystallogr., Sect. B 30*, 1979–1982 (1974).
964. M. T. Averbuch-Pouchot and A. Durif, *J. Solid State Chem. 60*, 13–19 (1985).
965. E. Steger, *Z. anorg. allg. Chem. 294*, 146–154 (1958).
966. E. Steger and A. Simon, *Z. anorg. allg. Chem. 294*, 1–9 (1958).
967. M. Beucher and J. C. Grenier, *Mater. Res. Bull. 3*, 643–648 (1968).
968. M. Laügt, A. Durif, and M. T. Averbuch-Pouchot, *Bull. Soc. fr. Minéral. Cristallogr. 96*, 383–385 (1973).
969. M. Laügt, J. C. Guitel, I. Tordjman, and G. Bassi, *Acta Crystallogr., Sect. B 28, 201–208 (1972).*
970. L. N. Shchegrov, N. M. Antraptseva, and I. G. Ponomareva, *Izv. Akad. Nauk SSSR, Neorg. Mater. 25*, 308–312 (1989).
971. D. Z. Serazetdinov, E. V. Poletsev, and Yu. A. Kushnikov, *Russ. J. Inorg. Chem. 12*, 1599–1600 (1967).
972. A. B. Bekturov, D. Z. Serazetdinov, Yu. A. Kushnikov, E. V. Poleaev, and S. M. Divnenko, *Russ. J. Inorg. Chem. 12*, 1242–1246 (1967).
973. G. A. Bukhalova, I. G. Rabkina, and I. V. Mardirosova, *Russ. J. Inorg. Chem. 20*, 332–334 (1975).
974. E. A. Genkina, N. S. Triodina, O. K. Mel'nikov, and B. A. Maksimov, *Izv. Akad. Nauk SSSR, Neorg. Mater. 24*, 1158–1162 (1988).
975. A. G. Nord and K. B. Lindberg, *Acta Chem. Scand. A29*, 1–6 (1975).
976. A. G. Nord, *Crystal Struct. Commun. 11*, 1467–1474 (1982).
977. A. G. Nord, *Acta Chem. Scand. A37*, 539–543 (1983).
978. A. G. Nord, *Mater. Res. Bull. 18*, 765–773 (1983).
979. E. A. Genkina, B. A. Maksimov, and O. K. Mel'nikov, *Sov. Phys. Crystallogr. 30*, 513–515 (1985).
980. A. G. Nord, T. Ericsson, and P. E. Werner, *Z. Kristallogr. 192*, 83–90 (1990).
981. M. T. Averbuch-Pouchot, A. Durif, and M. Bagieu-Beucher, *Acta Crystallogr., Sect. C 39*, 5–26 (1983).
982. W. Gunsser, D. Fruehauf, K. Rohwer, A. Zimmermann, and A. Wiedenmann, *J. Solid State Chem. 82*, 43–51 (1989).
983. M. Mohan Rao, M. Trojan, and L. Benes, *Mater. Res. Bull. 26*, 813–819 (1991).
984. M. Trojan and L. Benes, *Mater. Res. Bull. 26*, 693–700 (1991).
985. M. Schneider and K. H. Jost, *Z. anorg. allg. Chem. 580*, 175–180 (1990).
986. M. T. Averbuch-Pouchot, *Z. anorg. allg. Chem. 503*, 231–237 (1983).
987. G. Foumakoye, Thesis, University of Liège, Liège, Belgium (1986).
988. H. Nariai, I. Motooka, Y. Kanaji, and M. Tsuhako, *Bull. Chem. Soc. Jpn. 64*, 2912–2917 (1991).
989. A. V. Lavrov, T. A. Bykanova, and N. N. Chudinova, *Izv. Akad. Nauk SSSR, Neorg. Mater. 13*, 334–338 (1977).
990. H. Nariai, I. Motooka, and M. Tsuhako, *Bull. Chem. Soc. Jpn. 65*, 777–780 (1992).
991. M. Schneider, K. H. Jost, and H. Fichtner, *Z. anorg. allg. Chem. 500*, 117–122 (1983).
992. U. Schülke, personal communication, cited in Ref. 77.
993. M. Schneider and K. H. Jost, *Z. anorg. allg. Chem. 576*, 267–271 (1989).
994. A. Durif and M. T. Averbuch-Pouchot, *Acta Crystallogr. 503*, 927–928 (1986).
995. H. Worzala, *Z. anorg. allg. Chem. 421*, 122–128 (1976).
996. H. Worzala, *Z. anorg. allg. Chem. 445*, 27–35 (1978).
997. B. Klinkert and M. Jansen, *Z. anorg. allg. Chem. 556*, 85–91 (1988).
998. H. Nariai, I. Motooka, Y. Kanaji, and M. Tsuhako, *Bull. Chem. Soc. Jpn. 60*, 1337–1341 (1987).
999. M. T. Averbuch-Pouchot and A. Durif, *Acta Crystallogr., Sect. C 39*, 811–812 (1983).
1000. G. A. Bukhalova and I. A. Tokman, *Izv. Akad. Nauk SSSR, Neorg. Mater. 8*, 528–532 (1972).

1001. A. Durif, C. Martin, I. Tordjman, and D. Tranqui, *Bull. Soc. fr. Minéral. Cristallogr.* 89, 439–441 (1966).

1002. I. Tordjman, C. Martin, and A. Durif, *Bull. Soc. fr. Minéral. Cristallogr.* 90, 293–298 (1967).

1003. C. Cavéro-Ghersi and A. Durif, *J. Appl. Crystallogr.* 8, 562–564 (1975).

1004. J. B. Gill and R. M. Taylor, *J. Chem. Soc.* 1964, 5905–5906.

1005. M. Schneider, *Z. anorg. allg. Chem.* 503, 238–240 (1983).

1006. A. Takenaka, I. Motooka, and H. Nariai, *Chem. Soc. Jpn.* 62, 2819–2823 (1989).

1007. M. T. Averbuch-Pouchot and A. Durif, *Acta Crystallogr., Sect. C 41*, 1557–1558 (1985).

1008. M. T. Averbuch-Pouchot and A. Durif, *Acta Crystallogr., Sect. C 44*, 212–216 (1988).

1009. A. Durif, M. T. Averbuch-Pouchot, and J. C. Guitel, *Acta Crystallogr., Sect. C 39*, 812–813 (1983).

1010. A. Jouini, M. Dabbabi, and A. Durif, *J. Solid State Chem.* 60, 6–12 (1985).

1011. A. Jouini, M. Soua, and M. Dabbabi, *J. Solid State Chem.* 69, 135–144 (1987).

1012. I. Tordjman, R. Masse, and J. C. Guitel, *Acta Crystallogr., Sect. B 32*, 1643–1645 (1976).

1013. S. B. Hendricks and R. W. G. Wyckoff, *Am. J. Sci.* 13, 491–496 (1927).

1014. M. Bagieu-Beucher, *J. Appl. Crystallogr.* 9, 368–369 (1976).

1015. M. Bagieu-Beucher and J. C. Guitel, *Acta Crystallogr., Sect. B 34*, 1439–1442 (1978).

1016. G. Wappler, Diplom-Arbeit, Humbolt University, Berlin, Germany (1958).

1017. L. P. Mezentseva, A. I. Domanskii, and I. A. Bondar, *Russ. J. Inorg. Chem.* 22, 43–45 (1977).

1018. M. Beucher, Les Eléments des Terres Rares, International. Meeting, Paris, Grenoble (1969).

1019. R. Masse, J. C. Guitel, and A. Durif, *Acta. Crystallogr., Sect. B 33*, 630–632 (1977).

1020. M. Rzaigui, M. T. Averbuch-Pouchot, and A. Durif, *Acta Crystallogr., Sect. C 39*, 1612–1613 (1983).

1021. H. Koizumi and J. Nakano, *Acta Crystallogr., Sect. B 33*, 2680–2684 (1977).

1022. G. I. Dorokhova, O. S. Filipenko, L. O. Atovmyan, and B. N. Litvin, *Russ. J. Inorg. Chem.* 33, 1581–1585 (1988).

1023. K. K. Palkina, S. I. Maksimova, and N. T. Chibiskova, *Izv. Akad. Nauk SSSR, Neorg. Mater.* 17, 1248–1252 (1981).

1024. K. K. Palkina, S. I. Maksimova, and N. T. Chibiskova, *Sov. Phys. Dokl.* 26, 254–256 (1981).

1025. K. K. Palkina, A. K. Mustaev, S. I. Maksimova, R. Yu. Khusainova, and N. T. Chibiskova, *Russ. J. Inorg. Chem.* 34, 1533–1534 (1989).

1026. H. Nariai, A. I. Motooka, and M. Tsuhako, *Bull. Chem. Soc. Jpn.* 61, 2811–2815 (1988).

1027. S. A. Linde, Yu. E. Gorbunova, and V. Lavrova, *Russ. J. Inorg. Chem.* 28, 785–787 (1983).

1028. A. Durif, M. T. Averbuch-Pouchot, and J. C. Guitel, *J. Solid State Chem.* 41, 153–159 (1982).

1029. M. T. Averbuch-Pouchot and A. Durif, *Acta Crystallogr., Sect. C 43*, 1245–1247 (1987).

1030. M. T. Averbuch-Pouchot and A. Durif, *C. R. Acad. Sci., Ser. 2 304*, 269–271 (1987).

1031. M. T. Averbuch-Pouchot, A. Durif, and J. C. Guitel, *Acta Crystallogr., Sect. C 44*, 888–890 (1988).

1032. M. T. Averbuch-Pouchot, A. Durif, and J. C. Guitel, *Acta Crystallogr., Sect. C 44*, 1416–1418 (1988).

1033. M. T. Averbuch-Pouchot, A. Durif, and M. Bagieu-Beucher, *Z. Kristallogr.* 185, 506 (1988).

1034. M. T. Averbuch-Pouchot, A. Durif, and J. C. Guitel, *Acta Crystallogr., Sect. C 45*, 428–430 (1989).

1035. M. T. Averbuch-Pouchot and A. Durif, *Acta Crystallogr., Sect. C 45*, 46–49 (1989).

1036. M. T. Averbuch-Pouchot, A. Durif, and J. C. Guitel, *Acta Crystallogr., Sect. C 44*, 1189–1191 (1988).

1037. M. Bagieu-Beucher, A. Durif, and J. C. Guitel, *Acta Crystallogr., Sect. C 44*, 2063–2065 (1988).

1038. M. Bdiri and A. Jouini, *C. R. Acad. Sci., Ser. 2 308*, 1345–1348 (1989).

1039. A. Jouini, *Acta Crystallogr., Sect. C 45*, 1877–1879 (1989).

1040. M. Bdiri and A. Jouini, *J. Solid State Chem.* 83, 350–360 (1989).

1041. M. Bdiri and A. Jouini, *C. R. Acad. Sci., Ser. 2 309*, 881–885 (1989).

1042. M. Bdiri and A. Jouini, *Eur. J. Solid State Inorg. Chem.* 26, 585–592 (1989).

1043. E. Thilo and U. Schülke, *Z. anorg. allg. Chem.* 344, 293–307 (1965).

1044. K. H. Jost, *Acta Crystallogr., Sect. B 28*, 732–738 (1972).

1045. E. Thilo and U. Schülke, *Angew. Chem., Int. Ed. Engl.* 2, 742 (1963).

1046. E. Thilo and U. Schülke, *Angew. Chem.* 75, 1175–1176 (1963).

1047. A. McAdam, K. H. Jost, and B. Beagley, *Acta Crystallogr., Sect. B 28*, 2740–2743 (1972).

1048. E. J. Griffith and R. L. Buxton, *Inorg. Chem.* 4, 549–551 (1965).

1049.   M. Laügt and A. Durif, *Acta Crystallogr., Sect. B 30*, 2118–2121 (1974).

1050.   M. Bagieu-Beucher and J. C. Guitel, *Acta Crystallogr., Sect. B 33*, 2529–2533 (1977).

1051.   S. A. Linde, Yu. E. Gorbunova, A. V. Lavrov, and V. G. Kuznetsov, *Dokl. Akad. Nauk SSSR 241*, 1083–1085 (1978).

1052.   U. Schülke and R. Kayser, *Z. anorg. allg. Chem. 531*, 167–175 (1985).

1053.   M. T. Averbuch-Pouchot, *Z. Kristallogr. 189*, 17–23 (1990).

1054.   W. K. Trunov, L. N. Kholodkovskaya, L. A. Borodina, and N. N. Chudinova, *Kristallografiya 34*, 748–751 (1989).

1055.   M. Bagieu-Beucher and M. Rzaigui, *J. Solid State Chem. 108*, 11–17 (1994).

1056.   L. A. Borodina and N. N. Chudinova, *Izv. Akad. Nauk SSSR, Neorg. Mater. 24*, 345–347 (1988).

1057.   H. Nariai, I. Motooka, and M. Tsuhako, *Bull. Chem. Soc. Jpn. 64*, 2353–2355 (1991).

1058.   E. Thilo and U. Schülke, *Z. anorg. allg. Chem. 341*, 293–307 (1965).

1059.   N. N. Chudinova, L. A. Borodina, and U. Schülke, *Izv. Akad. Nauk SSSR, Neorg. Mater. 25*, 303–307 (1989).

1060.   M. T. Averbuch-Pouchot, *Acta Crystallogr., Sect. C 45*, 1273–1275 (1989).

1061.   L. N. Kholodkovskaya, L. A. Borodina, W. K. Trunov, and N. N. Chudinova, *Izv. Akad. Nauk SSSR, Neorg. Mater. 25*, 466–469 (1989).

1062.   L. N. Kholodkovskaya, L. A. Borodina, W. K. Trunov, and N. N. Chudinova, *Izv. Akad. Nauk SSSR, Neorg. Mater. 25*, 454–458 (1989).

1063.   N. N. Chudinova, L. A. Borodina, U. Schülke, and K. H. Jost, *Izv. Akad. Nauk SSSR, Neorg. Mater. 25*, 459–465 (1989).

1064.   M. T. Averbuch-Pouchot, *Acta Crystallogr., Sect. C 45*, 539–540 (1989).

1065.   A. Takenaka, I. Motooka, and H. Nariai, *Bull. Chem. Soc. Jpn. 60*, 4299–4303 (1987).

1066.   M. T. Averbuch-Pouchot and A. Durif, *C. R. Acad. Sci., Ser. 2 308*, 1699–1702 (1989).

1067.   L. N. Kholodkovskaya, L. A. Borodina, W. K. Trunov, and N. N. Chudinova, *Izv. Akad. Nauk SSSR, Neorg. Mater. 25*, 470–476 (1989).

1068.   M. T. Averbuch-Pouchot, *Z. anorg. allg. Chem. 574*, 225–234 (1989).

1069.   M. T. Averbuch-Pouchot, *Acta Crystallogr., Sect. C 47*, 930–932 (1991).

1070.   M. T. Averbuch-Pouchot and A. Durif, *Acta Crystallogr., Sect. C 47*, 1150–1152 (1991).

1071.   M. Elmokhtar, M. Rzaigui, and A. Jouini, *Acta Crystallogr., Sect. C 49*, 435–437 (1993).

1072.   M. T. Averbuch-Pouchot, personal communication.

1073.   M. T. Averbuch-Pouchot and A. Durif, *Acta Crystallogr., Sect. C 47*, 932–936 (1991).

1074.   E. V. Lazarevski, L. V. Kubasova, N. N. Chudinova, and I. V. Tananaev, *Izv. Akad. Nauk SSSR, Neorg. Mater. 16*, 120–125 (1980).

1075.   E. V. Lazarevski, L. V. Kubasova, N. N. Chudinova, and I. V. Tananaev, *Izv. Akad. Nauk SSSR, Neorg. Mater. 18*, 1550–1556 (1982).

1076.   E. V. Lazarevski, L. V. Kubasova, N. N. Chudinova, and I. V. Tananaev, *Izv. Akad. Nauk SSSR, Neorg. Mater. 18*, 1544–1549 (1982).

1077.   M. T. Averbuch-Pouchot, *Acta Crystallogr., Sect. C 45*, 1275–1277 (1989).

1078.   M. T. Averbuch-Pouchot, *Z. anorg. allg. Chem. 570*, 138–144 (1989).

1079.   M. T. Averbuch-Pouchot and A. Durif, *C. R. Acad. Sci., Ser. 2 309*, 535–537 (1989).

1080.   H. Nariai, I. Motooka, Y. Kanaji, and M. Tsuhako, *Phosphorus Res. Bull. 1*, 113–118 (1991).

1081.   E. V. Lazarevski, L. V. Kubasova, N. N. Chudinova, and I. V. Tananaev, *Izv. Akad. Nauk SSSR, Neorg. Mater. 17*, 486–491 (1981).

1082.   M. Rzaigui, M. T. Averbuch-Pouchot, and A. Durif, *Acta Crystallogr., Sect. C 48*, 241–243 (1992).

1083.   M. Laügt, A. Durif, and C. Martin, *J. Appl. Crystallogr. 7*, 448–449 (1974).

1084.   M. T. Averbuch-Pouchot, *Acta Crystallogr., Sect. C 45*, 1856–1858 (1989).

1085.   M. T. Averbuch-Pouchot and A. Durif, *Acta Crystallogr., Sect. C 46*, 968–970 (1990).

1086.   M. T. Averbuch-Pouchot and A. Durif, *Acta Crystallogr., Sect. C 46*, 10–13 (1990).

1087.   M. T. Averbuch-Pouchot and A. Durif, *Acta Crystallogr., Sect. C 47*, 1148–1150 (1991).

1088.   A. Takenaka, H. Kobayashi, K. Tsuchie, I. Motooka, and H. Nariai, *Bull. Chem. Soc. Jpn. 62*, 3808–3811 (1989).

1089. M. T. Averbuch-Pouchot, *Acta Crystallogr., Sect. C 46*, 2005–2007 (1990).

1090. M. T. Averbuch-Pouchot and A. Durif, *Eur. J. Solid State Inorg. Chem. 30*, 573–581 (1993).

1091. I. Tordjman,. M. T. Averbuch-Pouchot, and A. Durif, to be published.

1092. A. Takenaka, I. Motooka, and H. Nariai, *Bull. Chem. Soc. Jpn. 62*, 2819–2823 (1989).

1093. P. Rémy and A. Boullé, *Bull. Soc. Chim. Fr.* 1972, 2213–2221.

1094. Z. Y. Kanene, Z. A. Konstant, and V. V. Krasnikov, *Izv. Akad. Nauk SSSR, Neorg. Mater. 21*, 1552–1554 (1985).

1095. M. Rzaigui, *J. Solid State Chem. 89*, 340–344 (1990).

1096. M. Bagieu-Beucher, M. T. Averbuch-Pouchot, and M. Rzaigui, *Acta Crystallogr., Sect. C 47*, 1364–1366 (1991).

1097. M. Rzaigui, *J. Solid State Chem.*, in press.

1098. M. Bagieu-Beucher and M. Rzaigui, *Acta Crystallogr., Sect. C 47*, 1789–1791 (1991).

1099. V. K. Trunov, N. N. Chudinova, and L. A. Borodina, *Dokl. Akad. Nauk SSSR 300*, 1375–1379 (1988).

1100. M. Bagieu-Beucher and M. Rzaigui, *Acta Crystallogr., Sect. C 48*, 244–246 (1992).

1101. A. Durif and M. T. Averbuch-Pouchot, *Acta Crystallogr., Sect. C 46*, 2026–2028 (1990).

1102. A. Durif and M. T. Averbuch-Pouchot, *Acta Crystallogr., Sect. C 45*, 1884–1887 (1989).

1103. M. T. Averbuch-Pouchot and A. Durif, *Acta Crystallogr., Sect. C 47*, 1579–1583 (1991).

1104. M. T. Averbuch-Pouchot and A. Durif, manuscript in preparation.

1105. M. T. Averbuch-Pouchot and A. Durif, *Acta Crystallogr., Sect. C 46*, 179–181 (1990).

1106. M. T. Averbuch-Pouchot and A. Durif, *Acta Crystallogr., Sect. C 47*, 1576–1579 (1991).

1107. M. T. Averbuch-Pouchot and A. Durif, *Acta Crystallogr., Sect. C 46*, 2236–2238 (1990).

1108. M. T. Averbuch-Pouchot and A. Durif, *Acta Crystallogr., Sect. C 46*, 965–968 (1990).

1109. M. T. Averbuch-Pouchot and A. Durif, *Eur. J. Solid State Inorg. Chem. 29*, 1161–1172 (1992).

1110. M. T. Averbuch-Pouchot and A. Durif, *Eur. J. Solid State Inorg. Chem. 30*, 447–459 (1993).

1111. J. R. Van Wazer and E. Karl-Kroupa, *J. Am. Chem. Soc. 78*, 1772 (1956).

1112. U. Schülke, *Z. anorg. allg. Chem. 360*, 231–246 (1968).

1113. U. Schülke, *Angew. Chem. Int. Ed. Engl. 7*, 71 (1968).

1114. U. Schülke and N. N. Chudinova, *Izv. Akad. Nauk SSSR, Neorg. Mater. 10*, 1697–1703 (1974).

1115. U. Schülke, M. T. Averbuch-Pouchot, and A. Durif, *J. Solid State Chem. 98*, 213–218 (1992).

1116. U. Schülke, M. T. Averbuch-Pouchot, and A. Durif, *Z. anorg. allg. Chem. 619*, 374–380 (1993).

1117. B. Brühne and M. Jansen, *Z. anorg. allg. Chem. 619*, 1633–1638 (1993).

1118. M. Laügt, Thesis, University of Grenoble, Grenoble, France (1974).

1119. E. V. Lazarevski, L. V. Kubasova, N. N. Chudinova, and I. V. Tananaev, *Izv. Akad. Nauk SSSR, Neorg. Mater. 15*, 2180–2184 (1979).

1120. E. V. Lazarevski, L. V. Kubasova, and N. N. Chudinova, *Izv. Akad. Nauk SSSR, Neorg. Mater. 19*, 498–499 (1983).

1121. N. N. Chudinova, I. V. Tananaev, and M. A. Avaliani, *Izv. Akad. Nauk SSSR, Neorg. Mater. 13*, 2234–2237 (1977).

1122. N. N. Chudinova, M. A. Avaliani, L. S. Guzeeva, and I. V. Tananaev, *Izv. Akad. Nauk SSSR, Neorg. Mater. 14*, 2054–2057 (1978).

1123. I. Grunze, N. N. Chudinova, and K. K. Palkina, *Izv. Akad. Nauk SSSR, Neorg. Mater. 19*, 1943–1945 (1983).

1124. K. K. Palkina, S. I. Maksimova, V. G. Kusznetsov, and N. N. Chudinova, *Dokl. Akad. Nauk SSSR 245*, 1386–1389 (1979).

1125. H. Nariai, I. Motooka, and M. Tsuhako, *Bull. Chem. Soc. Jpn. 61*, 2811–2815 (1988).

1126. M. T. Averbuch-Pouchot, A. Durif, and U. Schülke, *Eur. J. Solid State Inorg. Chem. 30*, 741–750 (1993).

1127. M. T. Averbuch-Pouchot, A. Durif, and U. Schülke, *Eur. J. Solid State Inorg. Chem. 30*, 557–563 (1993).

1128. M. T. Averbuch-Pouchot and A. Durif, *Acta Crystallogr., Sect. C 48*, 1173–1176 (1992).

1129. U. Schülke, M. T. Averbuch-Pouchot, and A. Durif, *Z. Kristallogr. 204*, 143–152 (1993).

1130. M. T. Averbuch-Pouchot and A. Durif, *Acta Crystallogr., Sect. C 48*, 361–363 (1993).

1131. M. T. Averbuch-Pouchot and U. Schülke, Z. Kristallogr., in press.

1132. M. T. Averbuch-Pouchot and A. Durif, Eur. J. Solid State Inorg. Chem. 28, 9–22 (1991).

1133. M. Bagieu-Beucher and N. El-Horr, personal communication.

1134. M. Bagieu-Beucher, A. Durif, and J. C. Guitel, J. Solid State Chem. 40, 248 (1981).

1135. U. Schülke, unpublished results.

1136. U. Schülke, M. T. Averbuch-Pouchot, and A. Durif, Z. anorg. allg. Chem. 612, 107–112 (1992).

1137. M. T. Averbuch-Pouchot, unpublished results.

1138. M. T. Averbuch-Pouchot, A. Durif, and U. Schülke, J. Solid State Chem. 97, 299–304 (1992).

1139. C. I. Cabello and E. Baran, Spectrochim. Acta 41A, 1359–1360 (1985).

1140. U. Schülke and M. T. Averbuch-Pouchot, Z. anorg. allg. Chem. 620, 545–550 (1994).

1141. M. T. Averbuch-Pouchot and U. Schülke, Z. anorg. allg. Chem., in press.

1142. A. V. Lavrov, V. P. Nikolaev, G. G. Sadikov, and M. Y. Voitenko, Dokl. Akad. Nauk SSSR 259, 103–106 (1981).

1143. G. U. Wolf and M. Meisel, Z. anorg. allg. Chem. 509, 101–110 (1984).

1144. M. Meisel, Z. Chem. 23, 117–125 (1983).

1145. G. U. Wolf and M. Meisel, Z. Chem. 20, 451–452 (1980).

1146. K. K. Palkina, S. I. Maksimova, and G. U. Wolf, Izv. Akad. Nauk SSSR, Neorg. Mater. 16, 1466–1488 (1980).

1147. G. U. Wolf and M. Meisel, Z. Chem. 20, 452 (1980).

1148. G. U. Wolf and M. Meisel, Z. anorg. allg. Chem. 494, 49–54 (1982).

1149. M. Meisel, G. U. Wolf, and M. T. Averbuch-Pouchot, Acta Crystallogr., Sect. C 47, 1368–1370 (1991).

1150. K. K. Palkina, S. I. Maksimova, N. T. Chibiskova, M. S. Pautina, G. U. Wolf, and T. B. Kuvshinova, Izv. Akad. Nauk SSSR, Neorg. Mater. 28, 1486–1490 (1992).

1151. T. B. Kuvshinova, G. U. Wolf, and M. Meisel, Izv. Akad. Nauk SSSR, Neorg. Mater. 20, 1056–1062 (1984).

1152. V. V. Ilyukhin, V. R. Kalinin, T. B. Kuvshinova, and I. V. Tananaev, Dokl. Akad. Nauk SSSR 266, 1387–1391 (1981).

1153. V. V. Ilyukhin, V. R. Kalinin, T. B. Kuvshinova, and I. V. Tananaev, Dokl. Akad. Nauk SSSR 267, 85–88 (1982).

1154. V. V. Ilyukhin, V. R. Kalinin, and T. B. Kuvshinova, Izv. Akad. Nauk SSSR, Neorg. Mater. 24, 828–835 (1988).

1155. V. P. Nikolaev and T. B. Kuvshinova, Izv. Akad. Nauk SSSR, Neorg. Mater. 23, 622–625 (1987).

1156. E. A. Prodan, L. I. Petrovskaya, and T. B. Kuvshinova, Izv. Akad. Nauk SSSR, Neorg. Mater. 25, 1339–1343 (1989).

1157. T. S. Kubshinova, G. U. Wolf, T. N. Kuzmina, and M. S. Pautina, Izv. Akad. Nauk SSSR, Neorg. Mater. 25, 1349–1355 (1989).

1158. L. I. Petrvskaya, M. S. Pautina, E. A. Prodan, T. B. Kuvshinova, and S. I. Pytlev, Izv. Akad. Nauk SSSR, Neorg. Mater. 27, 1023–1027 (1991).

1159. M. Meisel, G. U. Wolf, and M. T. Averbuch-Pouchot, Acta Crystallogr., Sect. C 46, 2239–2241 (1990).

1160. D. M. Wiench and M. Jansen, Acta Crystallogr., Sect. C 39, 1613–1615 (1983).

1161. A. Larbot, J. Durand, S. Vilminot, and A. Norbert, Acta Crystallogr., Sect. B 37, 1023–1027 (1981).

1162. M. De la Rochère, A. Kahn, F. d'Yvoire, and E. Bretey, Mater. Res. Bull. 20, 27–34 (1985).

1163. B. Kojic-Prodic, M. Sljukic, and Z. Ruzic-Toros, Acta Crystallogr., Sect. B 38, 67–71 (1982).

1164. A. Larbot, A. Norbert, and J. Durand, Z. anorg. allg. Chem. 486, 200–206 (1982).

1165. G. Ladvig and H. Worzala, Z. Chem. 18, 374–375 (1978).

1166. K. K. Palkina and S. I. Maksimova, Dokl. Akad. Nauk SSSR 250, 1130–1134 (1980).

1167. J. J. Chen, K. H. Lii, and S. L. Lang, J. Solid State Chem. 76, 204–209 (1988).

1168. A. Leclaire, M. M. Borel, A. Grandin, and B. Raveau, Z. Kristallogr. 188, 77–83 (1989).

1169. A. Leclaire, J. C. Monier, and B. Raveau, J. Solid State Chem. 48, 147–153 (1983).

1170. K. H. Lii and C. C. Wang, J. Solid State Chem. 77, 117–123 (1988).

1171. A. Benmoussa, A. Leclaire, A. Grandin, and B. Raveau, *Acta Crystallogr., Sect. C 45,* 1277–1279 (1989).
1172. G. G. Sadikov, V. P. Nikolaev, A. V. Lavrov, and M. A. Porai-Koshits, *Dokl. Akad. Nauk SSSR 264,* 862–867 (1982).
1173. Yu. A. Ivanov, M. A. Simonov, and N. V. Belov, *Dokl. Akad. Nauk SSSR 242,* 599–602 (1978).
1174. R. C. Haushalter and F. W. Lai, *J. Solid State Chem. 76,* 218–223 (1988).
1175. V. P. Nikolaev, G. G. Sadikov, A. V. Lavrov, and M. A. Porai-Koshits, *Izv. Akad. Nauk SSSR, Neorg. Mater. 19,* 448–451 (1983).
1176. I. Grunze, N. N. Chudinova, and K. K. Palkina, *Izv. Akad. Nauk SSSR, Neorg. Mater. 20,* 1053–1055 (1984).
1177. I. Grunze, K. K. Palkina, S. I. Maksimova, and N. T. Chibiskova, *Dokl. Akad. Nauk SSSR 275,* 879–883 (1984).
1178. J. R. Van Wazer and K. A. Holst, *J. Am. Chem. Soc. 72,* 639–644 (1950).
1179. P. T. Clève, *Bull. Soc. Chim. 43,* 169 (1885).
1180. K. R. Johnsson, *Ber. dtsch. Chem. Ges.* 22, 976–980 (1889).
1181. A. V. Kroll, *Z. anorg. Chem. 76,* 387–418 (1912).
1182. A. V. Kroll, *Z. anorg. Chem. 77,* 1–40 (1912).
1183. A. V. Kroll, *Z. anorg. Chem. 78,* 95–133 (1912).
1184. S. Jaulmes, *C. R. Acad. Sci. 268C,* 935–937 (1969).
1185. D. Stachel, H. Paulus, C. Guenter, and H. Fuess, *Z. Kristallogr. 199,* 275–276 (1992).
1186. C. Baez-Doelle, D. Stachel, I. Svoboda, and H. Fuess, *Z. Kristallogr. 203,* 282–283 (1993).
1187. O. V. Yakubovich, O. V. Dimitrova, and A. I. Vidrevich, *Kristallografiya 38,* 77–85 (1993).
1188. M. Beucher, *Mater. Res. Bull. 4,* 15–18 (1969).
1189. I. Tordjman, M. Bagieu-Beucher, and R. Zilber, *Z. Kristallogr. 140,* 145–153 (1974).
1190. D. Stachel, H. Paulus, I. Svoboda, and H. Fuess, *Z. Kristallogr. 202,* 117–118 (1992).
1191. L. K. Minacheva, M. A. Porai-Koshits, A. S. Antsyshkina, V. G. Ivanova, and A. V. Lavrov, *Koord. Khim. 1,* 421–428 (1975).
1192. A. V. Lavrov, T. A. Bykanova, and Yu. M. Kessler, *Izv. Akad. Nauk SSSR, Neorg. Mater. 13,* 329–333 (1977).
1193. A. Olbertz, D. Stachel, I. Svoboda, and H. Fuess, *Z. Kristallogr.,* in press.
1194. A. S. Antsyshkina, M. A. Porai-Koshits, L. K. Minacheva, and V. G. Ivanova, *Koord. Khim. 5,* 268–275 (1979).
1195. A. S. Antsyshkina, M. A. Porai-Koshits, L. K. Minacheva, V. G. Ivanova, and A. V. Lavrov, *Koord. Khim. 4,* 448–454 (1978).
1196. A. S. Antsyshkina, M. A. Porai-Koshits, L. K. Minacheva, A. V. Lavrov, and V. G. Ivanova, *Dokl. Akad. Nauk SSSR 229,* 896–897 (1976).
1197. A. V. Lavrov, L. K. Minacheva, and L. S. Guzeeva, *Izv. Akad. Nauk SSSR, Neorg. Mater. 9,* 1466–1467 (1973).
1198. N. G. Chernorukov, N. P. Egorov, and T. A. Galanova, *Izv. Akad. Nauk SSSR, Neorg. Mater. 17,* 328–332 (1981).
1199. A. V. Lavrov, *Izv. Akad. Nauk SSSR, Neorg. Mater. 15,* 942–946 (1979).
1200. Yu. E. Gorbunova, S. A. Linde, and A. V. Lavrov, *Russ. J. Inorg. Chem. 26,* 383–386 (1981).
1201. N. N. Chudinova, K. K. Palkina, N. B. Komarovskaya, S. I. Maksimova, and N. T. Chibiskova, *Dokl. Akad. Nauk SSSR 306,* 635–638 (1989).
1202. K. K. Palkina, S. I. Maksimova, N. T. Chibiskova, N. N. Chudinova, and N. B. Karmanovskaya, *Neorg. Mater. 29,* 119–120 (1993).
1203. D. Tranqui, M. Bagieu, and A. Durif, *Acta Crystallogr., Sect. B 30,* 1751–1755 (1974).
1204. M. Bagieu, I. Tordjman, A. Durif, and G. Bassi, *Crystallogr. Struct. Commun. 3,* 387–390 (1973).
1205. D. Tranqui, M. Bagieu-Beucher, and A. Durif, *Bull. Soc. fr. Minéral. Cristallogr. 95,* 437–440 (1972).
1206. N. N. Chudinova and K. H. Jost, *Z. anorg. allg. Chem. 400,* 185–188 (1973).
1207. M. Rzaigui, N. Kbir Ariguib, M. T. Averbuch-Pouchot, and A. Durif, *J. Solid State Chem. 52,* 61–65 (1984).

1208. G. D. Dudko, V. V. Fedorov, and R. S. Shevelevich, *Izv. Akad. Nauk SSSR, Neorg. Mater.* 22, 456–459 (1986).

1209. A. Durif, *Bull. Soc. fr. Minéral. Cristallogr.* 94, 314–318 (1971).

1210. I. A. Bondar, L. P. Mezentseva, A. I. Domanskii, and M. M. Piryutko, *Russ. J. Inorg. Chem.* 20, 1448–1452 (1975).

1211. K. R. Albrand, R. Attig, J. Fenner, J. P. Jeser, and D. Mootz, *Mater. Res. Bull.* 9, 129–140 (1974).

1212. I. Szczygiel and T. Znamierowska, *J. Solid State Chem.* 82, 181–185 (1989).

1213. M. Yoshimura, K. Fujii, and S. Somiya, *Mater. Res. Bull.* 16, 327–333 (1981).

1214. H. Schulz, K. H. Thiemann, and J. Fenner, *Mater. Res. Bull.* 9, 1525–1530 (1974).

1215. R. Parrot, C. Barthou, B. Canny, B. Blanzat, and G. Collin, *Phys. Rev. B11*, 1001–1012 (1975).

1216. B. Jezowska-Trzebiatowska, Z. Mazurak, and T. Lis, *Acta Crystallogr., Sect. B 36*, 1639–1641 (1980).

1217. Z. Kangjing, G. Yitai, and H. Guangyan, *Crystal Struct. Commun. 11*, 1695–1699 (1982).

1218. H. Y-P. Hong and J. W. Pierce, *Mater. Res. Bull.* 9, 179–190 (1974).

1219. B. Borkowski, E. Grzesiak, F. Kaczmarek, Z. Kaluski, J. Karolczak, and M. Szymanski, *J. Crystal. Growth 44*, 320–324 (1978).

1220. B. C. Tofield, H. P. Weber, T. C. Damen, and G. H. Pasteur, *Mater. Res. Bull.* 9, 435–447 (1974).

1221. H. G. Danielmayer, J. P. Jeser, E. S. Schönherr, and W. Stetter, *J. Cryst. Growth 22*, 298–302 (1974).

1222. H. Kasano and Y. Furuhata, *J. Electrochem. Soc. 126*, 1567–1572 (1979).

1223. M. Marais, N. D. Chinh, H. Savary, and J. P. Budin, *J. Cryst. Growth 35*, 329–333 (1976).

1224. R. D. Plattner, W. W. Kruhler, W. K. Zwicker, T. Kovats, and S. R. Chinn, *J. Cryst. Growth 49*, 274–290 (1980).

1225. D. C. Miller, L. K. Shick, and C. D. Brandle, *J. Crystal. Growth 23*, 313–317 (1974).

1226. H. Kasano and Y. Furuhata, *J. Electrochem. Soc. 127*, 2464–2468 (1980).

1227. B. C. Tofield, P. M. Bridenbaugh, and H. P. Weber, *Mater. Res. Bull. 10*, 1091–1096 (1975).

1228. Y. Ito and S. Ashida, *J. Crystal. Growth 37*, 91–100 (1977).

1229. H. P. Weber, B. C. Tofield, and P. F. Liao, *Phys. Rev. B11*, 1152–1159 (1975).

1230. T. Kobayashi, T. Sawada, H. Ikeo, K. Muto, and J. Kai, *J. Phys. Soc. Jpn.* 40, 595–596 (1976).

1231. M. Rzaigui and N. Kbir Ariguib, *J. Solid State Chem.* 56, 122–125 (1985).

1232. H. Y. P. Hong and S. R. Chinn, *Mater. Res. Bull. 11*, 461–468 (1976).

1233. G. M. Loiacano, M. Delfino, and W. A. Smith, *Appl. Phys. Lett. 32*, 595–596 (1978).

1234. A. R. Grimmer and N. N. Chudinova, *Z. Chem. 12*, 149–150 (1972).

1235. D. W. Cruickshank, *J. Chem. Soc. 1972*, 5486–5505.

1236. V. I. Pakhomov, *Izv. Akad. Nauk SSSR, Neorg. Mater. 13*, 1341–1352 (1977).

Index